Environmental Exposure, Biomonitoring and Exposure Assessment

Environmental Exposure, Biomonitoring and Exposure Assessment

Editors

Maaike van Gerwen
Roel Vermeulen
Lauren Petrick

MDPI

Basel • Beijing • Wuhan • Barcelona • Belgrade • Novi Sad • Cluj • Manchester

Editors
Maaike van Gerwen
Icahn School of Medicine at Mount Sinai
New York
USA

Roel Vermeulen
Utrecht University
Utrecht
The Netherlands

Lauren Petrick
Icahn School of Medicine at Mount Sinai
New York
USA

Editorial Office
MDPI AG
Grosspeteranlage 5
4052 Basel, Switzerland

This is a reprint of articles from the Topic published online in the open access journals *International Journal of Environmental Research and Public Health* (ISSN 1660-4601), *Journal of Xenobiotics* (ISSN 2039-4713), *Pollutants* (ISSN 2673-4672), *Toxics* (ISSN 2305-6304), and *Toxins* (ISSN 2072-6651) (available at: https://www.mdpi.com/topics/exposure_risk).

For citation purposes, cite each article independently as indicated on the article page online and as indicated below:

Lastname, A.A.; Lastname, B.B. Article Title. *Journal Name* **Year**, *Volume Number*, Page Range.

ISBN 978-3-7258-1743-6 (Hbk)
ISBN 978-3-7258-1744-3 (PDF)
doi.org/10.3390/books978-3-7258-1744-3

Contents

About the Editors

Maaike van Gerwen

Maaike van Gerwen graduated with a Medical Degree from Radboud University (The Netherlands) in 2000 and a Ph.D. in Clinical Research from the Icahn School of Medicine at Mount Sinai (USA) in 2020. Maaike is an assistant professor and Director of Research in the Department of Otolaryngology—Head and Neck Surgery at the Icahn School of Medicine at Mount Sinai in New York where she leads the Thyroid Research Program to improve the care and outcomes of thyroid cancer patients by combining basic, clinical and epidemiological research. The research program focuses on environmental exposure contributing to thyroid cancer risk. Additionally, she is the principal investigator of a number of federally funded research projects. She has published multiple papers on the association between environmental exposure and thyroid health.

Roel Vermeulen

Roel Vermeulen is professor of Environmental Epidemiology at Exposome Science at Utrecht University and the University Medical Center (UMC) Utrecht (The Netherlands). He is the Scientific Director of the Institute for Preventive Health, part of the EUWW strategic alliance, as well as an MT member of the Institute for Risk Assessment Sciences (IRAS) at Utrecht University Faculty of Veterinary Sciences. He holds an adjunct professor position at the Julius Center, Department of Public Health Healthcare Innovation and Medical Humanities (UMC Utrecht), and a visiting professorship at Imperial College London, UK. His scientific research focuses on environmental risk factors for non-communicable diseases with a strong emphasis on integrating epidemiology, high-quality exposure assessment, and molecular biology into multidisciplinary investigations.

Lauren Petrick

Lauren Petrick obtained her Master of Science in Analytical Chemistry from the University of California in 2005 and graduated with a Ph.D. in Environmental Science from Technion - Israel Institute of Technology in 2011. Lauren is an associate professor in the Department of Environmental Medicine and Public Health at the Icahn School of Medicine at Mount Sinai in New York (USA) where she is the head of Untargeted Metabolomics and Exposomics at the Lautenberg Environmental Health Sciences Laboratory and performs urine, plasma, and serum analysis as the metabolomics lead of the ISMMS Children's/Human Health Exposure Analysis Resource (CHEAR/HHEAR) Untargeted Resource. She is also the Director of the Center for Metabolomics and Molecular Profiling at the Sheba Medical Center (Israel). Lauren is an exposome chemist and her research involves the use of high-resolution mass spectrometry and advanced biostatistics/bioinformatics techniques to discover biomarkers of disease.

International Journal of
Environmental Research and Public Health

MDPI

Article

Correlation between Lindane Use and the Incidence of Thyroid Cancer in the United States: An Ecological Study

Mathilda Alsen [1], Vikram Vasan [1], Eric M. Genden [1], Catherine Sinclair [1] and Maaike van Gerwen [1,2,*]

[1] Department of Otolaryngology-Head and Neck Surgery, Icahn School of Medicine at Mount Sinai, New York, NY 10029, USA
[2] Institute for Translational Epidemiology, Icahn School of Medicine at Mount Sinai, New York, NY 10029, USA
* Correspondence: maaike.vangerwen@mountsinai.org

Abstract: The increasing rate of thyroid cancer may be attributable to endocrine disruptive chemicals. Lindane is a persistent organochlorine pesticide with endocrine disruptive properties that has been classified as carcinogenic to humans. The aim of this ecological study was to evaluate potential correlation between lindane exposure and thyroid cancer incidence in the United States (US). Data on statewide age-adjusted thyroid cancer incidence rate (per 100,000 people) was obtained from the Centers for Disease Control and Prevention for all US states for 2019. Lindane use estimates per cropland (kg/acres cropland) were then overlaid on the map of age-adjusted thyroid cancer incidence rate using ArcGIS. The trend of lindane use in the US between 1992 and 2007 was calculated using the Mann Kendall correlation test. The correlation between statewide lindane use and age-adjusted thyroid cancer incidence rates was calculated using Spearman correlation. Lindane use in the US decreased significantly between 1992 and 2007 ($T = -0.617$; $p < 0.001$). There was no statistically significant correlation between lindane use in 1992 and thyroid cancer incidence rate for any of the years between 1999 and 2019. Our results suggest that restrictions clearly seem to be effective in reducing lindane use, however, more research is needed for individual pesticides and thyroid cancer.

Keywords: endocrine disruptive chemicals; EDCs; thyroid cancer; pesticides; lindane; environmental health; epidemiology

Citation: Alsen, M.; Vasan, V.; Genden, E.M.; Sinclair, C.; van Gerwen, M. Correlation between Lindane Use and the Incidence of Thyroid Cancer in the United States: An Ecological Study. *Int. J. Environ. Res. Public Health* **2022**, *19*, 13158. https://doi.org/10.3390/ijerph192013158

Academic Editor: Paul B. Tchounwou

Received: 2 September 2022
Accepted: 9 October 2022
Published: 13 October 2022

1. Introduction

The incidence rate of thyroid cancer has increased significantly in recent decades in the United States (US) [1]. The increased incidence may be partially explained by the early detection of small, mainly papillary thyroid cancers (PTC) due to more frequent use of advanced imaging technologies, so- called overdiagnosis [2]. Although overdiagnosis- by definition- leads to the detection of small PTCs in early stage and should thus result in a decline of thyroid cancer specific mortality, studies have reported that the overall thyroid cancer mortality remained stable and mortality of distant or advanced stage thyroid cancer even increased [2]. Overdiagnosis has been estimated to explain ~50% of the increase in PTC incidence, so research into other risk factors that would explain this increase is needed [2].

Although previous epidemiological studies suggested that approximately 40% of thyroid cancer diagnoses in the US may be attributable to environmental factors, research investigating the role of environmental factors in cancer risk is limited [2]. Chemicals with endocrine disruptive properties are potential risk factors, because of their ability to mimic the thyroid hormones in the body [2,3]. Organochlorine pesticides (OCPs) are known endocrine disrupting chemicals. OCPs were widely used for residential and agricultural purposes since the 1940s until some of them, including dichlorodiphenyltrichloroethane (DDT) and hexachlorobenzene (HCB), were banned or restricted in the 1960s and 1970s because of their toxicity and carcinogenic effects [4–6]. OCPs are known to alter thyroid hormone levels due to their similar structure to triiodothyronine (T3) and thyroxine (T4) [6,7].

Existing literature has shown that exposure to specific OCPs has been associated with various types of malignancies [8,9], as was exposure to chlordane exposure associated with rectal cancer, dieldrin exposure with lung cancer, toxaphene exposure with melanoma, and chlordane/heptachlor exposure with leukemia [8]. One study found endosulfan and 4,4,-DDE exposure to be significantly associated with breast and blood cancer [10]. Another study has reported a positive association between heptachlor, methoxychlor, DDT, and HCB exposure, and breast cancer [11]. Several epidemiological studies investigating the association between OCPs and thyroid cancer risk showed conflicting results, although some literature supports the hypothesis that pesticides could play a major role in the thyroid cancer etiology [3,12].

In OCP, lindane is the y-isomer of the hexachlorocyclohexane (HCH) [13]. Lindane was extensively used as an insecticide for wood, crops, livestock, and seed, starting in the 1950s; however lindane use was restricted in agriculture between 1970 and 1990 [13], and discontinued for agricultural purposes in the US in 2006 [14]. Lindane is, however, still being formulated and used in prescription lotions, cream, and shampoos for head lice and scabies [14]. The most common acute side effects of lindane include seizures, dizziness, and headache, although high doses may be fatal [15]. The International Agency for Research on Cancer (IARC) has classified lindane as carcinogenic to humans (Group 1), while the EPA has classified lindane as a possible human carcinogen (B2/C) [14,16]. Lindane is listed in the Stockholm convention under Annex A (elimination), concluding that lindane is likely to lead to significant adverse human health and environmental effects [17]. Exposure to lindane is significantly associated with non-Hodgkin's lymphoma [18]. Blair et al. reported a 50% increased risk of non-Hodgkin's lymphoma among agricultural workers exposed to lindane [19].

Lindane has been detected in air, soil, water, and sediment, and can be absorbed into the body through inhalation and dermal exposure, but mostly through diet [20]. It is a persistent type of OCP and concentrates in the environment mostly in water, but can also be found in soil and air [21]. The reported half-life of lindane in the environment is long and varies between 11 years in seawater and 100 years in the Arctic Ocean [22]. The half-life of lindane in soil is approximately 2 weeks [20]. The isomers of lindane are lipohilic, allowing accumulation in fatty tissues for long periods; the beta isomer (beta-HCH) has a half-life of about seven years in fatty tissues [20].

To the best of our knowledge, only one study investigated the association between lindane exposure and thyroid cancer risk, and found a significantly positive association between lindane exposure and thyroid cancer risk among male pesticide users (hazard ratio (HR) = 1.74, confidence interval (CI): 1.06–2.84) [9]. Understanding the potential association between OCP exposure and thyroid cancer risk is imperative as some OCPs are still being used to date [23]. Given that lindane is a carcinogen and exposure to lindane has been associated with thyroid dysfunction, this study assessed the potential correlation between lindane exposure and thyroid cancer in the US by performing an ecological study of thyroid cancer incidence and lindane use.

2. Materials and Methods

Data on statewide age-adjusted thyroid cancer incidence rate (per 100,000 people) was obtained by gender from the Centers for Disease Control and Prevention (CDC) for all US states for 2019, the most recent year with this data available. To examine any possible link between lindane use and thyroid cancer before year 2019 we looked at the correlation between lindane use (kg/acres of cropland) in 1992 and statewide thyroid cancer incidence rates for each year between 1999 and 2019 for males and females combined. This incidence data has been compiled from selected cancer registries meeting US Cancer Statistics data quality criteria covering 99% of the US population [24]. The data includes all ages, races/ethnicities, and sex. The data was imported into the ArcGIS, a mapping software developed by the Environmental Systems Research Institute (version 10.8; ESRI, Redlands, CA, USA) [25].

County-level lindane use was obtained from the United States Geological Survey (USGS). Lindane use was estimated using two estimation methods in the USGS survey: EPest-low and EPest-high. These estimation methods, developed by Thelin and Stone (2013) [26], calculate the estimated pesticide use (EPest) rate use for each crop acre by year using surveys from the Crop Reporting Districts (CRD) in conjunction with data reported by the US Department of Agriculture National Agricultural Statistics Service (NASS) for all states except California [26]. EPest-high includes more counties with estimated data compared to EPest-low: EPest-high estimates the use based on pesticide-by-crop use in surrounding areas whereas EPest-low assumes zero use for missing data. EP-pest high estimates were used in the present study as this represents a better estimate of potential lindane exposure of the population.

To the best of our knowledge, the latency period between lindane/pesticide exposure and thyroid cancer is unknown. To take latency into consideration, the earliest year available for the lindane data (1992) was included in the present study. Furthermore, lindane data for the year 2007 was explored because this was the first year following the discontinuation of lindane use in 2006. For the included years, annual lindane use estimates were reported as kilograms (kg) by county, which was then summarized by state and divided by total acres cropland/1000 (kg/acres cropland). Data on the cropland acreage per state was obtained from the US Department of Agriculture (USDA) for the included years.

Statistical Analysis

Trends of overall lindane use between the years 1992 and 2007 was calculated using the Mann Kendall correlation test. Lindane use, presented in quintiles for the years 1992 and 2007, was then overlaid on the map of age-adjusted thyroid cancer incidence rate (2019) using the ArcGIS. The correlation between statewide lindane use (kg/acres cropland) for the years 1992 and 2007 and age-adjusted thyroid cancer incidence rates for the year 2019 was calculated using Spearman correlation by gender. The correlation between lindane use (kg/acres of cropland) in 1992 and statewide thyroid cancer incidence rates for each year between 1999 and 2019 was also calculated for males and females combined. Analyses were performed using the R software (version 4.1.2; Vienna, Austria).

3. Results

The age-adjusted thyroid cancer incidence rates in 2019 ranged from 8.0 to 19.6 cases per 100,000 persons, with the lowest age-adjusted thyroid cancer incidence rate in Mississippi and the highest reported thyroid cancer incidence rate in New York. For males, the age-adjusted incidence rate ranged from 4.50 to 10.8 cases per 100,000 persons with the lowest age-adjusted thyroid cancer incidence rate in Mississippi and the highest age-adjusted reported thyroid cancer incidence rate in New York. The age-adjusted incidence rates ranged from 11.40 to 30.1 cases per 100,000 persons for females with the lowest age-adjusted thyroid cancer incidence rate in Mississippi and the highest reported age-adjusted thyroid cancer incidence rate in North Dakota. Data on thyroid cancer incidence rates were not available for the state of Nevada. (Figures 1–3) (Supplementary Table S1).

There was no significant correlation between lindane use in 1992 and thyroid cancer incidence rates between 1999–2019 for males and females combined (Table 1). Overall lindane use decreased significantly between 1992 and 2007 ($T = -0.617$; $p < 0.001$) (Table 2). In 1992, the lindane estimates ranged from 0.0003 to 2.33 kg/ cropland in Washington and Georgia, respectively. In 2007, the lindane estimates ranged from 0.0015–0.80 kg/cropland in Oklahoma and Delaware, respectively (Supplementary Table S1). There was no significant correlation between lindane use in 1992 and thyroid cancer incidence rate in 2019 for males and females combined ($r = 0.17$; $p = 0.364$), for males ($r = 0.17$; $p = 0.368$) or females ($r = 0.17$; $p = 0.368$) (Figure 4). There was no significant correlation between lindane use in 2007 and thyroid cancer incidence rates in 2019 for males and females combined (r = −0.17; p = 0.327), for males (r = −0.17.; p = 0.329) or females (r = −0.203; p = 0.234) (Supplementary Figures S1–S4).

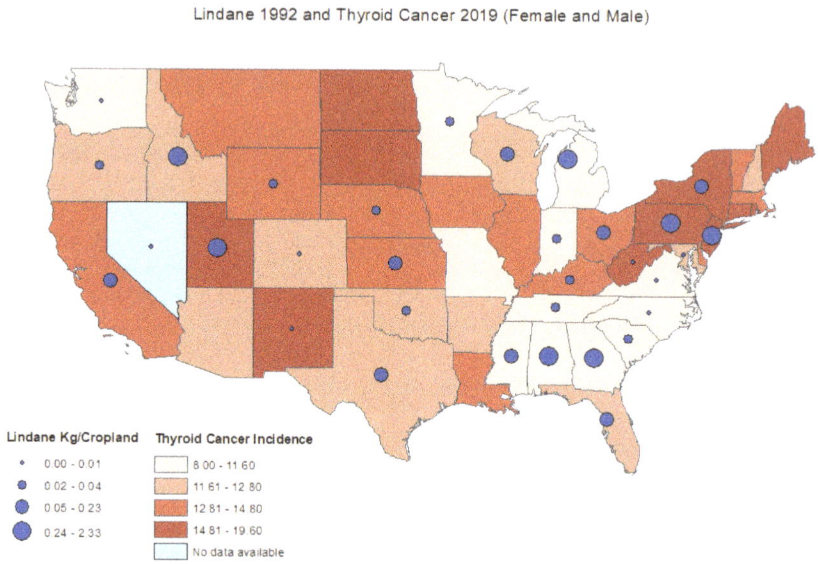

Figure 1. Distributions of age-adjusted thyroid cancer incidence rates per state and estimated lindane high-use (kg/Cropland) male and female.

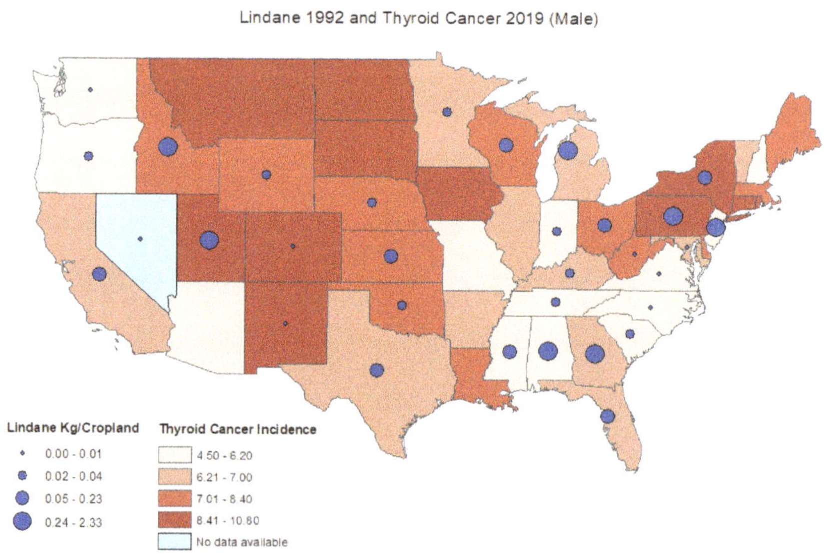

Figure 2. Distribution of age-adjusted thyroid cancer incidence rates per state and estimated lindane high-use (kg/Cropland) male.

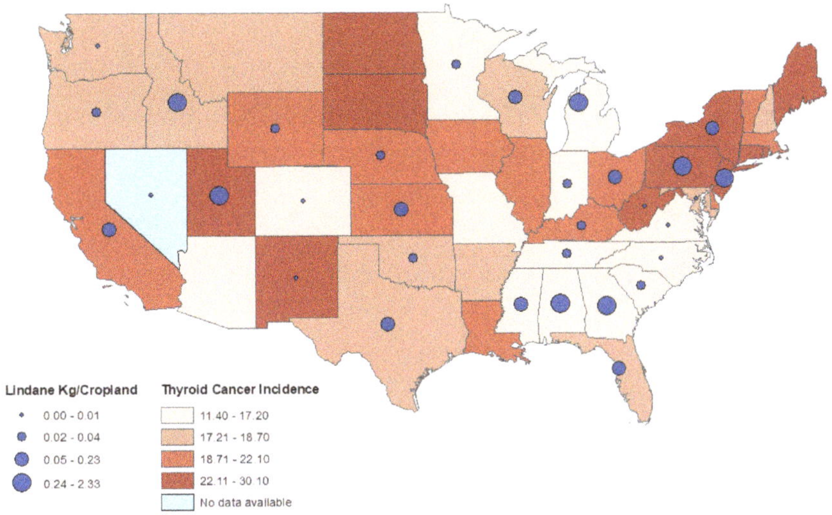

Figure 3. Distribution of age-adjusted thyroid cancer incidence rates per state and estimated lindane high-use (kg/Cropland) female.

Table 1. Lindane use in 1992 correlated with thyroid cancer incidence rates 1999–2019.

Year	Correlation Coefficient	*p*-Value
1999	0.021	0.911
2000	0.109	0.567
2001	0.059	0.753
2002	0.122	0.521
2003	0.133	0.482
2004	0.110	0.561
2005	0.06	0.764
2006	0.02	0.929
2007	0.05	0.791
2008	0.01	0.954
2009	0.06	0.756
2010	0.07	0.716
2011	0.07	0.725
2012	0.06	0.742
2013	−0.04	0.838
2014	0.13	0.482
2015	0.09	0.623
2016	0.14	0.453
2017	0.04	0.821
2018	0.09	0.611
2019	0.17	0.364

Table 2. Overall lindane use between 1992 and 2007.

Year	Mann-Kendall	*p*-Value
1992–2007	−0.617	<0.001

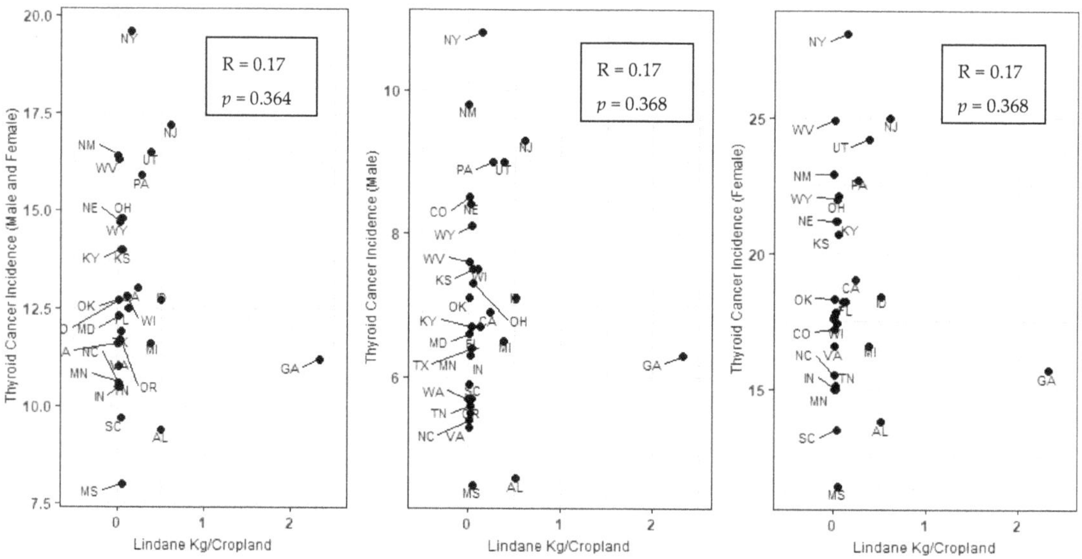

Figure 4. Spearman correlation of age-adjusted thyroid cancer incidence rates per state (2019) and statewide lindane use (kg/Cropland) (1992).

4. Discussion

This first ecological study investigating a potential correlation between lindane use and thyroid cancer incidence showed clear statewide differences in the patterns of lindane use as well as thyroid cancer incidence rates. The overall lindane use decreased significantly between 1992 and 2007, likely indicating the effect of implemented restrictions in lindane use. While there was no evident correlation between thyroid cancer and lindane use, it is worth noting that certain states had high lindane use and high thyroid cancer rates in the northeast area, including Delaware, Rhode Island, New York, and New Jersey.

The present study highlights the effectiveness of pesticide restrictions with lindane use trending downwards over time. The agricultural use of lindane has been restricted nationwide since the 1970s, but California was the first state to ban the pharmaceutical use of lindane in 2002, after detecting lindane in wastewater, which then entered lakes, rivers, and the ocean. California engineers calculated that one single lice or scabies treatment containing lindane would contaminate 6 million gallons of water [27]. After the ban in California in 2002, other states including New York, Washington, Maine, and Michigan, began similar discussions about a potential lindane ban [28]. In 2005, the WHO confirmed that the carcinogenicity of lindane [29], and in 2006 Lindane was banned in the US for agricultural purposes by the EPA [14]. Nevertheless, the use of lindane as a lice or scabies treatment is still approved by the US Food and Drug Administration (FDA) [14]. In the EU, all uses of lindane have been banned [22]. Overall, lindane is currently banned in 52 countries and restricted in 33 countries [29].

There are some environmental factors that have been shown to be directly linked to thyroid cancer risk [4]. Exposure to radiation has been proven carcinogenic, where the thyroid is particularly susceptible to radioiodine exposure [4]. Radioactive exposure can lead to DNA lesion and cell death [4]. One study reported that thyroid cancer has been more prevalent in children ages 0–4 years who lived in an area with high levels of ionizing radiations [30]. Other environmental exposures such as flame-retardants, phthalates, polyfluoroalkyl substances (PFAS) and polychlorinated biphenyls (PCBs) are known EDCs with thyroid-disruptuve properties, and interfere with endocrine signalling at cellular and molecular levels [3]. Metals such as cadmium, manganese and lead induce inflammation and the immune response to autoantigens. They also induce inflammatory reactions in the thyroid [4]. Air pollution has shown to have a similar inflammatory response to autoantigens and produce reactive oxygen species [4].

Lindane may exert thyroid-related carcinogenic effects through the generation of oxidative stress, [31], which has shown to be an important contributor to the formation and progression of cancer [32], and through increased mutations and DNA damage [33]. According to IARC, there is weak evidence that lindane alters cell proliferation or death, or induces chronic inflammation [13]. Furthermore, there is moderate evidence that lindane modulates receptor-mediated effects and that it is genotoxic [14]. Exposure to lindane treatment in human cells in vitro showed a chromosomal change, sister-chromatid exchange, and increased micronucleus formation, while results were mixed or negative for DNA-adduct formation and other types of DNA damage. There is no reliable data on genotoxicity in humans [14]. Nevertheless, there is strong evidence that lindane is immunosuppressive and that it induces oxidative stress [13]. An animal study reported excess thyroid C-cell adenoma among female rats after being exposed to lindane [34]. Lindane use in humans and the association with thyroid cancer was investigated using the Agricultural Health Study (AHS), a prospective cohort including 57,310 licensed private and commercial pesticide applicators. The study reported a positive association between lindane exposure and thyroid cancer among male participants (hazard ratio (HR) = 1.74, confidence interval (CI): 1.06–2.84) [9]. It is therefore of importance to further evaluate the effects of lindane exposure among those who live, or work, in high risk areas, such as farmers and their families, since occupational exposure poses a unique risk.

Although chronic, low dose exposure may be more difficult to measure and evaluate, some evidence suggests that chronic exposure to low dose lindane may be of greater health concern resulting in neurobehavioral, neurochemical, and electrophysiological effects, due to induced alteration in GABA receptor activity [35]. Multiple studies have investigated the effects of lindane on the thyroid gland. Prenatal exposure to beta-HCH, a transformable isomer of lindane, has been shown to alter thyroid hormones in newborns [36]. An increased risk of hypothyroidism was found among female spouses of pesticide applicators (adjusted odds ratio (ORadj) = 1.2 (95% CI: 1.0, 1.6)) after exposure to several OCPs combined, where an elevated but non-significant increased odds was found for hypothyroidism after being exposed to lindane (ORadj = 1.5 (95% CI: 0.93, 2.4)), further highlighting the importance of examining highly exposed populations including their family members [37]. Furthermore, a significant positive association between lindane exposure and hypothyroidism was found among pesticide applicators of 62 years and older (HR = 1.54; 95% CI: 1.23, 19.4) [38]. Thyroid dysfunction following lindane exposure has been confirmed in animal studies reporting decreased thyroid hormones after exposure to lindane [14,18,39].

Exposure to lindane and other OCPs during critical developmental periods in early life may negatively impact thyroid health, which has been assessed in some human and animal studies. A study reported that exposure to DDT metabolites before the age of 14 is associated with breast cancer in young women, confirming the need to evaluate the role of OCPs exposure during developmental periods [40]. Evidence has been provided that lindane transfers through the placenta. Prenatal lindane exposure in the offspring of exposed rats induced postnatal alterations in xenobiotic-metabolizing cytochrome P450s in the brain and liver [41]. Other OCPs also pass the placenta as a positive association was found between exposure to DDT and HCH during pregnancy and intra-uterine growth retardation [32,41,42]. Animal studies have found an increased risk of altered neurochemical and behavioral effects in developing rats after prenatal exposure to lindane, as well as long-lasting effects on reproduction later in life [43,44]. However, little is known about lindane exposure during critical developmental windows in utero and early life, and the risk of thyroid disease/cancer later in life, thus warranting further investigation.

To the best of our knowledge, this is the first ecological study investigating the association between lindane exposure and thyroid cancer risk in large representative populations. Therefore, we believe our report provides valuable new information in relation to the effects of lindane use. Strengths include the large number of states included in the study, despite some missing data for a few states. This study has several limitations. First, ecological studies are observational and look at entire populations in question, so no conclusions can be drawn at individual level nor on causality. Ecological studies are also not useful for hypothesis testing because of potential uncontrolled confounding. It is important to note that lindane application does not necessarily result in actual exposure at individual level. In addition, CDC only include all thyroid cancer histology types and no distinction of different types of thyroid cancer is made. It would be beneficial for future studies to examine lindane exposure and its association with different aggressive histological variants.

5. Conclusions

Although no correlation between statewide lindane use and thyroid cancer risk was found, and lindane use trended downwards following restrictions and bans, future studies are needed to explore the effects of lindane exposure on the thyroid gland, mainly given the carcinogenic and endocrine disruptive properties of lindane combined with the long environmental half-life of lindane leaving the population at risk of exposure long after usage restrictions. Research is particularly warranted in areas of high exposure and among highly exposed populations, including long-term, low-dose exposure to lindane, as well as in combination with exposure to other OCPs.

Supplementary Materials: The following supporting information can be downloaded at: https://www.mdpi.com/article/10.3390/jerph192013158/s1, Figure S1: distribution of age-adjusted thyroid cancer incidence rates per state and estimated lindane high-use (kg/Cropland) Male and Female (2007). Figure S2: distribution of age-adjusted thyroid cancer incidence rates per state and estimated lindane high-use (kg/Cropland) Male (2007). Figure S3: distribution of age-adjusted thyroid cancer incidence rates per state and estimated lindane high-use (kg/Cropland) Female (2007). Figure S4: Spearman correlation of age-adjusted thyroid cancer incidence rates per state (2019) and statewide lindane use (kg/Cropland) (2007). Table S1: statewide lindane use kg/cropland (earliest and most recent year available). Thyroid cancer incidence rate per/100,000 by sex.

Author Contributions: Conceptualization: M.v.G., C.S. and M.A.; methodology, M.v.G. and M.A.; software. M.A.; validation, M.v.G., C.S., M.A. and V.V.; formal analysis, M.A., investigation, M.v.G., M.A. and V.V.; resources. M.v.G. and E.M.G.; data curation, M.A.; writing—origianl draft preparation, M.v.G., M.A. and V.V.; writing—review and editing, M.v.G., M.A., V.V., C.S. and E.M.G.; visualization, M.v.G., M.A. and C.S.; supervision, M.v.G.; project administration, M.v.G. and M.A.; funding acquisition, E.M.G. All authors have read and agreed to the published version of the manuscript.

Funding: This research received no external funding.

Institutional Review Board Statement: Not applicable.

Informed Consent Statement: Not applicable.

Data Availability Statement: Publicly available datasets were analyzed in this study. This data can be found here: Data | U.S. Geological Survey at https://www.usgs.gov/products/data and USCS Data Visualizations—CDC at https://gis.cdc.gov/Cancer/USCS/#/AtAGlance/ (accessed on 1 September 2022).

Conflicts of Interest: The authors declare no conflict of interest.

References

1. Davies, L.; Welch, H.G. Increasing incidence of thyroid cancer in the United States, 1973–2002. *JAMA* **2006**, *295*, 2164–2167. [CrossRef] [PubMed]
2. Kitahara, C.M.; Sosa, J.A. The changing incidence of thyroid cancer. *Nat. Rev. Endocrinol.* **2016**, *12*, 646–653. [CrossRef] [PubMed]

3. Alsen, M.; Sinclair, C.; Cooke, P.; Ziadkhanpour, K.; Genden, E.; van Gerwen, M. Endocrine Disrupting Chemicals and Thyroid Cancer: An Overview. *Toxics* **2021**, *9*, 14. [CrossRef] [PubMed]
4. Fiore, M.; Conti, G.O.; Caltabiano, R.; Buffone, A.; Zuccarello, P.; Cormaci, L.; Cannizzaro, M.A.; Ferrante, M. Role of Emerging Environmental Risk Factors in Thyroid Cancer: A Brief Review. *Int. J. Environ. Res. Public Health* **2019**, *16*, 1185. [CrossRef] [PubMed]
5. Leemans, M.; Couderq, S.; Demeneix, B.; Fini, J.B. Pesticides with Potential Thyroid Hormone-Disrupting Effects: A Review of Recent Data. *Front. Endocrinol.* **2019**, *10*, 743. [CrossRef]
6. Deziel, N.C.; Warren, J.L.; Huang, H.; Zhou, H.; Sjodin, A.; Zhang, Y. Exposure to polychlorinated biphenyls and organochlorine pesticides and thyroid cancer in connecticut women. *Environ. Res.* **2021**, *192*, 110333. [CrossRef]
7. Freire, C.; Koifman, R.J.; Sarcinelli, P.N.; Rosa, A.C.S.; Clapauch, R.; Koifman, S. Long-term exposure to organochlorine pesticides and thyroid status in adults in a heavily contaminated area in Brazil. *Environ. Res.* **2013**, *127*, 7–15. [CrossRef]
8. Purdue, M.P.; Hoppin, J.A.; Blair, A.; Dosemeci, M.; Alavanja, M.C. Occupational exposure to organochlorine insecticides and cancer incidence in the Agricultural Health Study. *Int. J. Cancer* **2007**, *120*, 642–649. [CrossRef]
9. Lerro, C.C.; Freeman, L.E.B.; DellaValle, C.T.; Andreotti, G.; Hofmann, J.N.; Koutros, S.; Parks, C.G.; Shrestha, S.; Alavanja, M.C.; Blair, A.; et al. Pesticide exposure and incident thyroid cancer among male pesticide applicators in agricultural health study. *Environ. Int.* **2021**, *146*, 106187. [CrossRef]
10. Attaullah, M.; Yousuf, M.J.; Shaukat, S.; Anjum, S.I.; Ansari, M.J.; Buneri, I.D.; Tahir, M.; Amin, M.; Ahmad, N.; Ullah Khan, S. Serum organochlorine pesticides residues and risk of cancer: A case-control study. *Saudi J. Biol. Sci.* **2018**, *25*, 1284–1290. [CrossRef]
11. Mekonen, S.; Ibrahim, M.; Astatkie, H.; Abreha, A. Exposure to organochlorine pesticides as a predictor to breast cancer: A case-control study among Ethiopian women. *PLoS ONE* **2021**, *16*, e0257704.
12. Han, M.A.; Kim, J.H.; Song, H.S. Persistent organic pollutants, pesticides, and the risk of thyroid cancer: Systematic review and meta-analysis. *Eur. J. Cancer Prev.* **2019**, *28*, 344–349. [CrossRef] [PubMed]
13. Lindane–World Health Organization. Available online: https://monographs.iarc.who.int/wp-content/uploads/2018/07/mono113-02.pdf (accessed on 23 August 2022).
14. Lindane (Gamma-Hexachlorocyclohexane)–US EPA. Available online: https://www.epa.gov/sites/default/files/2016-09/documents/lindane.pdf (accessed on 23 August 2022).
15. Brown, V.J. *Life after Lindane in California: Water Concentrations, Poison Control Calls Drop Following Ban*; National Institute of Environmental Health Sciences: Durham, CA, USA, 2008.
16. IARC Monographs on the Identification of Carcinogenic Hazards to Humans. Available online: https://monographs.iarc.who.int/list-of-classifications (accessed on 23 August 2022).
17. Lindane-Stockholm Convention. Available online: http://chm.pops.int/Implementation/Alternatives/AlternativestoPOPs/ChemicalslistedinAnnexA/Lindane/tabid/5865/Default.aspx (accessed on 23 August 2022).
18. Schinasi, L.; Leon, M.E. Non-Hodgkin lymphoma and occupational exposure to agricultural pesticide chemical groups and active ingredients: A systematic review and meta-analysis. *Int. J. Environ. Res. Public Health* **2014**, *11*, 4449–4527. [CrossRef] [PubMed]
19. Blair, A.; Cantor, K.P.; Zahm, S.H. Non-Hodgkin's lymphoma and agricultural use of the insecticide lindane. *Am. J. Ind. Med.* **1998**, *33*, 82–87. [CrossRef]
20. Biomonitoring Summary CDC. Available online: https://www.cdc.gov/biomonitoring/Hexachlorocyclohexane_BiomonitoringSummary.html (accessed on 23 August 2022).
21. Misra, K.; Brar, K.S.; Verma, M.; Tyagi, R.; Trivedy, R.; Sharma, S. Lindane-A contaminant of global concern. *I Control Pollut.* **2007**, *23*, 169–187.
22. Lindane (Persistent Organic Pollutant) in the EU. Available online: https://www.europarl.europa.eu/RegData/etudes/STUD/2016/571398/IPOL_STU(2016)571398_EN.pdf (accessed on 23 August 2022).
23. Jayaraj, R.; Megha, P.; Sreedev, P. Organochlorine pesticides, their toxic effects on living organisms and their fate in the environment. *Interdiscip. Toxicol.* **2016**, *9*, 90–100. [CrossRef]
24. U.S. Cancer Statistics Data Visualizations Tool. Available online: www.cdc.gov/cancer/dataviz (accessed on 23 August 2022).
25. ArcGIS. Available online: https://www.esri.com/en-us/arcgis/about-arcgis/overview (accessed on 23 August 2022).
26. Thelin, G.P.; Stone, W.W. Estimation of annual agricultural pesticide use for countries of the conterminous United States, 1992–2009. In *U.S. Geological Survey Scientific Investigations Report*; US Department of the Interior, US Geological Survey: Sacramento, CA, USA, 2013; Volume 54.
27. Humphreys, E.H.; Janssen, S.; Heil, A.; Hiatt, P.; Solomon, G.; Miller, M.D. Outcomes of the California ban on pharmaceutical lindane: Clinical and ecologic impacts. *Environ. Health Perspect.* **2008**, *116*, 297–302. [CrossRef]
28. Schafer, K. Lindane; Partial Ban in California Spurs Action in Other States. *Glob. Pestic. Campaign.* **2004**, *14*, 12.
29. IARC Monographs Evaluate DDT, Lindane, and 2,4-D World Health Organization. Available online: https://www.iarc.who.int/wp-content/uploads/2018/07/pr236_E.pdf (accessed on 23 August 2022).
30. Spinelli, C.; Ghionzoli, M.; Oreglio, C.; Sanna, B.; De Napoli, L.; Morganti, R.; Antonelli, A.; Morabito, A.; Miccoli, P. Increased trend of thyroid cancer in childhood over the last 30 years in EU countries: A call for the pediatric surgeon. *Eur. J. Pediatr.* **2022**, *181*, 3907–3913. [CrossRef]

31. Salimi, F.; Asadikaram, G.; Abolhassani, M.; Nejad, H.Z.; Bagheri, F.; Kahnouei, M.M.; Abbasi-jorjandi, M.; Sanjari, M. Association between Organochlorine Pesticides and Thyroid Tumors; an in-silico and in-vivo Case Control Study. *Res. Sq.* **2020**, 1–38. [CrossRef]
32. Klaunig, J.E. Oxidative Stress and Cancer. *Curr. Pharm. Des.* **2018**, *24*, 4771–4778. [CrossRef] [PubMed]
33. Reuter, S.; Gupta, S.C.; Chaturvedi, M.M.; Aggarwal, B.B. Oxidative stress, inflammation, and cancer: How are they linked? *Free Radic. Biol. Med.* **2010**, *49*, 1603–1616. [CrossRef] [PubMed]
34. Program, N.T. Bioassay of lindane for possible carcinogenicity. *Natl. Cancer Inst. Carcinog. Tech. Rep. Ser.* **1977**, *14*, 1–99.
35. Anand, M.; Agrawal, A.; Rehmani, B.; Gupta, G.; Rana, M.; Seth, P. Role of GABA receptor complex in low dose lindane (HCH) induced neurotoxicity: Neurobehavioural, neurochemical and electrophysiological studies. *Drug Chem. Toxicol.* **1998**, *21*, 35–46. [CrossRef]
36. Alvarez-Pedrerol, M.; Ribas-Fitó, N.; Torrent, M.; Carrizo, D.; Garcia-Esteban, R.; Grimalt, J.O.; Sunyer, J. Thyroid disruption at birth due to prenatal exposure to beta-hexachlorocyclohexane. *Environ. Int.* **2008**, *34*, 737–740. [CrossRef]
37. Goldner, W.S.; Sandler, D.P.; Yu, F.; Hoppin, J.A.; Kamel, F.; Levan, T.D. Pesticide use and thyroid disease among women in the Agricultural Health Study. *Am. J. Epidemiol.* **2010**, *171*, 455–464. [CrossRef]
38. Shrestha, S.; Parks, C.G.; Goldner, W.; Kamel, F.; Umbach, D.M.; Ward, M.H.; Lerro, C.C.; Koutros, S.; Hofmann, J.N.; Freeman, L.E.B.; et al. Pesticide Use and Incident Hypothyroidism in Pesticide Applicators in the Agricultural Health Study. *Environ. Health Perspect.* **2018**, *126*, 97008. [CrossRef]
39. Schantz, S.L.; Widholm, J.J. Cognitive effects of endocrine-disrupting chemicals in animals. *Environ. Health Perspect.* **2001**, *109*, 1197–1206. [CrossRef]
40. Cohn, B.A.; Wolff, M.S.; Cirillo, P.M.; Sholtz, R.I. DDT and breast cancer in young women: New data on the significance of age at exposure. *Environ. Health Perspect.* **2007**, *115*, 1406–1414. [CrossRef]
41. Johri, A.; Dhawan, A.; Singh, R.L.; Parmar, D. Persistence in alterations in the ontogeny of cerebral and hepatic cytochrome P450s following prenatal exposure to low doses of lindane. *Toxicol. Sci.* **2008**, *101*, 331–340. [CrossRef]
42. Siddiqui, M.K.; Srivastava, S.; Srivastava, S.P.; Mehrotra, P.K.; Mathur, N.; Tandon, I. Persistent chlorinated pesticides and intra-uterine foetal growth retardation: A possible association. *Int. Arch. Occup. Environ. Health* **2003**, *76*, 75–80. [CrossRef] [PubMed]
43. Rivera, S.; Rosa, R.; Martínez, E.; Suñol, C.; Serrano, M.T.; Vendrell, M.; Rodríguez-Farré, E.; Sanfeliu, C. Behavioral and monoaminergic changes after lindane exposure in developing rats. *Neurotoxicol. Teratol.* **1998**, *20*, 155–160. [CrossRef]
44. Brannen, K.C.; Devaud, L.L.; Liu, J.; Lauder, J.M. Prenatal exposure to neurotoxicants dieldrin or lindane alters tert-butylbicyclophosphorothionate binding to GABA(A) receptors in fetal rat brainstem. *Dev. Neurosci.* **1998**, *20*, 34–41. [CrossRef] [PubMed]

International Journal of
Environmental Research and Public Health

MDPI

Review

Forty-Year Biomonitoring of Environmental Contaminants in Russian Arctic: Progress, Gaps and Perspectives

Alexey A. Dudarev [1] and Jon Oeyvind Odland [2,3,4,*]

1 Northwest Public Health Research Center, 191036 St. Petersburg, Russia
2 Department of Public Health and Nursing, NTNU, The Norwegian University of Science and Technology, 7034 Trondheim, Norway
3 Department of General Hygiene, I.M. Sechenov First Moscow State Medical University, Ministry of Health of Russia, 119992 Moscow, Russia
4 Institute of Ecology, National Research University Higher School of Economics, 101000 Moscow, Russia
* Correspondence: jon.o.odland@ntnu.no

Abstract: This article presents a comprehensive survey of the studies on the biomonitoring of persistent organic pollutants (POPs) and metals in biota and human tissues on the territory of the Russian Arctic. It is concluded that a relatively small number of studies were carried out during the last 40 years; for several Russian Arctic regions there is no data up to date, and for some regions the data are scarce, with most of the studies carried out in 1990s, followed by the large-scale GEF/AMAP/RAIPON project of 2001–2004 covering four regions. After that, single projects have been implemented in a few regions. Only the Nenets okrug and the Chukotka okrug (and hardly the Murmansk oblast) can be attributed as the regions where the biomonitoring of contaminants was carried out during last decades on several occasions, and for which the content of POPs and metals in biota and the human organism was assessed in 12–15-year dynamic trends (at least "at three points"). For the rest of the Russian Arctic territories, only fragmentary "cross-sections" of biomonitoring data is available, mainly obtained in the 1990s or early 2000s, which do not allow judging either the dynamics of the processes or the current state of affairs. The overwhelming majority of the studies in the Russian Arctic (more than 90%) were carried out within the framework of international projects, i.e., with cofinancing, assistance and contribution (including laboratory analyses) from the foreign colleagues and partners. The shortcomings of the Russian system of biomonitoring, including the weakness of the laboratory and research base, are considered. Perspectives of the Russian Arctic biomonitoring are discussed in detail, with the proposal of the elaboration of the national Russian Arctic Contaminants Program (RACP).

Keywords: biomonitoring; Russian Arctic; biota; wildlife species; traditional foods; indigenous people; human tissues; blood; breastmilk; contaminants; POPs; DDTs; PCBs; metals; exposure; health risks

Citation: Dudarev, A.A.; Odland, J.O. Forty-Year Biomonitoring of Environmental Contaminants in Russian Arctic: Progress, Gaps and Perspectives. *Int. J. Environ. Res. Public Health* **2022**, *19*, 11951. https://doi.org/10.3390/ijerph191911951

Academic Editors: Roel Vermeulen, Lauren Petrick and Maaike van Gerwen

Received: 14 July 2022
Accepted: 19 September 2022
Published: 21 September 2022

1. Introduction

1.1. Biomonitoring of Persistent Toxic Substances (PTS)

Within the scope of long-term international scientific collaboration in the circumpolar Arctic, the biomonitoring of environmental contaminants is understood as a systematic measurement of persistent toxic substances (PTS) in human bodily fluids and tissues, as well as in biota (wildlife species). Biomonitoring is the basis for assessing human exposure to contaminants, as well as for studying the biological effects and health risks associated with the exposure to PTS accumulated in the human organism.

Regarding human biological media, this review focuses on the levels of contaminants in blood (and breastmilk) because, in the context of circumpolar collaboration, all participating countries have collected blood samples (and less frequently breastmilk) for international comparison. Hair, nails, urine, feces and other matrices are not discussed here.

Along with human biological media, the objects of the biomonitoring of contaminants in the Arctic include various wildlife species used by the Arctic peoples as food products. Local species of flora and fauna, being included in food chains (terrestrial, freshwater or marine), are exposed to PTS through bioaccumulation and biomagnification processes. Consumption of local foods which contain elevated concentrations of contaminants leads to the increased exposure and body burdens of the population to these contaminants.

The biomonitoring of contaminants in the Arctic covers two main groups of PTS: persistent organic pollutants (POPs) and metals. POPs are represented mainly by polyhalogenated hydrocarbons, some of which are included in the list of the Stockholm Convention as the most dangerous compounds banned for production, use and transportation, and which are subjected to extermination. All POPs are xenobiotics, but not all of them are products of intentional chemical synthesis, due to which the pesticides and technical additives were produced and effectively used in agriculture and industry for decades all over the world, circulated and accumulated in the planet's environment, and gradually increasing the impact and danger to the living beings and humans. Partially, POPs have entered and continue to enter the planet's biosphere as byproducts of various industrial productions, with emissions from power plants and boilers running on fossil fuels, with waste incineration, fires, etc. Some banned POPs are still produced in a number of countries and continue to be used, in particular in agriculture and for the control of malaria and other vector-borne diseases.

Metals (including mercury, lead and cadmium) naturally occur widely in nature, and exist in rocks, soil, water and biota. Metals are transported over vast distances due to rock weathering, forest fires and volcanic eruptions. Industrial sources (e.g., mining, ferrous and nonferrous metallurgy, fuel and waste combustion) account for one to two thirds of the total metal contamination of the Arctic.

PTS, represented by substances and compounds of an inorganic or organic nature, natural origin or xenobiotics, are characterized by a combination of several specific features; they are:

- Stable in the environment (for years or decades), resistant to thermolysis, photolysis, hydrolysis and chemical and bacteriological degradation;
- Could be transferred globally (thousands of kilometers); long-range transport of PTS to the Arctic occurs with atmospheric flows, rivers, seas and ocean currents;
- Able to bioaccumulate, reaching high levels in the upper links of food chains due to biomagnification processes; concentrations of PTS in the organisms of top predators can be from thousands to millions times higher than those in the air, soil, water and organisms of the lower tiers of the food chains;
- Extremely slowly metabolized and excreted from organisms, which contributes to the long-term accumulation of significant doses of PTS, even at their low concentrations in food. This phenomenon has gained particular importance in assessing the exposure to PTS of the indigenous people of the Arctic, whose diet is mainly based on the consumption of local (country, traditional) foods obtained by hunting and fishing;
- Toxic and hazardous to health in low (and ultra-low) concentrations. The lipophilicity of many POPs contributes to their accumulation in adipose tissue and lipid-rich organs (including the brain and spinal cord). The high permeability of the blood–brain barrier for many POPs predetermines the risk of their direct toxic effect on the central nervous system and the endocrine sphere. An increased content of POPs in the body can provoke serious disorders in the neuroendocrine, immune and reproductive spheres, in the processes of fetal development and in antitumor resistance. Some POPs are "hormonal mimics" that inhibit the production of natural hormones and disrupt the normal processes regulated by the endocrine glands. Many POPs penetrate the placental barrier, entering the child's body during fetal development and in infancy—with mother's milk.

1.2. Circumpolar Biomonitoring of Contaminants: Historical Milestones

The biomonitoring of contaminants in the circumpolar Arctic began in 1960 with the first study in Alaska. The DDT and DDE levels were analyzed in the local foods and human body fat in Eskimos, who lived in isolated small settlements where there was little or no use of insecticides and who ate food of local origin with maximal intake of animal fat. The food samples were collected in the villages of Shungnak, Kotzebue, Gambell, Hooper Bay and Point Hope. The studied foods which made up the major portion of the Eskimos' diet included various fresh and dried fish, eider duck, white owl; fat, oil or meat from beaver, caribou, moose, seal, bearded seal, walrus, whale, beluga and polar bear; and miscellaneous foods including cranberries, salmonberries and wild rhubarb [1].

Starting from the 1960s, the first studies on POPs in birds, marine mammals and fish were carried out in some regions of the circumpolar Arctic. During the Alaskan expedition in summer 1966 along the Yukon River (from Circle, Alaska to Castle Rock, Y.T.), the samples of eggs, fat, muscle, liver, brain of Alaskan Peregrines and the prey of peregrines (boreal birds, seed-eating and insectivorous passerines, sandpipers) were collected and analyzed (by electron gas chromatography) for DDTs and dieldrin. The authors concluded that their data demonstrated that, even in so remote a region as interior Alaska, the peregrine's food chain is contaminated with significant, measurable quantities of persistent residues of the chlorinated hydrocarbons [2]. From 1967 to 1970, the data on Falco peregrinus eggs were collected from three geographic Alaskan populations: on Amchitka Island in the Aleutians, along the Yukon River in the interior taiga and along the Colville River in the foothill tundra of the Arctic Slope. It was concluded that changes in eggshell thickness and the pesticide residues reflected different degrees of exposure to DDT. There was a significant negative correlation between shell thickness and DDE content in peregrine eggs [3].

The first data on POPs in marine mammals were published by Holden in the *Nature* journal in 1967 [4], which demonstrated significant levels of DDT and dieldrin in the blubber of seals and porpoises in the Scottish and Canadian North Atlantic. Soon, the detailed report appeared on worldwide seals contamination by POPs. Seal blubber analyses have shown that Arctic, Antarctic and Pacific Ocean waters had a lower concentration of organochlorine pesticide residues than did the waters of the Baltic Sea, North Sea and Irish Sea; dieldrin and PCB levels were the highest around Great Britain; the highest DDT levels were recorded in the Baltic and the Gulf of the St. Lawrence [5]. The detailed reports on DDT-related compounds and PCBs in the ringed seal and beluga from Canada and Greenland were published in the early 1970s [6–8]. Also in the early 1970s, the first publications on elevated levels of POPs in northern freshwater and anadromous fish species appeared [9,10]; the latter publication was the first to report OCs in polar bears.

The biological effects of organochlorine compounds in Arctic ecosystems have not previously been studied, with the exception of the impact on birds of prey, the decline in populations of which in the Arctic and temperate zones of the Northern Hemisphere, due to the beginning of the widespread use of pesticides, has been documented since the mid-1960s. Concerns about the potential human health impacts of POPs in the Arctic first appeared in the 1980s, when it was discovered that the breastmilk of indigenous northern Quebec (Inuit) women contained extremely high PCB levels (3–5 times higher than that of white women of southern Quebec) [11]. The results of studies of POPs in the diet of the Inuit of northern Canada in the late 1980s and early 1990s [12] demonstrated that namely the dietary route of human exposure to POPs determines the high levels of contaminants in the breastmilk and blood of indigenous Arctic peoples. The revealed patterns caused the need for additional information about contaminants in "traditional food products" and gave a powerful stimulus for further research in the circumpolar Arctic, including the beginning of such research in Russia.

1.3. Establishment of AMAP and Human Health Assessment Group

At the meeting in Rovaniemi (Finland) in June 1991, the ministers of eight Arctic countries (Canada, Denmark, Finland, Iceland, Norway, Russia, Sweden and the USA) signed the Declaration on the Protection of the Arctic Environment. Moreover, the Arctic Environmental Protection Strategy (AEPS) was adopted at this meeting, which stated that "The Arctic is highly vulnerable to contaminants, and most of its people and culture are directly dependent on the health of the region's ecosystem." As part of the Declaration, the Arctic Monitoring and Assessment (AMAP) was established to monitor contaminants and assess their impact on all components of the Arctic environment.

Under the umbrella of the AMAP, in October 1992, the Human Health Assessment Group (HHAG) was formed in Nuuk (Greenland), which identified POPs and metals as priority contaminants subjected to biomonitoring in the Arctic. For almost 30 years, the AMAP HHAG has been the "headquarters for coordination" of international circumpolar biomonitoring of contaminants; it generalizes and systematizes the results of the national monitoring studies of POPs and metals in human media and biota food webs, examines the peculiarities of PTS bioaccumulation and biomagnifications in local food chains (terrestrial, freshwater, marine), assesses the intake of PTS by the humans (exposure levels) and evaluates the effects of exposure to contaminants and human health risks at various territories of the circumpolar Arctic.

The AMAP HHAG has prepared and published five reports of "Human Health in the Arctic": 1998, 2002, 2009, 2015 and 2021 [13–17]; the AMAP 2015 report was translated into Russian and published in 2018 [18]. The chronology of the AMAP reports up to 2015 are inclusive, and their quintessence was described in the Russian journal of *Human Ecology* [19].

1.4. Geography of Biomonitoring of Contaminants in the Russian Arctic

The Russian Arctic territory (called by the government as the "Arctic Zone of the Russian Federation") today, in 2022, consists of 11 separate territories: Murmansk Oblast, Karelia republic (northern districts), Arkhangelsk Oblast (northern districts), Nenets autonomous okrug, Komi republic (northern districts), Yamalo-Nenets autonomous okrug, Taimyr Dolgano-Nenets District of the Krasnoyarsk kraj, Turukhansky District of the Krasnoyarsk kraj, Evenkiysky District of the Krasnoyarsk kraj (northern areas), Yakutia (Sakha) republic (northern districts) and the Chukotka autonomous okrug.

Generally, the territory of the Russian Arctic is located above the latitude 64° N (Figure 1). In this review we do not consider data obtained from regions (or districts) located to the south of the designated territories. Below, we present a spatiotemporal "picture" of all the studies (which we managed to find) carried out in the Russian Arctic (since the beginning of 1977 and up to 2021) related to the biomonitoring of contaminants both in biota and in the human body. The summarized information is also presented in Tables 1 and 2.

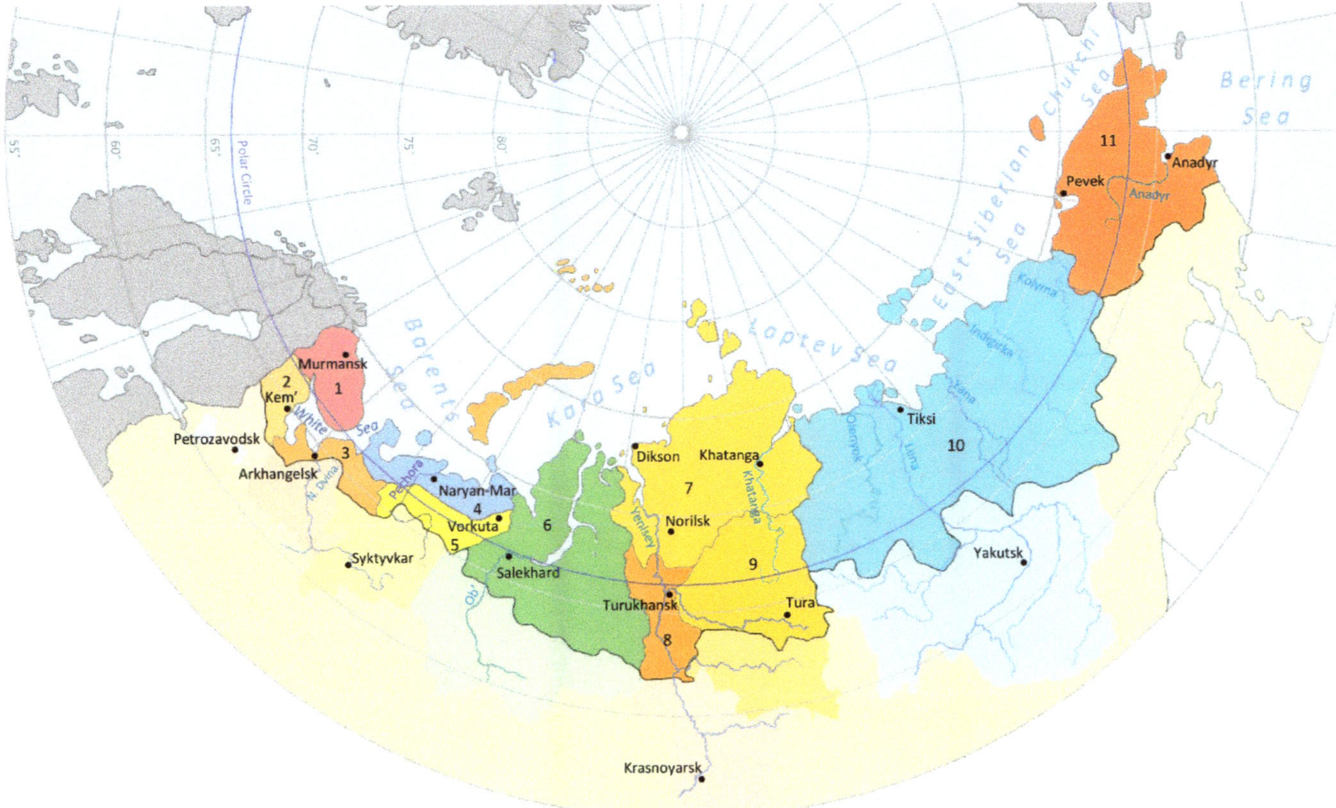

Figure 1. Map of the Russian Arctic territory ("Arctic Zone of the Russian Federation"): 1—Murmansk Oblast; 2—Karelia republic (northern districts); 3—Arkhangelsk Oblast (northern districts); 4—Nenets autonomous okrug; 5—Komi republic (northern districts); 6—Yamalo-Nenets autonomous okrug; 7—Taimyr Dolgano-Nenets District of the Krasnoyarsk kraj; 8—Turukhansky District of the Krasnoyarsk kraj; 9—Evenkiysky District of the Krasnoyarsk kraj (northern areas); 10—Yakutia (Sakha) republic (northern districts); 11—Chukotka autonomous okrug. (*The map was designed by Alexey V. Dozhdikov, research assistant from the Northwest Public Health Research Center, St-Petersburg, Russia*).

Table 1. Russian Arctic biomonitoring of contaminants in BIOTA.

Region	Areas, Districts, Cities, Settlements	Sampling Year	Samples	Contaminants	Reference	Foreign Collaborator
Murmansk Oblast	Southern Barents Sea (northern coast of Kola Peninsula)	1977	sea fish, invertebrates	DDTs	Savinova et al., 1981 [20]	-
		1979; 1983	sea fish, seabirds	DDTs, PCBs	Savinova, 1991 [21]	-
		1977–1989	sea fish, seabirds	DDTs, PCBs	Savinova et al., 1995 [22]	Norway
		1991–1992	Seabirds	Cd, Zn, Cu, As, Se, Hg	Savinov et al., 2003 [23]	Norway
	Upper and middle course of Ponoy river (eastern part of Kola peninsula)	1991	peregrine falcon eggs	DDTs, PCBs, PCDD/Fs, HCHs, HCB, CHLs, mirex, toxaphenes, dieldrin	Henny et al., 1994 [24]	USA
	Pechenga district—Kuetsjarvi lake (near Nickel town) in the Pasvik River system	1991–1992	freshwater fish	Cd, Cu, Cr, Hg, Ni, Zn	Amundsen et al., 1997 [25]	Norway
	Seven Islands Archipelago—southern Barents Sea (northern coast of Kola Peninsula)	1992; 2002	Seabirds	DDTs, HCHs, HCB, CHLs, PCBs	Savinov et al., 2007 [26]	Norway
	Rybach'ya bank—southern Barents Sea (northern coast of Kola Peninsula)	1994	sea fish	Cd, Pb, Hg, Ni, Cu, Zn	Zauke et al., 1999 [27]	Germany
	Southern extremity of Kola Peninsula, White Sea	1992; 1998	harp seals	DDTs, HCHs, CBs, CHLs, PCBs, mirex	Muir et al., 2003 [28]	Canada, Norway
	Lovozero district	2001–2002	reindeers, hares, birds, fish, berries, mushrooms	DDTs, HCHs, HCB, CHLs, mirex, toxaphenes, PCBs, PCDD/Fs + Hg, Pb, Cd	AMAP-2004 Report [29]	Norway, Canada
	Pechenga district (Nickel and Zapolyarny cities area)	2013–2014	reindeers, hares, birds, fish, berries, mushrooms, vegetables	DDTs, HCHs, HCB, PCBs (fauna) + Pb, Hg, Cd, As, Ni, Cu, Zn, Cr, Fe, Mn, Co, Sr, V (fauna and flora)	Dudarev et al., 2015 [30]; Dudarev et al., 2015 [31]; Hansen et al., 2017 [32]	Norway, Finland

Table 1. *Cont.*

Region	Areas, Districts, Cities, Settlements	Sampling Year	Samples	Contaminants	Reference	Foreign Collaborator
Karelia republic (northern districts)	Sorokskaya Guba of Onega Bay (Belomorsk district), White Sea	1993	ringed seals	DDTs, HCHs, HCB, CHLs, PCBs	Kostamo et al., 2000 [33]	Finland
	Chupa Guba of Kandalaksha Bay (Loukhsky district), White Sea	1999–2000	sea fish	DDTs, HCHs, CBs, CHLs, PCBs, mirex	Muir et al., 2003 [28]	Canada, Norway
	Northwestern Onega Bay (Kemsky district) at the junction with Kandalaksha Bay (Loukhsky district), White Sea	2001	ringed seals	DDTs, HCHs, CBs, CHLs, PCBs, mirex, PCDD/Fs, toxaphenes	Muir et al., 2003 [28]; Savinov et al., 2011 [34]	Canada, Norway
	Four small lakes (Krivoye, Sredneye, Krugloye and Zhemchuzhnoe) in the White Sea lowland (Loukhsky district)	2000s?	freshwater perch	Hg	Nemova et al., 2014 [35]	-
Arkhangelsk Oblast (northern districts)	Franz Josef Land	1991	seabirds	DDTs, HCHs, HCB, CHLs, PCBs	Savinova et al., 1995 [36]	Norway
	Franz-Josef Land; southern extremity of Novaya Zemlya	1991–1992	seabirds	Cd, Zn, Cu, As, Se, Hg	Savinov et al., 2003 [23]	Norway
	Outer Dvina Bay, White Sea	1998; 1999–2000	ringed seals, bearded seals; sea fish	DDTs, HCHs, CBs, CHLs, PCBs, mirex	Muir et al., 2003 [28]	Canada, Norway
Nenets autonomous okrug	Islands Vaygach, Dolgy, Kolguev	1991–1992	seabirds	Cd, Zn, Cu, As, Se, Hg	Savinov et al., 2003 [23]	Norway
	Southern Barents Sea (East Ice) to the north of Cape Kanin	1993	harp seal	DDTs, HCHs, HCB, CHLs, PCBs	Kleivane et al., 1997 [37]	Norway
	Southeastern Barents Sea; Yugor Peninsula; lower reaches of the Pechera River	1994–1995	birds, fish, hare, reindeer	DDTs, HCHs, CBs, CHLs, PCBs + Cd, Pb, Hg	RCMA data in AMAP-1998 Report [13]	-
	Southeastern Barents Sea (near Vaygach Island)	2002	ringed seals	DDTs, HCHs, CBs, CHLs, PCBs, PCDD/Fs, mirex, toxaphenes	Savinov et al., 2011 [34]	Canada, Norway
	Nelmin-Nos area	2001–2002	reindeers, hares, birds, fish, berries, mushrooms	DDTs, HCHs, HCB, CHLs, mirex, toxaphenes, PCBs, PCDD/Fs + Hg, Pb, Cd	AMAP-2004 Report [29]	Norway, Canada
	Krasnoe, Nelmin-Nos and Indiga villages	2017–2018	fish	Hg, Pb, Cd, As, Ni, Co, Cu, Zn	Sobolev et al., 2019 [38]	Norway
	Indiga village	2017–2018	fish	DDTs, CBs, CHLs, PCBs, mirex	Lakhmanov et al., 2020 [39]	Norway

Table 1. *Cont.*

Region	Areas, Districts, Cities, Settlements	Sampling Year	Samples	Contaminants	Reference	Foreign Collaborator
Komi republic (northern districts)—NO data						
Yamalo-Nenets autonomous okrug	Not specified	2000s	freshwater fish	Cd, Pb, Ni	Agbalyan, 2012 [40]	-
	Southern part of Kara Sea; Yamal and Gydan peninsula; lower reaches of the Ob' River	1994–1995	birds, fish, hare, reindeer	DDTs, HCHs, CBs, CHLs, PCBs + Cd, Pb, Hg	RCMA data in AMAP-1998 Report [13]	-
Taimyr Dolgano-Nenets District of the Krasnoyarsk kraj	Four lakes at southern foothills of Byrranga mountains, northern Taimyr	1993	freshwater fish	As, Hg, Pb, Cd, Ni, Cu, Se, Zn	Allen-Gil et al., 2003 [41]	USA
	Southeastern part of Kara Sea; Taimyr peninsula; lower reaches of the Yenisey River	1994–1995	birds, fish, hare, reindeer	DDTs, HCHs, CBs, CHLs, PCBs + Cd, Pb, Hg	RCMA data in AMAP-1998 Report [13]	-
	Yenisei River estuary near Dikson Island (Kara Sea)	1995	ringed seals	DDTs, HCHs, HCB, CHLs, PCBs	Nakata et al., 1998 [42]	Japan
	Yenisei River estuary near Dikson Island (Kara Sea)	2002	ringed seals	DDTs, HCHs, CBs, CHLs, PCBs, PCDD/Fs, mirex, toxaphenes	Savinov et al., 2011 [34]	Canada, Norway
	Khatanga area, Dudinka area	2001–2002	reindeers, hares, birds, fish, berries, mushrooms	DDTs, HCHs, HCB, CHLs, toxaphenes, mirex, PCBs, PCDD/Fs + Hg, Pb, Cd	AMAP-2004 Report [29]	Norway, Canada
Turukhansky district of the Krasnoyarsk kraj—NO data						
Evenkiysky district of the Krasnoyarsk kraj—NO data						
Yakutia (Sakha) republic (northern districts)	Lower Lena River	1992	ptarmigan, freshwater muksun	DDTs, HCHs	Rovinsky et al., 1995 [43]	-
	Southern part of Laptev Sea and East-Siberian Sea; New Siberian Islands, lower reaches of the Lena River and Indigirka River	1994–1995	birds, fish, hare, reindeer	DDTs, HCHs, CBs, CHLs, PCBs + + Cd, Pb, Hg	RCMA data in AMAP-1998 Report [13]	-
	Vilyuisky and Momsky districts of northern Yakutia	1990s–2000s (not specified)	fish	Hg, Pb, Cd	Nyukkanov, Kolesnikov, 2004 [44]	-
	Lower reaches of the rivers Khroma, Indigirka and Kolyma	2006–2010	fish	Cd, Pb, Hg	Tyaptirgyanov and Tyaptirgyanova, 2014, 2015 [45–47]	-

Table 1. *Cont.*

Region	Areas, Districts, Cities, Settlements	Sampling Year	Samples	Contaminants	Reference	Foreign Collaborator
Chukotka autonomous okrug	Chaunsky district, coastal Chukotka	1993	seabirds	Cd, Hg, Fe, Mn, Zn, Cu	Kim et al., 1996 [48]	Japan
	Anadyrsky district (Kanchalan area); Chukotsky district (Lavrentiya, Lorino, Uelen)	2001–2002	reindeers, hares, birds, fish, berries, mushrooms, marine mammals (in coastal areas)	DDTs, HCHs, HCB, CHLs, mirex, toxaphenes, PCBs, PCDD/Fs + Hg, Pb, Cd	AMAP-2004 Report [29]	Norway, Canada
	Chukotsky district (Lavrentiya area)	2002	gray whales, walruses, ringed seals, bearded seals	PCBs, PCDD/Fs	Amirova et al., 2004 [49]	-
	Anadyrsky district (Kanchalan area); Chukotsky district (Lavrentiya, Lorino, Uelen)	2001–2003	reindeers, hares, birds, fish + marine mammals (Chukotsky district)	DDTs, HCHs, HCB, PCBs + Hg, Pb, Cd	Dudarev, 2012 [50]	Norway, Canada
	Southwestern coast of the Chukchi Sea (near Vankarem village)	2005	ringed seals	DDTs, HCHs, CBs, CHLs, PCBs, PCDD/Fs, mirex, toxaphenes	Savinov et al., 2011 [34]	Canada, Norway
	Mechigmensky Bay of the Bering Sea (Lorino village, coastal eastern Chukotka)	2010–2011	gray whales, walruses	HCHs, DDTs (*lipid weight for all tissues!*) + Pb, Cd, Hg, As (*dry weight for all tissues!*)	Tsygankov, 2012 [51]; Tsygankov et al., 2014 [52], Tsygankov et al., 2015 [53]	Doubtful results!
		2011	walruses	Pb, Cd	Trukhin et al., 2013 [54]	-
	Providensky district (Enmelen, Nunligran, and Sireniki villages)	2016	fish (marine, migratory, freshwater), meat of land mammals (reindeer, hare), meat and blubber of marine mammals (whale, walrus, seals), mushrooms, berries, wild plants, seafood (weeds, mollusks, ascidians)	DDTs, HCHs, CBs, CHLs, PCBs, mirex (fauna) +Pb, As, Cd, Hg, Cu, Zn, Ni, Cr, Al, Mn, Ba, Sr, Co, V, Be, Mo, Sn, Sb (fauna and flora)	Dudarev et al., 2019 [55]; Dudarev et al., 2019 [56];	USA

Table 2. Russian Arctic HUMAN biomonitoring of contaminants.

Region	Areas, Districts, Cities, Settlements	Sampling Year	Medium (Number of Samples); Indigenous or Not	Contaminants	Reference	Foreign Collaborator
Murmansk Oblast	Murmansk and Monchegorsk cities	1993	Breastmilk (30)—nonindigenous	DDTs, HCHs, HCB, CHLs, PCBs, PCDD/Fs	Polder et al., 1998 [57]	Norway
	Nickel and Monchegorsk cities	1993–1994	Maternal blood serum (50 + 51)—nonindigenous	Cu, Zn;	Odland et al., 1999 [58]	Norway
			Maternal and cord whole blood (50 mother–child pairs)—nonindigenous	Cd, Pb; Hg	Odland et al., 1999 [59,60]	Norway
	Lovozero village, Krasnoschelye village, Apatity city	1995	Whole blood of children (63 + 47 + 14)—indigenous and nonindigenous	Pb	Odland et al., 1999 [61]	Norway
	Murmansk city	2000	Breastmilk (14)—nonindigenous	DDTs, HCHs, HCB, CHLs, mirex, PCBs, PCDD/Fs, PBDEs	Polder et al., 2008 [62]	Norway
	Lovozero district (Lovozero and Krasnoschelye villages)	2001–2002	Blood serum and whole blood: general population (20 + 20), mother–child pairs (7)—indigenous	DDTs, HCHs, PCBs + Hg, Pb	Dudarev et al., 2004 [63]	Norway, Canada
				DDTs, HCHs, HCB, CHLs, mirex, toxaphenes, PCBs, PCDD/Fs, PBDEs + Hg, Pb, Cd	AMAP-2004 Report [29]; Holoubek et al., 2008 [64] (extended data); AMAP-2009 Report [15] (extended data)	Norway, Canada
	Lovozero and Krasnoschelye villages; Kola town (near Murmansk)	2001–2002	Blood serum: women (47 + 17 + 0), men (4 + 15 + 0)—indigenous; pregnant women (0 + 0 + 16)—nonindigenous	DDTs, HCHs, HCB, PCBs	Sandanger et al., 2009 [65]	Norway, Canada
	Pechenga district (Nickel and Zapolyarny cities)	2013–2014	Blood serum and whole blood: general population (50), pregnant women (50)—nonindigenous	DDTs, HCHs, HCB, PCBs + Hg, Pb, Cd, As, Ni, Mn, Co, Cu, Zn	Dudarev et al., 2016 [66,67]; Adlard et al., 2021 [68] (fragment on maternal Hg)	Norway, Finland
Karelia republic (northern districts)—NO data						

Table 2. *Cont.*

Region	Areas, Districts, Cities, Settlements	Sampling Year	Medium (Number of Samples); Indigenous or Not	Contaminants	Reference	Foreign Collaborator
Arkhangelsk Oblast (northern districts)	Arkhangelsk city	1993	Maternal blood serum (50)—nonindigenous	Cu, Zn	Odland et al., 1999 [58]	Norway
			Maternal and cord whole blood (50 mother–child pairs)—nonindigenous	Cd, Pb, Hg	Odland et al., 1999 [59–61]	Norway
		1996	Maternal blood serum (27)—nonindigenous	DDTs, HCHs, HCB, CHLs, mirex, PCBs	Sandanger et al., 2003 [69]	Norway
	Arkhangelsk and Severodvinsk cities	1996–1997	Breastmilk (51 + 50)—nonindigenous	DDTs, HCHs, HCB, CHLs, PCBs	Polder et al., 2003 [70]	Norway
	Arkhangelsk city	2000	Breastmilk (23)—nonindigenous	DDTs, HCHs, HCB, CHLs, mirex, PCBs, PCDD/Fs, PBDEs	Polder et al., 2008 [62]	Norway
Nenets autonomous okrug	Naryan-Mar city	1996–1997	Breastmilk (20)—indigenous and nonindigenous	DDTs, HCHs, HCB, CHLs, PCBs	Polder et al., 2003 [70]	Norway
	Lower Pechora River area, mainly Nelmin-Nos village	2001–2002	Blood serum and whole blood: general population (32), mother–child pairs (21)—indigenous	DDTs, HCHs, PCBs + Hg, Pb	Dudarev et al., 2004 [63]	Norway, Canada
				DDTs, HCHs, HCB, CHLs, mirex, toxaphenes, PCBs, PCDD/Fs, PBDEs + Hg, Pb, Cd	AMAP-2004 Report [29]; Holoubek et al., 2008 [64] (extended data); AMAP-2009 Report [15] (extended data)	Norway, Canada
	Naryan-Mar city	2002	Maternal blood serum (12)—indigenous and nonindigenous	PBDEs, PFOS	Odland et al., 2006 [71]	Norway
	Lower Pechora River area	2001–2002	Blood serum: pregnant women (38), women (31), men (13)—indigenous	DDTs, HCHs, HCB, PCBs	Sandanger et al., 2009 [65]	Norway, Canada
	Nelmin-Nos village	2009–2010	Blood serum and whole blood: women (87), men (22)—indigenous	DDTs, HCHs, HCB, CHLs, mirex, PCBs + Hg, Pb, Cd	Rylander et al., 2011 [72]	Norway
	Seven rural coastal and inland settlements: Bugrino, Varnek, Amderma, Indiga, Shoina, Nelmin-Nos, Krasnoe	2018	Blood serum of women (204)—indigenous and nonindigenous	HCHs, DDTs, CHLs, aldrin, mirex, CBs, PCBs	Varakina et al., 2021 [73]	Norway
			Whole blood of general population (297)—indigenous and nonindigenous	As, Cd, Hg, Pb, Mn, Co, Cu, Zn	Sobolev et al., 2019 [38]	Norway

Table 2. *Cont.*

Region	Areas, Districts, Cities, Settlements	Sampling Year	Medium (Number of Samples); Indigenous or Not	Contaminants	Reference	Foreign Collaborator
Komi republic (northern districts)	Izhma and Usinsk settlements	2009–2010	Blood serum and whole blood of general population: women (25 + 25), men (25 + 25)—indigenous	DDTs, HCHs, HCB, CHLs, mirex, PCBs + Hg, Pb, Cd	Rylander et al., 2011 [72]	Norway
Yamalo-Nenets autonomous okrug	Salekhard city	1995	Maternal and cord whole blood; breastmilk (21 mother–child pairs)—nonindigenous	Hg, Pb, Cd, Ni	Klopov, 1998 [74]	Norway
		1996–1998	Maternal blood serum; breastmilk (31)—nonindigenous	DDTs, HCHs, HCB, CHLs, PCBs	Klopov et al., 1998 [75]	Norway
	Tazovsky village (and Tazovsky tundra), Gyd village (and Gydan tundra), Antipayut village (and Antipayutinskaya tundra), Nakhodka village (and Nakhodka tundra), Kutopyugan village (and Kutopyugan tundra) and Kharsaim village	2016	Whole blood of general population (151)—indigenous	Pb, Cd, Ni, Hg; As, Cu, Mg, Ca, Fe, Mn, Zn, Cr, Co	Agbalyan, 2012 [40]; Agbalyan and Shinkaruk, 2018 [76]; Agbalyan, 2020 [77]	- -
Taimyr Dolgano-Nenets District of the Krasnoyarsk kraj	Norilsk city	1995	Maternal and cord whole blood; breastmilk (21 mother–child pairs)—nonindigenous	Hg, Pb, Cd, Ni	Klopov, 1998 [74]	Norway
	Norilsk and Dudinka cities	1995–1996	Maternal blood serum; breastmilk (49 + 27)—nonindigenous	DDTs, HCHs, HCB, CHLs, PCBs	Klopov et al., 1998 [75]	Norway
	Khatanga area, Dudinka area, Norilsk city	2001–2002	Blood serum and whole blood: mother–child pairs (29 + 38 + 10), general population (5 + 0 + 0)—indigenous and nonindigenous	DDTs, HCHs, PCBs + Hg, Pb	Dudarev et al., 2004 [63]	Norway, Canada
				DDTs, HCHs, HCB, CHLs, mirex, toxaphenes, PCBs, PCDD/Fs, PBDEs + Hg, Pb, Cd	AMAP-2004 Report [29]; Holoubek et al., 2008 [64] (extended data); AMAP-2009 Report [15] (extended data)	Norway, Canada

Table 2. *Cont.*

Region	Areas, Districts, Cities, Settlements	Sampling Year	Medium (Number of Samples); Indigenous or Not	Contaminants	Reference	Foreign Collaborator
Taimyr Dolgano-Nenets District of the Krasnoyarsk kraj	Lower Yenisei River area; Norilsk city	2001–2002	Blood serum: pregnant women (69 + 59), women (40 + 0), men (12 + 0)—indigenous and nonindigenous	DDTs, HCHs, HCB, PCBs	Sandanger et al., 2009 [65]	Norway, Canada
	Taimyr (not specified)	2002	Maternal blood serum (12)	PBDEs, PFOS	Odland et al., 2006 [71]	Norway
	Norilsk city	2001	Maternal and cord whole blood and plasma (7 mother–child pairs)	PFASs (PFOS, PFOA, PFNA, FOSA, PFHxS, PFUnDA)	Hanssen et al., 2013 [78]	Norway
Turukhansky District of the Krasnoyarsk kraj—NO data						
Evenkiysky District of the Krasnoyarsk kraj—NO data						
Yakutia (Sakha) republic (northern districts)	Yuryung-Khaya village (Anabar district)	2017	Blood *serum* (35 men and 72 women = 107)—indigenous Dolgans	Pb, Ni, Cu, Zn, Cr, Mn, Fe, Sc, Ti, Rb, Sr, Cs	Sivtseva et al., 2020 [79]	-
Chukotka autonomous okrug	Uelen village (Chukotsky district)	2001	Blood serum (50)—indigenous	DDTs, HCHs, HCB, CHLs, mirex, toxaphenes, PCBs	Sandanger et al., 2003 [80]	Norway
	Chukotsky district, Iultinsky district, Anadyrsky district, Anadyr city	2001–2003	Blood serum and whole blood: mother–child pairs (47 + 5+39 + 12); general population (30 + 0 + 50 + 0)—indigenous and nonindigenous	DDTs, HCHs, PCBs + Hg, Pb	Dudarev et al., 2004 [63]	Norway, Canada
				DDTs, HCHs, HCB, CHLs, mirex, toxaphenes, PCBs, PCDD/Fs, PBDEs + Hg, Pb, Cd	AMAP-2004 Report [29]; Holoubek et al., 2008 [64] (extended data); AMAP-2009 Report [15] (extended data)	Norway, Canada
	Chukotsky district, Anadyrsky district, Anadyr city.	2001–2003	Breastmilk (27 + 21 + 7)—indigenous	DDTs, HCHs, HCB, CHLs, mirex, toxaphenes, PCBs, PCDD/Fs, PBDEs	AMAP-2004 Report [29]; Holoubek et al., 2008 [64]	Norway, Canada

Table 2. *Cont.*

Region	Areas, Districts, Cities, Settlements	Sampling Year	Medium (Number of Samples); Indigenous or Not	Contaminants	Reference	Foreign Collaborator
Chukotka autonomous okrug	17 settlements in coastal Chukotsky district and inland Anadyrsky district	2001–2002	Maternal blood, cord blood and breastmilk (48 mother–child pairs)—indigenous	DDTs, HCHs, HCB, CHLs, mirex, toxaphenes, PCBs + Hg, Pb, Cd	Anda et al., 2007 [81]	Norway
	Uelen village (Chukotsky district), Kanchalan village (Anadyrsky district)	2001–2002	Blood serum: pregnant women (59 + 67), women (26 + 28), men (24 + 14)—indigenous	DDTs, HCHs, HCB, PCBs	Sandanger et al., 2009 [65]	Norway, Canada
	Lorino and Lavrentiya villages (Chukotsky district)	2007 vs. 2001–2002	Blood serum and whole blood: 17 mother–child pairs (13 + 4)—indigenous (personalized follow-up)	DDTs, HCHs, HCB, PCBs, CHLs + Hg, Pb	Dudarev et al., 2010 [82]	Norway
	Kanchalan village (Anadyr district) and Novoye Chaplino village (Providensky district)	not specified	Whole blood (98 + 36)—indigenous	Pb, Ni, Cu, Zn, Fe, K, Ca, Ga, Ge, Se, Br, Rb, Sr, Zr	Gyrgolkau et al., 2015 [83]	-
	Inland and coastal districts, incl. Chukotsky district	2014–2015	Maternal blood serum (250), incl. (63)	DDTs, HCHs, HCB, PCBs, PeCB, mirex, PBDEs	Bravo et al., 2019 [84]	Spain, Norway

2. Beginning of the Sea Fish and Seabirds Biomonitoring in the Russian Arctic

The biomonitoring of contaminants in the territory of the Russian Arctic started in Murmansk oblast in 1977, when, in the southern part of the Barents Sea near the northern coast of the Kola Peninsula, employees of the Murmansk Institute of Marine Biology (Tatiana Savinova and colleagues) for the first time analyzed the samples of several species of sea fish and invertebrates for DDT contamination. The results of these studies were first published in 1981 as a short article in the *Russian Journal of Hydrobiology* [20]. In 1979 and 1983, samples of various types of marine fish and birds were collected at several spots on the northern coast of the Kola Peninsula; the results of the analysis for DDTs and PCBs (in comparison with the circumpolar regions) were published in 1990 in the form of a scientific report by the Kola Science Center of the USSR Academy of Sciences, "Chemical pollution of the northern seas". In 1991, the report was translated into English and published by the Canadian Institute of Scientific and Technical Information in Ottawa [21].

Norway and the Soviet Union signed The Agreement on Environmental Cooperation in 1988, and later signed a similar agreement between Norway and Russia in 1992. The Joint Norwegian–Russian Commission on Environmental Cooperation was established, with annual meetings. Working groups were initiated to increase collaboration on different environmental topics, e.g., air pollution, marine environment and radioactive pollution. This collaboration contributed to the harmonization and development of scientific methodology and databases [22].

The beginning of Norwegian–Russian seabird expert group cooperation was marked by the preparation and publication of the joint report, "Chemical pollution in the Arctic and sub-Arctic marine ecosystems: an overview of current knowledge" with Tatyana Savinova as the lead author [22]. The report was focused on the data obtained by that time (literature review and own results) concerning POP and metal levels in circumpolar Arctic biota, including Russian data on seabirds and sea fish sampled in the period 1977–1989 in several spots on the northern coast of the Kola peninsula (southern part of the Barents Sea).

As part of the Norwegian–Russian project "Environmental Contaminants in Arctic Seabirds" (with participation of Tatyana Savinova and Vladimir Savinov from the Murmansk Marine Biological Institute), which was launched in 1991 with the aim of providing reliable mapping of contamination levels in seabirds, two international expeditions to seabird colonies in the Barents Sea were carried out in 1991 and 1992. The expedition of 1991 was aimed at the analysis of legacy POPs in bird samples and covered Norwegian islands and part of Norwegian northern mainland coastline, but also included the Russian Franz-Josef Land. The results of the study were published in 1995 [36].

During the other expedition of 1992 to various seabird colonies in the Barents Sea, thirteen species of seabirds belonging to five families were collected on the Norwegian territory and in several Russian spots in the Barents Sea: Franz-Josef Land, the southern extremity of Novaya Zemlya, the Nenets okrug islands (Vaygach, Dolgy, Kolguev) and the northern coast of the Kola peninsula. The aim of this study was the analysis of six metals (Cd, Zn, Cu, As, Se and Hg) in the tissues of seabirds, and the results were published in 2003 [23].

Several species of seabirds were caught on islands of the Seven Islands Archipelago (northern coast of Kola peninsula) in the summer of 1992 and 2002. Hepatic tissue samples collected in 1992 were analyzed for POPs at the Institute of Pharmacology and Toxicology, Norwegian Veterinary College (Oslo, Norway); samples collected in 2002 were analyzed for POPs at the Centre of Environmental Chemistry "Typhoon" (Obninsk, Russia). During the decade period, a statistically significant decrease in the residue levels of the analyzed POPs was found in a majority of the sampled birds [26].

The Japan–Russian project (Ehime University, Matsuyama, Japan and the Institute of Biological Problems of the North, Russian Academy of Science, Magadan, Russia) was carried out in Chaunsky district of coastal Chukotka, where seabirds belonging to three orders, four families and eleven species were obtained during June–July 1993 to determine concentrations of four essential metals (Fe, Mn, Zn and Cu) and two toxic metals (Cd and Hg) in the selected bird tissues; the results were published in 1996 [48].

In June 1994, the Russian–German expedition on the research vessel "Dalniye Zelentsy" collected samples of 15 marine fish species in the southern part of the Barents Sea (Rybach'ya bank near the northern coast of Kola peninsula). Six metals (Cd, Pb, Hg, Ni, Cu and Zn) in fish tissues were analyzed in the University of Oldenburg laboratory (Germany); the results were published in 1999 [27] including three orders, four families and eleven species of fish.

3. Beginning of Inland Biota Biomonitoring in the Russian Arctic

In the 1970s, David Peakall discovered that OCs could be extracted from the lipids of the membranes of dried bird eggshells. This made it possible to relate eggshell thinning to pesticide levels, even when the egg contents were not available [85]. Specimens were collected from museums and private collectors globally. In 1979, a large dataset of eggshell index and DDE levels in egg membranes were reported, including data on eggs (from the Moscow Zoological Museum) of *Falco peregrinus leukogenus* from the Russian Arctic, namely from the mouth of the Indigirka river (Yakutia; 71° N, 146° E) collected annually in 1961–1965, and from the Pyasina river (Taimyr; 73° N, 91° E) collected in 1966. The Yakutian DDE levels in the eggshells were in the range of 80–180 ppm (lipid), while the Taimyr levels were 180–200 ppm (lipid) [85].

Studies on a population of peregrine falcons on the Kola Peninsula have been ongoing since 1977. Within the US–Russian collaboration, an international expedition in 1991 along the upper and middle course of Ponoy river (eastern part of Kola peninsula, Murmansk oblast) was organized and aimed at visiting most of peregrine eyrie sites during the egg stage and "sampling" the eggs. The eggs were brought to the US for analysis of DDTs, PCBs, PCDD/Fs, HCHs, HCB, toxaphenes, chlordanes, dieldrin and mirex. The eggs revealed relatively low concentrations of DDE (geomean 3.5 mcg/g) and of other OCPs. The DDE levels and eggshell thinning (11.4%) were similar to those reported in the peregrine eggs from Alaska. Levels of PCBs were higher than DDE concentrations, comparable to Fennoscandian falcons, but higher than those found in Alaska (3–21 mcg/g ww; geomean of 7.3 mcg/g ww). Concentrations of PCDD/Fs and planar PCBs in eggs were also relatively high, with combined TEQ levels of 86–640 pg/g ww [24].

In the frame of the Norwegian–Russian collaboration (University of Tromso with two Institutes of the Russian Academy of Sciences: Institute of Problems of Industrial Ecology of the North, Apatity, Murmansk oblast, and the Severtzov Institute of Ecology and Animal Morphology, Moscow, Russia), the studies of 1991 and 1992 on the fish of three lakes (Kuetsjarvi, Bjarnevatn and Vaggatem) in a watercourse of the Pasvik River system in the border region between Norway and Russia, in the vicinity of the Nickel town mining and metallurgic activity, were organized. The metal content from the muscle and liver were studied in different fish tribes at the laboratory of the Norwegian Institute for Nature Research. Concentrations of Cd and Ni in fish tissue increased with increasing proximity to the smelters, whereas the other elements showed similar concentrations at the three localities. The levels were within the ranges reported for other areas, but higher than unpolluted study sites [25].

During investigations of the Lower Lena River region in August–September 1992 by the Institute of Global Climate and Ecology of the Russian Federal Service for Hydrometeorology and Natural Environmental Monitoring (Russian Academy of Sciences, Moscow, Russia), the levels of background pollution of the atmosphere, soil, vegetation and biota were studied. The levels of DDT, DDE, α-HCH and γ-HCH were analyzed in two samples of ptarmigan (*Lagopus lagopus*) and four samples of freshwater muksun (*Coregonus muksun*) [43].

Within the US–Russian project (with participation of the Extreme North Research Institute, Norilsk, Russia) during the summer of 1993, 50 samples of freshwater fish (char, whitefish and burbot) were collected from four lakes located at southern foothills of the Byrranga mountains in the northern part of Taimyr peninsula, along with samples of lemmings taken at two sites, and soils, sediments and vegetation taken at seven sites. All samples were stored frozen and then shipped by air to the Battelle Marine Sciences Laboratory in the US, where the fish samples were analyzed for eight metals (As, Hg, Pb, Cd, Ni, Cu, Se and Zn) [41].

In 1991, the "Regional Center Monitoring of the Arctic" (RCMA) was established in St-Petersburg, Russia, on the basis of the Arctic and Antarctic Research Institute of the USSR State Committee for Hydrometeorology (in 2006, the RCMA was reassigned to the North-Western Branch of the SPA "Typhoon"). In the early 1990s, the RCMA was the key organizer and implementer of the first large-scale assessment of the contamination of the Russian Arctic seas coastline; this investigation was a part of the Russian AMAP National Implementation Plan. During the period of 1994–1995, samples of vegetation and biota were collected by the RCMA in many spots along the whole of the Russian Arctic coastline (of the Barents, Kara, Laptev, East Siberian and Chukchi seas), mainly at the deltas of large rivers and on the nearby islands. The results of the chemical analysis of the POPs and metals (in RCMA laboratory) in the sampled biota (terrestrial and seabirds, fish and mammals, including reindeer) were published as the reports of the RCMA and Rosgydromet in 1994–1996 [86–90] and were included in the first AMAP-1998 report.

The RCMA provided levels for Cd, Pb and Hg in the muscle and liver tissues of birds across Arctic Siberia (done during 1995). In the muscle tissue, concentrations of all metals decreased, with highest levels in the carnivores, but liver tissues provided no differences. Some geographical differences were observed, with higher Pb levels in the east. The same trend was observed for Cd in the muscle and liver of omnivores, and for Cd and Pb in the liver tissues. Moreover, the data on Cd, Pb and Hg in several fish species (sea, anadromous and freshwater) from various locations (sampling 1994–1995), and in liver, kidney and muscle of reindeer (*Rangifer tarandus*), were presented. Concentrations of Pb in the muscle and liver of reindeer were generally higher in the eastern part of the Russian Arctic; there was no similar trend for Cd [13].

PCBs, DDTs, HCHs, CBs and CHLs were analyzed by the RCMA in the tissues of reindeer, waterfowl and other terrestrial birds. The Russian data are sparse due to small numbers. An interesting observation was the difference between two years (1994 and 1995), with only one year difference, thus making it difficult to compare with other countries. However, the 1994 results were generally higher than for Canadian caribou and Svalbard reindeer. It should be noted that the PCBs in the Russian reindeer were based on only seven congeners (PCBs 28, 52, 101, 118, 153, 138 and 180), which were major congeners in Canadian caribou fat [13].

4. Beginning of Marine Mammals Biomonitoring in the Russian Arctic

Studies on the POP contamination of marine mammals in three Russian Arctic seas (Barents Sea, White Sea and Kara Sea) began almost simultaneously in the early 1990s.

Blubber samples from a few dozen of the Eastern harp seal (*Phoca groenlandica*) were obtained during the Norwegian expedition in April and May 1993 in the southern Barents Sea (East Ice) to the north of Cape Kanin (Nenets okrug). Significantly higher concentrations of the three major pollutants ($\sum$PCB, $\sum$DDT and $\sum$CHL) were found in adults as compared to the juveniles, while no such relations were found for $\sum$HCH or HCB [37].

In the frame of the Finnish–Russian project, ringed seals (*Phoca hispida*) were sampled in the winter of 1993 in Sorokskaya Guba (near Belomorsk city in northern Karelia) of Onega Bay of the White Sea, and also in Lake Ladoga (Russia) and Lake Saimaa (Finland). The blubber samples were taken from seals that had died naturally or drowned in fishing gear. DDTs, HCHs, HCB, CHLs and PCBs were analyzed in the samples. It was concluded that, of the three areas studied, the highest concentrations of PCBs and DDTs in ringed seal blubber were found in Lake Saimaa; Lake Ladoga samples were contaminated moderately; the Arctic White Sea ringed seals indicated the lowest pollution due to long-range transport [33].

Under a cooperative Japan–Russian project in April 1995, 38 ringed seals were collected in Taimyr near Dikson Island (Kara Sea). The collected samples of blubber were analyzed for POPs. The levels of DDTs and PCBs were significantly higher than comparable results from the Canadian and Norwegian Arctic. It was predicted that significant local sources were present. Concentrations of CHLs and HCHs in ringed seals were similar, probably because of the global transportation to the Arctic region [42].

Within the Canadian–Russian project, the collection of tissues of several species of seals was conducted in three areas of the White Sea Basin during the 1998 and 2001 expeditions by the Northern Department of the Polar Research Institute of Marine Fishery and Oceanography (Arkhangelsk). In 1998, about 50 samples of harp seals, ringed seals and bearded seals were collected in the White Sea: ringed and bearded seals in the outer Dvina Bay (Arkhangelsk oblast) and harp seals near the southern extremity of the Kola Peninsula (Murmansk oblast). In autumn 2001, 23 ringed seals were collected in northwest part of the Onega Bay (Kemsky district of northern Karelia). In addition, blubber samples from 10 harp seal pups collected near the southern extremity of the Kola Peninsula (Murmansk oblast) in March 1992 were included to examine the temporal trends of OCs [28].

Moreover, several marine fish prey species (navaga, herring, bullrout, smelt and cod), as well as invertebrates, were collected in 1999–2000 from the research vessel "Professor Vladimir Kuznetsov" (Zoological Institute, St-Petersburg) in Chupa Guba of Kandalaksha Bay (Loukhsky district of northern Karelia) and the outer Dvina Bay (Arkhangelsk oblast) of the White Sea. The collected samples were analyzed for POPs in the "Typhoon" laboratory (Obninsk, Russia). It was shown that ∑PCB and p,p-DDE concentrations in ringed seals from eastern Svalbard and the southern Barents Sea were 2–3 times higher than in this study. PCB and ∑DDT concentrations in blubber were similar to levels from Svalbard and East Greenland. All major OCs, except HCH isomers, were higher compared to previous results in the Canadian and Alaskan Arctic, but lower than studies from the Kara Sea. A reduction of approximately 33% for ∑DDT to 60% for ∑PCB in the blubber of harp seal pups collected in 1992 compared with 1998 was expected, as declining concentrations ∑DDT in seawater were reported in inflowing rivers to the White Sea in the 1980s and early 1990s [28].

Under the auspices of the Norwegian–Canadian–Russian collaboration, samples of the blubber of adult ringed seals from the four areas of the Russian Arctic (White Sea, Barents Sea, Kara Sea and Chukchi Sea) were collected in 2001–2005 and analyzed for POPs at the Centre of Environmental Chemistry, S.P.A. Typhoon (Obninsk, Russia). The highest concentrations were found in the Kara Sea compared with those found in the other three areas of the Russian Arctic, while the highest mean concentrations of ΣHCHs and PCDD/Fs were found in the blubber of ringed seals from the Chukchi Sea and the White Sea, respectively. No significant differences between ΣPBDE concentrations were found in the blubber of ringed seals from the three study areas of the European part of the Russian Arctic, but the PBDE contamination level in ringed seals from the Chukchi Sea was 30–50 times lower [34].

5. Beginning of Human Biomonitoring in the Russian Arctic

The first data on the content of POPs and metals in human biological media were obtained in 1993–1996, also within the framework of the Russian–Norwegian cooperation under the auspices of AMAP, with participation and the active promotion of Jon Oyvind Odland from Tromso University. In the period 1996–1999, the first publications appeared on the results of these studies: on the content of essential trace elements and toxic metals in maternal and cord blood sampled in the Murmansk Oblast and Arkhangelsk Oblast [58–61,91], on the content of POPs in the breastmilk of mothers from the cities of Murmansk and Monchegorsk [57], Arkhangelsk, Severodvinsk (Arkhangelsk oblast) and Naryan-Mar (Nenets okrug) [70] (data on 5–7 year dynamics of POPs content in breastmilk from the Murmansk and Arkhangelsk oblast were published later [62]), on the content of POPs and metals in maternal blood, breastmilk and cord blood taken in Salekhard

(Yamal-Nenets okrug), Norilsk and Dudinka (Taimyr okrug) [74,75] and of the POPs in the maternal blood from Arkhangelsk [69]. The preliminary results of these studies were included by fragments in the AMAP-1998 report [13], mainly with reference to "personal communication". The AMAP-2002 report [14] contained the already-published materials of the above-mentioned Russian–Norwegian studies on the biomonitoring of contaminants in human biological media in the enumerated regions of the Russian Arctic.

6. GEF/AMAP/RAIPON Project "Persistent Toxic Substances, Food Security and Indigenous Peoples of the Russian North"

In 2001, Russia's first large-scale GEF/AMAP/RAIPON project "Persistent Toxic Substances, Food Security and Indigenous Peoples of the Russian North" was launched with the participation of the Russian Polar Fund, the Russian Association of Indigenous Peoples of the North, Siberia and the Far East (RAIPON), the Russian Ministry of Natural Resources, the Ministry of Health and Social Affairs (Northwest Public Health Research Center—NWPHRC), the Federal Service for Hydrometeorology and Environmental Monitoring, Roshydromet (RCMA and SPA Typhoon) and multiple Russian Arctic regional organizations of various departmental affiliations, as well as foreign colleagues from Norway and Canada.

The GEF/AMAP/RAIPON project of 2001–2004 initially covered four regions of the Russian Arctic: Murmansk Oblast, Nenets, Taimyr and Chukotka autonomous okrugs. Within the framework of the project, the sampling and analysis of hundreds of various biota samples (terrestrial, freshwater and marine) was carried out by the RCMA and SPA Typhoon, and biota samples were partially taken by specialists from the NWPHRC. The sampling of human biological media (blood, cord blood, breastmilk) was the full prerogative of the NWPHRC. Subsequent analysis of human samples was carried out in SPA Typhoon, and partly at the RCMA, Unilab Analyze AS (Tromsø, Norway) and the Center de Toxicologue du Quebec (Quebec, QC, Canada). All the collected samples were analyzed for three metals (Hg, Pb and Cd) and the wide list of legacy POPs.

For the first time during the expeditions of 2001–2003 (headed by Alexey Dudarev from the NWPHRC) to several coastal and mainland settlements in Chukotka autonomous okrug, hundreds of samples of human biological media of indigenous people (Chukchi and Eskimos) were collected. It turned out that, right in the coastal Chukotka (settlements of Uelen, Lorino and Lavrentiya), the highest levels of exposure of indigenous people to legacy POPs were observed (among the studied Russian Arctic regions). This phenomenon in coastal Chukotka was related to the ancient indigenous tradition of marine mammal hunting, and therefore to the regular consumption by the local people of marine mammal blubber highly contaminated by lipophilic POPs. The coastal Chukotka POP exposure levels were comparable to the highest circumpolar levels inherent to the natives of northern Canada, Greenland and the Faroe Islands.

The first results of the GEF/AMAP/RAIPON project were firstly presented at the International Congress of Circumpolar Health in 2003 in Nuuk, Greenland [63]. Some data on the exposure to POPs of coastal Chukchi and Eskimos of Uelen village [80] and data on PCB and PCDD/Fs levels in marine mammals of coastal Chukotka [49] were published before the release of the AMAP-2004 report [29]. The materials of the report published at the end of 2004 were later included in the First Regional Report of 2008 on the implementation of the "Global Monitoring Plan for POPs under the Stockholm Convention" [64] in the AMAP Report of 2009 [15] and were discussed in multiple publications in Russian and international journals [50,65,81,82,92–95].

Since 2002, Alexey Dudarev, being included in the AMAP Human Health Assessment Group as the key national Russian expert, has been a lead coauthor of a number of chapters of the three AMAP Reports "Human Health in the Arctic" of 2009, 2015 and 2021 [15–17]. The AMAP-2009 report [15] contained the first detailed comparison of biomonitoring data on contaminants in biota and human tissues in all circumpolar countries, including five subjects of the Russian Federation. The fifth subject (added to Murmansk Oblast, Nenets,

Taimyr and Chukotka autonomous okrugs) was Kamchatka and the Commander Islands, where, in 2003–2004, within the framework of the Russian–US cooperation between the NWPHRC and the Aleutian International Association, the project "PTS, food safety and the indigenous peoples of Kamchatka and the Commander Islands" was implemented. Materials on the biomonitoring of POPs in human blood serum in five subjects of the Russian Arctic were also analyzed and compared in a joint article with Norwegian and Canadian colleagues in 2009 [65].

The first pilot follow-up personalized biomonitoring study of the content of legacy POPs and metals in the blood of indigenous people of coastal Chukotka (Lorino and Lavrentiya settlements) was carried out in 2007 (5 years after the first survey in 2001–2002), when the same mothers and their grown-up 5-year-old children were reexamined for comparison "puerperal $\rightarrow$ mother" and "fetus (cord blood) $\rightarrow$ child". Maternal blood levels of POPs during the 5-year period decreased significantly (by 33–74%); blood levels of Pb decreased by 21%, while the Hg levels remained the same. The infant blood serum levels of most POPs during the five-year period increased considerably; the blood Pb levels did not change, while the Hg levels decreased by 31%. Initially lower (compared to maternal) levels of contaminants in cord blood (due to the placenta barrier) significantly increased during the 5 years in the blood of children (due to breastfeeding and further consumption of local food), reaching the initial maternal levels [82].

The AMAP-2015 report [16] contained the latest Chukotka published data, as well as the results of the Norwegian–Russian biomonitoring in the Komi Republic (Izhma and Usinsk settlements) and the Nenets Autonomous Okrug (Nelmin-Nos settlement), where the concentrations of POPs in the blood of residents of a Nenets village, first measured in 2001–2003, were compared with the levels of POPs identified in 2009–2010 (after 7–9 years): an almost twofold decrease in DDE plasma concentrations and a significant decrease levels of HCB in both women and men was found; however, there were no clear trends in PCB levels [72].

One more unique pilot study, the results of which have not been included in the AMAP-2015 report, should be mentioned. Exposure to perfluoroalkyl substances (PFASs) experienced (blood sampling of 2001–2002) by delivering women and their newborns was assessed in the industrial city of Norilsk (Taimyr Peninsula) compared to the rural towns of Urgench and Khazarasp (Aral Sea region of Uzbekistan) [78]. The observed Norilsk PFOS plasma concentrations were comparable to those in samples collected in 2001–2002 from indigenous delivering women from remote rural areas of Taimyr and Naryan-Mar (Nenets okrug) [71]; similar levels were also observed for Inuit adults of Nunavik in the Canadian Arctic. It should be noted that these two studies of PFASs were the only ones in the Russian Arctic biomonitoring history.

7. Recent Biomonitoring Studies in the Russian Arctic

The latest AMAP-2021 report [17] has become "loaded" with the new Russian biomonitoring data on POPs and metals in both biota and human blood from three regions of the Russian Arctic: Murmansk Oblast, Chukotka okrug and Nenets okrug.

The three lateral EU project Kolarctic KO467 "Food and health safety in the border regions of Norway, Finland and Russia: linking industry, communities and socio-economic consequences" was implemented in 2013–2016. The Russian part of the project was carried out by the NWPHRC (headed by Alexey Dudarev) in the Pechenga district of the Murmansk Oblast, namely in the Nickel and Zapolyarny towns located close to the Ni-Cu smelter of the Pechenganickel enterprise. Practically all kinds of local biota were sampled in 2013 (fish, poultry, game, mushrooms, wild and garden berries and garden vegetables). Blood sampling involved the participation of 50 adult men and women in 2013 and 50 pregnant women in 2013–2014. A broad suite of legacy POPs and 13 metals (Pb, As, Cd, Hg, Cu, Zn, Ni, Cr, Fe, Mn, Co, Sr and V) were analyzed in the collected samples in the laboratory of the NW branch of SPA Typhoon (former RCMA) in St-Petersburg. Below, we present more detailed results obtained during the project to acquaint an English-speaking reader with Russian-language publications on the project.

In the biota of the Pechenga district, ΣDDT levels were highest in king salmon (4.5 µmg/kg ww), followed by whitefish and burbot (3.7 µg/kg ww). All other species had lower levels. As for the DDT group, the concentrations were far below the Russian maximum permissible concentrations (MPC). Concentrations of DDE and DDT in moose and birds were low. Mean concentrations of HCB in all samples of game, freshwater fish and marine cod were low. This was also the situation for the PCB group, below the Russian maximum permissible concentration for fish. Concentrations of POPs were very low in all food samples measured [30].

Analysis of metals in local foods of the Pechenga district demonstrated that the mean Pb concentrations in all Russian samples of the local fauna and flora were low (<0.05 mg/kg ww). Significant excesses of the Russian maximum permissible concentrations (MPC) of some metals in many local foods were revealed (for Ni in mushrooms—up to 30 MPC). Mushroom analyses demonstrated high concentrations of toxic elements. This points out the importance of using mushrooms as indicators of metal contamination, as well as the main food contributor to human exposure. For the first time, it has been determined that highly toxic Ni, the presence of which was found in high concentrations in mushrooms, berries and vegetables, should be considered as the most important factor of dietary exposure (and health risk) of the population of the Pechenga district [31]. The levels of metals in local foods collected in the border areas of the three participating countries of the Kolarctic project (Russia, Norway and Finland) were compared in a later joint publication [32], where the Russian data were the most comprehensible and the concentrations of metals were the highest.

Another important task of the Russian part of the Kolarctic project was the analysis of drinking water in Nickel and Zapolyarny. Samples of drinking water in both towns were highly contaminated by Ni, probably due to the neighboring mining and processing of the metals. The Ni concentration in the drinking water was 2–3-fold higher than the Russian MPC [96].

Of the forty POPs analyzed in human plasma samples collected in the Pechenga district, only HCB, HCHs, DDTs and PCBs were detectable. Among the HCHs, the β-HCH dominated; α-HCH was detected in 28% of women's blood samples; γ-HCH in 50% of women and in 22% of pregnant women. The presence of HCH isomers in the blood of the examined residents was obviously associated with other (than local food) sources, since local food did not contain HCH. Among the DDTs, 4.4DDE and 4.4DDT predominated. At the same time, 4.4 DDD was found in 22% of pregnant women, which indicated the probable presence of a household "fresh" source of DDT in the maternity ward of the Nickel town. The average levels of HCB, β-HCH, 4.4DDE and 4.4DDT were in the range of 0.13–1.4 µg/L of blood serum. Of the fifteen PCB congeners analyzed in the blood of the examined persons, the largest "contribution" to the total PCB was made by the following congeners: PCB-118 (17–25%), PCB-138 and PCB-153 (11–17% each), PCB-52 (10–14%) and PCB-101 (8–9%). In general, the PCB congener "profile" in the blood of the surveyed population corresponded to that in local foods. The ∑PCBs levels in the surveyed population were low: about 1 µg/L serum (120 ng/g lipids) in pregnant women and about 1.5 µg/L serum (320 ng/g lipids) in the general population. None of the examined blood serum samples (including pregnant women) exceeded the international recommended levels for ∑PCB in the blood. The levels of HCB, 4,4DDE and PCB-153 in the blood of the population of the Pechenga region were the lowest in comparison with the other Russian Arctic regions studied previously (from the Murmansk Oblast to Chukotka in 2001–2003), and among pregnant women it was similar to that in neighboring northern Norway [66].

Analysis of metals in the blood of the surveyed men, women and pregnant women in the Pechenga district demonstrated the following ranges (min–max) of metal concentrations (mcg/L whole blood)—Pb: 3.5–108; Hg: 0.3–25.2; Cd: 0.16–2.4; As: 0.41–28.7; Mn: 7.3–305.0; Ni: 1.7–95.9; Cu: 912–2800; Zn: 3830–10,600. The highest concentrations of Mn, Co, Ni, Cu,

Zn, As and Pb were observed among women, while the highest observed concentrations of Cd and Hg were among men. For the geometric mean concentrations of Hg and Pb, the male levels were up to three-fold higher compared to pregnant women. In contrast, men had lower levels of Mn, Co and As compared to women, but pregnant women still had the lowest mean concentrations. For the first time, it was reported that 50% of the surveyed men and women, and 100% of pregnant women, in the Pechenga district demonstrated a significant excess of the WHO "reference levels" of Ni in the blood (5 mcg/L). When compared with the data obtained in 2001–2003 for other Russian Arctic territories, it should be noted that the population of the Pechenga district is characterized by the lowest levels of exposure to Pb, and average levels of exposure to Hg and Cd [67].

In 2021, a joint article by members of the AMAP Human Health Assessment Group was published, summarizing the results of the MercuNorth project on the biomonitoring of Hg in the blood of pregnant women in the circumpolar Arctic (samples collected in 2010–2016). In the article, Russian data on the Pechenga district of the Murmansk oblast were used, along with materials from all eight Arctic countries that participated in the project [68].

A community-based dietary and lifestyle survey, with sampling of local terrestrial, freshwater and marine biota for analysis of environmental POPs and metals, was undertaken in spring 2016 in the Providensky district of coastal eastern Chukotka (settlements of Enmelen, Nunligran and Sireniki). This study, headed by Alexey Dudarev from the NWPHRC, was a part of the US–Russian project "Food Security and Lactic Bacteria Use in Alaska and the Bering Strait Region" (2016–2019), administered by the Institute of Northern Engineering, University of Alaska Fairbanks. Four articles in a special issue of the *IJERPH* were published in 2019, presenting the results of the study [55,56,97,98].

Concentrations of POPs in the Providensky district were very low in fish, land and marine mammal meat. However, none of the POPs analyzed exceeded the Russian MPC.

Temporal changes in the levels of POPs in local foods in the Providensky district indicated that there was a tendency for rising levels of HCB in marine mammals. These observations were consistent with the global decline in the levels of the majority of POPs [55].

Analysis of additional sources of in-home food contamination (home-brewed alcohol, domestic insecticides and wash-outs from the kitchen walls) was performed in the Providensky district. The results showed high levels of HCHs, DDTs and PCBs, still representing a share of the dietary exposure of local people to POPs [55].

An ongoing study in the communities of the Chukchi Peninsula was the first to report contamination levels of wild plants and seafood (seaweed, blue mussels and ascidians) by 18 metals in the coastal region of the northern Bering Sea [56]. The consumption of marine mammal blubber was shown to be the main exposure route for POPs, and dietary exposure to metals was primarily from fish, seafood, wild plants, mammal meat and fowl. An algorithm for calculating the Recommended Food Daily Intake Limits (RFDILs) for certain types of local foods was developed and applied for the practical recommendations for local Chukotka natives, based on Russian and foreign standards (ADIs and TDIs oral) for POPs and metals. The developed methodology for calculating RFDILs provides the ability to take into account the simultaneous presence of several contaminants in several food products [98].

With regard to the human biomonitoring of contaminants in the body of indigenous peoples of Chukotka, a Spanish–Russian–Norwegian study showed a significant decrease (by 1.5–3 times) in the concentrations of most POPs in the blood of pregnant women in coastal Chukotka (sampling in 2014–2015) compared with the results obtained in the same region of Chukotka in 2001–2002 [84].

In 2017, the Arctic Biomonitoring Laboratory was established on the basis of NArFU (Northern (Arctic) Federal University, named after M.V. Lomonosov) in Arkhangelsk. The newly formed laboratory focused its efforts on the Nenets autonomous okrug (NAO), adjacent to the Arkhangelsk region, where, in 2017–2018, Russian–Norwegian research was conducted as a sequel of the initial biomonitoring studies of the contaminants in the biota and human tissues of local people, carried out during the implementation of the GEF/AMAP/RAIPON project in 2001–2002.

The main POPs in fish from the area were DDTs and PCBs. However, no health risks were identified, and most of the substances showed a decreasing trend. Hg concentrations were low and were comparable to other Arctic studies. Northern pike was the only species for which Hg bioaccumulated. Cd and Pb were present in lower concentrations compared to Hg [38].

Human blood sampling in NAO was carried out in 2018 with the participation of 204 women from seven rural settlements, and the collected samples were investigated for PCBs and 17 organochlorine pesticides in blood serum [73]. Concentrations of all POPs in human serum were low, and most POPs decreased in the last 15 years (compared to that of 2001–2002). The results indicated a clear downward trend in the serum concentrations of p,p-DDE, HCB, PCB138 and PCB153 from 2002 to 2009 [72] and further to 2018. Interestingly, the most pronounced decline was observed between 2009 and 2018 [73]. It can be stated that, in the European part of the Russian Arctic, the 15-year POP temporal trends became available for the Nelmin-Nos settlement in the NAO.

A total of 297 whole-blood samples collected in 2018 in seven rural settlements of NAO were analyzed for As, Cd, Hg, Pb, Mn, Co, C and, Zn, which were generally low. The highest concentrations were found in the coastal populations, except for Co, Cd and Se, which were higher in the inland population. Pb concentrations were, however, at exceeding levels, with increased risk of nephrotoxicity [99].

8. Russian Regional "Hotbeds of Science" Not Involved in the AMAP Activities

When speaking about biomonitoring in the Russian Arctic during last decade, it is necessary to mention some regional "hotbeds of science" which are not involved in the AMAP activities or other international collaboration, and which are lacking unified international approaches and protocols.

One of these scientific centers is located in the Yamalo-Nenets autonomous okrug (YaNAO). The "Scientific Research Centre of the Arctic" (SRCA) was established in 2010 in the Salekhard city of the YaNAO to conduct a wide range of research. The opening in 2011 of a branch of the SRCA in the Nadym settlement of the YaNAO allowed for keeping experienced specialists from the Research Institute of Medical Problems of the Far North of the Russian Academy of Medical Sciences, which was liquidated in Nadym two years earlier. Currently, the SRCA has seven research sectors, including environmental protection, ecological and biological research. The latter was headed during last decade by Dr. Elena Agbalyan, chief researcher.

More than 130 samples of the YaNAO local fish (muscles and liver of muksun, nelma, broad whitefish and humpback whitefish) and 40 samples of reindeer meat (place and date of sampling not specified in the publication) were examined for contents of Cd, Pb and Ni by the atomic absorption method in an acetylene–air flame using spectrophotometer Spectr AA-50B (Varian, Sydney, Australia) at the Research Institute of Medical Problems of the Far North (Nadym, YaNAO). The highest levels of Cd (0.078 mg/kg ww) were detected in nelma liver, the highest levels of Pb (0.121 mg/kg ww) were detected in muksun liver and the highest levels of Ni (0.254 mg/kg ww) were detected in venison [40]. We also disposed some scattered data on the concentrations of Hg in the YaNAO local fish and reindeer meat; however, these materials are unpublished, and again, the place and date of sampling are not specified.

In 2016, 151 blood samples from the indigenous and nonindigenous residents of the YaNAO were collected in the Tazovsky settlement (and Tazovsky tundra), Gyd settlement (and Gydan tundra), Antipayut settlement (and Antipayutinskaya tundra), Nakhodka settlement (and Nakhodka tundra), Kutopyugan settlement (and Kutopyugan tundra) and Kharsaim settlement. A majority of the collected samples were from the Tazovsky district (89 samples) of the YaNAO. Chemical analysis of Hg in the blood samples was carried out in the Laboratory for Biotic Medicine (Moscow, Russia) using inductively coupled plasma mass spectrometry (ICP-MS) on a Nexion 300D quadrupole mass spectrometer (Perkin Elmer, Waltham, MA, USA). High levels of Hg were revealed in a significant portion of the samples taken. In particular, in the villages of Kharsaim and Kutopyugan, the portion of residents with an excess of blood Hg levels over 15 mcg/L were 37% (mean value 24 mcg/L) and 30% (mean value 26 mcg/L), respectively. The average levels of Hg in the blood from indigenous men (11.2 mcg/L) was almost five times higher than in nonindigenous men (2,3 mcg/L); in indigenous women (9.6 mcg/L), the difference was almost three times in comparison with the nonindigenous women (3.8 mcg/L) [76]. The revealed high Hg levels in the blood of the YaNAO indigenous people are extraordinary because, in the whole Russian Arctic (from the Kola Peninsula to Chukotka), the Hg levels over 20 mcg/L have been detected only sporadically.

Additionally, for the Hg analysis, the same 151 blood samples from the YaNAO residents were also analyzed (in the same laboratory using the same mass spectrometer) for the content of metals (Pb, Cd, As, Ni, Cu, Mg, Ca, Fe, Mn, Zn, Cr and Co). The average blood levels of toxic metals were relatively low: Pb—24 mcg/L; As—1.6 mcg/L; Cd—0.9 mcg/L; Ni—3.0 mcg/L. It should be noted that the latter results were published in conference proceedings, not in a peer-reviewed journal [77].

We have identified several publications on the contamination of some biota species sampled in the southern and central regions of the Karelia republic (below 64°N), which cannot be attributed to the Arctic regions. We found a small fragment from the aquatic ecology research, namely the analysis of Hg in perch sampled from four small lakes in the northern Karelian Loukhsky district (the White Sea lowland) by a team of specialists from two institutions of the Russian Academy of Sciences—Institute of Biology of the Karelian Research Center (Petrozavodsk, Karelia Republic) and the Papanin Institute for Biology of Inland Waters (Borok, Yaroslavl Oblast), which were carrying out a "long-term monitoring" (obviously since the early 2000s) of Hg in freshwater fish. Hg concentrations in perch muscles, determined by the atomic absorption spectroscopy at the Papanin Institute, were the following: in the lake Krivoye, 0.09 mg/kg ww; in the lake Sredneye, 0.14 mg/kg ww; in the lake Krugloye, 0.15 mg/kg ww; in the lake Zhemchuzhnoe, 0.37 mg/kg ww. It was concluded that the content of Hg in the muscles of perch from these small lakes was "relatively stable over the past 10–15 years" [35].

Biota and human blood concentrations of metals have also been studied in the Yakutia (Sakha) Republic. The assessment of metal levels in freshwater fish from the northern Yakutia began in the 1990s. Ajan Nyukkanov from the Yakut State Agricultural Academy (Yakutsk city) defended his candidate dissertation, "The content of heavy metals in freshwater fish of Yakutia", in 1996, and doctoral dissertation, "Impact of natural ecotoxicants on hydrobionts in the Sakha (Yakutia)", in 2004 (in the same year he published a monograph of the same name [44]). Unfortunately, we could not find any of Nyukkanov's informative articles related to the biomonitoring term in any domestic or foreign scientific journals. Nyukkanov's monograph and the dissertation contain the data on the levels of Hg, Pb and Cd in samples of muscles, liver, intestine, bones and gills from perch, crucian carp, roach and chukuchan collected in the Vilyuisky and Momsky districts of northern Yakutia. However, the given tables do not indicate the number of samples analyzed, or the specific location or the calendar year of fish sampling.

In 1972, Matvey Tyaptirgyanov started working in the laboratory of ichthyology at the Institute of Biology of the Yakutian Department of the Siberian Branch of the USSR Academy of Sciences (SBAS); in 1986, he became the head of the laboratory of aquatic ecosystems. In 1988–1991, he was the head of the ichthyology sector at the newly created Institute of Ecological Problems of the North of the Academy of Sciences of the Republic of Sakha (Yakutia). In 1994–2010, he worked as the director of the Department of Biological Resources of the Ministry of Nature Protection of the Republic of Sakha (Yakutia), managing and implementing a number of scientific projects. In 2017, he defended his doctoral thesis, "Fish of freshwater reservoirs of Yakutia (systematics, ecology, impact of anthropogenic factors)". Being the docent of the North-Eastern Federal University named after Ammosov

(Yakutsk city) before defending the dissertation, from 2014 to 2015, Matvey Tyaptirgyanov published three separate articles in the *Yakut Medical Journal* (in Russian) containing the data on Cd [45], Pb [46] and Hg [47] in fish samples. These articles were aimed at summarizing the results of multiyear expedition studies carried out in several Yakutian rivers. Again, we face the same problem: the given tables do not contain information on sampling locations (only the names of rivers without the indication of upper, middle or lower courses), on the numbers of samples analyzed and on the calendar year of fish sampling. The dissertation of Matvey Tyaptirgyanov helped us to find out that these articles combine the results of 2006–2010 studies (and earlier period) of the lower reaches of the rivers Khroma, Indigirka and Kolyma, which flow into the Arctic Ocean, and also the Vilyui and Amga rivers, which lapse into the Lena River at the latitudes of 64° N and 62° N, respectively.

Chemical analysis of metals (Cd, Pb, Cd) was carried out in the sanitary–hygienic laboratory of the Center for Hygiene and Epidemiology in the Republic of Sakha (Yakutia) (Yakutsk city) using the atomic absorption spectrophotometer (AAS-3, Perkin-Elmer) in an air–propane flame. Similar to Nyukkanov's results, the Tyaptirgyanov data is presented for muscles, liver, intestine, bones and gills sampled from perch, crucian carp, roach, chukuchan and also from pike, dace and broad whitefish. The authors concluded that the exceedances of Cd concentrations in Yakutian freshwater fish constitute 1–2 Russian maximum permissible concentrations (MPC), more often in polluted basins, localized in fish gills and liver [45]. Studies indicated insignificant levels of Pb in the tissues and organs of herbivorous and carnivorous fish, and few exceedances of Pb over MPC were detected in the regions of industrial mining [46]. The data obtained on Hg show 1–3 MPC exceedances in predatory fish (pike and perch) and in nonpredatory (crucian carp, roach and dace) fish in the industrial areas of diamond and gold mining [47]. It should be noted that all studies of metals in Yakutian fish were carried out by biologists–ichthyologists, who set the main goal of their research for studying the impact of contaminants on the fish population and health in conditions of their habitat pollution, but not for performing the biomonitoring of metals.

In 2017, the M.K. Ammosov North-Eastern Federal University and the Institute for Humanities Research and Indigenous Studies of the North (Yakutsk city, Russia), together with the Institute of Physiology and Basic Medicine (Novosibirsk city, Russia), carried out a human biomonitoring study in northern Yakytia. The study included 107 indigenous people (35 men and 72 women) belonging to the ethnic group of Dolgans living in the village of Yuryung-Khaya (Anabar district, Yakutia) situated on the shore of the Anabar river, 100 km from the Arctic Ocean coast. Blood serum concentrations of 13 elements (P, Sc, Ti, Cr, Mn, Fe, Ni, Cu, Zn, Rb, Sr, Cs and Pb) were analyzed by the inductively coupled plasma mass spectrometry (MS-ICP) using the Elan 9000 (Perkin Elmer, USA) at the Institute of Tectonics and Geophysics (Khabarovsk). The median levels of metals in the blood serum of the Dolgans were the following: Pb—9.5 μg/L (12.8 μg/L in men and 8.6 μg/L in women); Cr—277 μg/L; Mn—133 μg/L; Ni—57 μg/L. The authors concluded that their study revealed enhanced levels of several elements in blood serum which may affect the development of cardiovascular and other diseases among Dolgans under the conditions of the industrialization of the Arctic and expressed a hope that their study was the starting point in "the study of the microelement composition of the blood of the indigenous peoples of the Russian North" [79].

It should be noted that the above study is among several studies of the microelement status (mainly for clinical patients) which use the blood serum as a human biological medium. This phenomenon contradicts the long-established worldwide practice of studying toxic and essential elements in whole blood (which has been practiced by the AMAP for 30 years) and hampers the correct comparisons of the results of various biomonitoring studies at the circumpolar and global scales.

The Research Institute of Therapy and Preventive Medicine and the Institute of Chemical Kinetics and Combustion named after V.V. Voievodsky of the Siberian Branch of the Russian Academy of Sciences (both from Novosibirsk) carried out a study of the "elemental composition" of the whole blood of indigenous people of the coastal and inland Chukotka (the results were published in 2015, but it was not specified when the blood sampling was performed). A total of 98 blood samples of inland Chukchy from the Kanchalan village (Anadyr district, inland Chukotka) and 36 blood samples of coastal Eskimo from the Novoye Chaplino village (Providensky district, coastal Chukotka) were analyzed for 14 elements (Pb, K, Ca, Fe, Ni, Cu, Zn, Ga, Ge, Se, Br, Rb, Sr and Zr) by X-ray fluorescence analysis using synchrotron radiation (XRF SR) at the elemental analysis station of the Siberian Center for Synchrotron and Terahertz Radiation of the Institute of Nuclear Physics named after G.I. Budker of the Siberian Branch of the Russian Academy of Sciences [83]. The results presented in the article raise very serious doubts about the correctness of the performed analysis. For example, the geometric mean values of the Pb blood concentrations were 270 mcg/L in Chukchi men, 250 mcg/L in Chukchi women and 300 mcg/L in Eskimo men and women. Based on multiyear biomonitoring studies in the circumpolar Arctic (including coastal and inland Chukotka), it should be stated that the Pb concentrations presented by the authors of the above study are the highest ever found, and the average levels should be 10 times lower! In such a situation, it is unacceptable to take seriously all the work done by the authors.

The last in our list of the Russian regional "independent hotbeds of science" which are not involved in the AMAP activities, and therefore are lacking unified international approaches and protocols, but which are trying to deal with the biomonitoring of contaminants in marine biota (mainly in the water areas of the Sea of Okhotsk, the Sea of Japan and the northwestern part of the Pacific ocean), is the Vladivostok "alliance" of the Far Eastern Federal University (FEFU) and the Pacific Research Fisheries Center (TINRO-Center); the third Vladivostok's institute (V.I. Il'ichev Pacific Oceanological Institute of the Far Eastern Branch of the Russian Academy of Sciences, POI FEB RAS) seems to work separately. These institutions during last decade have produced a dozen articles (in Russian and international journals) containing the results of biomonitoring studies of POPs and metals in marine mammals from the Chukotka coastal waters of the Bering Sea. It turned out that all these articles are based on the results of two expeditions in the summers of 2010 and 2011 to one single place—Mechigmensky Bay of the Bering Sea (Lorino village, coastal eastern Chukotka). The first article [51] presented the data on the chemical analysis of HCHs, DDTs and four metals (Pb, Cd, Hg, As) in the muscles and liver of seven gray whales (four males and three females) hunted in the summer of 2010. The second article [52] showed the results of the analysis of the same contaminants (except As) in the muscles and liver of eight walruses (five males and three females) hunted in the summer of 2011. The third [53], and several subsequent articles, contains the combined POPs results of two previous publications, i.e., data on HCHs and DDTs in the muscles and livers of gray whales and Pacific walruses.

In the above studies, the chemical analysis of α-, β- and γ-HCH, DDT and DDE was performed using the gas chromatograph Shimadzu with an ECD (electron capture detector). The authors demonstrate unusually high levels of HCHs and DDTs in whale and walrus muscles and liver (blubber was not analyzed; unknown why), with the results presented on the lipid weight basis. It is unclear why, but the authors display the "total OCP concentration" ($\sum$HCHs + $\sum$DDTs) in marine mammal tissues, apparently forgetting that the global OCPs are numerous and are not limited to HCHs and DDTs. The author's "total OCP concentration" in the whale liver was up to 13,808 ng/g lipids, and in the walrus liver it was up to 90,263 ng/g lipids. These enormous concentrations (90,000 ng/g lipids = 90 mcg/g lipids) do not inspire confidence. The levels of $\sum$HCHs and $\sum$DDTs up to 100 mcg per kilogram wet weight each (200 mcg/kg ww together) have been detected in the blubber of whales and walruses sampled in the coastal Chukotka in 2016 [55] and in the early 2000s [29]. For the blubber tissue (up to 95% of fat, unlike 2–7% of fat in liver tissue), the wet weight is almost the same as the lipid weight, meaning that the author's levels of POPs in the liver are 500 times higher compared to that in the blubber of Chukotka marine mammals! It just cannot be.

Chemical analysis of metals (Pb, Cd, Hg and As) in the gray whale muscles and liver [51], and in the Pacific walrus muscles and liver [52], was carried by atomic absorption spectrophotometry using the Shimadzu AA 6800. The same equipment and method was applied for analysis of Pb and Cd in the organs (kidneys, liver, spleen, intestine, testicles, lungs, heart and bones) sampled from 22 walruses (11 males and 11 females) from the same place (Mechigmensky Bay of the Bering Sea) in the summer of 2011 [54]. All metal concentrations in these three publications are presented on a dry weight basis; in this review, we will not consider these results as unacceptable. The biomonitoring of metals in the biota in the circumpolar Arctic, for many years, has used the calculations of metal concentrations on a wet weight basis, not on a dry weight basis, as was done by the authors of these articles. No unified correction factors exist; the value of the metal concentrations in the same sample, calculated in different ways (for wet and dry weight), will differ by several times.

One last thing worth mentioning regarding the Vladivostok authors is their selective approach to the comparison of their own results with the results of other researchers. Speculations and the manipulation of data are obvious in this regard; it is not clear which contaminants and in which tissues are compared. The concentrations of contaminants in the tissues of walruses and whales of the Bering Sea, obtained in the above studies, were compared by the authors with the levels of contaminants in the tissues of whales and dolphins, reported from other regions of the planet: metals data are from the Gulf of Genoa, Mediterranean Sea, Argentine and Brazil, and POPs data are from Korean waters, Australian waters, the Black Sea, California waters, the Sea of Japan, but not from the Arctic. There is a strong belief that the authors from Vladivostok are living in "Another World" (perhaps the Eastern world) and simply ignore the abundance of results reported from different regions of the circumpolar Arctic, where, for at least 30 years, POPs and metals in different biota species have been analyzed on a wet weight basis (not on dry weight) in raw tissues (not in lipids). This contradiction excludes the possibility of comparing Vladivostok's results with the international Arctic data obtained in the frame of circumpolar activities.

Thus, the Russian regional "hotbeds of science", which are not involved in the AMAP activities or other international collaboration, could be characterized by the lack of unified international approaches and protocols, and therefore by the incomparability and frequent incorrectness of the results obtained.

9. Summarizing the Results of Russian Arctic Biomonitoring

Completing the description of the accumulated 40-year results of the biomonitoring of contaminants in the Russian Arctic, we will try to summarize the work done (Tables 1 and 2), outlining the main achievements, gaps, problems and perspectives. The main characteristics of the Russian Arctic biomonitoring are the following:

- Relatively small number of studies carried out generally;
- Small number of studies carried out in each region;
- Regarding the biota contamination, there is no data from Komi, Turukhansky and Evenkiysky districts, and few results from Karelia, Arkhangelsk Oblast, Yamal, Taimyr and Yakutia;
- Regarding the human biomonitoring, there is no data from Karelia, Turukhansky and Evenkiysky districts, and few results from Arkhangelsk Oblast, Komi, Yamal and Yakutia;
- Most of the studies were carried out in the 1990s, followed by the large-scale GEF/AMAP/RAIPON project of 2001–2004, which covered four regions (Murmansk oblast, Nenets okrug, Taimyr and Chukotka); after that, only single projects have been implemented in a few regions;

Thus, only the Nenets okrug and the Chukotka okrug (and hardly the Murmansk Oblast) can be attributed to the regions where the biomonitoring of contaminants was carried out during last decades on several occasions, and for which the 12–15-year dynamic trends (at least "at three points") of the content of POPs and metals in the biota and human organism were obtained. For the rest of the Russian Arctic territories, only fragmentary "cross-sections" of biomonitoring data is available, mainly obtained in the 1990s or early 2000s, which do not allow to judge either the dynamics of the processes or the current state of affairs.

Another important issue is that the majority of the studies in the Russian Arctic (more than 90%) were carried out within the framework of international projects, i.e., with cofinancing, assistance and contribution (including laboratory analyses) from foreign colleagues and partners.

In addition, it should be noted that, with the exception of a single pilot personalized cohort study in coastal Chukotka in 2007 [82], studies similar to those permanently conducted, for example, in Norway, Canada or the Faroe Islands (when the same individuals are examined for decades as they grow older, and hundreds or even thousands of people are involved in the studies), have never been performed in the Russian Arctic. Moreover, most biomonitoring studies in the Russian Arctic are characterized by a small number of observations (samples analyzed), which does not allow drawing statistically valid conclusions, comparisons or generalizations.

As for the spectrum of the analyzed POPs, in contrast to the modern circumpolar Arctic, where, in parallel with the "old" ("legacy") POPs, the "new" ("emerging") contaminants (polybrominated, polyfluorinated, etc.) are on the agenda of biomonitoring studies, in the Russian Arctic, only "legacy" POPs are the focus of research, as a rule. The reason for that is the limited chemical–analytical potential in the Russian Arctic and generally in Russia. There is a lack of modern laboratories with modern equipment and methodologies needed for the analysis of ultra-low concentrations of POPs and metals in biological tissues, as well as a lack of qualified specialists. In the entire set of the Russian Arctic regions, there is not a single laboratory accredited for the analysis of POPs in human biological media (particularly in blood) which are able to perform research at the modern international level. Under the current internal ban on the export of biological samples abroad, the implementation of biomonitoring studies in Russia, even in cooperation with foreign colleagues who dispose the necessary laboratory facilities, becomes problematic for Russian researchers. Very few Russian laboratories integrated into international scientific collaboration, which are able to provide high-quality results and to confirm their research quality by the regular participation in international interlaboratory comparisons, are located outside the Russian Arctic—in Moscow, Obninsk, St. Petersburg, Ufa and some other scientific centers. The Russian Arctic regional "hotbeds of science", which are not involved in international collaboration, and which sometimes use very "special" laboratory equipment and apply some "exotic" laboratory methods, are doomed to incorrectness and the incomparability of the results obtained.

Summing up all of the above, it must be recognized that the Arctic biomonitoring of contaminants in Russia has not yet been formed; by now, it is only a certain set of scattered results from different studies conducted in different years, which so far cannot be called a functioning system. The available Russian biomonitoring dataset may serve as a basis for the system formation in the future, which will require reasonable elaboration, substantial efforts and significant financing.

10. Perspectives of the Russian Arctic Biomonitoring

First of all, the elaboration of a concept and a strategy for the development of Russian Arctic biomonitoring is needed. It is clear that the priority is the updating of available scientific capacities, including the extension and radical modernization of existing chemical–analytical laboratories focused on the studying and monitoring of contaminants in the Arctic environment, and environmental health research centers dealing with the science-based assessment and prediction of exposure burdens and adverse health risks in Arctic populations.

How to provide scientific capacities and how to "cover" (by the studies) the Russian Arctic territory with a coastline of 20,000 km in length?

As we already pointed, there are only few chemical labs in Russia, primarily the "Typhoon" in Obninsk city and the northwestern branch of "Typhoon" in St-Petersburg (former RCMA), which are historically (for more than three decades) dealing with Arctic contaminants issues, which are accredited for the analysis of POPs and metals in biological tissues and which are regularly participating in the international (including AMAP) ring tests.

There are also few research centers in Russia dealing with Arctic environmental health issues, primarily the Northwest Public Health Research Center in St-Petersburg, which, during almost three decades, has participated in international Arctic environmental health projects focusing on contaminants, and which is deeply integrated in the AMAP system.

The Arkhangelsk NArFU Arctic Biomonitoring Laboratory (ABL) was established in 2017. The ABL presents itself as a "specialized multidisciplinary laboratory (biomonitoring centre) for monitoring POPs in the Russian Arctic ("from the bottom")". They are planning "to accredit in the National Accreditation System the methodologies for the identification of inorganic and organic pollutants in food samples and human blood samples" and intend to develop a "methodology for conducting biological monitoring in the Russian Arctic" [100].

The ABL "within the governmental megagrant project has developed proposals to change the existing legal acts regulating biomonitoring". They "revealed gaps in the current Russian legislation" in the sphere of the national (state) monitoring system regarding the contamination of traditional food consumed by indigenous peoples in the Arctic. The ABL staff put forward an idea to include into the list of objects for the mandatory systematic control (in the frame of sociohygienic monitoring headed by the Federal Service for Supervision of Consumer Rights Protection and Human Welfare, Rospotrebnadzor) the following item: "Chemical contamination of migratory species of birds, fish and wild animals consumed by indigenous peoples of the North, Siberia and the Far East of the Russian Federation", anticipating that the "collection of the necessary data and information is becoming mandatory". The ABL declares that "the state monitoring system is a very effective mechanism for a country as large as Russia", and that "such changes will also help to harmonize monitoring activities in Russia with other Arctic States and to fill in the gaps in the Global Monitoring Reports and the Arctic Monitoring and Assessment Programme (AMAP) reports on persistent organic pollutants in traditional indigenous food" [100].

One clear problem is in the Russian system of sociohygienic monitoring (SHM), which, in the Arctic, is far from perfect.

Indeed, the Rospotrebnadzor in the Arctic has to control a few chemical contaminants (including some PTS) in a certain few types of local foods produced by licensed companies for further sale in the retail network. Local (traditional) foods of individual hunting–fishing–gathering–growing, consumed by the local population, remain out of control by any regional or federal authorities. Thus, the Arctic SHM database does not contain any data on POPs and metals in traditional foods in the Russian Arctic regions, especially the monitoring data.

It is important to note that the Rospotrebnadzor branches in the regions of the Russian Arctic were reduced to a minimum during last two decades and are now experiencing an acute shortage of personnel, material resources (including the poor state of its laboratories) and financing. The SHM system in the Russian Arctic is flawed; the information on environmental factors (chemical contamination of water, air and soil) currently being sampled is characterized by a clear insufficiency both in quantitative and qualitative terms: scanty coverage of the controlled territories (small remote communities are almost out of control), irregular and infrequent data collection, limited list of analyzed contaminants, low objectivity and comparability (between regions) of the collected materials and obtained results and a general lack of valid information. Thus, it would be extremely difficult to oblige the Arctic regional Rospotrebnadzor branches to carry out the regular additional work of sampling numerous types of local foods in numerous settlements (particularly in the remote and hard-to-reach communities), and it would be impossible to perform a chemical analysis of the hypothetically collected samples on the spot.

In this situation, the question arises: how to develop biomonitoring in the Russian Arctic? Is it evident that there are no opportunities (financial–economic resources, equipment, specialists, etc.) to establish a wide net of chemical labs and research centers in each region of the "Arctic Zone of Russian Federation"? A scenario of strengthening those few scientific platforms available in the Arctic, and the active involvement of the above-mentioned Russian professional institutions, should be considered. If so, is there a need to understand how to build the national biomonitoring system, leaning on these few national scientific platforms called to cope with serious challenges and to deal with large volumes of research in numerous regions of the Russian Arctic?

In our opinion, the only way to create the Russian Arctic biomonitoring system is to elaborate and launch the national Russian Arctic Contaminants Program (RACP), where the biomonitoring of contaminants will be an important structural element. The RACP should function on a permanent basis (for decades onward), it should be interdisciplinary and it should consist of several research and monitoring subprograms, managed by several constantly interacting expert groups:

- Environmental monitoring of contaminants (air, soil, water, bottom sediments, etc.) in three major ecosystem types: terrestrial, freshwater and marine;
- Biomonitoring of contaminants in biota (different wildlife species), primarily those consumed as traditional/local foods by indigenous peoples;
- Traditional diets and nutrition of indigenous peoples (dietary surveys) and dietary exposure assessments;
- Human biomonitoring (contaminants in human tissues);
- Human health assessments (health effects related to human exposure);
- Human health risk assessment, management and communication.

Each expert group in the first phase of their work should explore in detail all the appropriate information (spatial–temporal) accumulated to date, release a summarized comprehensive report and, after that, develop a perspective plan containing a regional-specific blueprint for the Russian Arctic.

New information for each subprogram will be collected by research expeditions in the frames of scientific projects, which could involve different specialists from different institutions, according to their plans. Funding of the projects should be provided by the research institutions themselves by obtaining various grants, including direct state financial support.

Community-based monitoring is preferable, as it allows covering (in dynamics) all the scope of the RACP subprograms (environment, biota, human) in the selected places in parallel. In each region of the Russian Arctic, it is necessary to define the limited number of these "target communities" ("focal ecosystems") and to determine the regional priorities for the contaminant measurements in certain wildlife species that represent important traditional/local foods which are frequently harvested by local (particularly indigenous) people. The list of such "target communities" in each region should be limited but "representative"; it must include the places of compact residences of indigenous peoples, the remote hard-to-reach settlements, the coastal villages (bordering the ocean/sea and large rivers/lakes) as well as the inland settlements. The choice should fall on "typical" locations where the inhabitants are engaged in sea hunting and fishing, reindeer herding, game hunting, freshwater fishing, gathering mushrooms, berries and wild plants, growing vegetables, etc. It is also important to include in the list the vicinities of industrial enterprises that pollute the environment (and the local food chains) by contaminants of local origin.

The RACP should be focused, first of all, on legacy POPs, new POPs, chemicals of emerging concern, toxic metals and essential elements. The health studies could be supplemented by the assessment of vitamins, hormones and other biochemical components in the human organism. The priorities for the human biomonitoring, health and risk assessments should be mother–child pairs, children and women of reproductive age (as vulnerable groups), but adult populations of both genders are very important to be included in the analysis as highly exposed groups for which health effects may become noticeable.

Monitoring plans (frequency of samplings and surveys) should take into account the time intervals required for the monitoring of different contaminants in various mediums in certain regions, but, of course, should be based on actual research opportunities. The formation of biobanks (archives of frozen biological samples stored for future analysis) is of great importance when regular financing is hardly possible.

11. Conclusions

Apart from the single studies of the late 1970s and 1980s in the southern part of the Barents Sea, the Russian Arctic biomonitoring of contaminants only actively started in the northwest part of the country (first of all, in Murmansk Oblast and Arkhangelsk Oblast) in the early 1990s, much later than the beginning of the global and circumpolar biomonitoring. The biomonitoring in the territory of the Russian Arctic has made a certain contribution to understanding the patterns of biota contamination and population exposure to POPs and metals in the circumpolar Arctic. However, it should be concluded that only a relatively small number of studies were carried out during the last 40 years; for several Russian Arctic regions, there is no up-to-date data, and for some regions, the data are scarce; most of the studies were carried out in the 1990s, followed by the large-scale GEF/AMAP/RAIPON project of 2001–2004 covering four regions; after that, single projects have been implemented in only a few regions. Only the Nenets okrug and the Chukotka okrug (and hardly Murmansk oblast) can be attributed to the regions where the biomonitoring of contaminants was carried out during last decades on several occasions, and for which the content of POPs and metals in biota and the human organism may be assessed in 12–15-year dynamic trends (at least "at three points"). For the rest of the Russian Arctic territories, only fragmentary "cross-sections" of biomonitoring data is available, mainly obtained in the 1990s or early 2000s, which do not allow judging either the dynamics of the processes or the current state of affairs. The overwhelming majority of the studies in the Russian Arctic (more than 90%) were carried out within the framework of international projects, i.e., with cofinancing, assistance and contribution (including laboratory analyses) from foreign colleagues and partners. The shortcomings of the Russian system of biomonitoring, including the weaknesses of the laboratory and research base, are considered. Perspectives of the Russian Arctic biomonitoring belong to the proposal of elaboration of the national Russian Arctic Contaminants Program (RACP).

Author Contributions: Conceptualization, A.A.D.; methodology, A.A.D.; writing—original draft preparation, A.A.D.; review and editing, A.A.D. and J.O.O.; validation, J.O.O.; supervision, project administration, J.O.O. All authors have read and agreed to the published version of the manuscript.

Funding: This research received no external funding.

Conflicts of Interest: The authors declare no conflict of interest.

References

1. Durham, W.F.; Armstrong, J.F.; Upholt, W.M.; Heller, C. Insecticide content of diet and body fat of Alaskan natives. *Science* **1961**, *134*, 1880–1881. [CrossRef] [PubMed]
2. Cade, T.J.; White, C.M.; Haugh, J.R. Peregrines and Pesticides in Alaska. *Condor* **1968**, *70*, 170–178. [CrossRef]
3. Cade, T.J.; Lincer, J.L.; White, C.M.; Roseneau, D.G.; Swartz, L.G. DDE Residues and Eggshell Changes in Alaskan Falcons and Hawks. *Science* **1971**, *172*, 955–957. [CrossRef]
4. Holden, A.V.; Marsden, K. Organochlorine pesticides in seals and porpoises. *Nature* **1967**, *216*, 1274–1276. [CrossRef] [PubMed]
5. Holden, A.V. Monitoring organochlorine contamination of the marine environment by the analysis of residues in seals. In *Marine Pollution and Sea Life*; Ruivo, M., Ed.; Fishing News Books Ltd.: Totnes, UK, 1972; pp. 266–272.
6. Addison, R.F.; Brodie, P.F. Occurrence of DDT residues in beluga whales (Delphinapterus leucas) from the Mackenzie Delta, N.W.T. *J. Fish. Res. Board Can.* **1973**, *30*, 1733–1736. [CrossRef]
7. Addison, R.F.; Smith, T.G. Organochlorine residue levels in Arctic ringed seals: Variation with age and sex. *Oikos* **1974**, *25*, 335–337. [CrossRef]

8. Clausen, J.; Braestrup, L.; Berg, O. The content of polychlorinated hydrocarbons in Arctic mammals. *Bull. Environ. Contam. Toxicol.* **1974**, *12*, 529–534. [CrossRef]
9. Reinke, J.; Uthe, J.F.; Sabatino, T. Organochlorine residues in commercial fish caught in Canada–1970. *Pestic. Monit. J.* **1972**, *6*, 43–49.
10. Bowes, G.W.; Jonkel, C.J. Presence and distribution of polychlorinated biphenyls (PCB) in arctic and subarctic marine food chains. *J. Fish. Res. Board Can.* **1975**, *32*, 2111–2123. [CrossRef]
11. Dewailly, E.; Nantel, A.; Weber, J.-P.; Meyer, F. High levels of PCBs in breast milk of women from Arctic Québec. *Bull. Environ. Contam. Toxicol.* **1989**, *43*, 641–646. [CrossRef]
12. Kinloch, D.; Kuhnlein, H.; Muir, D. Inuit foods and diet. A preliminary assessment of benefits and risks. *Sci. Total Environ.* **1992**, *122*, 247–278. [CrossRef]
13. AMAP. *AMAP Assessment Report: Arctic Pollution Issues*; Arctic Monitoring and Assessment Programme (AMAP): Oslo, Norway, 1998; 859p.
14. AMAP. *AMAP Assessment 2002: Human Health in the Arctic*; Arctic Monitoring and Assessment Programme (AMAP): Oslo, Norway, 2003; 137p.
15. AMAP. *AMAP Assessment 2009: Human Health in the Arctic*; Arctic Monitoring and Assessment Programme (AMAP): Oslo, Norway, 2009; 254p.
16. AMAP. *AMAP Assessment 2015: Human Health in the Arctic*; Arctic Monitoring and Assessment Programme (AMAP): Oslo, Norway, 2015; 165p.
17. AMAP. *AMAP Assessment 2021: Human Health in the Arctic*; Arctic Monitoring and Assessment Programme (AMAP): Tromso, Norway, 2021; 240p.
18. *Report of AMAP-2015: Human Health in the Arctic—SPb*; Publishing and Printing Company "Kosta": Ohrid, Republic of Macedonia, 2018; 176p. (In Russian)
19. Dudarev, A.A.; Odland, J.O. Human health in connection with Arctic pollution—Results and perspectives of international studies under the aegis of AMAP. *Ekologiya cheloveka* **2017**, *9*, 3–14. (In Russian) [CrossRef]
20. Savinova, T.N.; Ugryumova, L.E.; Andryuzchenko, V.V. Levels of DDT and its metabolites in fish and invertebrates of the Barents Sea shelf. *Gidrobiologycal J.* **1981**, *17*, 93–96. (In Russian)
21. Savinova, T.N. *Chemical Pollution of the Northern Seas*; Canadian translation of Fisheries and Aquatic Sciences No. 5536; National Research Council of Canada: Ottawa, ON, Canada, 1991; p. 174.
22. Savinova, T.N.; Gabrielsen, G.W.; Falk-Petersen, S. *Chemical Pollution in the Arctic and Sub-Arctic Marine Ecosystems: An Overview of Current Knowledge*; Report no. 3: 1994/1995. NINA-fagrapport 1; The joint Norwegian-Russian Commission on Environmental Cooperation: Moscow, Russia, 1995; pp. 1–68.
23. Savinov, V.M.; Gabrielsen, G.W.; Savinova, T.N. Cadmium, zinc, copper, arsenic, selenium and mercury in seabirds from the Barents Sea: Levels, inter-specific and geographical differences. *Sci. Total Environ.* **2003**, *306*, 133–158. [CrossRef]
24. Henny, C.J.; Ganusevich, S.A.; Ward, F.P.; Schwartz, T.R. Organochlorine pesticides, chlorinated dioxins and furans, and PCBs in Peregrine Falcon eggs from the Kola Peninsula, Russia. In *Raptor Conservation Today*; Meyburg, B.-U., Chancellor, R.D., Eds.; Pica Press: London, UK, 1994; pp. 739–749.
25. Amundsen, P.-A.; Staldvik, F.J.; Lukin, A.A.; Kashulin, N.A.; Popova, O.A.; Reshetnikov, Y.S. Heavy metal contamination in freshwater fish from the border region between Norway and Russia. *Sci. Total Environ.* **1997**, *201*, 211–224. [CrossRef]
26. Savinov, V.; Gabrielsen, G.W.; Savinova, T.; Alekseeva, L.; Polder, A.; Skaare, J.U.; Koryakin, A. POPs in marine birds from the eastern Barents Sea (Russia): Comparative analysis of data set from 1992 and 2002. *Organohalogen Compound.* **2007**, *69*, 595–598.
27. Zauke, G.P.; Savinov, V.M.; Ritterhoff, J.; Savinova, T. Heavy metals in fish from the Barents Sea (summer 1994). *Sci. Total Environ.* **1999**, *9*, 161–173. [CrossRef]
28. Muir, D.; Savinova, T.; Savinov, V.; Alexeeva, L.; Potelov, V.; Svetochev, V. Bioaccumulation of PCBs and chlorinated pesticides in seals, fishes and invertebrates from the White Sea, Russia. *Sci. Total Environ.* **2003**, *1*, 111–131. [CrossRef]
29. AMAP. *Persistent Toxic Substances, Food Security and Indigenous Peoples of the Russian North. Final Report*; AMAP Report 2004:2; Arctic Monitoring and Assessment Programme (AMAP): Oslo, Norway, 2004; p. 192.
30. Dudarev, A.A.; Dushkina, E.V.; Sladkova, Y.N.; Burova, D.V.; Gushchin, I.V.; Talykova, L.V.; Nikanov, A.N.; Lukicheva, L.A. Persistent organic pollutants (POPs) in local food products of the Pechenga district of the Murmansk region. *Toxicol. Rev.* **2015**, *4*, 18–25. (In Russian)
31. Dudarev, A.A.; Dushkina, E.V.; Chupakhin, V.S.; Sladkova, Y.N.; Burova, D.V.; Gushchin, I.V.; Talykova, L.V.; Nikanov, A.N.; Lukicheva, L.A. Metal content in local food products of the Pechenga district of the Murmansk region. *Russ. J. Occup. Health Ind. Ecol.* **2015**, *2*, 35–40. (In Russian)
32. Hansen, M.D.; Nøst, T.H.; Heimstad, E.S.; Evenset, A.; Dudarev, A.A.; Rautio, A.; Myllynen, P.; Dushkina, E.V.; Jagodic, M.; Christensen, G.N.; et al. The impact of a Nickel-Copper smelter on concentrations of toxic elements in local wild food from the Norwegian, Finnish, and Russian border regions. *Int. J. Environ. Res. Public Health* **2017**, *14*, 694. [CrossRef] [PubMed]
33. Kostamo, A.; Medvedev, N.; Pellinen, J.; Hyvarinen, H.; Kukkonen, V.K. Analysis of organochlorine compounds and extractable organic halogen in three subspecies of ringed seal from northeastern Europe. *Environ. Toxicol. Chem.* **2000**, *19*, 848–854. [CrossRef]
34. Savinov, V.; Muir, D.C.; Svetochev, V.; Svetocheva, O.; Belikov, S.; Boltunov, A.; Alekseeva, L.; Reiersen, L.O.; Savinova, T. Persistent organic pollutants in ringed seals from the Russian Arctic. *Sci. Total Environ.* **2011**, *409*, 2734–2745. [CrossRef] [PubMed]

35. Nemova, N.N.; Lysenko, L.A.; Meshcheryakova, O.V.; Komov, V.T. Mercury in fish: Biochemical indication. *J. Biosphere* **2014**, *6*, 176–186. (In Russian)
36. Savinova, T.N.; Polder, A.; Gabrielsen, G.W.; Skaare, J.U. Chlorinated hydrocarbons in seabirds from the Barents Sea area. *Sci. Total Environ.* **1995**, *160*, 497–504. [CrossRef]
37. Kleivane, L.; Espeland, O.; Fagerheim, K.A.; Hylland, K.; Polder, A.; Skaare, J.U. Organochlorine pesticides and PCBs in the east ice harp seal (*Phoca groenlandica*) population. *Mar. Environ. Res.* **1997**, *43*, 117–130. [CrossRef]
38. Sobolev, N.; Aksenov, A.; Sorokina, T.; Chashchin, V.; Ellingsen, D.G.; Nieboer, E.; Varakina, Y.; Veselkina, E.; Kotsur, D.; Thomassen, Y. Essential and non-essential trace elements in fish consumed by indigenous peoples of the European Russian Arctic. *Environ Pollut.* **2019**, *253*, 966–973. [CrossRef]
39. Lakhmanov, D.; Varakina, Y.; Aksenov, A.; Sorokina, T.; Sobolev, N.; Kotsur, D.; Plakhina, E.; Chashchin, V.; Thomassen, Y. Persistent Organic Pollutants (POPs) in Fish Consumed by the Indigenous Peoples from Nenets Autonomous Okrug. *Environments* **2020**, *7*, 3. [CrossRef]
40. Agbalyan, E.V. The content of heavy metals and the public health risk in the Yamal North. *Hyg. Sanit.* **2012**, *1*, 14–16. (In Russian)
41. Allen-Gil, S.M.; Ford, J.; Lasorsa, B.; Monetti, M.; Vlasova, T.; Landers, D.H. Heavy metal contamination in the Taimyr peninsula, Siberian Arctic. *Sci. Total Environ.* **2003**, *301*, 119–138. [CrossRef]
42. Nakata, H.; Tanabe, S.; Tatsukawa, R.; Koyama, Y.; Miyazaki, N.; Belikov, S.; Boltunov, A. Persistent organochlorine contaminants in ringed seals (*Phoca hispida*) from the Kara Sea, Russian Arctic. *Environ. Toxicol. Chem.* **1998**, *17*, 1745–1755. [CrossRef]
43. Rovinsky, F.; Pastukhov, B.; Bouyvolov, Y.; Burtseva, L. Present day state of background pollution of the natural environment in the Russian Arctic in the region of the Ust-Lena Reserve. *Sci. Total Environ.* **1995**, *160*, 193–199. [CrossRef]
44. Nyukkanov, A.N.; Kolesnikov, V.A. *Impact of Natural Ecotoxicants on Hydrobionts in the Sakha (Yakutia) Republic;* Krasnoyarsk State Agrarian University Publishing: Krasnoyarsk, Russia, 2004; p. 240. (In Russian)
45. Tyaptirgyanov, M.M.; Tyaptirgyanova, V.M. Ecological and hygienic assessment of the accumulation and distribution of cadmium compounds in the organs and tissues of freshwater fish in Yakutia. *Yakut Med. J.* **2014**, *3*, 69–73. (In Russian)
46. Tyaptirgyanov, M.M.; Tyaptirgyanova, V.M. Ecological and hygienic assessment of the accumulation and distribution of lead compounds in the organs and tissues of freshwater fish in Yakutia. *Yakut Med. J.* **2014**, *4*, 47–50. (In Russian)
47. Tyaptirgyanov, M.M.; Tyaptirgyanova, V.M. Ecological and hygienic assessment of mercury accumulation in the organs and tissues of freshwater fish in Yakutia. *Yakut Med. J.* **2015**, *1*, 34–38. (In Russian)
48. Kim, E.-Y.; Ichihashi, H.; Saeki, K.; Atrashkevich, G.; Tanabe, S.; Tatsukava, R. Metal accumulation in tissues of seabirds from Chaun, Northeast Siberia, Russia. *Environ. Poll.* **1996**, *92*, 247–252. [CrossRef]
49. Amirova, Z.; Kruglov, E.; Vlasov, S.; Loshkina, E.; Khalilov, R. PCBs and PCDD/Fs distribution in tissues and organs of marine animals in Russian Arctic. *Organohalog Compd.* **2004**, *66*, 1533–1541.
50. Dudarev, A.A. Dietary exposure to persistent organic pollutants and metals among Inuit and Chukchi in Russian Arctic Chukotka. *Int. J. Circumpolar Health* **2012**, *71*, 18592. [CrossRef]
51. Tsygankov, V.Y. Organochlorine pesticides and heavy metals in organs of gray whale from the Bering Sea. *Izvestiya TINRO (J. Trans. Pac. Res. Inst. Fish. Oceanogr.)* **2012**, *170*, 202–209. (In Russian)
52. Tsygankov, V.Y.; Boyarova, M.D.; Lukyanova, O.N. Persistent toxic substances in the muscles and liver of the pacific walrus Odobenus rosmarus divergens Illiger, 1815 from the Bering Sea. *Russ. J. Mar. Biol.* **2014**, *40*, 147–151. [CrossRef]
53. Tsygankov, V.Y.; Boyarova, M.D.; Lukyanova, O.N. Bioaccumulation of persistent organochlorine pesticides (OCPs) by gray whale and Pacific walrus from the western part of the Bering Sea. *Mar. Pollut. Bull.* **2015**, *99*, 235–239. [CrossRef]
54. Trukhin, A.M.; Kolosova, L.F.; Slinko, E.N. Toxic metals in walruses (Odobenus rosmarus divergens Linnaeus, 1785) of the Bering Sea. *J. Stud. Aquat. Biol. Resour. Kamchatka Northwest. Part Pac. Ocean Petropavlovsk-Kamchatsky* **2013**, *28*, 140–146. (In Russian)
55. Dudarev, A.A.; Chupakhin, V.S.; Vlasov, S.V.; Yamin-Pasternak, S. Traditional Diet and Environmental Contaminants in Coastal Chukotka II: Legacy POPs. *Int. J. Environ. Res. Public Health* **2019**, *16*, 695. [CrossRef] [PubMed]
56. Dudarev, A.A.; Chupakhin, V.S.; Vlasov, S.V.; Yamin-Pasternak, S. Traditional Diet and Environmental Contaminants in Coastal Chukotka III: Metals. *Int. J. Environ. Res. Public Health* **2019**, *16*, 699. [CrossRef] [PubMed]
57. Polder, A.; Becher, G.; Savinova, T.N.; Skaare, J.U. Dioxins, PCBs and some chlorinated pesticides in human milk from the Kola Peninsula, Russia. *Chemosphere* **1998**, *37*, 1795–1806. [CrossRef]
58. Odland, J.O.; Nieboer, E.; Romanova, N.; Thomassen, Y.; Brox, J.; Lund, E. Concentrations of essential trace elements in maternal serum and the effect on birth weight and newborn body mass index in sub-arctic and arctic populations of Norway and Russia. *Acta Obstetricia et Gynecologica Scandinavica* **1999**, *78*, 605–614.
59. Odland, J.O.; Nieboer, E.; Romanova, N.; Thomassen, Y.; Lund, E. Blood lead and cadmium and birth weight among sub-arctic and arctic populations of Norway and Russia. *Acta Obstetricia et Gynecologica Scandinavica* **1999**, *78*, 852–860.
60. Odland, J.O.; Nieboer, E.; Romanova, N.; Thomassen, Y.; Brox, J.; Lund, E. Self reported ethnic status of delivering women, newborn body mass index, and blood, serum and urine concentrations of toxic and essential elements in Norwegian and Russian Arctic populations. *Int. J. Circumpolar Health* **1999**, *58*, 4–13.
61. Odland, J.O.; Perminova, I.; Romanova, N.; Thomassen, Y.; Tsuji, L.J.S.; Brox, J. Elevated blood lead concentrations in children living in isolated communities of the Kola Peninsula, Russia. *Ecosystem Health* **1999**, *5*, 75–81. [CrossRef]

62. Polder, A.; Gabrielsen, G.W.; Odland, J.O.; Savinova, T.N.; Tkachev, A.; Loken, K.B.; Skaare, J.U. Spatial and temporal changes of chlorinated pesticides, PCBs, dioxins (PCDDs/PCDFs) and brominated flame retardants in human breast milk from Northern Russia. *Sci. Total Environ.* **2008**, *25*, 41–54. [CrossRef]
63. Dudarev, A.A.; Konoplev, A.V.; Sandanger, T.M.; Vlasov, S.V.; Miretsky, G.I.; Samsonov, D.P.; Chernik, G.V.; Morshina, T.N.; Pasynkova, E.M.; Pervunina, R.I.; et al. Blood concentrations of persistent toxic substances in the indigenous communities of the Russian Arctic. *Int. J. Circumpolar Health* **2004**, *63* (Suppl. S2), 179–182. [CrossRef] [PubMed]
64. Holoubek, I.; Klanova, J.; Kocan, A.; Cupr, P.; Dudarev, A.; Boruvkova, J.; Chroma, K. *Global Monitoring Plan for Persistent Organic Pollutants under the Stockholm Convention Article 16 on Effectiveness Evaluation*; Recetox-Tocoen Reports No. 339; First Regional Monitoring Report Central and Eastern European and Central Asian Region; Recetox mu Brno: Brno, Czechia, 2008.
65. Sandanger, T.M.; Anda, E.E.; Dudarev, A.A.; Nieboer, E.; Konoplev, A.V.; Vlasov, S.V.; Weber, J.P.; Odland, J.O.; Chashchin, V.P. Combining data sets of organochlorines (OCs) in human plasma for the Russian Arctic. *Sci. Total Environ.* **2009**, *407*, 5216–5222. [CrossRef] [PubMed]
66. Dudarev, A.A.; Dushkina, E.V.; Sladkova, Y.N.; Chupakhin, V.S.; Lukicheva, L.A. Exposure levels to persistent organic pollutants (POPs) of the population of the Pechenga district of the Murmansk region. *Toxicol. Rev.* **2016**, *3*, 2–9. (In Russian) [CrossRef]
67. Dudarev, A.A.; Dushkina, E.V.; Sladkova, Y.N.; Chupakhin, V.S.; Lukicheva, L.A. Levels of exposure to metals of the population of the Pechenga district of the Murmansk region. *Russ. J. Occup. Health Ind. Ecol.* **2016**, *6*, 11–16. (In Russian)
68. Adlard, B.; Lemire, M.; Bonefeld-Jørgensen, E.C.; Long, M.; Olafsdottir, K.; Odland, J.O.; Rautio, A.; Myllynen, P.; Sandanger, T.M.; Dudarev, A.A.; et al. MercuNorth–monitoring mercury in pregnant women from the Arctic as a baseline to assess the effectiveness of the Minamata Convention. *Int. J. Circumpolar Health* **2021**, *80*, 1881345. [CrossRef]
69. Sandanger, T.M.; Odland, J.O.; Tkachev, A.; Burkow, I.C. Persistent organic pollutants in plasma of delivering women from Arkhangelsk. *Sci. Total Environ.* **2003**, *306*, 171–178. [CrossRef]
70. Polder, A.; Odland, J.O.; Tkachev, A.; Foreid, S.; Savinova, T.N.; Skaare, J.U. Geographic variation of chlorinated pesticides, toxaphenes and PCBs in human milk from sub-arctic and arctic locations in Russia. *Sci. Total Environ.* **2003**, *1*, 179–195. [CrossRef]
71. Odland, J.O.; Sandanger, T.M.; Heimstad, E.S. *Screening and assessment of contaminants in human blood samples from Northern Norway, North-West Russia and Siberia*; TA-2184/2006; Statens Forurensningstilsyn: Oslo, Norway, 2006.
72. Rylander, C.; Sandanger, T.M.; Petrenya, N.; Konoplev, A.; Bojko, E.; Odland, J.O. Indications of decreasing human PTS concentrations in North West Russia. *Glob. Health Action* **2011**, *4*, 8427. [CrossRef]
73. Varakina, Y.; Lahmanov, D.; Aksenov, A.; Trofimova, A.; Korobitsyna, R.; Belova, N.; Sobolev, N.; Kotsur, D.; Sorokina, T.; Grjibovski, A.M.; et al. Concentrations of Persistent Organic Pollutants in Women's Serum in the European Arctic Russia. *Toxics* **2021**, *9*, 6. [CrossRef]
74. Klopov, V. Levels of heavy metals in women residing in the Russian Arctic. *Int. J. Circumpolar Health* **1998**, *57*, 582–585.
75. Klopov, V.; Odland, J.O.; Burkow, I.C. Persistent organic pollutants in maternal blood plasma and breast milk from Russian Arctic populations. *Int. J. Circumpolar Health* **1998**, *57* (Suppl. S1), 239–248. [PubMed]
76. Agbalyan, E.V.; Shinkaruk, E.V. The mercury content in the blood of inhabitants of the Yamalo-Nenets autonomous district. *Hyg. Sanit.* **2018**, *97*, 799–802. (In Russian) [CrossRef]
77. Agbalyan, E.V. The prevalence and content of elements in the blood of the aboriginal population and migrants of the Yamalo-Nenets Autonomous Okrug. In Proceedings of the Ecosystem Services and Management of Natural Resources: Materials of the International Scientific and Practical Conference, Tyumen, Russia, 28–30 November 2019; Gashev, S.N., Ed.; Vector Book: Tyumen, Russia, 2020; pp. 262–267.
78. Hanssen, L.; Dudarev, A.A.; Huber, S.; Odland, J.O.; Nieboer, E.; Sandanger, T.M. Partition of perfluoroalkyl substances (PFASs) in whole blood and plasma, assessed in maternal and umbilical cord samples from inhabitants of arctic Russia and Uzbekistan. *Sci. Total Environ.* **2013**, *447*, 430–437. [CrossRef] [PubMed]
79. Sivtseva, A.I.; Sivtseva, E.N.; Shadrina, S.S.; Melnikov, V.N.; Boyakova, S.I.; Dokhunaeva, A.M. Microelement composition of serum in Dolgans, indigenous inhabitants of the Russian Arctic, in the conditions of industrial development of territories. *Int. J. Circumpolar Health* **2020**, *79*, 1. [CrossRef] [PubMed]
80. Sandanger, T.M.; Brustad, M.; Odland, J.O.; Doudarev, A.A.; Miretsky, G.I.; Chaschin, V.; Burkow, I.C.; Lund, E. Human plasma levels of POPs, and diet among native people from Uelen, Chukotka. *J. Environ. Monitoring.* **2003**, *5*, 689–696. [CrossRef] [PubMed]
81. Eik Anda, E.; Nieboer, E.; Dudarev, A.A.; Sandanger, T.M.; Odland, J.O. Intra- and intercompartmental associations between levels of organochlorines in maternal plasma, cord plasma and breast milk, and lead and cadmium in whole blood, for indigenous peoples of Chukotka, Russia. *J. Environ. Monit.* **2007**, *9*, 884–893. [CrossRef]
82. Dudarev, A.A.; Chupakhin, V.S.; Odland, J.O.; Reiersen, L.O.; Chashchin, V.P. A follow-up study of blood levels of persistent toxic substances (PTS) among indigenous peoples of the coastal Chukotka, Russia, 2001–2007. *Int. J. Circumpolar Health* **2010**, *69* (Suppl. S7), 263–268. [CrossRef]
83. Gyrgolkau, L.A.; Zhuravskaya, E.Y.; Savchenko, T.I.; Chankina, O.V. Chemical elements of the blood of the indigenous inhabitants of Chukotka and their relationship with anthropometric indicators. *News Russ. Acad. Sciences. Phys. Ser.* **2015**, *79*, 69–72. (In Russian)
84. Bravo, N.; Grimalt, J.O.; Chashchin, M.; Chashchin, V.P.; Odland, J.O. Drivers of maternal accumulation of organohalogen pollutants in Arctic areas (Chukotka, Russia) and 4,4′-DDT effects on the newborns. *Environ. Int.* **2019**, *124*, 541–552. [CrossRef]

85. Peakall, D.B.; Kiff, L.F. Eggshell thinning and DDE residue levels among peregrine falcons Falco Peregrinus: A global perspective. *Ibis* **1979**, *121*, 200–204. [CrossRef]
86. RCMA. *Data Listing of 1994 Chemical-Analytical Studies*; Regional Centre Monitoring of the Arctic, Russian Federal Service for Hydrometeorology and Environmental Monitoring: St Petersburg, Russia, 1994.
87. RCMA. *Data Listing of 1995 Chemical-Analytical Studies*; Regional Centre Monitoring of the Arctic, Russian Federal Service for Hydrometeorology and Environmental Monitoring: St sPetersburg, Russia, 1996.
88. Rosgidromet. *Results of Chemical-Analytical Studies of the Russian AMAP National Implementation Plan*; Regional Centre "Monitoring of the Arctic": St. Petersburg, Russia, 1995.
89. Melnikov, S.A.; Gorshkov, A.N.; Vlasov, S.V. *Results of Chemical-Analytical Studies*; Regional Centre Monitoring of the Arctic, Russian Federal Service for Hydrometeorology and Environmental Monitoring: St. Petersburg, Russia, 1995.
90. Melnikov, S.A.; Gorshkov, A.N.; Vlasov, S.V. *Results of Chemical-Analytical Studies*; Regional Centre Monitoring of the Arctic (RCMA), Russian Federal Service for Hydrometeorology and Environmental Monitoring: St. Petersburg, Russia, 1996.
91. Odland, J.O.; Romanova, N.; Sand, G.; Thomassen, Y.; Brox, J.; Khotova, E.; Duriagin, A.; Lund, E.; Nieboer, E. Preliminary report of trace elements in mothers and newborns living in the Kola Peninsula and Arkhangelsk region of Russia compared to Norwegian populations. *Arctic. Med. Res.* **1996**, *55* (Suppl. S1), 38–46. [PubMed]
92. Dudarev, A.A. Persistent polychlorinated hydrocarbons and heavy metals in the Arctic biosphere—The main patterns of exposure and reproductive health of indigenous people. *Biosphere* **2009**, 2, 186–202. (In Russian)
93. Dudarev, A.A.; Chupakhin, V.S.; Mizernyuk, V.N.; Lebedev, G.B.; Chashchin, V.P. Heavy metals in the blood of women of indigenous ethnic groups in the Far North. *Hyg. Sanit.* **2010**, *4*, 31–34. (In Russian)
94. Dudarev, A.A.; Chupakhin, V.S.; Ivanova, Z.S.; Lebedev, G.B. Specifics of exposure to dichlorodiphenyltrichloroethane of indigenous peoples of coastal and inland Chukotka. *Hyg. Sanit.* **2012**, *2*, 15–20. (In Russian)
95. Dudarev, A.A.; Chupakhin, V.S.; Ivanova, Z.S.; Lebedev, G.B. Peculiarities of exposure to polychlorinated biphenyls of indigenous peoples of coastal and inland Chukotka. *Hyg. Sanit.* **2012**, *4*, 22–28. (In Russian)
96. Doushkina, E.V.; Dudarev, A.A.; Sladkova, Y.N.; Zachinskaya, I.Y.; Chupakhin, V.S.; Goushchin, I.V.; Talykova, L.V.; Nikanov, A.N. Metallic content of water sources and drinkable water in industrial cities of Murmansk region. *Russ. J. Occup. Health Ind. Ecol.* **2015**, *2*, 29–34. (In Russian)
97. Dudarev, A.A.; Yamin-Pasternak, S.; Pasternak, I.; Chupakhin, V.S. Traditional Diet and Environmental Contaminants in Coastal Chukotka I: Study Design and Dietary Patterns. *Int. J. Environ. Res. Public Health* **2019**, *16*, 702. [CrossRef]
98. Dudarev, A.A.; Yamin-Pasternak, S.; Pasternak, I.; Chupakhin, V.S. Traditional Diet and Environmental Contaminants in Coastal Chukotka IV: Recommended Intake Criteria. *Int. J. Environ. Res. Public Health* **2019**, *16*, 696. [CrossRef]
99. Sobolev, N.; Ellingsen, D.G.; Belova, N.; Aksenov, A.; Sorokina, T.; Trofimova, A.; Varakina, Y.; Kotsur, D.; Grjibovski, A.M.; Chashchin, V.; et al. Essential and non-essential elements in biological samples of inhabitants residing in Nenets Autonomous Okrug of the Russian Arctic. *Environ. Int.* **2021**, *152*, 106510. [CrossRef]
100. Sorokina, T.Y. A national system of biological monitoring in the Russian Arctic as a tool for the implementation of the Stockholm Convention. *Int. Environ. Agreem.* **2019**, *19*, 341–355. [CrossRef]

International Journal of
Environmental Research and Public Health

MDPI

Article

DNA Barcoding of Chironomid Larvae (Diptera: Chironomidae) from Large Rivers in South Korea to Facilitate Freshwater Biomonitoring and Public Health Surveillance

Hyo Jeong Kang [1], Min Jeong Baek [2], Ji Hyoun Kang [3] and Yeon Jae Bae [3,4,*]

1 Department of Life Science, Graduate School, Korea University, Seoul 02841, Korea
2 National Institute of Biological Resources, Incheon 22689, Korea
3 Korean Entomological Institute, Korea University, Seoul 02841, Korea
4 Department of Environmental Science and Ecological Engineering, College of Life Sciences, Korea University, Seoul 02841, Korea
* Correspondence: yjbae@korea.ac.kr; Tel.: +82-2-3290-3408

Abstract: Chironomid larvae are among the dominant benthic macroinvertebrates in all types of water systems in South Korea. They may pass through pipes in rivers (raw water) and occur in drinking water, thus creating public health issues. However, little is known about the larval stages of chironomids in large South Korean rivers. Therefore, we examined larval–adult associations in chironomids inhabiting major rivers used as water sources. The larvae were collected in 2015 and 2016 from nine locations along the four largest rivers in South Korea using a Ponar grab. Cytochrome oxidase subunit I (*COI*) sequences were generated from the larval specimens, and the species were identified by comparing these sequences to those in a newly constructed DNA barcode library of Chironomidae in South Korea. The samples from the four rivers yielded 61 mitochondrial *COI* sequences belonging to 18 species, including *Hydrobaenus kondoi* Saether, 1989, which was reported for the first time in the Korean Peninsula. Further, morphological identification of the larvae was conducted, and a pictorial taxonomic key to Chironomidae species in large rivers in South Korea was developed to facilitate freshwater biomonitoring research. Finally, an action flow chart was created for the rapid identification of chironomid larvae in infested drinking water or water purification facilities.

Keywords: large river Chironomidae; DNA barcode library; larval-adult association; rapid identification protocol; South Korean rivers

Citation: Kang, H.J.; Baek, M.J.; Kang, J.H.; Bae, Y.J. DNA Barcoding of Chironomid Larvae (Diptera: Chironomidae) from Large Rivers in South Korea to Facilitate Freshwater Biomonitoring and Public Health Surveillance. *Int. J. Environ. Res. Public Health* **2022**, *19*, 12035. https://doi.org/10.3390/ijerph191912035

Academic Editors: Roel Vermeulen, Lauren Petrick and Maaike van Gerwen

Received: 3 August 2022
Accepted: 21 September 2022
Published: 23 September 2022

1. Introduction

Rivers are an important source of freshwater for humans. In South Korea, the main sources of industrial and domestic water supply are the country's four largest rivers: the Han (HR), Geum (GR), Yeongsan (YR), and Nakdong Rivers (NR) [1]. However, as the country develops, anthropogenic activities (e.g., dam construction) have negatively impacted South Korea's river ecosystems through increased channelization, eutrophication, and the increase in lentic areas [2]. These activities, in combination with rising water temperatures due to climate change, have promoted the mass emergence of sporadic/nuisance insect species inhabiting rivers and streams. Among these, chironomids (Diptera: Chironomidae) have become one of the dominant invertebrate groups in South Korean large rivers, and along with oligochaete worms, they account for over 90% of the benthic macroinvertebrate fauna in terms of richness and abundance, particularly in the lentic areas of large river systems [3–5]. Most cases of sporadic/nuisance insect emergence in South Korean rivers can be attributed to riverine chironomids.

Owing to the massive occurrence and abundance of riverine chironomids in South Korea, the invasion of tap water and drinking water purification facilities by their larvae has recently become a key social issue [6]; numerous cases of larval infestation of tap water have been reported in many large and small cities, including Incheon, Seoul, Busan, and Jeju [6,7], raising national concerns regarding water safety. However, in severe cases of larval infestation in tap water and water purification facilities, most local governments have not been able to appropriately respond due to lack of information on chironomid larvae.

Chironomid larvae are widely used in freshwater biomonitoring due to their high abundance, rich species diversity, high pollution tolerance, overlapping life cycles, low larval mobility, and ease of sampling [8,9]. Moreover, they are valuable models for toxicity tests under specific environmental conditions, such as assays of toxic sediment stress and microplastic concentrations in freshwater [10–13]; however, despite their rich species diversity, chironomids are often neglected in aquatic biodiversity and biomonitoring studies due to the lack of knowledge on larval identification.

As is the case with other dipteran groups, many chironomid species are difficult to discriminate at the larval stage, and chironomid taxonomy is based primarily on adult male morphology, such as the characteristics of the male genital hypopygium [14]. Furthermore, conventional approaches to establish adult–larval associations in Diptera typically involve rearing larvae or collecting pupal exuviae to associate with adults, but these methods also have significant technical limitations: larval rearing can be inefficient in certain groups, and generally requires specific conditions (e.g., temperature, food, and photoperiod), whereas pupal sampling is limited by short emergence periods [15–17].

However, the identification of chironomid larvae is critical for rapidly responding to public health issues such as larval infestation in drinking water as well as for supporting freshwater biomonitoring research. In this context, recently developed DNA barcodes for adult chironomids using mitochondrial *COI* genes [18] have rendered adult–larval association practicable. Therefore, in this study, we aimed (i) to correctly identify chironomid larvae inhabiting large rivers in South Korea using morphology and DNA barcodes, (ii) to create a pictorial key to facilitate the identification of chironomid larvae by researchers and the public, and (iii) to develop a protocol for public agencies and stakeholders, which will allow the rapid and accurate examination of invertebrate species found in drinking water.

2. Materials and Methods

2.1. Larval Sampling and Morphological Identification

In 2015 and 2016, chironomid larvae were collected from nine locations distributed across South Korea's four largest rivers (HR, three sites; GR, three sites; NR, two sites; and YR, one site; Figure 1). The larvae were collected from lentic areas (2–10 m deep; 20 samples in total per site) approximately 0.3–1 km upstream of weirs using a Ponar grab (15 × 15 cm) on a boat. River substrates were relatively homogenous mixtures of sand (20–40%) and silt (60–80%) at all sites, except those along the HR, where the proportion of sand was higher (50–70%). Each sample was passed through a 0.5 mm sieve, and the recovered larvae were preserved in 80% ethanol. The preserved larval specimens were separated into morpho-species, photographed (Zeiss Stereo Discovery V12 microscope; Carl Zeiss Microscope GmbH, Jena, Germany) to examine thoracic and abdominal setae and tubules, and then identified to the genus or subfamily level using available identification keys and references [19–22]. For each specimen, the head and abdominal segments VII–IX were carefully dissected under a stereomicroscope (SZX7; Olympus, Tokyo, Japan) and placed in Hoyer's medium for slide preparation [23]. Additionally, the larvae's mouthparts were drawn using a drawing tube attached to an optical microscope (BX53; Olympus, Tokyo, Japan), and larval images were digitized using Adobe Illustra-tor CS6 (Adobe Systems Inc., San Jose, CA, USA). The remaining body parts were preserved in absolute ethanol at −20 °C for subsequent DNA extraction.

Figure 1. Sampling sites along the four largest rivers in South Korea. Sampling was conducted at 0.3–1 km upstream of weirs (pink marks) at each site.

2.2. Molecular Analysis

To reduce the risk of contamination, the specimens were placed in Petri dishes containing 1 mL of 80% ethanol. We dissected, from the thorax I–III to the abdomen, using a needle to reveal the gut, which was carefully removed from the specimen using forceps under a stereomicroscope (SZX7; Olympus, Tokyo, Japan). Total DNA was extracted from thorax segments I–III or abdominal tissue using the DNeasy Blood and Tissue Kit (Qiagen, Hilden, Germany) following the manufacturer's instructions. Moreover, a 658-bp fragment of the mitochondrial cytochrome oxidase I (*COI*) gene was amplified by polymerase chain reaction (PCR) using the AccuPower PCR premix (Bioneer, Daejeon, South Korea) and previously published universal primers (LCO1490 and HCO2198) [24]. Amplification was performed in 20 μL reactions under the following conditions: 94 °C for 5 min, 35 cycles at 94 °C for 30 s, 48 °C for 1 min, and 72 °C for 1 min 30 s, and 72 °C for 10 min [25]. The resulting PCR products were purified and sequenced according to the methodology described by Kang et al. [18], and the sequences were deposited in GenBank (accession numbers OP381663–OP381723).

To identify the chironomids at the species level, the *COI* sequences obtained from larval specimens were compared against the adult reference sequences from NCBI and our previous studies [18], and the sequences were then aligned using the CLC Main Workbench (version 7.8.1; CLC bio, Aarhus, Denmark). The results were crosschecked using the ClustalW algorithm in MEGA 7.0., and a neighbor-joining (NJ) analysis was conducted using MEGA 7.0 to examine the relationships among specimens [26,27].

3. Results

3.1. Specimen Identification

A total of 61 *COI* sequences were successfully generated from the larval specimens (Figure 2), and in the NJ analysis, most of the gene's conspecific sequences were grouped with 100% bootstrap support. Of the 61 sequences, 56 were assigned to 14 species, and one (H20L) was identified at a presumed genus level (*Stenochironomus* Kieffer, 1919). The remaining four sequences (H01L, N28L, N29L, and N31L) were assigned to three undescribed species. Of the 14 identified species, 11 belonged to nine genera in the Chironominae subfamily, two belonged to two genera in the Orthocladiinae subfamily, and one belonged to a genus in the Tanypodinae subfamily (Table 1). Furthermore, the *COI* sequences of specimens identified as *Benthalia carbonaria* (Meigen, 1804) formed two clades which differed by a maximum of 4.2% (K2P-distances) or 77 nucleotide sites. These diverging specimens were collected from the GR and NR populations (Figure 2).

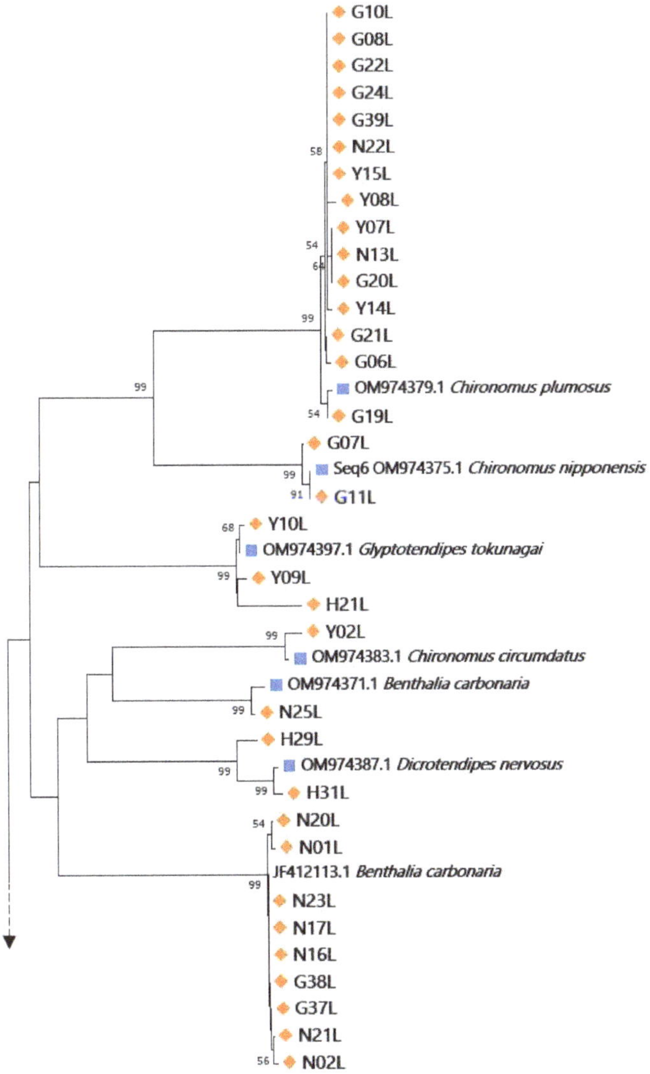

Figure 2. *Cont.*

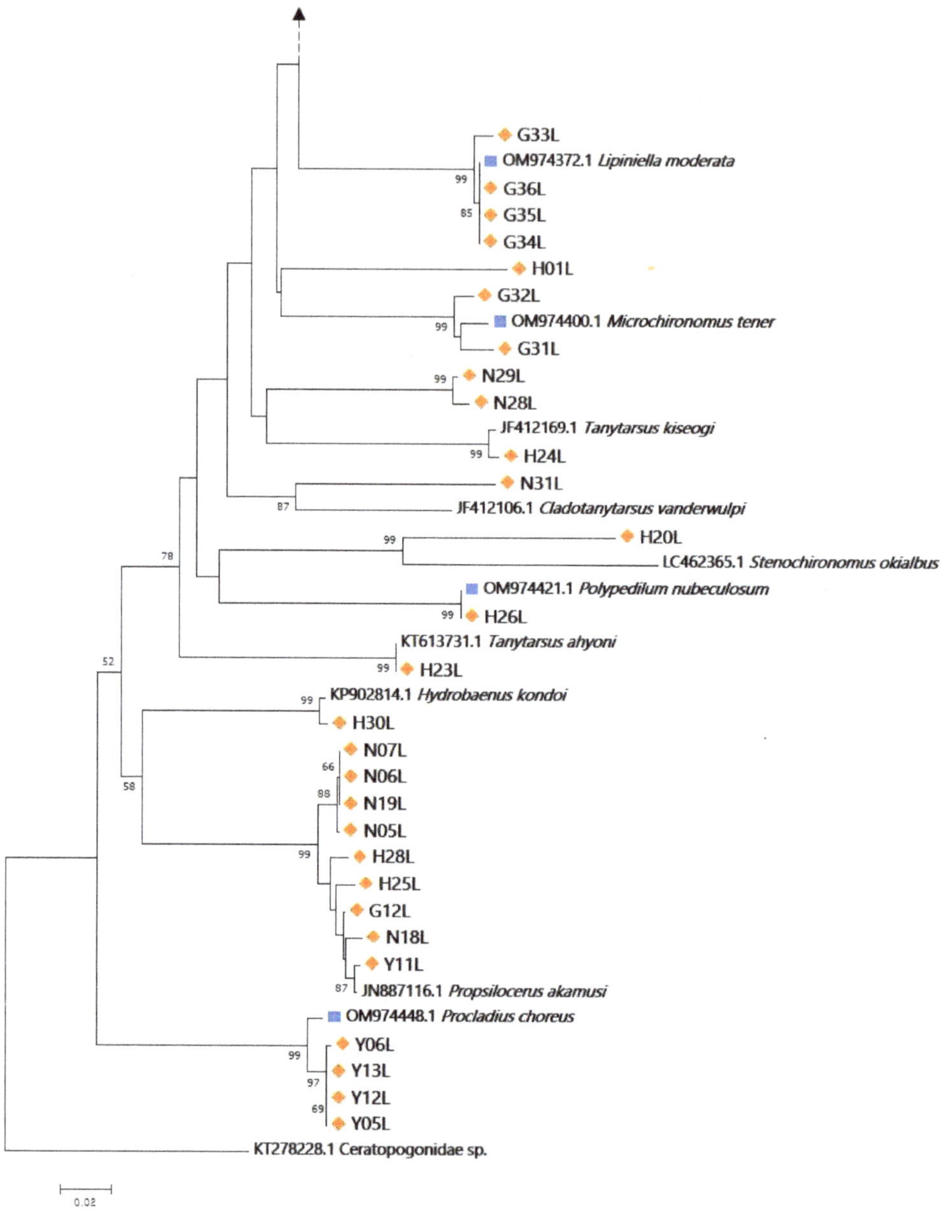

Figure 2. Neighbor-joining tree of 79 cytochrome oxidase I (*COI*) sequences from larval and adult chironomid specimens. Sequences from larval and adult specimens are marked using orange diamonds and blue squares, respectively, and the other of adult reference sequences from NCBI are unmarked.

Table 1. List of larval Chironomidae taxa collected from four large rivers in South Korea and identified using a DNA barcode library in this study.

Subfamily	Species	Accession Number
Tanypodinae	*Procladius choreus* (Meigen, 1804)	OM974448 †, OP381670, OP381671, OP381677, OP381678
Orthocladiinae	*Hydrobaenus kondoi* Saether, 1989 *	KP902814 †, OP381696
	Propsilocerus akamusi (Tokunaga, 1938)	JN887116 †, OP381672, OP381685, OP381686, OP381690, OP381691, OP381692, OP381698, OP381700, OP381718
Chironominae	*Benthalia carbonaria* (Meigen, 1804)	OM974371 †, JF412113 †, OP381680, OP381681, OP381682, OP381683, OP381684, OP381687, OP381688, OP381693, OP381694, OP381704, OP381706
	Chironomus circumdatus (Kieffer, 1916)	OM974383 †, OP381679
	Chironomus nipponensis Tokunaga, 1940	OM974375 †, OP381719, OP381722
	Chironomus plumosus (Linnaeus, 1758)	OM974379 †, OP381668, OP381669, OP381675, OP381676, OP381689, OP381705, OP381713, OP381714, OP381715, OP381716, OP381717, OP381720, OP381721, OP381723
	Dicrotendipes nervosus (Staeger, 1839)	OM974387 †, OP381695, OP381697
	Glyptotendipes tokunagai Sasa, 1979	OM974397 †, OP381673, OP381674, OP381703,
	Microchironomus tener (Kieffer, 1918)	OM974400 †, OP381711, OP381712
	Lipiniella moderata Kalugina, 1970	OM974372 †, OP381707, OP381708, OP381709, OP381710
	Polypedilum nubeculosum (Meigen, 1804)	OM974421 †, OP381699
	Stenochironomus sp.	OP381666
	Tanytarsus ahyoni Ree & Jeong, 2010	KT613731 †, OP381702
	Tanytarsus kiseogi Ree & Jeong, 2010	JF412169 †, OP381701
	Chironominae sp. 1	OP381663
	Chironominae sp. 2	OP381664, OP381665
	Chironominae sp. 3	OP381667

Asterisks (*) indicate species recorded as new to the South Korean fauna. GenBank accession numbers with daggers (†) indicate the specimens whose sequences were retrieved from GenBank.

3.2. Morphological Key and Action Protocol

The larvae were morphologically identified and a pictorial key was created based on their diagnostic characteristics (Figure 3), including head capsule features such as eye spots, antennae, and mouthparts; tubules on abdominal segments VII–VIII; and armature of segment IX. Based on these key characteristics, taxa were identified at subfamily, tribe, genus, and species levels. Furthermore, an action protocol was formulated for public use in order to enable a rapid response to environmental issues, such as larval infestation of drinking water (Figure 4). The mechanism underlying the protocol was as follows: when small insects such as midges are detected in drinking water, public agencies are notified, and stakeholders then collect samples that are sent to identification experts. Specimen collection is important for quickly identifying species according to the provided protocol, and therefore the flow chart of the protocol includes a simple morphological identification process that uses the pictorial key (Figure 3) and a DNA barcode analysis should larval material be damaged or incomplete.

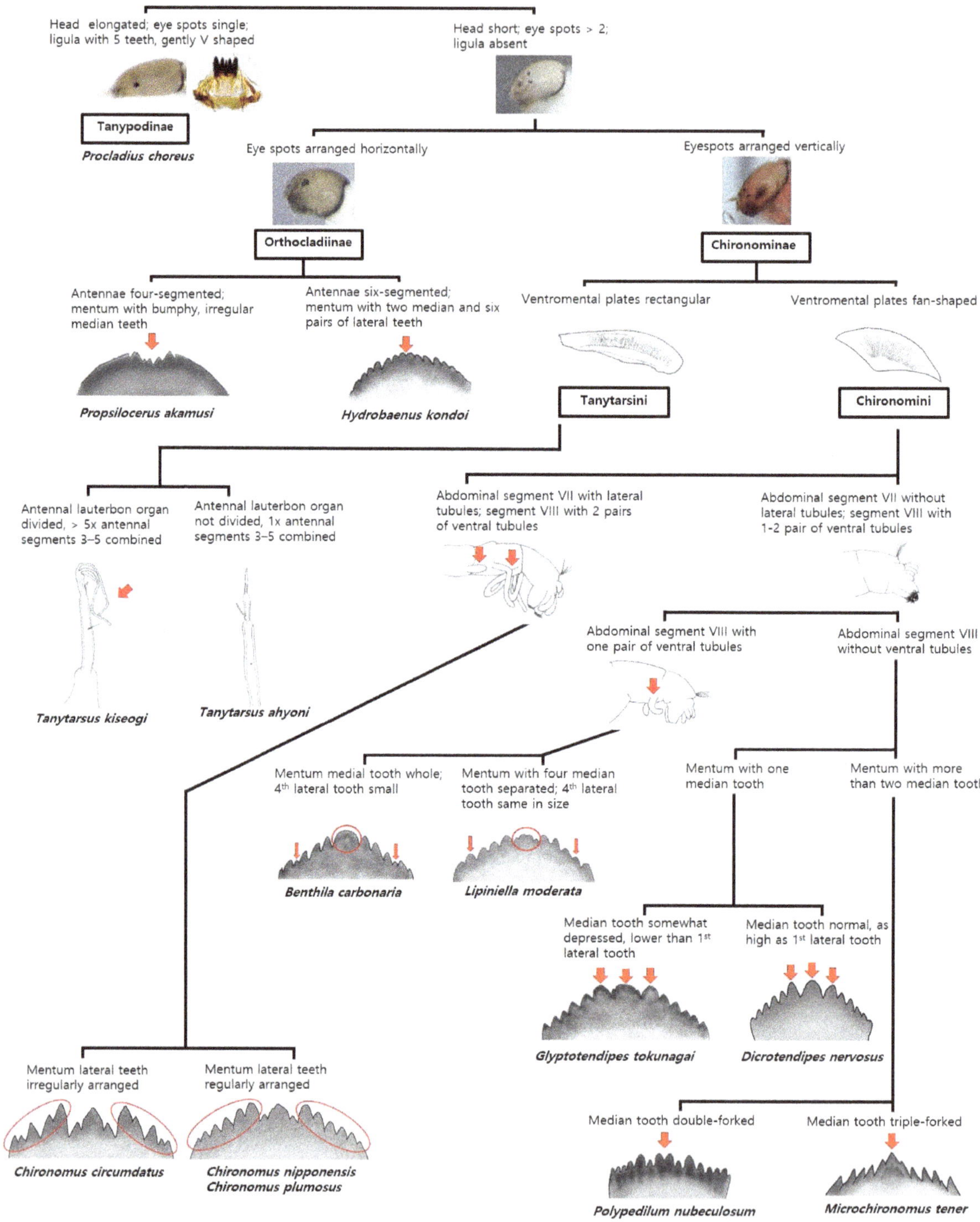

Figure 3. Pictorial key to chironomid larvae inhabiting large rivers in South Korea.

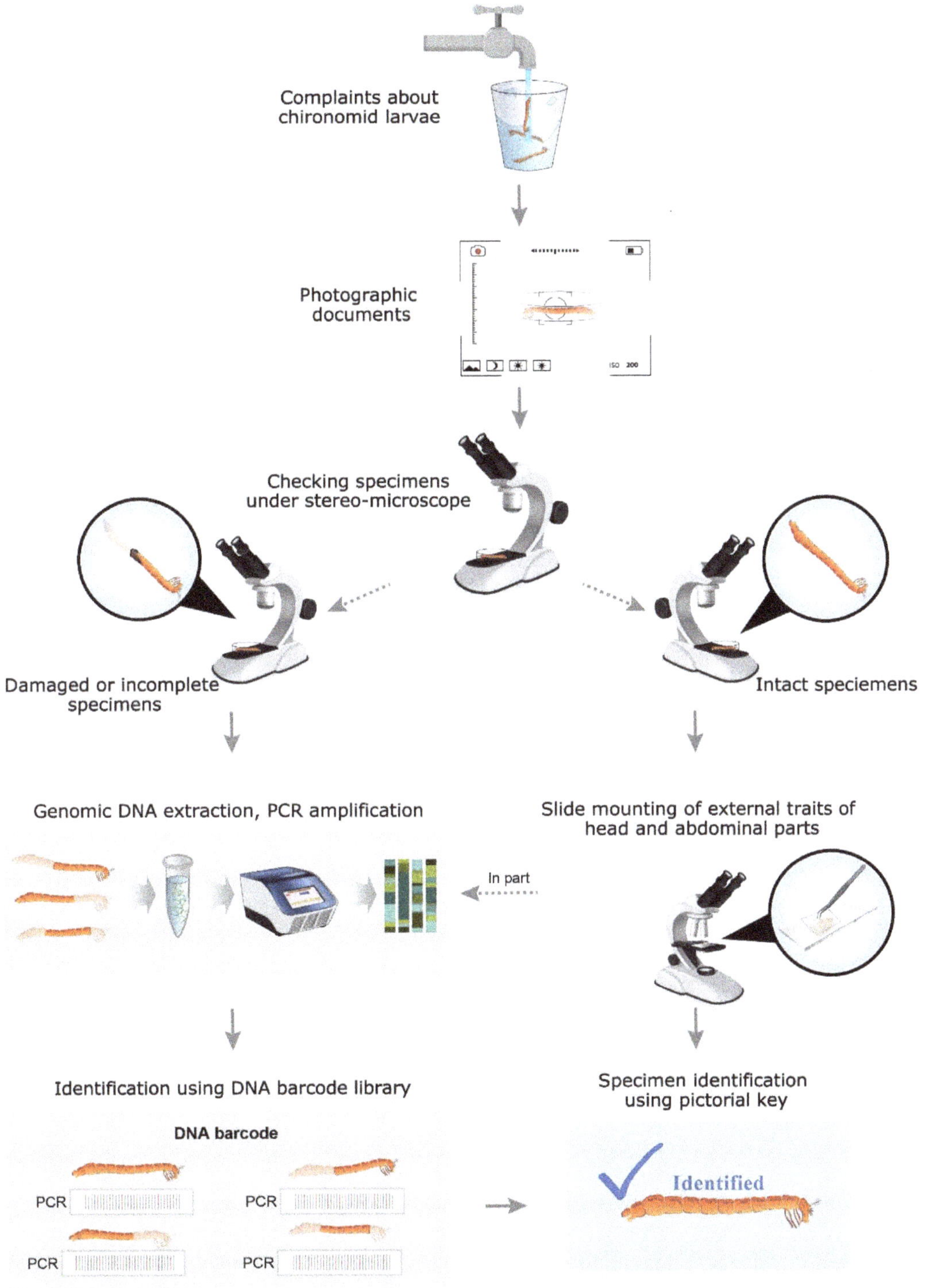

Figure 4. Action protocol for rapid identification of chironomid larvae to respond to environmental issues.

4. Discussion

The chironomid fauna of large rivers in South Korea, which are for the source of most of the freshwater supply in the country, remains virtually unknown due to difficulties in the sampling of deep river beds and identification of specimens because of limited knowledge on larval morphology. Therefore, we used a previously developed chironomid DNA barcode library [18] in this study to identify, for the first time, larval specimens of 18 chironomid species inhabiting four major rivers in South Korea, which account for approximately 45% of all known adult chironomid species found in these rivers [18].

Currently, DNA barcoding is a well-established tool in taxonomy and ecology, which has been widely applied for the identification of pests, invasive species, and foreign insect materials in food, in addition to various other applications in the public health sector [28–31]. Although the present study incorporated sequences from GenBank, we used our own species identification data based on adult morphology, as well as DNA sequences of adult and larval specimens from the same sampling sites, to improve the accuracy of species identification [18]. In this study, larval and adult specimens were correctly differentiated by (1) identifying adult males based on morphological traits, specifically features of their genital structures; (2) constructing phylogenetic trees based on *COI* sequences of adult male specimens to clearly delineate the species category; and (3) assigning larval sequences to species categories, established according to the phylogenetic analysis. Nonetheless, NCBI reference data should be used with caution as sequences may be derived from different geographic regions or misidentified specimens [32,33].

In addition, as shown in the present study, the high-resolution identification of larval specimens can be achieved using a combination of DNA barcoding and morphological analyses. For instance, the approach we adopted allowed us to identify for the first time a chironomid species complex that was previously unidentifiable from larval specimens and lacked a clearly delineated species category, which was *Hydrobaenus kondoi* Saether, 1989. Moreover, we noted morphological deviations from the data recorded in the Holarctic key during genus identification using the general taxonomic key [19]; for instance, a record for a larva belonging to the genus *Dicrotendipes* in the Holarctic key indicated that it possessed a pair of ventral tubes, although two specimens collected from HR in this study did not exhibit this feature. These differences may represent intraspecific or intrageneric variations in larval morphology; therefore, further comprehensive evaluation of larval and adult material, as well as DNA sequencing data, will most likely help define chironomid taxa.

Furthermore, in line with our observations, Kim et al. [34] reported that the genera *Chironomus*, *Dicrotendipes*, *Glyptotendipes*, *Michrochironomus*, *Polypedilum*, *Stictochironomus*, *Tanytarsus*, and *Propsilocerus* were most abundant in large South Korean rivers. Individuals in these genera can thrive in large rivers because of their adaptive larval and pupal traits such as hemoglobin production, and specific behaviors such as the construction of tubes and burrows in fine sediments [35–37]. Moreover, most of the species identified in our study are resistant to eutrophication, and some genera are highly resistant to hypoxia, a frequent phenomenon occurring in polluted streams and benthic zones in lentic areas of large rivers in Northeast Asia. Furthermore, our results suggest that larval species inhabiting large South Korean rivers may be detected in tap water and water purification facilities, because water resources are supplied from the impoundment areas of dams and weirs of these rivers.

Infestation of tap water by chironomid larvae was first reported in Incheon, South Korea in July 2020 [6], which escalated to a nationwide hot topic. Since then, local governments throughout the country have devoted close attention to drinking water purification facilities and supply systems, with the inclusion of "larval infestation" as a new criterion in the drinking water quality checklist. Indeed, larval infestation of tap water has been reported recurrently since the 1940s in other countries, such as Germany, Great Britain, the United States, and Israel [38–42].

The new pictorial key and protocol developed for the identification of chironomid species found in large South Korean rivers presented herein (Figures 3 and 4) are expected to facilitate the identification of larvae associated with infestation problems and, along with the attached taxonomic key (Table S1), they may enable the biomonitoring of large rivers in East Asia [43] as practiced for larval chironomids in Europe, North America, and Australia [19–22]. Furthermore, the action protocol formulated here facilitates rapid identification of chironomid larvae in infested drinking water (Figure 4).

5. Conclusions

Through the comprehensive sampling of chironomid larvae from four large rivers in South Korea and the use of a recently constructed DNA barcode library of Korean Chironomidae, this study identified 18 species within this family and successfully associated larval morphotypes with known adult species for the first time. Moreover, our newly developed pictorial key to chironomid species in large South Korean rivers may facilitate biomonitoring of large rivers in Northeast Asia, and our action protocol for larval identification can be used by the public to enable a rapid response to environmental issues such as larval infestation of drinking water, helping alleviate a problem that has greatly affected countries such as Korea in the past.

Supplementary Materials: The following supporting information can be downloaded at https://www.mdpi.com/article/10.3390/jerph191912035/s1, Table S1: Taxonomic key to species of Chironomidae larvae in large South Korean rivers.

Author Contributions: Conceptualization: H.J.K., M.J.B. and Y.J.B.; methodology: H.J.K., J.H.K. and Y.J.B.; software: H.J.K. and J.H.K.; validation: H.J.K. and Y.J.B.; formal analysis: H.J.K., M.J.B., J.H.K. and Y.J.B.; investigation: H.J.K., M.J.B. and Y.J.B.; data curation: H.J.K., M.J.B. and J.H.K.; writing (original draft preparation): H.J.K. and Y.J.B.; writing (review and editing): H.J.K., M.J.B., J.H.K. and Y.J.B.; visualization: H.J.K., J.H.K. and Y.J.B.; supervision: Y.J.B.; project administration: H.J.K. and M.J.B.; funding acquisition: Y.J.B. All authors have read and agreed to the published version of the manuscript.

Funding: This work was funded by K-Water within the research on benthic macroinvertebrates in four major rivers and major inflow rivers and by the National Institute of Biological Resources (NIBR) (NIBR202203201, NIBR202211101) under the Ministry of Environment (MOE) of the Republic of Korea. Additional funding was provided through a grant from Korea University.

Institutional Review Board Statement: Not applicable.

Informed Consent Statement: Not applicable.

Data Availability Statement: The data presented in this manuscript are available from the authors upon reasonable request.

Acknowledgments: We are grateful to Myoung Chul Kim, Seung Phil Chun (SOKN Institute of Ecology and Conservation, Seoul, South Korea), and Hwang Goo Lee (Sang Ji University, Wonju, South Korea) for sampling during field investigations. We are also grateful to late Leonard C. Ferrington, Jr. (Minnesota University, Minneapolis, MN, USA), Djuradj Milošević (Niš University, Niš, Serbia), and Dubravka Čerba (Josip Juraj Strossmayer University, Osijek, Croatia) for providing literature and technical advice on the identification of chironomid larvae.

Conflicts of Interest: The authors declare no conflict of interest.

References

1. Park, H.; Choo, G.; Kim, H.; Oh, J.E. Evaluation of the current spill status of PFASs and OPFRs in South Korean tap water associated with its origin. *Sci. Total Environ.* **2018**, *634*, 1505–1512. [CrossRef] [PubMed]
2. Bae, Y.J.; Lee, B.H. Human impacts on stream ecosystems and freshwater arthropods in Korea. *Korean J. Entomol.* **2001**, *31*, 63–76.
3. K-Water. *Research of Benthic Macroinvertebrates in 4 Major Rivers and Major Inflow Rivers*; K-Water: Deajeon, Korea, 2016.
4. Kwak, I.; Lee, D.; Hong, C.; Park, Y. Distribution patterns of benthic macroinvertebrates in streams of Korea. *Korean J. Ecol. Environ.* **2018**, *51*, 60–70. [CrossRef]
5. Min, J.; Kong, D. Distribution patterns of benthic macroinvertebrate communities based on multispatial-scale environmental variables in the river systems of Republic of Korea. *J. Freshw. Ecol.* **2020**, *35*, 323–347. [CrossRef]

6. Kwak, I.; Park, J.; Kim, W.; Park, K. Morphological and genetic species identification in the *Chironomus* larvae (Diptera: Chironomidae) found in domestic tap water purification plants. *Korean J. Ecol. Environ.* **2020**, *53*, 286–294. [CrossRef]
7. Kwak, I.; Park, J.; Kim, W.; Park, K. Morphological and genetic species identification in the Chironomidae larvae found in tap water purification plants in Jeju. *Korean J. Ecol. Environ.* **2021**, *54*, 240–246. [CrossRef]
8. Rosenberg, D.M.; Resh, V.H. Introduction to freshwater biomonitoring and benthic macroinvertebrates. In *Freshwater Biomonitoring and Benthic Macroinvertebrates*; Rosenberg, D.M., Resh, V.H., Eds.; Chapman & Hall: New York, NY, USA, 1993; pp. 1–9.
9. Rosenberg, D.M. Freshwater biomonitoring and Chironomidae. *Neth. J. Aquat. Ecol.* **1992**, *26*, 101–122. [CrossRef]
10. Kawai, K.; Yamagishi, T.; Kubo, Y.; Konishi, K. Usefulness of chironomid larvae as indicators of water quality. *Med. Entomol. Zool.* **1989**, *40*, 269–283. [CrossRef]
11. Vermeulen, A.C. Elaborating chironomid deformities as bioindicators of toxic sediment stress: The potential application of mixture toxicity concepts. *Ann. Zool. Fenn.* **1995**, *32*, 265–285.
12. Baek, M.J.; Kang, H.J.; Yoon, T.J.; Bae, Y.J. Biological and genetic characteristics of *Glyptotendipes tokunagai* (Diptera: Chironomidae) on the basis of successive rearing of forty-two generations over seven years under laboratory conditions. *Environ. Entomol.* **2014**, *43*, 1406–1418. [CrossRef]
13. Lin, C.T.; Chiu, M.C.; Kuo, M.H. Effects of anthropogenic activities on microplastics in deposit-feeders (Diptera: Chironomidae) in an urban river of Taiwan. *Sci. Rep.* **2021**, *11*, 1–8. [CrossRef]
14. Ekrem, T.; Stur, E.; Hebert, P.D.N. Females do count: Documenting Chironomidae (Diptera) species diversity using DNA barcoding. *Org. Divers. Evol.* **2010**, *10*, 397–408. [CrossRef]
15. Maier, K.J.; Kosalwat, P.; Knight, A.W. Culture of *Chironomus decorus* (Diptera: Chironomidae) and the effect of temperature on its life history. *Environ. Entomol.* **1990**, *19*, 1681–1688. [CrossRef]
16. Ristola, T.; Pellinen, J.; Ruokolainen, M.; Kostamo, A.; Kukkonen, J.V.K. Effect of sediment type, feeding level, and larval density on growth and development of a midge (*Chironomus riparius*). *Environ. Toxicol. Chem.* **1999**, *18*, 756–764. [CrossRef]
17. Baek, M.J.; Yoon, Y.J.; Bae, Y.J. Development of *Glyptotendipes tokunagai* (Diptera: Chronomidae) under different temperature conditions. *Environ. Entomol.* **2012**, *41*, 950–958. [CrossRef]
18. Kang, H.J.; Baek, M.J.; Kang, J.H.; Bae, Y.J. Diversity and DNA barcode analysis of chironomids (Diptera: Chironomidae) from large rivers in South Korea. *Insects* **2022**, *13*, 346. [CrossRef]
19. Wiederholm, T. *Chironomidae of the Holarctic Region. Keys and Diagnoses. Part 1: Larvae*; Entomologica Scandinavica: Stockholm, Sweden, 1983; pp. 1–457.
20. Epler, J.H. *Identification Manual for the Larval Chironomidae (Diptera) of Florida*; Bureau of Surface Water Management, Florida Department of Environmental Protection: Tallahassee, FL, USA, 1995.
21. Madden, C.P. Key to genera of larvae of Australian Chironomidae (Diptera). *Mus. Vic. Sci. Rep.* **2010**, *13*, 1–31. [CrossRef]
22. Cranston, P.S. Identification guide to genera of aquatic larval Chironomidae (Diptera) of Australia and New Zealand. *Zootaxa* **2019**, *4706*, 71–102. [CrossRef]
23. Sasa, M.; Kikuchi, M. *Chironomidae [Diptera] of Japan*, 1st ed.; University of Tokyo Press: Tokyo, Japan, 1995; pp. 1–333.
24. Folmer, O.; Black, M.; Hoeh, W.; Lutz, R.; Vrijenhoek, R. DNA primers for amplification of mitochondrial cytochrome c oxidase subunit I from diverse metazoan invertebrates. *Mol. Mar. Biol. Biotechnol.* **1994**, *3*, 294–299. [PubMed]
25. Kim, S.; Song, K.H.; Ree, H.I.; Kim, W. A DNA barcode library for Korean Chironomidae (Insecta: Diptera) and indexes for defining barcode gap. *Mol. Cells* **2012**, *33*, 9–17. [CrossRef]
26. Kumar, S.; Stecher, G.; Tamura, K. MEGA7: Molecular evolutionary genetics analysis version 7.0 for bigger datasets. *Mol. Biol. Evol.* **2016**, *33*, 1870–1874. [CrossRef]
27. Tamura, K.; Nei, M.; Kumar, S. Prospects for inferring very large phylogenies by using the neighbor-joining method. *Proc. Natl. Acad. Sci. USA* **2004**, *101*, 11030–11035. [CrossRef]
28. Armstrong, K.F.; Ball, S.L. DNA barcodes for biosecurity: Invasive species identification. *Philosoph. Trans. Roy. Soc. B Biol. Sci.* **2005**, *360*, 1813–1823. [CrossRef] [PubMed]
29. Chu, K.H.; Xu, M.; Li, C.P. Rapid DNA barcoding analysis of large datasets using the composition vector method. *BMC. Bioinform.* **2009**, *10*, S8. [CrossRef]
30. Ashfaq, M.; Hebert, P.D.N.; Mirza, J.H.; Khan, A.M.; Zafar, Y.; Mirza, M.S. Analyzing mosquito (Diptera: Culicidae) diversity in Pakistan by DNA barcoding. *PLoS ONE* **2014**, *9*, e97268. [CrossRef] [PubMed]
31. Cho, S.Y.; Suh, K.I.; Bae, Y.J. DNA barcode library and its efficacy for identifying food-associated insect pests in Korea. *Entomol. Res.* **2013**, *43*, 253–261. [CrossRef]
32. Meiklejohn, K.A.; Damaso, N.; Robertson, J.M. Assessment of BOLD and GenBank-Their accuracy and reliability for the identification of biological materials. *PLoS ONE* **2019**, *14*, e0217084. [CrossRef]
33. Meyer, C.P.; Paulay, G. DNA Barcoding: Error rates based on comprehensive sampling. *PLoS Biol.* **2005**, *3*, e422. [CrossRef]
34. Kim, W.; Park, J.; Hong, C.; Choi, B.; Kim, H.; Park, Y.; Park, J.; Song, H.; Kwak, I. Changes in community structure of Chironomidae caused by variability of environmental factors among weir sections in Korean rivers. *Korean J. Ecol. Environ.* **2020**, *53*, 46–54. [CrossRef]
35. Nagell, B.; Landahl, C.C. Resistance to anoxia of *Chironomus plumosus* and *Chironomus anthracinus* (Diptera) larvae. In *Holarctic Ecology*; Vik, R., Ed.; Munksgaard: Copenhagen, Danmark, 1978; Volume 1, pp. 333–336. [CrossRef]

36. Zou, W.; Cai, Y.; Tolonen, K.T.; Zhu, G.; Qin, B.; Peng, K.; Gong, Z. The adaptations to tube-dwelling life of *Propsilocerus akamusi* (Diptera: Chironomidae) larvae and its eutrophication-tolerant mechanisms. *Limnologica* **2019**, *77*, 125684. [CrossRef]
37. Walshe, B.M. The function of haemoglobin in *Chironomus plumosus* under natural conditions. *J. Exp. Biol.* **1950**, *27*, 73–95. [CrossRef]
38. Krüger, V.F. Eine parthenogenetische Chironomide als Wasserleitungsschädling. In *Die Naturwissenschaften*; Süffert, F., Ed.; Springer: Berlin/Heidelberg, Germany, 1941; Volume 36, pp. 556–558. [CrossRef]
39. Kelly, S.N. Infestation of the Norwich England water system. *J. Am. Water Works Assoc.* **1955**, *47*, 330–334. [CrossRef]
40. Williams, D.A. An infestation by a parthenogenetic chironomid. *Water Treat. Exam.* **1974**, *23*, 215–229.
41. Bay, E.C. Chironomid (Diptera: Chironomidae) larval occurrence and transport in a municipal water system. *J. Am. Mosq. Control Assoc.* **1993**, *9*, 275–284. [PubMed]
42. Alexander, M.K.; Merrit, R.W.; Berg, M.B. New strategic for the control of the parthenogenetic chironomid (*Paratanytarsus grimmii*) (Diptera: Chironomidae) infesting water systems. *J. Am. Mosq. Control Assoc.* **1997**, *13*, 189–192. [PubMed]
43. Morse, J.; Bae, Y.J.; Sangpradub, N.; Tanida, K.; Vshivkova, T.S.; Yang, L.; Yule, C. Freshwater biomonitoring with benthic macroinvertebrates in East Asia. *Front. Ecol. Environ.* **2007**, *5*, 33–42. [CrossRef]

International Journal of
Environmental Research and Public Health

MDPI

Article

Determination of Carbonyl Compounds in Different Work Environments: Comparison between LC-UV/DAD and LC–MS/MS Detection Methods

Federica Castellani [1,2], Arianna Antonucci [1], Ivano Pindinello [1], Carmela Protano [1] and Matteo Vitali [1,*]

[1] Department of Public Health and Infectious Diseases, Sapienza University of Rome, P.le Aldo Moro, 5, 00185 Rome, Italy
[2] Department of Ecological and Biological Sciences, Tuscia University, Largo dell'Università snc, 01100 Viterbo, Italy
* Correspondence: matteo.vitali@uniroma1.it; Tel.: +39-0649914560

Abstract: There were two analytical methods for the determination of 12 carbonyl compounds (CCs) by using liquid chromatography (LC) coupled with mass spectrometry (MS/MS) and diode array detector (UV/DAD) that were developed and applied to 52 samples that were collected in 10 workplaces. Linearity ($0.996 < R^2 < 0.999$), intra-day repeatability (0.7 < RSD% < 10), and inter-day repeatability (5 < RSD% < 16) were acceptable for both techniques, but the highest sensibility of the MS/MS method allowed us to correctly quantify 98% of the samples (versus 32% by UV/DAD). The comparison of the concentrations that were obtained by quantifying the same sample with both techniques showed good agreement for acetaldehyde and formaldehyde (0.1 < % deviation < 30) but much higher for the less abundant congeners. In real samples, formaldehyde was the most abundant congener (concentrations between 2.7 and 77 $\mu g\ m^{-3}$), followed by acetaldehyde (concentrations between 1.5 and 79 $\mu g\ m^{-3}$) and butyraldehyde (concentrations between 0.4 and 13 $\mu g\ m^{-3}$). In all the beauty salon samples, instead, the most abundant congener was acetaldehyde (concentrations between 19 and 79 $\mu g\ m^{-3}$), probably associated with the use of beauty products. Principal components analysis (PCA) confirms the ubiquitous character of formaldehyde and highlights the influence of minority CCs on different workplaces.

Keywords: airborne formaldehyde; airborne carbonyl compounds; occupational exposure; analytical method; liquid chromatography

Citation: Castellani, F.; Antonucci, A.; Pindinello, I.; Protano, C.; Vitali, M. Determination of Carbonyl Compounds in Different Work Environments: Comparison between LC-UV/DAD and LC–MS/MS Detection Methods. *Int. J. Environ. Res. Public Health* **2022**, *19*, 12052. https://doi.org/10.3390/ijerph191912052

Academic Editors: Maaike van Gerwen, Roel Vermeulen and Lauren Petrick

Received: 8 August 2022
Accepted: 20 September 2022
Published: 23 September 2022

1. Introduction

Carbonyl compounds (CCs), such as aldehydes and ketones, are ubiquitous pollutants that are among the most widespread in the environment [1,2]. Airborne CCs are primarily emitted by anthropogenic sources, such as the incomplete combustion of fossil fuels or biomass burning activities [3]. When released into the environment, CCs play an important role in the atmospheric chemistry as the precursors to several species are involved in photochemical smog (e.g., free radicals, ozone and peroxyacyl nitrates) [4].

In the last years, CCs have received more normative, scientific, and public attention both for their active role in atmospheric chemistry and for their association with several adverse effects on human health [5]. Among CCs, formaldehyde and acetaldehyde have particular health concerns; indeed, formaldehyde is a well-known sensory irritant compound, especially for sensitive individuals and it has been associated with allergies and negative outcomes for the respiratory system, while acetaldehyde can irritate the skin, eyes, and nose [6]. Besides, the International Agency for Research on Cancer (IARC) classified formaldehyde and acetaldehyde as human carcinogens (Group 1) and possibly carcinogenic to humans (Group 2B), respectively [7–9]. Indoor environments increase the risk of adverse health effects that are related to the exposure to formaldehyde and acetaldehyde because it is well-known that the indoor concentrations of pollutants can be two to five times higher than the respective outdoor concentrations [10,11]. In addition, the quality of indoor air is of particular importance because the general population spends the majority of the time in indoor environments, both residential or non-residential [12]. Particularly relevant is the exposure of workers that are employed in the manufacture of formaldehyde or formaldehyde-containing products, healthcare professionals, firefighters, beauticians and hairdressers, printing-rooms workers, professionals of gross anatomy and pathology laboratories, veterinarians, and embalmers [13]. Given the ubiquitous nature of formaldehyde, the World Health Organization (WHO) indicated that occupational exposure occurs in all workplaces [5], mainly through the inhalation route, but also through the dermal route [14]. Acetaldehyde is the other majority aldehyde that was revealed in the workplace, whose presence is related to the use of cosmetic products, combustion appliances, cigarette smoking, and special adhesives [15,16]. In addition to formaldehyde and acetaldehyde, other CCs that are associated with different emission sources may be present in work environments [17]. For example, crotonaldehyde is synthesized and used in the field of industry [18], propionaldehyde is emitted by housing and furniture, and butyraldehyde, hexanal, heptanal octanal, and nonanal are associated with the use of solvents, paints, perfumes, and flavoring agents [15]. Several studies focused their attention on the appearance of adverse effects on human health due to combined exposure to different CCs [19,20], that can lead both to the formation of DNA adducts and to DNA damaging [21].

Due to their impact on human health, a simple, fast, and highly effective collection sampling method is required for CCs routine determination. Generally, the analytical approaches are limited to formaldehyde and acetaldehyde determination; however, given the potential toxicity of other CCs and the synergic effects deriving from the exposure to mixture of CCs, an analytical procedure for simultaneously measuring the levels of the most common airborne CCs would be very useful. The aims of this study were: 1. to optimize an analytical procedure useful for determining the concentrations of 12 CCs; and 2. to study the distribution of 12 CCs in ten different work environments. For this purpose, the air levels of these 12 CCs were measured in ten different work environments among those that were recognized as determining potential exposure to these compounds: six wards of four different Italian hospitals, a copy shop, a beauty salon, a pharmacy, and a chemistry laboratory. The sampling was carried out by portable sampling pumps and adsorbing and derivatizing cartridges, then treated and analyzed by liquid chromatography (LC) coupled both to a spectroscopic detector (diode array detector, DAD) and tandem mass spectrometer (MS/MS) in order to optimize the two different detection methods. The performances and the results of the two detection methods were compared to define their applicability in different exposure scenarios.

2. Materials and Methods

2.1. Chemical and Materials

LC-MS grade water, acetonitrile (ACN), acetic acid, and ammonium formate were supplied by Carlo Erba (Milan, Italy), while the 12 Carbonyl-DNPH Derivatives standard solution (Formaldehyde-DNPH, Fo-DNPH; Acetaldehyde-DNPH, Ace-DNPH; Propionaldehyde-DNPH, Pro-DNPH; Crotonaldehyde-DNPH, Cro-DNPH; Butyraldehyde-DNPH, Bu-DNPH; Cyclohexanone-DNPH, Cy-DNPH; Valeraldehyde-DNPH, Val-DNPH; Hexanal-DNPH, Hex-DNPH; Heptanal-DNPH, Hep-DNPH; Octanal-DNPH, Oct-DNPH; Nonanal-DNPH, Non-DNPH; Decanal-DNPH, Dec-DNPH) by Agilent Technologies S.r.l. (Milan, Italy). Working standard solutions were prepared daily by dilution in ACN and kept in amber vials at +4 °C. Sample filtration was carried out by Choice™ PTFE Syringe Filters (0.22 μm × 13 mm, Thermo Scientific, Waltham, MA, USA). The used dual-bed sampling cartridges, coated with 2,4-dinitrophenylhydrazine (DNPH) and 1,2-bis(2-pyridyl) ethylene (BPE) for the CCs derivatization to form stable hydrazone derivatives and the removal of negative ozone interference (130 mg 2-BPE coated silica, 270 mg DNPH-coated silica), were purchased from Supelco Analytical (Bellefonte, PA, USA). A total of six SKC AirChek® TOUCH sampling pumps (SKC, Eighty Four, PA, USA) were used, calibrated by a DryCal DC-lite primary flowmeter (Bios International Corporation, Butler, NJ, USA). The used analytical column was an Acclaim Carbonyl C18 RSLC (150 × 3 mm, 3 μm), acquired by Thermo Scientific (Waltham, MA, USA).

2.2. Instrumentation

LC analyses were carried out on an HPLC 1260 Infinity II system (Agilent Technologies Italy S.p.A. Milan, Italy) that was equipped with high pressure mixing and an autosampler Agilent G7129A. The MS/MS determinations were performed by using a QTRAP 5500 mass spectrometer with an ESI source, operating in MRM (multiple reaction monitoring) mode (AB SCIEX S.r.l. Forster City, CA, USA). Analyses were achieved in negative mode. The DAD determinations were performed by UV-DAD 1290 DAD FS (Agilent Technologies Italy S.p.A. Milan, Italy), set at 360 nm. The used measuring cell had an optical path of 1 mm and a volume of 1 μL.

2.3. Sampling Sites and Strategy

The indoor carbonyl levels were measured in ten different workplaces, selected based on a recent systematic review [13]. The general information and characteristics of each sampling are reported in Table 1. Briefly, for each working environment, four to six sample pumps were used both as personal samplers, in order to control employees' exposure, and as environmental samplers, in order to estimate indoor contamination. In every sampling site, the assessment of CCs' background levels was made by placing a sampling pump outside and near the monitoring site. Air samples were collected into the dual-bed cartridges at a flow rate of 0.14 L min^{-1}. Sampling was carried out during normal working hours, with sampling time ranging from 51 to 406 min (Table 1). The chosen flow rate and sampling times ensure that the collected CCs would not consume more than 30% of the DNPH that was coated on the cartridges; this was checked for every sample by LC-UV/DAD analysis at 360 nm and subsequent quantification of unreacted DNPH (Supplementary Material S1). A preliminary test was performed in order to assess that the chosen sampling flow rate did not consume more than 30% of DNPH. Analysis of the blank samples (i.e., 5 unsampled cartridges) was performed to verify possible contaminations during extraction procedure. The portable pumps' flow rates were checked before and after every sampling campaign using a calibrated flowmeter. The sampled DNPH-coated cartridges were stored in a refrigerator at +4 °C until the analysis, carried out within two weeks from the sampling campaign.

Table 1. Specifications of the sampling sites and conditions.

	Sampling Date	Sampling SITE	Sampling Type	Main Task/Activity	Sampling Time (min)	Volume (m^3)
PL1-lab	28 January 2021	Pathology Laboratory	Laboratory Environment	Preparation of anatomical specimens in formalin	264	0.037
PL2-out			Environmental—External window ledge	Urban environment	266	0.037
PL3-tech			Staff—Laboratory Technician	Preparation of anatomical specimens in formalin	261	0.037
PL4-tech			Staff—Laboratory Technician	Preparation of anatomical specimens in formalin	262	0.037
PL5-doc			Staff—Ward doctor	Preparation of anatomical specimens in formalin	202	0.028
PL6-solv			Environmental—Solvent storage	Storage of formalin tanks	273	0.038
GA1-nurs	2 February 2021	Gastroenteroloy Ambulatory	Staff—Nurse	Clinical examinations—gastroscopy and colonoscopy	149	0.021
GA2-nurs			Staff—Nurse	Clinical examinations—gastroscopy and colonoscopy	146	0.020
GA3-doc			Staff—Ward doctor	Clinical examinations—gastroscopy and colonoscopy	125	0.018
GA4-lab			Laboratory Environment	Clinical examinations—gastroscopy and colonoscopy	159	0.022
GA5-out			Environmental—External window ledge	Urban environment	140	0.020
CS1-desk	30 April 2021	Copy Shop	Staff—desk	Cashier	300	0.042
CS2-glue			Staff—Gluing machine	Gluing of backs of books	300	0.042
CS3-photo			Staff—Photocopier	Use of laser photocopier	300	0.042
BS1-hair	31 April 2021	Beauty Salon	Staff—Hairdresser	Hairstyling, hair dye	351	0.049
BS2-nail			Staff—Nail Technician	Gel nails extension	352	0.049
BS3-sol			Environmental—Solarium	Tanning bed	347	0.049
BS4-rec			Staff—Receptionist	Secretarial work	343	0.048
BS5-bea			Staff—Beautician	Waxing, manicure, pedicure	334	0.047
BS6-out			Environmental—External window ledge	Urban environment	335	0.047

Table 1. *Cont.*

	Sampling Date	Sampling SITE	Sampling Type	Main Task/Activity	Sampling Time (min)	Volume (m^3)
PH1-out	3 May 2021	Pharmacy	Environmental—External window ledge	Urban environment	375	0.053
PH2-ware			Environmental—Warehouse	Storage of pharmaceutical products	315	0.044
PH3-store			Staff—Storeman	Storage of pharmaceutical products	313	0.044
PH4-dercon			Staff—Dermocosmetic consultant	Employed to the sale	314	0.044
PH5-phar			Staff—Pharmacist	Employed to the sale	321	0.045
PH6-der			Environmental—Dermocosmetic	Presence of wooden shelves	314	0.044
HL1-lab	22 September2021	Histology Laboratory	Laboratory Environment	Preparation of anatomical specimens in formalin	406	0.057
HL2-tech			Staff—Laboratory Technician	Preparation of anatomical specimens in formalin	406	0.057
HL3-tech			Staff—Laboratory Technician	Preparation of anatomical specimens in formalin	397	0.056
HL4-sec			Staff—Secretary	Acceptance of histological samples	394	0.055
HL5-tech			Staff—Laboratory Technician	Preparation of anatomical specimens in formalin	391	0.055
HL6-out			Environmental—External window ledge	Urban environment	393	0.055
HRL1-tech	5 April 2022	Histology Research Laboratory	Staff—Laboratory and Enclosure Technician	Preparation of animal anatomical specimens in formalin and management of the enclosure	154	0.022
HRL2-lab			Laboratory Environment	Preparation of animal anatomical specimens in formalin	155	0.022
HRL3-out			Environmental—External window ledge	Urban environment	155	0.022
HRL4-bio			Staff—Biologist Head of the Enclosure	Histological sampling of animal parts and management of the enclosure	81	0.011
HRL5-lab			Laboratory environment—Enclosure	Histological sampling of animal parts	51	0.007

Table 1. *Cont.*

	Sampling Date	Sampling SITE	Sampling Type	Main Task/Activity	Sampling Time (min)	Volume (m^3)
HpL1-out	6 April 2022	Histopathology Laboratory	Environmental—External window ledge	Urban environment	244	0.034
HpL2-tech			Staff—Laboratory Technician	Preparation of anatomical specimens in formalin	311	0.044
HpL3-lab			Laboratory Environment	Preparation of anatomical specimens in formalin	306	0.043
HpL4-lab			Laboratory Environment	Storage of anatomical specimens in formalin	302	0.042
HpL5-doc			Staff—Ward doctor	Preparation of anatomical specimens in formalin	163	0.023
HpL6-doc			Staff—Ward doctor	Preparation of anatomical specimens in formalin	251	0.035
CL1-lab	12 April 2022	Chemical Laboratory	Laboratory Environment	Environmental Samples Preparation and Storage	273	0.038
CL2-lab			Laboratory Environment	Environmental Samples Preparation and Analysis	270	0.038
CL3-lab			Laboratory Environment	Quantitative analytical chemistry	271	0.038
CL4-off			Office	Data elaboration	271	0.038
CL5-out			Environmental—External window ledge	Urban environment	270	0.038
ParL1-lab	13 April 2022	Parasitology Laboratory	Laboratory environment	Sample preparation and analysis	222	0.031
ParL2-lab			Laboratory environment	Sample preparation and analysis	221	0.031
ParL3-corr			Corridor Outside the Parasitology Laboratory	-	222	0.031
ParL4-lab			Laboratory Environment	Sample preparation and analysis	220	0.031

2.4. Sample Analysis

The 12 analyzed carbonyl compounds (Fo-DNPH, Ace-DNPH, Pro-DNPH, Cro-DNPH, Bu-DNPH, Cy-DNPH, Val-DNPH, Hex-DNPH, Hep-DNPH, Oct-DNPH, Non-DNPH, Dec-DNPH) were eluted from the DNPH-coated cartridge with 4.0 mL of ACN, with an elution rate of 1 mL min^{-1} as suggested by the manufacturer of the cartridges. The eluate was filtered and then injected in the LC system and detected by using both spectrometric and spectroscopic detectors.

2.5. Optimization of the Analytical Determinations

Both for the LC-MS/MS and LC-UV/DAD determinations, the calibration curves were constructed by using an external standard method, in the range 1.56–100 μg L^{-1} for MS/MS and 8–1000 μg L^{-1} for UV/DAD. For the quantification of real samples, the external standard quantification method was used after verifying that there was no matrix effect for both the determination techniques. Limits of detection (LODs) and quantification (LOQs) were calculated as the minimum concentration of the analyte that produces a signal-to-noise ratio greater than 3 and 10, respectively. Intra-day reproducibility, evaluated as the relative standard deviation (RSD%), was calculated on 10 repeated injections in the same day at three concentration levels:

(1) Lowest Level (L_L): Concentration level representing an agreement between the LOQs of the detected analytes;
(2) Medium Level (M_L): Concentration level corresponding to the highest LOQ of the detected analytes;
(3) Highest Level (H_L): Concentration corresponding to 10 times the L_L.

Inter-day reproducibility, instead, was obtained by performing triplicate analysis of the same standard solution for seven consecutive days. The result was expressed as RSD%.

2.6. Statistical Elaboration

The statistical software R (R-project for statistical computing, Ver. 3.0, 32-bit) was used to perform principal component analysis (PCA) on the 12 individual CCs' congeners in order to illustrate the distribution pattern and the possible emission sources. Before the statistical analysis, the matrix of the data was transformed by column mean centering and row and column autoscaling in order to correct for different variable scaling and units.

3. Results and Discussion

3.1. Optimization of the Chromatographic Separation

To find the best chromatographic condition for CCs separation, we used a C18 Acclaim Carbonyl column, testing different elution gradients and flow rates. The best solution was a mobile phase consisting of (A) ammonium acetate 2 mM and (B) ACN at a flow rate of 0.8 mL min^{-1}. The gradient elution started from A:B 53:47 (v/v %) for 5 min and linearly increasing to 90% B in −5 min, then isocratic elution for 5 min. The initial elution conditions were then restored for 5 min. The total duration of the analysis was 20 min. The injection volume was 5 μL. The same optimized chromatographic separation was coupled both with MS/MS and UV/DAD detectors. The retention times of the selected analytes are reported in Table 2.

Table 2. Optimized spectrometric conditions for each studied analyte. The letter "Q" indicates the ion that was chosen as quantifier, the letter "q" indicates the ion that was chosen as qualifier.

	Retention Times (min)	Molecular Mass	Precursor Ion		Product Ion	Collision Energy (eV)	Collision Cell Exit Potential (V)
Fo-DNPH	5.83	210	209	Q	151	−11	−15
				q	163	−11	−15
Ace-DNPH	8.05	224	223	Q	151	−12	−6
				q	163	−12	−10
Pro-DNPH	9.68	238	237	Q	163	−15	−15
				q	151	−15	−15
Cro-DNPH	10.36	250	249	Q	172	−16	−15
				q	151	−16	−15
Bu-DNPH	10.65	252	251	Q	163	−15	−15
				q	151	−15	−15
Cy-DNPH	11.29	278	277	Q	247	−15	−15
				q	231	−24	−15
Val-DNPH	11.40	266	265	Q	163	−15	−15
				q	152	−25	−15
Hex-DNPH	12.06	280	279	Q	152	−25	−15
				q	163	−15	−15
Hep-DNPH	12.73	294	293	Q	163	−15	−15
				q	152	−25	−15
Oct-DNPH	13.53	308	307	Q	163	−19	−15
				q	205	−19	−15
Non-DNPH	14.53	322	321	Q	163	−20	−15
				q	205	−20	−15
Dec-DNPH	15.85	336	335	Q	163	−19	−15
				q	171	−19	−15

3.2. Optimization of the Spectrometric Determination and Data Quality

To choose the precursor and product ions that are useful to quantitative and qualitative determinations and to find the best parameters for ionic transmission in mass spectrometry, the infusion of standard solution of Ccs (10 μg L^{-1} in ACN) at a flow rate of 7 μL min^{-1} were carried out in negative mode. Nebulizer, turbo, and curtain gases were respectively set at 30, 55, and 60 psi; the temperature was set at 500°C; and declustering and entrance potential were set at −60 and −10, respectively. The optimized parameters for each analyte are reported in Table 2, together with precursor and product ions.

The validation parameters of the LC-MS/MS optimized method are shown in Table 3. The method linearity in the investigated concentration range (1.56–100 μg L^{-1}), expressed as R^2, was between 0.999 and 0.996 for all the studied analytes. The values of intra-day repeatability (RSD%) calculated at the lowest level (L_L, 0.20 μg L^{-1}) ranged from 3.0% to 7.8%. The RSD% were found to be lower at higher concentrations (M_L, 0.78 μg L^{-1} and H_L, 20 μg L^{-1}) with values between 1.5% and 3.3% for M_L and between 1.2% and 2.7% for H_L. The obtained results were much lower than those that were obtained by Chi et al. [22]. The intra-day repeatability values (RSD%) were between 8.6% and 15% for L_L, between 6.3% and 14% for M_L, and between 7.7% and 16% for H_L. LODs and LOQs were comparable or lower than those that were found by Chi et al. [22] and acceptable for the analytical determinations. The analytical data that were considered for the optimization of the spectrometric determination have confirmed this method to be sensitive, precise, and accurate for the evaluation of CCs in indoor air. Supplementary Material S2 shows the LC-MS/MS chromatogram that was obtained by injecting 5 μL of aldehydes-DNPH standard solution at 0.2 μg L^{-1}.

Table 3. LODs (μg L^{-1}), LOQs (μg L^{-1}), linearity (R^2), and the intra- and inter-day repeatability (RSD%) that were obtained from the LC-MS/MS optimized method. L_L, M_L, and H_L correspond respectively to 0.20, 0.78, and 20 μg L^{-1}.

			Linearity	Intra-Day Repeatability (n = 10)			Inter-Day Repeatability (n = 7)		
	LOD	LOQ	R^2	L_L	M_L	H_L	L_L	M_L	H_L
Fo-DNPH	0.39	0.78	0.998	7.8	3.0	1.3	14	6.5	6.5
Ace-DNPH	0.39	0.78	0.999	6.8	2.9	1.2	10	6.3	7.7
Pro-DNPH	0.049	0.20	0.998	3.0	2.2	1.4	11	6.6	8.5
Cro-DNPH	0.049	0.20	0.998	3.6	3.3	1.4	8.8	7.3	9.4
Bu-DNPH	0.098	0.39	0.998	5.8	2.1	1.5	11	7.1	9.3
Cy-DNPH	0.012	0.024	0.997	3.5	2.5	2.3	15	12	13
Val-DNPH	0.024	0.098	0.998	4.6	2.7	2.6	11	9.2	11
Hex-DNPH	0.012	0.049	0.997	3.5	2.3	2.2	11	10	13
Hep-DNPH	0.012	0.098	0.997	2.3	1.9	1.8	13	14	16
Oct-DNPH	0.098	0.39	0.997	3.4	2.7	2.3	12	13	15
Non-DNPH	0.024	0.098	0.996	5.4	3.3	2.7	12	14	15
Dec-DNPH	0.049	0.20	0.997	3.7	1.5	2.6	8.6	9.8	11

3.3. *Optimization of the Spectroscopic Determination and Data Quality*

The method linearity in the investigated concentration range (8–1000 μg L^{-1}), expressed as R^2, was 0.999 for all the analytes that were under study, as reported in Table 4. Although the range of the investigated concentrations was rather wide, the results were comparable with those that were obtained by Feng et al. [23]. The values of intra-day repeatability (RSD%), calculated at lowest level (L_L, 25 μg L^{-1}), were between 2.3% and 7.8%. The RSD% were found to be lower at higher concentrations (M_L, 62 μg L^{-1} and H_L, 250 μg L^{-1}) with values between 1.5% and 3.3% for M_L and between 1.2% and 2.7% for H_L (Table 4). The intra-day repeatability values (RSD%) were between 8.6% and 15% for L_L, between 6.3% and 14% for M_L, and between 7.7% and 16% for H_L (Table 4). The values that were obtained both for intra- and inter-day repeatability were comparable with those that were obtained for MS/MS determination, despite the different concentrations of the three examined levels. As expected, LODs and LOQs were higher than those that were obtained by MS/MS analysis but comparable or lower than those that were obtained in other works [24,25]. Supplementary Material S3 shows the LC-UV/DAD chromatogram that was obtained by injecting 5 μL of aldehydes-DNPH standard solution at 25 μg L^{-1}.

Table 4. LODs (μg L^{-1}), LOQs (μg L^{-1}), linearity (R^2), and the intra- and inter-day repeatability (RSD%) that were obtained from the LC-UV/DAD optimized method. L_L, M_L, and H_L correspond respectively to 25, 62, and 250 μg L^{-1}.

			Linearity	Intra-Day Repeatability (n = 10)			Inter-Day Repeatability (n = 7)		
	LOD	LOQ	R^2	L_L	M_L	H_L	L_L	M_L	H_L
Fo-DNPH	16	62	0.999	10	4.9	1.6	9.3	11	9.5
Ace-DNPH	12	62	0.999	6.0	4.0	1.8	6.5	8.3	8.8
Pro-DNPH	8	31	0.999	5.8	2.7	1.0	5.6	7.6	9.3
Cro-DNPH	8	31	0.999	7.8	3.2	1.7	6.1	7.8	8.8
Bu-DNPH	8	31	0.999	3.8	1.3	0.80	5.6	7.2	9.2
Cy-DNPH	8	25	0.999	2.3	2.0	0.85	5.2	7.6	9.4
Val-DNPH	8	25	0.999	4.4	1.9	0.74	5.4	7.2	9.1
Hex-DNPH	8	25	0.999	5.6	1.8	1.3	5.0	6.8	9.2
Hep-DNPH	8	31	0.999	7.0	3.2	1.3	7.1	6.8	8.1
Oct-DNPH	8	31	0.999	6.1	3.7	1.2	6.3	6.5	9.3
Non-DNPH	12	62	0.999	8.2	5.6	1.0	10	7.1	9.3
Dec-DNPH	16	62	0.999	9.5	2.5	1.5	11	6.7	9.0

3.4. Comparison between LC-MS/MS and LC-UV/DAD

The MS/MS detection method was compared with the UV/DAD detection at 360 nm method. All the real samples were analyzed by using both detection methods. Supplementary Material S4–S6 show the LC-MS/MS and LC-UV/DAD chromatograms that were obtained by injecting 5 µL of extracted real samples (PL5-doc, BS1-hair, and ParL4-lab). The results of three carbonyls (Fo-DNPH, Ace-DNPH, and Pro-DNPH) are listed in Table 5. Unfortunately, the comparison was possible only for three congeners because, due to the lower sensibility of the UV/DAD detection method, only lower molecular weight analytes were detected in more than 50% of the samples that were analyzed by using UV/DAD detector. Table 5 shows that for formaldehyde and acetaldehyde existed good agreement between the two detection methods with percentage deviations between 0.2 and 29% and 0.1 and 30%, respectively. For pro-DNPH, since the determined concentrations are much lower than those that were obtained for fo-DNPH and ace-DNPH, with values between 0.32 and 5 µg m^{-3}, small differences can cause a considerable deviation (percentage deviations between 1.2% and 50%). The obtained percentage deviations were in line with those that were found by Chi et al. [22]. The deviations of cy-DNPH and cro-DNPH were in the range of 443–2004% and 16576–1601%, respectively; the obtained results were not acceptable, and this trend could be explained by considering the lack of selectivity of the spectroscopic method. In real samples, in fact, there could be an interference that co-elutes with the target compounds causing an increase in the signal with a consequent overestimation of the concentrations when samples were analyzed by using UV/DAD detector. This problem was overcome with the use of a high selectivity spectrometer detector operating in MRM mode. Dure to the lower LODs (Tables 3 and 4), the quantification of low concentration's CCs in air samples (mainly medium and high molecular weight carbonyls) was much more accurate by using the MS/MS detector than by using UV/DAD. For this reason, the real results will be described by using the concentrations that were obtained by MS/MS analysis.

Table 5. Percentage deviations of real samples that were obtained from the comparison of ESI-MS/MS and UV/DAD detection methods.

	Formaldehyde	Acetaldehyde	Propionaldehyde
PL1-lab	18	18	-
PL2-out	21	20	1.2
PL3-tech	3.9	0.5	-
PL4-tech	11	4.7	
PL5-doc	2.3	1.4	-
PL6-solv	0.5	0.6	-
GA1-nurs	12	16	-
GA2-nurs	18	24	-
GA3-doc	15	19	-
GA4-lab	13	14	-
GA5-out	7.0	22	-
CS1-desk	2.1	19	24
CS2-glue	3.8	16	14
CS3-photo	16	24	28
BS1-hair	14	13	31
BS2-nail	2.2	25	35
BS3-sol	12	26	31
BS4-rec	16	30	31
BS5-bea	29	30	31
BS6-out	21	11	23
PH1-out	11	6.0	35
PH2-ware	7.8	10	40
PH3-store	19	19	42

Table 5. *Cont.*

	Formaldehyde	Acetaldehyde	Propionaldehyde
PH4-dercon	18	21	48
PH5-phar	23	22	44
PH6-der	12	11	47
HL1-lab	25	13	29
HL2-tech	3.4	9.7	5.3
HL3-tech	1.0	5.0	8.5
HL4-sec	4.2	16	22
HL5-tech	0.7	4.2	5.0
HL6-out	5.6	25	-
HRL1-tech	13	2.3	-
HRL2-lab	20	6.4	-
HRL3-out	18	30	-
HRL4-bio	7.1	18	-
HRL5-lab	7.1	1.6	-
HpL1-out	25	20	-
HpL2-tech	12	0.1	39
HpL3-lab	5.2	0.7	9.2
HpL4-lab	2.0	4.0	-
HpL5-doc	0.3	15	-
HpL6-doc	3.8	9.1	-
CL1-lab	1.2	0.1	-
CL2-lab	7.3	7.7	-
CL3-lab	0.2	3.1	49
CL4-off	8.1	3.7	50
CL5-out	12	6.3	47
ParL1-lab	26	0.4	-
ParL2-lab	6.1	0.3	-
ParL3-corr	7.3	15	-
ParL4-lab	8.3	3.2	-

3.5. Real Samples

Table 6 reports the concentration of the 12 CCs that were assessed in ten different work environments. The average blank concentrations (Supplementary Material S7) were subtracted from sample concentrations. Analyzing the samples with LC-MS/MS technique, only 2.4% of the samples were <LOQ (Table 6); this percentage rises to 68% when the samples were analyzed by using LC-UV/DAD. As we can see from Table 6, the highest concentration of hexanal (13 μg m^{-3}) was recorded in one of the samples that was collected in the beauty salon (BS5-bea) and associated with a personal sampler of a beautician. Unfortunately, a comparison with the literature is not possible, due to the lack of data. The highest concentrations of cyclohexanone, crotonaldehyde, heptanal, and octanal (2.7 μg m^{-3}, 1.8 μg m^{-3}, 3. μg m^{-3}, and 3.2 μg m^{-3}, respectively) were recorded in the sample HL1-lab, collected in the histology laboratory (environmental sampler). In the same workplace, was also recorded the highest concentration of decanal (4.0 μg m^{-3}) in a personal sampler (HL5-tech). The concentrations that were obtained in this study were comparable or much higher than those that were obtained by Lu et al. [26,27] in special rooms and different wards of Chinese hospitals. The highest concentrations of butyraldehyde, nonanal, propionaldehyde, and valeraldehyde (13 μg m^{-3}, 9.9 μg m^{-3}, 5.0 μg m^{-3}, and 8.0 μg m^{-3}, respectively) were recorded in the enclosure environment (HRL5-lab). Unfortunately, a comparison with the literature is not possible because, to our knowledge, no study was conducted in a similar work environment.

Table 6. Concentration of the 12 Ccs ($\mu g\ m^{-3}$) that were determined in the 10 different work environments.

	Acetaldehyde	Butyraldehyde	Cyclohexanone	Crotonaldehyde	Decanal	Heptanal	Hexanal	Formaldehyde	Nonanal	Octanal	Propionaldehyde	Valeraldehyde
PL1-lab	9.5	0.46	0.01	<LOQ	0.08	0.03	0.19	55	0.49	0.08	0.67	0.17
PL2-out	4.7	0.57	0.08	<LOQ	0.12	0.06	0.06	6.1	0.73	0.15	0.95	0.23
PL3-tech	9.4	0.89	0.01	<LOQ	0.15	0.03	0.21	71	0.55	0.09	0.92	0.21
PL4-tech	7.7	1.4	0.11	<LOQ	0.11	0.06	0.19	66	0.62	0.12	0.99	0.71
PL5-doc	8.4	0.80	0.02	<LOQ	0.18	0.02	0.18	64	0.63	0.11	0.89	0.22
PL6-solv	3.5	0.76	0.05	<LOQ	0.17	0.06	0.08	4.5	0.82	0.14	0.87	0.29
GA1-nurs	6.2	0.81	0.11	<LOQ	0.19	0.04	<LOQ	5.3	0.81	0.13	0.74	0.25
GA2-nurs	7.9	0.61	0.12	<LOQ	0.14	0.00	0.02	6.1	0.63	0.09	0.84	<LOQ
GA3-doc	7.9	0.68	0.26	0.02	0.70	0.11	0.19	8.8	1.7	0.43	1.1	0.03
GA4-lab	4.2	0.41	0.13	<LOQ	0.25	0.01	<LOQ	5.0	0.68	0.08	0.63	0.04
GA5-out	3.7	0.38	0.16	0.03	0.46	0.14	0.14	5.2	1.9	0.23	0.72	0.04
CS1-desk	8.5	1.2	0.06	0.08	1.0	0.52	2.2	19	2.2	0.56	1.2	0.95
CS2-glue	13	0.82	0.15	0.06	1.1	0.50	2.2	22	2.2	0.57	1.2	0.71
CS3-photo	14	0.98	0.07	0.05	0.72	0.33	1.4	12	1.7	0.38	1.1	0.58
BS1-hair	57	1.9	0.11	0.01	1.0	0.51	6.5	25	2.4	0.01	4.0	0.64
BS2-nail	52	1.7	0.12	0.01	1.7	0.88	7.8	26	3.5	0.91	4.4	2.3
BS3-sol	59	1.8	0.14	0.01	1.1	0.75	8.5	26	3.2	0.58	4.3	2.4
BS4-rec	46	1.8	0.07	0.01	0.79	0.57	7.4	23	2.2	0.44	3.9	2.6
BS5-bea	79	2.4	0.07	0.03	1.6	1.1	13	29	4.5	1.0	4.7	2.9
BS6-out	19	1.7	0.05	0.01	0.62	0.67	2.8	14	2.5	0.58	1.7	1.5
PH1-out	2.7	0.44	0.12	0.02	0.39	0.12	0.32	4.5	1.1	0.20	1.8	0.19
PH2-ware	8.1	1.1	0.27	0.02	0.70	0.34	2.0	17	2.5	0.39	1.6	0.87
PH3-store	8.4	1.7	0.15	0.02	0.73	0.31	2.2	14	1.9	0.40	1.8	1.3
PH4-dercon	12	1.2	0.09	0.02	0.65	0.42	2.7	13	2.5	0.42	1.9	0.97
PH5-phar	8.7	1.6	0.18	0.02	0.93	0.33	2.2	13	2.0	0.45	1.6	1.5
PH6-der	11	1.4	0.10	0.02	0.49	0.24	1.8	15	1.5	0.29	1.8	0.96
HL1-lab	4.2	2.8	2.7	1.8	3.4	3.0	3.1	13	4.7	3.2	2.1	2.6
HL2-tech	4.4	0.76	0.18	0.01	1.1	0.42	0.80	29	2.4	2.2	0.60	0.37
HL3-tech	3.7	0.73	0.14	0.004	2.3	0.82	1.2	77	5.4	1.8	0.64	0.51
HL4-sec	2.7	0.93	0.14	0.004	1.3	0.36	0.98	18	2.5	0.74	0.61	0.53
HL5-tech	3.9	1.17	0.22	0.005	4.0	0.36	1.0	42	3.9	1.6	0.69	0.72
HL6-out	1.5	0.45	0.04	0.002	0.50	0.15	0.32	3.3	1.0	0.25	0.32	0.15
HRL1-tech	8.4	3.8	0.52	0.02	0.75	0.50	1.80	5.7	3.9	0.64	2.0	2.4
HRL2-lab	3.6	1.7	0.19	0.01	0.68	0.22	0.20	4.9	4.0	0.63	1.5	0.54
HRL3-out	4.3	2.5	0.55	0.02	0.60	0.63	2.5	2.7	3.3	0.63	1.8	2.9
HRL4-bio	13	8.1	1.3	0.02	1.4	1.2	4.2	4.4	6.5	1.1	3.1	7.0
HRL5-lab	16	13	1.30	0.04	1.6	1.2	4.6	10	9.9	1.6	5.0	8.0
HpL1-out	3.6	2.6	0.29	0.01	0.30	0.20	0.61	3.3	1.5	0.30	1.8	1.3
HpL2-tech	4.7	2.0	0.29	0.01	0.46	0.28	0.85	12	1.9	0.35	1.7	1.0
HpL3-lab	6.3	3.1	0.42	0.01	0.59	0.31	0.93	16	1.9	0.34	2.3	1.5
HpL4-lab	4.7	2.1	0.40	0.01	0.78	0.32	0.76	13	3.0	0.46	1.6	1.3

Table 6. *Cont.*

	Acetaldehyde	Butyraldehyde	Cyclohexanone	Crotonaldehyde	Decanal	Heptanal	Hexanal	Formaldehyde	Nonanal	Octanal	Propionaldehyde	Valeraldehyde
HpL5-doc	5.3	2.0	0.17	0.01	0.54	0.36	0.97	7.7	3.0	0.47	1.7	1.1
HpL6-doc	5.2	3.3	0.20	0.01	0.41	0.32	1.5	17	2.1	0.36	2.2	1.8
CL1-lab	3.8	<LOQ	<LOQ	0.45	0.07	1.2	0.18	13	2.1	0.35	1.7	1.6
CL2-lab	3.9	2.2	0.34	0.01	0.69	0.23	0.42	13	2.4	0.41	1.8	0.92
CL3-lab	4.2	2.5	0.43	0.02	0.75	0.26	0.56	9.6	2.4	0.38	1.9	1.2
CL4-off	4.6	3.1	0.33	0.01	0.39	0.22	0.51	7.9	2.2	0.38	2.0	1.5
CL5-out	3.2	3.1	0.44	<LOQ	0.47	0.20	0.54	4.0	2.0	0.27	1.9	1.7
ParL1-lab	4.9	2.3	0.25	0.01	0.33	0.19	0.59	5.0	1.9	0.30	1.9	1.1
ParL2-lab	5.4	2.5	0.26	0.01	0.30	0.19	0.50	5.8	1.7	0.28	2.0	1.2
ParL3-corr	5.9	1.5	0.16	0.01	0.29	0.15	0.37	5.2	1.9	0.31	1.7	0.61
ParL4-lab	5.7	2.6	0.27	0.01	0.32	0.18	0.46	5.7	1.8	0.29	2.0	1.2

Formaldehyde and acetaldehyde concentrations in work environments' indoor air ranged from 2.7 to 77 μg m^{-3} and from 1.5 to 79 μg m^{-3}, respectively (Table 6). For all the sampling sites, the indoor concentrations of the two pollutants were higher than the corresponding outdoor values (Table 6). For all the sampling sites, the concentrations of formaldehyde were comparable or higher than the respective concentrations of acetaldehyde, except for the beauty salon. In all the five beauty salon's samples, in fact, the concentrations of acetaldehyde were much higher than the respective concentrations of formaldehyde (Figure 1). This trend was also found in the beauty salon of Tehran by Hadei et al. [10], and could be due to the composition of cosmetic products that are used in the beauty salon. Evtyugina et al. [28] measured CCs' concentrations in a beauty salon in Spain, finding formaldehyde and acetaldehyde mean concentrations of 12 and 9.0 μg m^{-3}, respectively. These values are much lower than those that were determined in this work (mean concentration of 24 and 52 μg m^{-3} for formaldehyde and acetaldehyde, respectively).

The highest concentrations of formaldehyde were detected in the hospital environments (sampling sites HL3, PL3, PL4, PL5, PL1, HL5, and HL2) where formalin solutions were used. Lower concentrations were detected in the hospital environments of gastroenterology and parasitology; in the first ward formaldehyde was not deliberately used and in the second ward a closed system was used to contain formaldehyde emissions. The maximum and minimum formaldehyde contamination levels that were detected in different hospital wards in this study were compared with the literature. The comparison showed that formaldehyde contamination levels in this work were much lower than those that were found in other Italian and international hospitals [29–33]. Concerning the acetaldehyde contamination levels, the values that were found in this work were comparable or much lower than the values that were found by Sousa et al. [6] and Lu et al. [26] in Brazilian and Chinese hospitals, respectively. For formaldehyde and acetaldehyde contamination levels in the copy shop, the concentrations that were found in this study were in line or lower than those that were found by Saraga et al. [34] in a printery in the center of Athens but higher than those that were found by Ho et al. [1] in a photocopy center at a Chinese university. Formaldehyde and acetaldehyde contamination levels in the selected pharmacy were in line with those that were found by Loh et al. [35] in eight different drug stores in the USA. Unfortunately, to our knowledge, there are no studies that were carried out in chemistry laboratories. However, the contamination levels that were found in this selected work environment were comparable with those that were found in the pharmacy and in some wards of the investigated hospitals (HpL and CA). Unfortunately, for the concentrations of the other CCs, a comparison with the literature is not possible given the lack of data. However, to further investigate the distribution patterns of target analytes in different workplaces, PCA has been conducted on the 12 individual CCs' congeners and 52 samples. The PCA results are summarized in the biplot that is shown in Figure 2, while scores and loadings are shown in Tables 7 and 8, respectively.

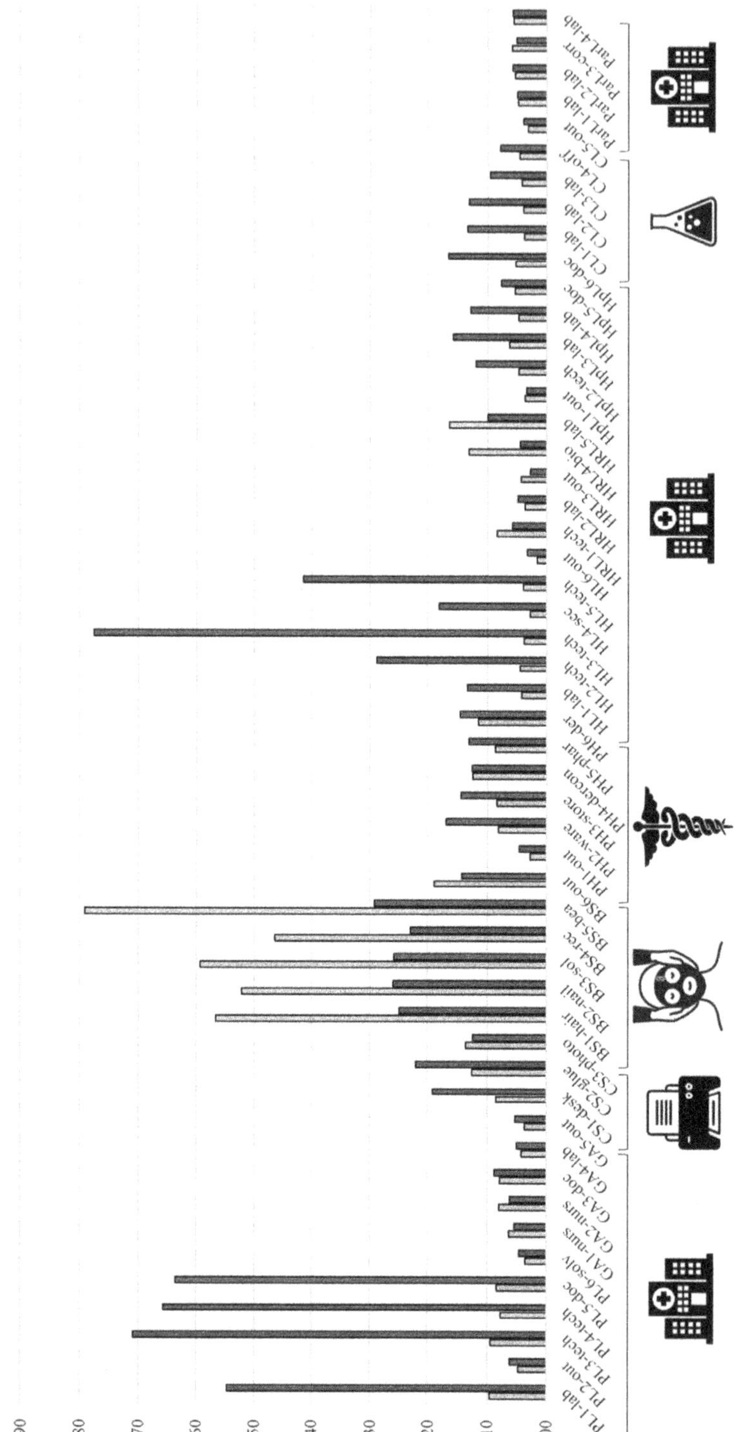

Figure 1. Distribution (concentrations expressed as µg m^{-3}) of formaldehyde and acetaldehyde in the previously described work environments.

Table 7. Scores and variance % of the five components that were obtained by the PCA that was performed on the concentration data that were yielded at each work place.

	PC1	PC2	PC3	PC4	PC5
Variance%	47.4	20.9	14.5	8.5	4.5
PL1−lab	−2.49	−0.09	1.10	−0.82	−1.55
PL2−out	−2.22	−0.29	−0.39	0.65	0.12
PL3−tech	−2.32	0.02	1.45	−1.40	−2.13
PL4−tech	−2.01	−0.03	1.00	−1.33	−2.13
PL5−doc	−2.31	−0.04	1.25	−1.22	−1.83
PL6−solv	−2.20	−0.32	−0.49	0.60	0.22
GA1−nurs	−2.19	−0.33	−0.44	0.59	0.18
GA2−nurs	−2.33	−0.24	−0.32	0.70	0.13
GA3−doc	−1.39	−0.59	−0.01	0.28	0.54
GA4−lab	−2.37	−0.48	−0.34	0.64	0.27
GA5−out	−1.77	−0.66	−0.24	0.42	0.53
CS1−desk	−0.37	−0.32	0.56	−0.07	0.34
CS2−glue	−0.34	−0.26	0.82	−0.05	0.30
CS3−photo	−1.00	−0.11	0.29	0.34	0.34
BS1−hair	1.08	3.20	1.62	1.04	−0.19
BS2−nail	2.92	2.81	1.99	0.43	0.49
BS3−sol	2.49	3.45	1.75	0.85	−0.09
BS4−rec	1.60	3.00	1.16	0.84	−0.28
BS5−bea	4.42	4.60	2.88	0.80	0.46
BS6−out	0.13	0.50	0.24	0.25	0.18
PH1−out	−1.74	−0.19	−0.33	0.74	0.31
PH2−ware	−0.48	−0.05	0.08	0.06	0.13
PH3−store	−0.46	0.19	−0.08	0.12	0.14
PH4−dercon	−0.31	0.40	0.14	0.30	0.32
PH5−phar	−0.29	0.07	−0.05	0.05	0.32
PH6−der	−0.89	0.38	−0.04	0.33	−0.03
HL1−lab	8.91	−7.13	2.20	2.94	−1.04
HL2−tech	0.03	−1.74	1.32	−1.17	0.93
HL3−tech	1.21	−1.73	2.88	−3.43	0.18
HL4−sec	−0.68	−0.96	0.47	−0.58	0.83
HL5−tech	1.26	−1.92	2.33	−2.71	2.24
HL6−out	−2.12	−0.73	−0.22	0.46	0.64
HRL1−tech	1.15	0.02	−1.32	−0.27	0.20
HRL2−lab	−0.41	−0.47	−0.67	−0.29	0.89
HRL3−out	0.98	−0.15	−1.17	0.19	0.16
HRL4−bio	5.69	0.41	−2.97	−1.04	−0.43
HRL5−lab	8.23	1.38	−4.08	−2.41	−0.52
HpL1−out	−0.83	−0.07	−1.13	0.44	−0.04
HpL2−tech	−0.72	−0.16	−0.59	0.14	−0.06
HpL3−lab	−0.10	0.09	−0.81	−0.03	−0.40
HpL4−lab	−0.14	−0.34	−0.60	−0.23	0.20
HpL5−doc	−0.33	−0.08	−0.67	0.01	0.35
HpL6−doc	−0.15	0.36	−0.88	−0.16	−0.44
CL1−lab	−0.22	−0.98	0.18	1.41	−0.87
CL2−lab	−0.50	−0.29	−0.62	−0.12	0.08
CL3−lab	−0.25	−0.26	−0.83	0.00	0.11
CL4−off	−0.34	0.02	−1.19	0.06	−0.13
CL5−out	−0.36	−0.08	−1.42	0.17	−0.06
ParL1−lab	−0.78	0.05	−0.99	0.36	0.03
ParL2−lab	−0.78	0.08	−1.05	0.35	−0.09
ParL3−corr	−1.20	−0.05	−0.67	0.41	0.23
ParL4−lab	−0.72	0.10	−1.08	0.33	−0.08

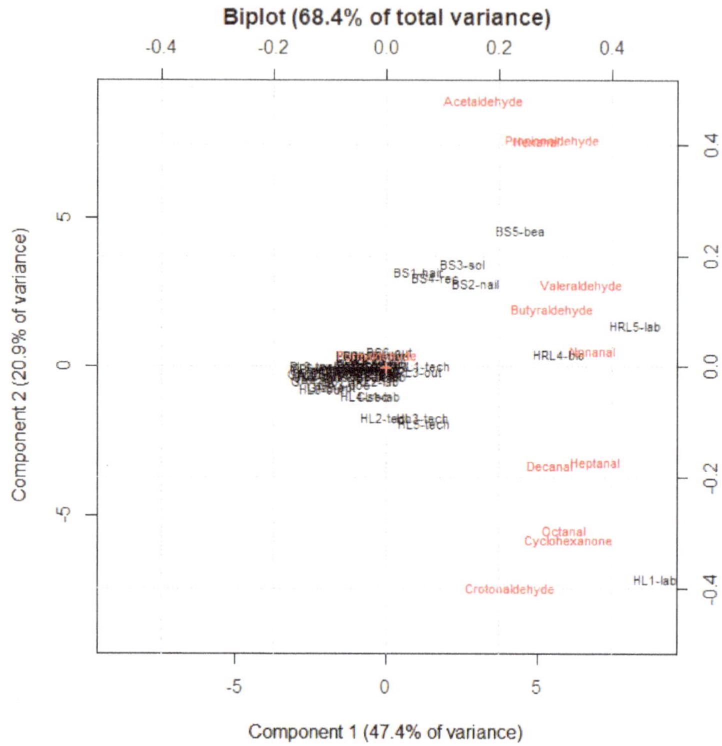

Figure 2. Biplot of the PCA (PC1 and PC2) that was performed on the concentration data that were yielded at each work place.

Table 8. Loadings of the five components that were obtained by the PCA that were performed on the concentration data that were yielded at each workplace.

	PC1	PC2	PC3	PC4	PC5
Acetaldehyde	0.17	0.48	0.34	0.21	0.00
Butyraldehyde	0.29	0.10	−0.46	−0.26	−0.18
Cyclohexanone	0.32	−0.31	−0.17	0.15	−0.23
Crotonaldehyde	0.22	−0.40	0.17	0.45	−0.30
Decanal	0.29	−0.18	0.32	−0.24	0.50
Heptanal	0.37	−0.17	0.14	0.22	−0.14
Hexanal	0.27	0.41	0.27	0.17	0.08
Formaldehyde	−0.02	0.02	0.50	−0.58	−0.63
Nonanal	0.37	0.03	−0.14	−0.34	0.23
Octanal	0.32	−0.29	0.21	−0.17	0.24
Propionaldehyde	0.29	0.41	−0.05	0.17	−0.11
Valeraldehyde	0.35	0.15	−0.32	−0.14	−0.18

Two significant components (PC1 and PC2), accounting for 68.4% of the total variance, were obtained. The biplot in Figure 2 well separates four clusters of work environments, each characterized by the presence of different CCs. The first cluster that was identified on the central part of the biplot, was predominated by formaldehyde, which did not show high variability across the 44 samples. This is in agreement with the fact that formaldehyde is considered a ubiquitous indoor pollutant [36,37]. The second cluster, in the upper right of the biplot, is characterized by the presence of all the indoor beauty salon samples and three carbonyl compounds (acetaldehyde, propionaldehyde, and hexanal) that are probably associated with the use of beauty products. The third cluster to the right of the biplot is characterized by the presence of the two samples (one personal and the other environmental) that are related to the enclosure. This environment was characterized by the presence of valeraldehyde, nonanal, and butyraldehyde, maybe associated with the presence of living animals. The fourth cluster, at the bottom right of the biplot, is characterized by the presence of one environmental sample that was collected from a histology laboratory. This work environment is characterized by the presence of five carbonyl compounds (decanal, heptanal, octanal, cyclohexanone, and crotonaldehyde) that are probably associated with the use of chemicals for different purposes, such as cleaning and disinfection.

4. Conclusions

This study demonstrated that LC-MS/MS detection is a suitable method for the determination of carbonyl compounds in different work environments. Good selectivity, sensibility, linearity, and reproducibility were all obtained. The proposed method, compared with LC–UV/DAD detection, resulted to be the most suitable strategy for the analysis of CCs in work environments given the high sensibility and the lowest LODs. The application of the optimized method to real samples also achieved excellent results: only 2.4% of the samples resulted to be lower than LOQs. When the same samples were analyzed by using LC-UV/DAD, this percentage rises to 68%. Among most of the work environments, formaldehyde was the most abundant detected carbonyl, followed by acetaldehyde and butyraldehyde. On the contrary, the most abundant congener that was detected in the beauty salon was acetaldehyde; this trend could be attributed to the composition of cosmetic products that are used in beauty salon. The PCA that was performed on the dataset confirms the ubiquitous character of formaldehyde and highlights the different influence of minority CCs on different work environments.

Supplementary Materials: The following supporting information can be downloaded at: https://www.mdpi.com/article/10.3390/jerph191912052/s1, S1. LC-UV/DAD chromatogram of blank cartridge; S2. LC-MS/MS chromatogram of aldehydes-DNPH standard solution; S3. LC-UV/DAD chromatogram of aldehydes-DNPH standard solution; S4. LC-MS/MS and LC-UV/DAD chromatograms of PL5-doc; S5. LC-MS/MS and LC-UV/DAD chromatograms of BS1-hair; S6. LC-MS/MS and LC-UV/DAD chromatograms of PaL4-lab; S7. Aldehyde concentrations in blank samples.

Author Contributions: Conceptualization, M.V.; Methodology, F.C. and C.P.; Software, F.C.; Validation, C.P.; Formal Analysis, F.C., A.A., and I.P.; Investigation, F.C., A.A., and I.P.; Writing—Original Draft Preparation, F.C. and A.A.; Writing—Review and Editing, C.P. and M.V.; Supervision: M.V. All authors have read and agreed to the published version of the manuscript.

Funding: This research was funded by INAIL, BRiC-2018, grant number B86C19000070001.

Institutional Review Board Statement: The investigations that were carried out do not require the approval of an ethics committee.

Informed Consent Statement: Not applicable.

Data Availability Statement: All data are available upon request to the corresponding author.

Conflicts of Interest: The authors declare no conflict of interest. The funders had no role in the design of the study; in the collection, analyses, or interpretation of data; in the writing of the manuscript; or in the decision to publish the results.

References

1. Ho, S.S.H.; Cheng, Y.; Bai, Y.; Ho, K.F.; Dai, W.T.; Cao, J.J.; Lee, S.C.; Huang, Y.; Ip, H.S.S.; Deng, W.J.; et al. Risk Assessment of Indoor Formaldehyde and Other Carbonyls in Campus Environments in Northwestern China. *Aerosol Air. Qual. Res.* **2016**, *16*, 1967–1980. [CrossRef]
2. De Carvalho, A.B.; Kato, M.; Rezende, M.M.; de Pereira, P.A.; de Andrade, J.B. Determination of carbonyl compounds in the atmosphere of charcoal plants by HPLC and UV detection. *J. Sep. Sci.* **2008**, *31*, 1686–1693. [CrossRef]

3. Lewtas, J. Air pollution combustion emissions: Characterization of causative agents and mechanisms associated with cancer, reproductive, and cardiovascular effects. *Mutat. Res.* **2007**, *636*, 95–133. [CrossRef]
4. Atkinson, R.; Arey, J. Atmospheric degradation of volatile organic compounds. *Chem. Rev.* **2003**, *103*, 4605–4638. [CrossRef]
5. WHO (World Health Organization). *WHO Guidelines for Indoor Air Quality. Selected Pollutants*; WHO Regional Office for Europe: Copenhagen, Denmark, 2010.
6. Sousa, F.W.; Caracas, I.B.; Nascimento, R.F.; Cavalcante, R.M. Exposure and cancer risk assessment for formaldehyde and acetaldehyde in the hospitals, Fortaleza-Brazil. *Build Environ.* **2011**, *46*, 2115–2120. [CrossRef]
7. IARC (International Agency for Research on Cancer). Monographs on the Evaluation of Carcinogenic Risks to Humans. In *Re-Evaluation of some Organic Chemicals, Hydrazine and Hydrogen Peroxide*; IARC Working Group on the Evaluation of Carcinogenic Risks to Humans; IARC: Lyon, France, 1999.
8. IARC (International Agency for Research on Cancer). Monographs on the Evaluation of Carcinogenic risks to Humans. In *Formaldehyde, 2-Butoxyethanol and 1-Tert-Butoxypropan-2-ol*; IARC Working Group on the Evaluation of Carcinogenic Risks to Humans; IARC: Lyon, France, 2006.
9. Protano, C.; Buomprisco, G.; Cammalleri, V.; Pocino, R.N.; Marotta, D.; Simonazzi, S.; Cardoni, F.; Petyx, M.; Iavicoli, S.; Vitali, M. The Carcinogenic Effects of Formaldehyde Occupational Exposure: A Systematic Review. *Cancers* **2022**, *14*, 165. [CrossRef]
10. Hadei, M.; Hopke, P.K.; Shahsavani, A.; Moradi, M.; Yarahmadi, M.; Emam, B.; Rastkari, N. Indoor concentrations of VOCs in beauty salons; association with cosmetic practices and health risk assessment. *J. Occup. Med. Toxicol.* **2018**, *13*, 1–9. [CrossRef]
11. Jafari, M.J.; Khajevandi, A.A.; Mousavi Najarkola, S.A.; Yekaninejad, M.S.; Pourhoseingholi, M.A.; Omidi, L.; Kalantary, S. Association of sick building syndrome with indoor air parameters. *Tanaffos* **2015**, *14*, 55–62.
12. EPA (United States Environmental Protection Agency). *Questions about Your Community: Indoor Air*; EPA: Washington, DC, USA, 2015.
13. Cammalleri, V.; Pocino, R.N.; Marotta, D.; Protano, C.; Sinibaldi, F.; Simonazzi, S.; Petyx, M.; Iavicoli, S.; Vitali, M. Occupational scenarios and exposure assessment to formaldehyde: A systematic review. *Indoor Air* **2022**, *32*, e12949. [CrossRef]
14. Scarselli, A.; Corfiati, M.; Di Marzio, D.; Iavicoli, S. National Estimates of Exposure to Formaldehyde in Italian Workplaces. *Ann. Work Expo. Health* **2017**, *61*, 33–43. [CrossRef]
15. Jurvelin, J.; Vartiainen, M.; Jantunen, M.; Pasanen, P. Personal Exposure Levels and Microenvironmental Concentrations of Formaldehyde and Acetaldehyde in the Helsinki Metropolitan Area, Finland. *J. Air Waste Manag. Assoc.* **2001**, *51*, 17–24. [CrossRef]
16. Maroni, M.; Seifert, B.; Lindvall, T. *Indoor Air Quality a Comprehensive Reference Book*, 1st ed.; Elsevier Science Ltd.: Amsterdam, The Netherlands, 1995.
17. Calisti, R.; Chiaverini, A.F.; Mei, R. Aldehydes others than formaldehyde: Exposure patterns and their significance in a set of Italian indoor workplaces, 2011–2020. *J. Occup. Environ. Hyg.* **2020**, *11*, 102–146. [CrossRef]
18. TOXNET. *Crotonaldehyde*; National Library of Medicine HSDB Database: Bethesda, MD, USA, 2020.
19. Zhang, S.; Chen, H.; Wang, A.; Liu, Y.; Hou, H.; Hu, Q. Assessment of genotoxicity of four volatile pollutants from cigarette smoking based on the in vitro γH2AX assay using high content screening. *Environ. Toxicol. Pharmacol.* **2017**, *55*, 30–36. [CrossRef] [PubMed]
20. Zhang, S.; Chen, H.; Wang, A.; Liu, Y.; Hou, H.; Hu, Q. Combined effect of co-exposure to formaldehyde and acrolein mixtures on cytotoxicity and genotoxicity in vitro. *Environ. Sci. Pollut. Res. Int.* **2018**, *25*, 25306–25314. [CrossRef] [PubMed]
21. Xie, M.Z.; Shoulkamy, M.I.; Salem, A.M.; Oba, S.; Goda, M.; Nakano, T.; Ide, H. Aldehydes with high and low toxicities inactivate cells by damaging distinct cellular targets. *Mutat. Res.* **2016**, *786*, 41–51. [CrossRef]
22. Chi, Y.; Feng, Y.; Wen, S.; Lü, H.; Yu, Z.; Zhang, W.; Sheng, G.; Fu, J. Determination of carbonyl compounds in the atmosphere by DNPH derivatization and LC-ESI-MS/MS detection. *Talanta* **2007**, *72*, 539–545. [CrossRef]
23. Zeng, Y.; Wen, S.; Chen, Y.; Wang, X.; Lü, H.; Bi, X.; Sheng, G.; Fu, J. Ambient levels of carbonyl compounds and their sources in Guangzhou, China. *Atmos. Environ.* **2005**, *39*, 1789–1800. [CrossRef]
24. Wang, H.W.; Tong, X.Y.; Yan, L.Q.; Sheng, J.Y.; Liu, S.M. Determination of Volatile Carbonyl Compounds in Cigarette Smoke by LC-DAD. *Chromatographia* **2005**, *62*, 631–636. [CrossRef]
25. Zhang, T.; Jiang, G.; Guazzotti, S.; Elmashni, D. *Quantitative Analysis of Carbonyl-DNPH Derivatives by UHPLC/UV*; Thermo Fisher Scientific Application note; Thermo Fisher: Waltham, MA, USA, 2010; p. 52007.
26. Lü, H.; Wen, S.; Feng, Y.; Wang, X.; Bi, X.; Sheng, G.; Fu, J. Indoor and outdoor carbonyl compounds and BTEX in the hospitals of Guangzhou, China. *Sci. Total Environ.* **2006**, *368*, 574–584. [CrossRef]
27. Lü, H.; Cai, Q.Y.; Wen, S.; Chi, Y.; Guo, S.; Sheng, G.; Fu, J.; Katsoyiannis, A. Carbonyl compounds and BTEX in the special rooms of hospital in Guangzhou, China. *J. Hazard Mater.* **2010**, *178*, 673–679. [CrossRef]
28. Evtyugina, M.; Vicente, E.D.; Vicente, A.M.; Nunes, T.; Lucarelli, F.; Calzolai, G.; Nava, S.; Blanco-Alegre, C.; Calvo, A.I.; Castro, A.; et al. Air quality and particulate matter speciation in a beauty salon and surrounding outdoor environment: Exploratory study. *Atmos. Pollut. Res.* **2021**, *12*, 101174. [CrossRef]
29. Perdelli, F.; Spagnolo, A.M.; Cristina, M.L.; Sartini, M.; Dallera, M.; Ottria, G.; Orlando, P. Occupational exposure to formaldehyde in three pathology departments. *Ann. Ig.* **2006**, *18*, 481–490. [PubMed]

30. De Ochs, M.; Grotz, L.O.; Factorine, L.S.; Rodrigues, M.R.; Pereira Netto, A.D. Occupational exposure to formaldehyde in an institute of morphology in Brazil: A comparison of area and personal sampling. *Environ. Sci. Pollut. Res. Int.* **2011**, *19*, 2813–2819. [CrossRef]
31. Bellisario, V.; Mengozzi, G.; Grignani, E.; Bugiani, M.; Sapino, A.; Bussolati, G.; Bono, R. Towards a formalin-free hospital. Levels of 15-F2t-isoprostane and malondialdehyde to monitor exposure to formaldehyde in nurses from operating theatres. *Toxicol. Res.* **2016**, *5*, 1122–1129. [CrossRef]
32. Higashikubo, I.; Miyauchi, H.; Yoshida, S.; Tanaka, S.; Matsuoka, M.; Arito, H.; Araki, A.; Shimizu, H.; Sakurai, H. Assessment of workplace air concentrations of formaldehyde during and before working hours in medical facilities. *Ind. Health* **2017**, *55*, 192–198. [CrossRef] [PubMed]
33. Lee, E.G.; Magrm, R.; Kusti, M.; Kashon, M.L.; Guffey, S.; Costas, M.M.; Boykin, C.J.; Harper, M. Comparison between active (pumped) and passive (diffusive) sampling methods for formaldehyde in pathology and histology laboratories. *J. Occup. Environ. Hyg.* **2017**, *14*, 31–39. [CrossRef] [PubMed]
34. Saraga, D.; Pateraki, S.; Papadopoulos, A.; Vasilakos, C.; Maggos, T. Studying the indoor air quality in three non-residential environments of different use: A museum, a printery industry and an office. *Build Environ.* **2011**, *46*, 2333–2341. [CrossRef]
35. Loh, M.M.; Houseman, E.A.; Gray, G.M.; Levy, J.I.; Spengler, J.D.; Bennett, D.H. Measured Concentrations of VOCs in Several Non-Residential Microenvironments in the United States. *Environ. Sci. Technol.* **2006**, *40*, 6903–6911. [CrossRef]
36. Liang, W.; Zhao, B.; Liu, J.; Pei, J. Can carbon dioxide be a good indicator for formaldehyde in residences?—Monte Carlo modeling for a whole year. *Sci. Technol. Built. Environ.* **2020**, *26*, 749–762. [CrossRef]
37. Shao, Y.; Wang, Y.; Zhao, R.; Chen, J.; Zhang, F.; Linhardt, R.J.; Zhong, W. Biotechnology progress for removal of indoor gaseous formaldehyde. *Appl. Microbiol. Biotechnol.* **2020**, *104*, 3715–3727. [CrossRef]

International Journal of
Environmental Research and Public Health

MDPI

Article

An Advanced Tape-Stripping Approach for High-Efficiency Sampling on Non-Absorbent Surfaces

Pengcheng Zhao [1], Pak-To Chan [1], Nan Zhang [1,2] and Yuguo Li [1,3,*]

[1] Department of Mechanical Engineering, The University of Hong Kong, Hong Kong SAR 999077, China
[2] Beijing Key Laboratory of Green Built Environment and Energy Efficient Technology, Beijing University of Technology, Beijing 100124, China
[3] School of Public Health, The University of Hong Kong, Hong Kong SAR 999077, China
* Correspondence: liyg@hku.hk

Abstract: Surface sampling is a frequent task in laboratory work and field studies. Simple methods usually have low efficiency in collecting target substances from surfaces. This study developed an advanced tape-stripping approach for efficient sampling on non-absorbent surfaces. A film-forming solution, prepared using polyvinyl alcohol, is applied to the target surface, where it covers and engulfs the surface deposits and then solidifies into an elastic membrane as it is exposed to air. The deposits are collected by stripping off the membrane and re-dissolving it in water. This new approach exhibited an efficiency of 100% in collecting uniform-size microspheres from glass surfaces and extremely high efficiencies (>96.6%) in detecting selected target DNA materials from glass and stainless steel surfaces. In comparison, the common swab-rinse method exhibited an efficiency of 72.6% under similar measuring conditions. The viability of *S. aureus* during sampling using the new approach decreased as the ethanol concentration in the applied solution increased. Using a solution with a mass ratio of ethanol of 17.6% balanced the effects of multiplication and degradation of the *S. aureus* on glass surfaces during sampling. Overall, the proposed approach exhibits high efficiency in collecting living and abiotic matter from non-absorbent surfaces, complementing existing sampling methods.

Keywords: film-forming solution; polyvinyl alcohol; qPCR; re-dissolvable membrane; surface microbes; swab

Citation: Zhao, P.; Chan, P.-T.; Zhang, N.; Li, Y. An Advanced Tape-Stripping Approach for High-Efficiency Sampling on Non-Absorbent Surfaces. *Int. J. Environ. Res. Public Health* **2022**, *19*, 12571. https://doi.org/10.3390/ijerph191912571

Academic Editors: Roel Vermeulen, Lauren Petrick and Maaike van Gerwen

Received: 3 August 2022
Accepted: 28 September 2022
Published: 1 October 2022

1. Introduction

Surface sampling is an important step in the detection of surface deposits [1,2]. Sampling efficiency generally refers to the ratio of detected target substances to total target substances each time a surface is sampled. High sampling efficiency and consistency in surface sampling are crucial for research [3–5]. However, a sampling efficiency of 100% is difficult to achieve due to losses during the collecting, transporting and detecting processes [6].

It is difficult to collect microbes from surfaces because some microbes are likely to undergo irreversible adhesion to substrates [7]. Research on stable and efficient techniques for surface sampling is ongoing [8–10]. Table 1 lists recent studies of surface sampling, presenting detailed data on the sampling efficiencies of specific sampling methods. The most widely used sampling methods for collecting surface deposits have efficiencies of <80%, depending on the measuring conditions (see Figure 1). Some commercial sampling kits can achieve higher efficiencies when collecting surface deposits [11,12]. However, efficient sampling techniques usually involve complex operations—it is difficult to simultaneously achieve high sampling efficiency and operational simplicity.

Most surface sampling methods belong to one of the following categories: wipe, contact plate, tape strip and elution [6]. Each method has some limitations: the destructive elution and glove-juice methods are highly efficient in detaching surface deposits but often require complicated post-collection treatments [13]; in contrast, the swab-rinse and contact-plate methods are simple, but their efficiencies are highly variable [14–17]. Tape-stripping is an efficient approach for collecting surface deposits (Figure 1); however, it suffers from a design conundrum. Strengthening the tape's adhesiveness increases the collection efficiency but also creates difficulties in detaching the collected material from the adhesive, unless the material on the tape is examined in situ [18,19]. In addition, existing methods can be affected by other factors, such as operational interference [20], surface features [16,21] and the characteristics of the deposits [22].

Table 1. Details of 16 studies on sampling methods. Their measured sampling efficiencies are presented in the rightmost column.

Studies	Sampling Targets	Surfaces	Sampling Methods	Sampling Efficiencies (%)
Whyte et al. [23]	Skin bacteria	Skin	Swab RODAC (contact-plate) Cylinder (eluting)	19 9 31
Perkins et al. [24]	Cytokines	Skin	Sebutape (tape-stripping)	≈50
Yamaguchi et al. [25]	*S. epidermidis*	Plastic	Swab Adhesive sheet	27 ± 7.2 [a] 54 ± 4.6 [a]
Buttner et al. [12]	*B. atrophaeus* spores	Metal	BiSKit (a foam material)	11.3–18.4 [b]
Daley et al. [26]	Respiratory epithelial cells	Nasal cavity; nasopharyngeal cavity	Flocked swab Rayon swab	20.1–75.1 [b] 9.1–43.5 [b]
Buttner et al. [27]	*Erwinia herbicola*	Metal; glass; wood; vinyl; plastic; concrete; nylon	Sponge (+qPCR) Swab (+qPCR)	0.7–44.8 [b] 0.8–52.2 [b]
Moore and Griffith [28]	*Escherichia coli; S. aureus*	SS	Swab	31–75 [b]
Brown et al. [29]	*B. atrophaeus* spores	SS; painted wallboard	Wipe (by polyester-rayon blend gauze)	8.1–67.4 [b]
Bisha and Brehm-Stecher [30]	*Salmonella*	Tomato	Adhesive tape	>99
Hong-Geller et al. [22]	*B. anthracis; Yersinia pestis*	Plastic; SS; glass; vinyl	Swab/wipe	<1–94 [b]
Rubuza et al. [11]	*S. aureus; Klebsiella pneumoniae*	Fabric pieces	Swab RODAC (contact-plate) Destructive elution Morapex device	<0.01 <0.01 0.07–0.85 [b] 0.10–0.36 [b]
Khamisse et al. [31]	Bacterial DNA	Polyvinyl chloride; SS	Swab	2–27 [b]
Lutz et al. [32]	*S. aureus*	SS	Electrostatic wipe Swab Roller sampler Contact-plate	0–33 [b] 0–46 [b] 0.01–15 [b] 0–0.09 [b]
Exum et al. [15]	*Escherichia coli*	Surfaces in households	Dry cloth	105
Madsen et al. [9]	*S. aureus*	SS	Dipslide eSwab Viscose swab Cotton swab	0.08–0.10 [b] 0–0.16 [b] 0–0.02 [b] 0–0.02 [b]
Lyons et al. [10]	*Clostridioides difficile* spores; *Klebsiella pneumoniae; S. aureus; Acinetobacter baumannii*	Nitrile glove	Wipe (by cellulose sponge) Contact-plate	4–53 [b] 9–28 [b]

Abbreviations: RODAC—replicate organism detection and counting; *S.*—*Staphylococcus*; *B.*—*Bacillus*; SS—stainless steel. [a] Mean ± standard deviation. [b] The range represents the measured values in different measuring conditions.

In this study, a novel approach for surface sampling is proposed. A polyvinyl alcohol (PVA) solution was prepared and applied to the target surface. The solution collects surface deposits as it solidifies into an elastic membrane that can be stripped off. The membrane releases the collected material when it is re-dissolved in water. The efficiencies of this new approach in sampling DNA materials and uniform-size microspheres from different surfaces were measured. In addition, the effect of the PVA solution on bacterial viability was tested. Although the proposed method, an advanced tape-stripping technique, is not suitable in all situations, it provides an option with high sampling efficiency and low cost that can complement existing methods for surface sampling.

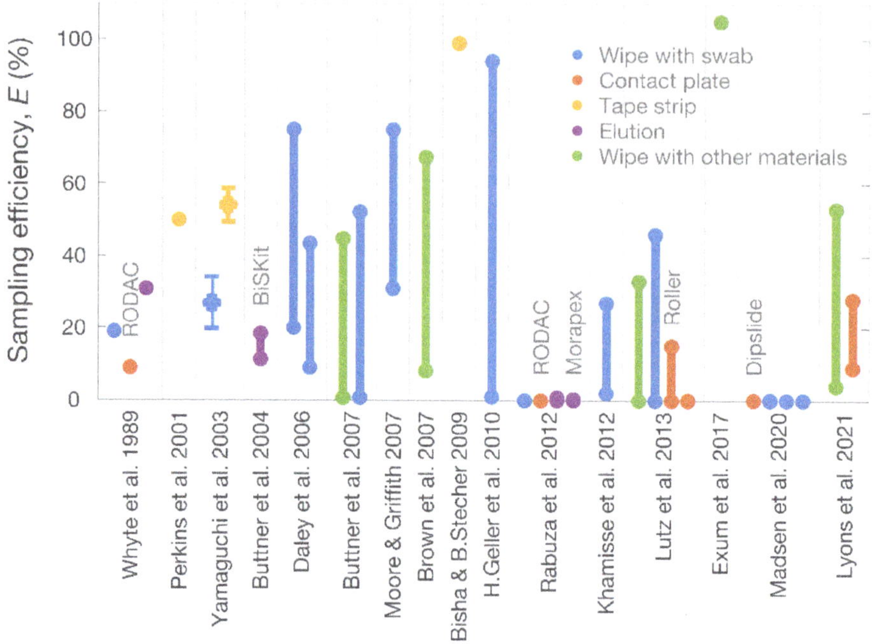

Figure 1. Efficiencies of surface sampling methods evaluated in the 16 studies listed in Table 1. According to the principles of sampling, the studies can be categorized into five groups: wipe with swab (blue), contact plate (red), tape strip (yellow), elution (purple) and wipe with other materials (i.e., sponge, clothing, and gauze; green). Each error bar represents the standard deviation of the data; each vertical line denotes the range of measured values in different measuring conditions. Specific kits or commercial materials used in the samplings are annotated alongside the datapoints.

2. Materials and Methods

2.1. Literature Review of Surface Sampling Methods

Sixteen articles on surface sampling, published in the 1989–2021 period, were retrieved by searching for the keywords "surface sampling", "sampling efficiency", "swab", "contact plate", "tape strip", "elute" and "wipe". Another criterion for selection was that the article should describe a systematic study on one or more sampling methods and clearly report the corresponding sampling efficiencies. It was difficult to retrieve all studies that met these criteria. Nevertheless, the chosen articles were not deliberately selected (e.g., articles reporting low sampling efficiencies) to emphasize the good qualities of the high sampling efficiency of the approach proposed in this study.

2.2. Materials

Four reference strains with double-stranded DNA as genetic materials, comprising a Gram-positive bacterium, *Staphylococcus aureus* ATCC 25923, a Gram-negative bacterium, *Escherichia coli* ATCC 25922, a fungus, *Saccharomyces cerevisiae* ATCC 9763, a bacteriophage, P22 ATCC 97540, and a type of fluorescence microsphere were applied to different surfaces. *Salmonella enterica serovar* Typhimurium ATCC 14028 was incubated to be the host for culturing phage P22. Lyophilized powders of *S. aureus*, *E. coli*, *S. cerevisiae* and *S. Typhimurium* were purchased from HKM, Guangdong, China. The bacteriophage P22 stock was purchased from Laval University, Canada. Fluoro-Max Green Particulate Marker Particles, fluorescence microspheres sold in powder form (catalogue number: 35–3; material: polystyrene–divinylbenzene; average diameter = 8 μm; coefficient of variation < 18%) were purchased from ThermoFisher, Hong Kong.

Two types of flat surfaces were prepared. Glass surfaces were obtained by directly using typical glass petri dishes with an average surface roughness (*Ra*) of ≈ 0.01 μm, measured using a surface roughness tester (TR200, Beijing Kaida Technology Development Co., Ltd., Beijing, China). Stainless steel surfaces were obtained by cutting stainless steel 301 plate (with a thickness of 1 mm and a measured *Ra* of ≈ 1 μm) into 10 cm × 10 cm coupons.

Film-forming solutions were prepared with four ingredients: (I) PVA powder (#1788; degree of polymerization = 1700; degree of alcoholysis = 88%), acquired from Shandong Usolf Chemical Technology Co., Ltd., linyi, China; (II) ethanol absolute for analysis (purity ≥ 99.5%; CAS: 64-17-5), acquired from Anaqua Global International Inc., Ltd., Hong Kong; (III) glycerol aqueous solution for analysis (86–88% wt%; CAS: 56-81-5), acquired from Acros Organics; and (IV) nuclease-free water for molecular biology (prepared from double distilled and deionized water; CAS: 7732-18-5), acquired from Sigma-Aldrich, Hong Kong.

The glycerol solution was also mixed with appropriated mass of nuclease-free water to obtain "50% glycerol" (i.e., glycerol aqueous solution; ≈50% wt%) for aliquot storage. Flocked swabs, FLOQSwabs® 552C, COPAN, were used to sample the surfaces using the swab-rinse method and used as an auxiliary tool in the proposed approach. Lysogeny broth (LB) and agar powders were purchased from Hangzhou BAISI Biotechnology Co., Ltd., Hangzhou, China. Powders of yeast extract peptone dextrose (YPD) medium were obtained from Shanghai RUICHU BioTech Co., Ltd., Shanghai, China.

2.3. *Surface Preparation*

2.3.1. Inoculum Preparation

Four reference strains were prepared following the protocol described in our previous study [33]. Each bacterial lyophilized powder was mixed with 1 mL of the reviving broth (i.e., a brain heart infusion broth) provided in the product package, and transferred into 5 mL LB for overnight culture at 37 °C. Frozen aliquots were obtained by mixing 5 mL of the culture broth with 5 mL of 50% glycerol and aliquoting 500 μL into each tube for storage at −80 °C. The *S. cerevisiae* lyophilized powder was revived using a similar process; however, the reviving broth was not used and the culture was kept in YPD medium at 25 °C. P22 stock was directly mixed with an equal volume of 50% glycerol, and 500 μL aliquots were transferred into each tube for storage at −80 °C.

Frozen aliquots of *S. aureus*, *E. coli* and *S. Typhimurium* were transferred into separate flasks containing 25 mL of LB, incubated overnight at 37 °C on an orbital shaker (130 rpm) and streaked for isolation on agar plates (LB with 1.5% agar). Then, a colony of each bacterium was sub-cultured in a flask with 25 mL LB and incubated for 15 h at 37 °C in an orbital shaker (130 rpm). Frozen aliquots of *S. cerevisiae* were transferred into a flask containing 25 mL YPD medium and incubated for 8 h at 25 °C in an orbital shaker (130 rpm). A 10 μL P22 stock sample was added to 25 mL of the prepared *S. Typhimurium* suspension; the newly mixed suspension was cultured for 24 h at 37 °C.

Suspensions of *S. aureus*, *E. coli* and *S. cerevisiae* were diluted to a concentration of 10^4–10^5 CFU/μL. Phage P22 was isolated from the host by centrifuging at 5000 rpm for 15 min, obtaining a supernatant in which the phages had a concentration of 10^7–10^8 PFU/μL, quantified by plaque assay. The prepared suspensions of each strain were aliquoted into separate tubes at 4 °C for less than 24 h and vortexed for 5 s before use.

2.3.2. Surface Seeding

The glass and stainless steel materials were autoclaved at 121 °C for 20 min and air-dried in a biosafety cabinet overnight before use. For each surface, a circular area with a diameter of 5 cm was defined as the target area for microbial inoculation and sampling. For each glass petri dish, the target area was marked on the outer side of the dish using a marker pen; for each stainless steel coupon, the sampling area was estimated with the naked eye. For each type of reference strain, a 200 μL microbial suspension was dispensed within the target area on a prepared surface (a stainless steel coupon or the inner side of a glass petri dish) and air-dried to visible dryness. For the microspheres, approximately 5 mg of microsphere powder was directly spread within the target area on a glass surface.

2.4. Surface Sampling

2.4.1. Preparation of PVA Solutions

A liquid solution for surface sampling was prepared as follows. First, 2.5 g of PVA powder was dissolved in 50 g of nuclease-free water by stirring at 95 °C for 60 min; 1 g of glycerol was added, and then the mixture was autoclaved at 121 °C for 20 min; after it cooled to room temperature, the sterilized mixture was further mixed evenly with 25 g of ethanol absolute [34,35]. Thus, a PVA solution was obtained, with a mass ratio of ethanol in the entire solution (m_e) of 32%, which is considered as the default mass ratio of ethanol in the following discussion unless m_e is otherwise specified. PVA solutions with a series of other values of m_e, i.e., 0, 3%, 6%, 10%, 18% and 25%, were prepared by adjusting the masses of water and ethanol during preparation.

2.4.2. Surface Sampling Using PVA Solution

The new surface sampling approach is illustrated step-by-step in Figure 2. First, 1 mL of the prepared PVA solution was dispensed onto the target area (Figure 2a,b), and was then spread evenly using a sterile dry swab to cover the surface deposits in the target area (Figure 2c). The swab was used to continually wipe the target area for approximately 30 s to facilitate the mixing of the surface deposits with the PVA solution (Figure 2c). The PVA solution was allowed to air-dry and solidify into an elastic membrane, which required 1–2 h under the laboratory conditions. During this drying process, the swab tip was kept static and in contact with the drying solution (Figure 2d). Then, the elastic membrane was stripped from the surface by pulling on the swab tip (Figure 2e,f). The PVA membrane along with the swab tip was cut off using sterile pliers and moved into a tube (Figure 2g). Finally, 1 mL of nuclease-free water was added to the tube, and the tube was vortexed for 60 s (Figure 2h). The collected materials were released and resuspended as the solidified membrane re-dissolved in water (Figure 2i).

The PVA solution applied on surfaces had an evaporating speed of approximately 10 mg/min under the laboratory measuring conditions. In the laboratory, air temperature was maintained at 22 ± 1 °C, relative humidity was maintained between 65% and 75%, and a slightly greater than normal air flow was generated from air-conditioning. The evaporating speed was slower for PVA solutions containing less ethanol.

2.4.3. Surface Sampling Using Swab

The common swab-rinse technique was also used to sample the surface. A sterile swab rinsed with 1 mol/L of phosphate-buffered saline solution was used to sample a surface by repeatedly wiping the target area over a 10 s period; then, the swab tip was cut off and moved into a tube. Then, 1 mL of nuclease-free water was added to the tube, and the tube was vortexed for 60 s.

2.5. Measurement of Sampling Efficiency

Sampling efficiency (E) is defined as the ratio of the quantity of target substances detected on a surface at a specific time of sampling (ΔC) to the quantity of target substances on the site just before their collection (C) (Equation (1)). For the new sampling approach, the collection was defined to occur at the moment at which the formed membrane was stripped off.

$$E = \frac{\Delta C}{C} \times 100\% \quad (1)$$

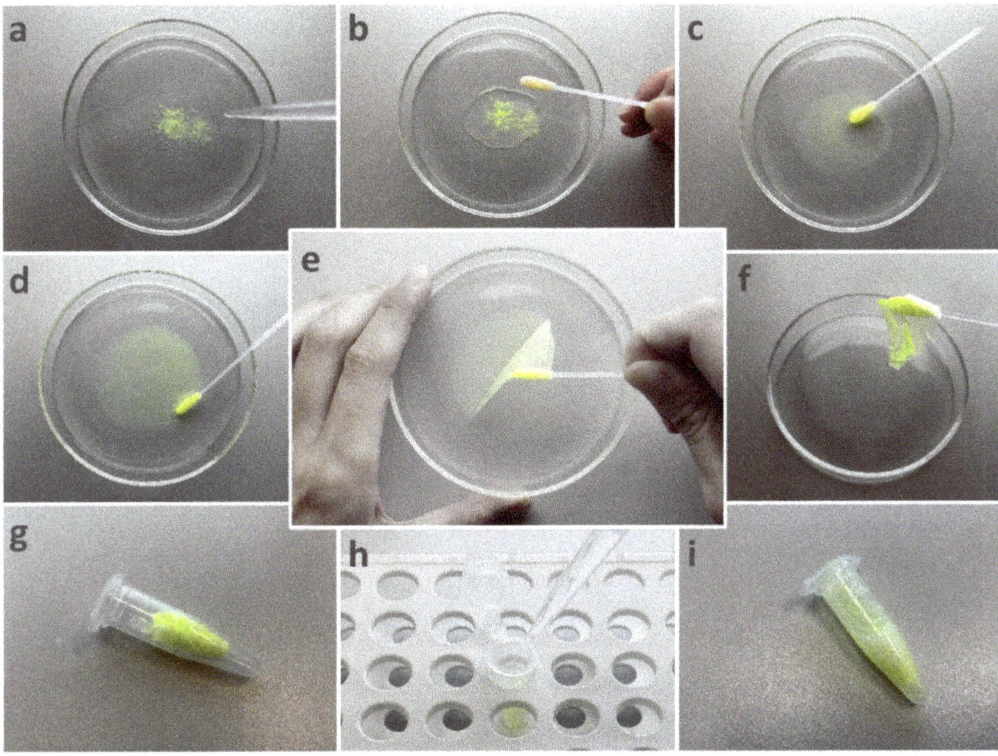

Figure 2. Demonstration of the new sampling approach. Fluorescence microspheres were sampled from a glass plate using the PVA solution. The sampling process is illustrated step-by-step as follows: (**a**) the PVA solution is dispensed into a contaminated glass surface; (**b**) a swab is used to wipe the glass surface; (**c**) the swab tip remains in contact with the solution, as the solution air-dries; (**d**) the solution solidifies into an elastic membrane; (**e**) the attached swab is lifted; (**f**) the membrane is stripped off; (**g**) the swab tip and the attached membrane are moved to a tube; (**h**) water is added into the tube; (**i**) the membrane is re-dissolved, releasing the sample. A short video demonstrating the sampling process is provided in Video S1.

In general, if the real E is very close to 100%, measurement errors can result in an estimated E exceeding 100% [15,22]. Thus, the E value is measured using the following sequential-sampling method [23,36]: a given surface is sampled N times, and the E is calculated using the quantities collected in the N samples, i.e., $\Delta C_1, \Delta C_2, \ldots, \Delta C_N$. According to Equation (1), after the first sampling, the quantity of sampling target substances remaining on the surface is

$$C - \Delta C_1 = C - EC = (1 - E)C$$

Assuming that (I) the sampling efficiency remains constant and (II) the sampling target substance does not multiply or degrade during the consecutive N samplings, it can be inferred that after the second sampling, the quantity of sampling target substances remaining on the surface is

$$C - \Delta C_1 - \Delta C_2 = C - EC - E(C - \Delta C_1) = (1 - E)^2 C$$

As such, after the n-th sampling, the quantity of sampling target substances remaining on the surface is

$$C - \sum_{i=1}^{n} \Delta C_i = (1 - E)^n C$$

Then, Equation (2) is derived from the above relationship.

$$\sum_{i=1}^{n} \Delta C_i = \left[1 - (1 - E)^n\right] C \tag{2}$$

According to Equation (2), the quantity of sampling target substances in the n-th sampling (ΔC_n, $1 \leq n \leq N$) can be derived as in Equation (3).

$$\Delta C_n = \sum_{i=1}^{n} \Delta C_i - \sum_{i=1}^{n-1} \Delta C_i = \left[E(1 - E)^{n-1}\right] C \tag{3}$$

Then, a relationship between ΔC_n and n is determined by combining Equations (2) and (3), as in Equation (4).

$$\frac{\Delta C_n}{\sum_{i=1}^{N} \Delta C_i} = \frac{E(1 - E)^{n-1}}{1 - (1 - E)^N} \tag{4}$$

where, E, as the only unknown parameter, can be evaluated by fitting a series of measured data (n, $\Delta C_n / \sum_{i=1}^{N} \Delta C_i$) with Equation (4). Thus, E can be calculated using the quantity of sampling target substances detected in the N samples (i.e., $\Delta C_1, \Delta C_2, \ldots, \Delta C_N$) based on this assumed relationship.

After the N sequential sampling procedures, one extra sampling procedure on the same site was performed using the swab-rinse method to validate the effectiveness of the new sampling approach—if the amount of sampling target substances on the final swab was not significantly more than the amount in each of the former samples, this would suggest that (I) almost all of the target substances had been removed from the surface and (II) the sampling target substances collected in the N samples did not significantly degrade.

2.6. Experimental Design

Fifteen sets of experiments were performed under different conditions, as listed in Table 2. The sets of experiments were categorized by two objectives: (I) to measure E under different sampling conditions; and (II) to measure *S. aureus* viability in the PVA solutions with different m_e. Before each set of tests, a negative-control test was performed by sampling a clean surface using the PVA solution to ensure that the prepared solution was free of the sampling target substances. Subsequent tests were valid only if the quantity of sampling target substances in the prepared solution in this validation test was below the detection limit. For each set of experiments, the results of 5 or 6 replicates were averaged (see Table 2).

Table 2. Experiment design in 15 sets of experiments.

No.	Aim	Sampling Target	Surface	Sampling Tools × Times	Quantification	Replicates
1	To measure E under different conditions	*S. aureus* DNA	Glass	PVA solution (m_e = 32%) × 5 + Swab × 1	qPCR	5
2		*E. coli* DNA	Glass	PVA solution (m_e = 32%) × 5 + Swab × 1	qPCR	5
3		Phage P22 DNA	Glass	PVA solution (m_e = 32%) × 5 + Swab × 1	qPCR	5
4		*S. cerevisiae* DNA	Glass	PVA solution (m_e = 32%) × 5 + Swab × 1	qPCR	5
5		*S. aureus* DNA	Stainless steel	PVA solution (m_e = 32%) × 5 + Swab × 1	qPCR	5
6		*S. aureus* DNA	Glass	PVA solution (m_e = 0) × 5 + Swab × 1	qPCR	5
7		*S. aureus* DNA	Glass	Swab × 6	qPCR	5
8		Microspheres	Glass	PVA solution (m_e = 32%) × 6	Microscopy	5
9	To measure *S. aureus* viability in the PVA solutions	*S. aureus*	Glass	PVA solution (ethanol-free) × 1	Plate count	6
10		*S. aureus*	Glass	PVA solution (m_e = 3%) × 1	Plate count	6
11		*S. aureus*	Glass	PVA solution (m_e = 6%) × 1	Plate count	6
12		*S. aureus*	Glass	PVA solution (m_e = 10%) × 1	Plate count	6
13		*S. aureus*	Glass	PVA solution (m_e = 18%) × 1	Plate count	6
14		*S. aureus*	Glass	PVA solution (m_e = 25%) × 1	Plate count	6
15		*S. aureus*	Glass	PVA solution (m_e = 32%) × 1	Plate count	6

2.6.1. Sampling Efficiencies under Different Conditions (Sets 1–8)

In sets 1–7 (Table 2), specific short DNA fragments were selected (listed in Table 3) for measuring the sampling efficiency in sampling the DNA of different microbes, as short

DNA fragments are relatively stable in the environment and do not replicate in the PVA solution (m_e = 32%) [37]. The DNA concentration in each sample was quantified by carrying out DNA extraction, followed by qPCR [33]. A 600 µL sample from the collecting tube was transferred to the PowerBead tube provided in the DNA isolation kit (DNeasy PowerSoil Kit, Qiagen) for DNA extraction. Before DNA extraction, 100 µL of phenol–chloroform–isoamyl alcohol solution (25:24:1, *v*/*v*) was added to the PowerBead tube to promote the removal of proteins from nucleic acids. The purified DNA was then quantified using a qPCR detection system (CFX Connect, Bio-Rad). iTaq Universal SYBR Green Supermix (CFX Connect, Bio-Rad) was used as the reagent for the 10 µL reaction volume. The DNA fragment and the cycling conditions in qPCR assay for each type of microorganisms are listed in Table 3. Calibration of the qPCR amplification efficiency for each pair of primers is provided in Supplementary File S2.

Table 3. The source, length and primers of the chosen DNA fragment and the cycling conditions in qPCR for each type of microorganism.

S. aureus and *E. coli* [38]	
DNA fragment	16s ribosomal DNA (rDNA)
Number of base pairs	178 bp
Primers (5′–3′)	(*341-F*) CCT ACG GGA GGC AGC AG (*518-R*) GTA TTA CCG CGG CTG CTG
Cycling conditions	95 °C × 1 min + [95 °C × 10 s + 60 °C × 30 s] × 40 cycles
Amplification efficiency	95.3%
S. cerevisiae [39,40]	
DNA fragment	Internal Transcribed Spacer-2 (between 5.8S rDNA and 28S rDNA)
Number of base pairs	≈350 bp
Primers (5′–3′)	(*ITS3-F*) GCA TCG ATG AAG AAC GCA GC (*ITS4-R*) TCC TCC GCT TAT TGA TAT GC
Cycling conditions	95 °C × 1 min + [95 °C × 30 s + 60 °C × 30 s + 72 °C × 1 min] × 40 cycles
Amplification efficiency	82.6%
P22 [41]	
DNA fragment	A segment from the 14,567th to the 14,617th base pair in the complete genome sequence
Number of base pairs	51 bp
Primers (5′–3′)	(*14567-F*) CTT AAC AAG CTC TGA CTG CTC ATC A (*14617-R*) CCA TCG CCT GTG ACT GGA T
Cycling conditions	95 °C × 1 min + [95 °C × 10 s + 60 °C × 30 s] × 40 cycles
Amplification efficiency	97.7%

After subjecting a sample to qPCR, a C_q value for the target DNA fragment was obtained. According to the working principle of qPCR, the DNA concentration in a sample (ΔC) is proportional to a function of C_q (Equation (5)).

$$\Delta C \propto (1+e)^{-C_q} \tag{5}$$

where *e* is the amplification efficiency in the qPCR process, and the value for each piece of DNA fragment is calibrated (see Supplementary File S2). Thus, the left-hand side of Equation (4) can be calculated from the raw qPCR data, as derived in Equation (6).

$$\begin{aligned} \frac{\Delta C_n}{\sum_{i=1}^{N} \Delta C_i} &= \frac{\Delta C_n}{\Delta C_1+\Delta C_2+\cdots+\Delta C_N} \\ &= \frac{(1+e)^{-C_{q(n)}}}{(1+e)^{-C_{q(1)}}+(1+e)^{-C_{q(2)}}+\cdots+(1+e)^{-C_{q(N)}}} \\ &= \frac{1}{\sum_{i=1}^{N}(1+e)^{[C_{q(n)}-C_{q(i)}]}} \end{aligned} \tag{6}$$

In set 8 (Table 2), the number of surface microspheres was quantified by directly counting the surface microspheres using a digital microscope (i.e., an optical microscope connected to a high-resolution camera and a monitor). A glass petri dish was fixed on the mechanical stage of the microscope. The microspheres were dispensed to the inner side of a glass petri dish and sampled multiple times on-site. A rectangular area of ≈100 µm × 56 µm on which microspheres were deposited was magnified and displayed on the monitor (Figure 3b,d). The number of microspheres removed from this area in one sampling process (ΔC_n) is the difference in the number of microspheres before and after the sampling, as shown in Figure 3b,d.

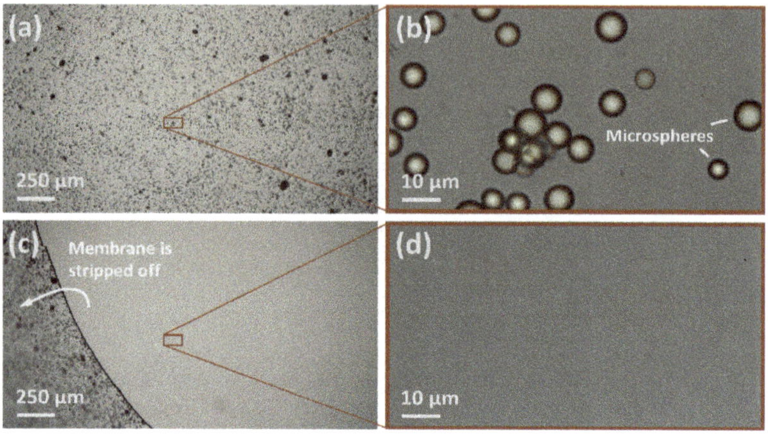

Figure 3. Digital microscope images demonstrating the process of sampling surface microspheres using the proposed approach. (**a**) Microspheres on a glass surface covered by the PVA solution, air-dried to a solid membrane; (**b**) a magnified area of 100 µm × 56 µm; (**c**) removal of the surface microspheres by stripping off the membrane; (**d**) the magnified area after stripping off the membrane.

2.6.2. *S. aureus* Viability in the PVA Solutions (Sets 9–15)

In the laboratory environment, the 1 mL of PVA solution on a 5-cm diameter circular area took 1–2 h to dry completely, during which the microbes engulfed in the solution could multiply or degrade. Thus, the number of microbial cells present at the moment the membrane is stripped off (C) is different from the number of microbial cells present when the PVA solution is applied to the surface (C_0), with a multiplication rate of $R = C/C_0$.

In the case of *S. aureus* on glass surfaces, it is possible to determine an optimal mass ratio of ethanol in the PVA solution to achieve $R = 1$. A 200 µL *S. aureus* suspension was inoculated on the inner side of a glass petri dish (i.e., a glass surface), air-dried and coated with the PVA solution with a specific m_e (sets 9–15 in Table 2). After the PVA solution solidified into a membrane, 20 mL of nuclease-free water was dispensed into the glass petri dish to re-dissolve the membrane on-site. Then, 1 mL of this solution was taken to perform plate counting to estimate the number of living *S. aureus* cells on the surface at the moment the PVA solution was completely solidified, thus obtaining the value of C.

Simultaneously, *S. aureus* (from the same tube of *S. aureus* used in the above trial) inoculated on a glass surface was air-dried and directly mixed on-site with 20 mL of nuclease-free water. Then, 1 mL of this solution was taken to perform plate counting to estimate the number of living cells of *S. aureus* on the surface before sampling, obtaining the value of C_0.

Then, the multiplication rate for the *S. aureus* was calculated as $R = C/C_0$, with $R > 1$ (or < 1) meaning that the surface *S. aureus* multiplied (or degraded) in the PVA solution during its solidification, and $R = 1$ means that the effects of multiplication and degradation counteracted each other.

3. Results

3.1. *Sampling Efficiency of the New Approach*

The efficiencies of the new approach in sampling DNA materials on surfaces were >96% under all the measuring conditions (sets 1–6 in Figure 4), with an average of 98.8%.

After five sequential sampling processes on a surface, minute DNA concentrations could be detected by the swab-rinse method, indicating that the new sampling approach effectively collected microbes from both types of surfaces, and the DNA in the PVA solutions exhibited no significant degradation.

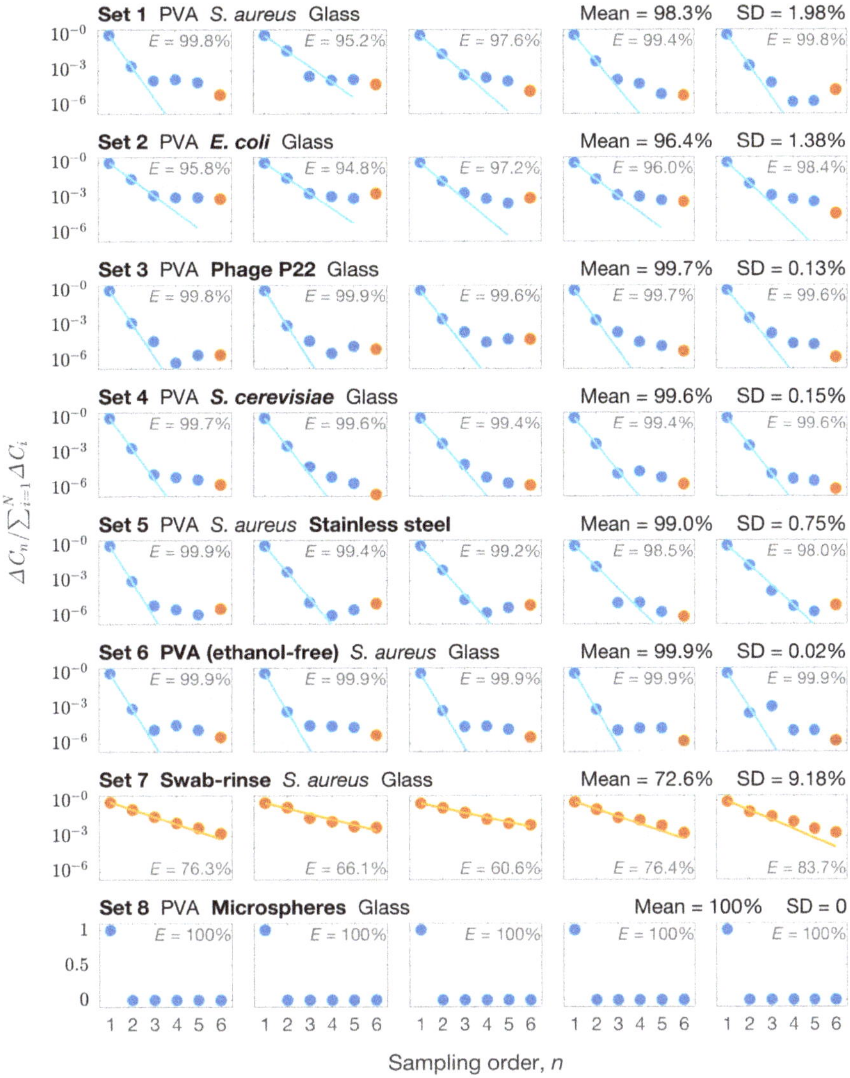

Figure 4. Evaluation of the sampling efficiencies (E) for the new approach (sets 1–6 and 8) and the swab-rinse method (set 7). E was evaluated through N-times sequential sampling, denoted by circles in a plot (blue: new approach; orange: swab-rinse method). Sampling efficiency (E) was evaluated by fitting the datapoints with Equation (4) and is shown as lines. In each set, the mean and SD of E were obtained from five E measurements.

As shown in Figure 4, most of the E values measured by the sequential sampling method were greater than 99%. Sampling *E. coli* on glass surfaces (set 2 in Figure 4) was slightly less efficient, but still exhibited an average E of 96.4%. When using ethanol-free PVA solution to sample *S. aureus* on glass surfaces (set 6 in Figure 4), the average E of 99.9% might be overestimated, because living *S. aureus* can multiply in an ethanol-free PVA solution. The relationship between the *S. aureus* multiplication and the overestimation of the *E* value is derived in Supplementary File S2. Nevertheless, the decreasing trend of the datapoints in each plot in sets 1–6 (Figure 4) indicates an extremely high efficiency of the new technique in detaching target substances from the surfaces.

In Figure 4, the fourth and fifth points in some plots deviated slightly upward from the fitting lines, and this deviation appears to be amplified on the logarithmic-scale *y*-axes. This could be due to errors in DNA quantification when there are excessively low concentrations of the target DNA fragments. For example, false positives may occur in qPCR tests due to the incorrect annealing of primers [42].

For comparison, the swab-rinse method was used to consecutively sample a surface (set 7 in Figure 4), obtaining an average *E* of 72.6% with a standard deviation (SD) of 9.18%. Both the sampling efficiency and the repeatability of the conventional swab-rinse method in the measuring condition were worse than those of the new approach.

Glass surfaces with uniform-size microspheres were sampled using the PVA solution (set 8 in Figure 4). A 100% sampling efficiency was obtained, as in each set of sequential sampling, all of the microspheres in the magnified area examined (as illustrated in Figure 3) were removed from the surface in the first sampling.

3.2. *S. aureus* Viability in the PVA Solutions

The effects of the PVA solutions with different m_e values on *S. aureus* viability are shown in Figure 5. A PVA solution of m_e = 32% inactivated 30–60% of *S. aureus*; in contrast, the ethanol-free PVA solution (m_e = 0) provided a favorable aquatic environment for bacteria, so the *S. aureus* multiplied to reach 3–5 times the number in the original inoculation. According to the data fitting (red line in Figure 5), the PVA solution with an $m_e \approx 17.6\%$ led to a constant concentration of *S. aureus* (i.e., *R* = 1). In that sample, the inhibition of microbial growth due to ethanol and its promotion by the aquatic environment counteracted each other.

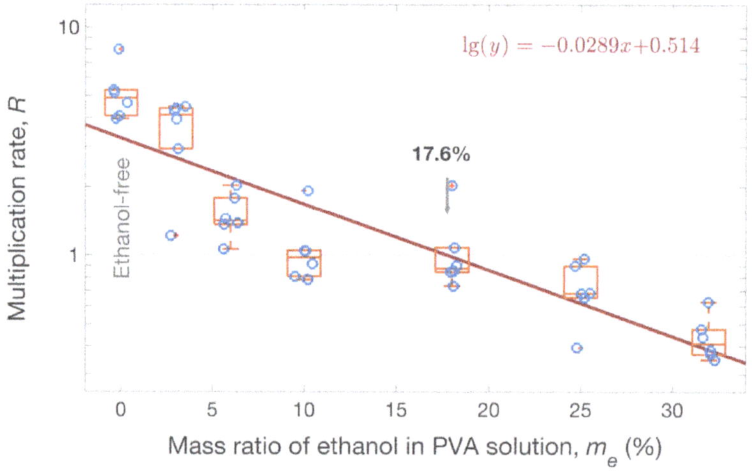

Figure 5. Effect of PVA solutions on *S. aureus* viability. *S. aureus* multiplication rate (*R*) in the PVA solutions with different m_e was evaluated. Each *R* value, the average of six replicates, is illustrated in the boxplot along with a jittering scatter plot. An m_e of 17.6% was obtained by data-fitting (red line) to balance the effects of multiplication and inhibition of *S. aureus*.

It is known that the multiplication of *S. aureus* can lead to an increase in DNA concentration when the inactivation of *S. aureus* is faster than the degradation of DNA. Thus, a PVA solution with an $m_e > 17.6\%$ can be used to sample DNA materials. In practice, to maintain a constant quantity of sampled organic or living substances during the sampling process, the optimum ethanol concentration in the PVA solution should be varied according to the target materials (e.g., different microbial species and genetic materials).

4. Discussion

4.1. Significance of This Study

In this study, we proposed a new approach to surface sampling using a film-forming agent, i.e., PVA solution. This solution, with low toxicity, high water solubility and chemical resistance, is suitable for collecting a variety of substances from surfaces [43]. The PVA solution can be considered a "soluble tape", as it can solidify to an elastic membrane when exposed to air, which can in turn be stripped off and re-dissolved in water. The cost of all of the ingredients in the prepared solution per milliliter is negligible in comparison with the cost of each swab used in this study.

The new approach offers a high and stable efficiency for collecting target substances from surfaces, which is independent of the surface area, the stripping movements, the existence of impurities and the selection of the auxiliary tools (e.g., sterile swabs and forceps). It minimizes the errors introduced by operational interference, such as wetting the swab head with appropriate buffer, squeezing the excessive buffer and swabbing surfaces in the swab-rinse technique. Compared with normal tape-stripping methods, applying the PVA solution to surfaces can efficiently collect multiple layers of deposits and completely release the collected materials in water. Despite some limitations, the new approach complements existing sampling methods and could be developed as a sampling tool for specific jobs such as collecting trace substances from ancient remains [44], cleaning instrument surfaces [45] and collecting DNA at crime scenes [46].

4.2. Sampling Mechanism

Existing sampling methods (see Table 1) can be categorized as "wipe", "contact", "adhere" or "elute" types. Microorganisms on a surface can form strong adhesions [47], which can be overcome by sampling tools, such as rinsed swabs and elution buffers. However, in conventional sampling methods, the "collection" of surface deposits occurs in a two-dimensional interface—the sampling tool and the surface are competing to adsorb the deposits. In contrast, in the new approach, the PVA solution engulfs surface deposits, as it is applied to a surface. In this scenario, most of the deposits, comprising both sampling target substances and other impurities, will already be detached from the surfaces before the solution solidifies to a membrane and is stripped from the surface. This fundamentally enhances sampling efficiency and eliminates the effect of surface/deposit material and impurities on the surface. In addition, the volume of the solution applied to a surface can be adjusted according to the sampling area, which avoids a decrease in sampling efficiency as the sampling area increases (e.g., which tends to occur when using swabs).

Water, ethanol and glycerol were used in the preparation of the PVA solution. These ingredients are chemically stable and exhibit no effect on most biological targets, including DNA molecules [48]. PVA has been widely used in the medical and pharmaceutical industries (e.g., in the preparation of eye drops and drugs) due to its low toxicity [43,49]. Glycerol has been widely used in cosmetics [50]. In the solutions proposed here, glycerol significantly increases the elasticity of the membrane.

4.3. Effect of the PVA Solutions on Microbial Viability

The sampling efficiencies exceeded 96% in all the measuring conditions (see Figure 4). Nevertheless, the 1 mL volume of PVA solution used during our laboratory sampling required 1–2 h for solidification. Surface microbes may multiply or degrade during this time.

Ethanol has biocidal efficacy against microbes. In contrast, short DNA fragments in surface sediments are relatively stable for hours [37,51] and can be precipitated in the presence of ethanol. Therefore, using the PVA solution (m_e = 32%) to sample surfaces will not change the DNA concentration of the samples.

If the sampling target is living bacteria or fungi, a high mass ratio of ethanol in the PVA solution has antimicrobial effects on the living cells, whereas a solution with a very low mass ratio of ethanol may provide an aquatic environment for cell multiplication [52]. In this study, the optimal mass ratio of ethanol in the PVA solution for sampling surface *S. aureus* was determined. This ratio balanced the effects of bacterial multiplication and bacterial degradation; however, this mass ratio is only valid under the specific experimental conditions. In future studies, shortening the sampling time may simultaneously reduce the influence of the PVA solution on microbial activity, and a solution with a higher mass ratio of ethanol would accordingly be required. Thus, the optimal mass ratio of ethanol in the PVA solution should be adjusted according to the specific sampling conditions.

4.4. Limitations and Improvements

The proposed PVA solution-based sampling approach has some limitations. First, it is not suitable for sampling from water-absorbing materials, such as clothing, paper, food, concrete, human skin and most porous surfaces. In the case of surfaces with rough texture, the solidified membrane may break when stripped off. Second, some time is required for the drying of the solution applied to the surfaces. Applying less solution per unit area can accelerate its drying, but the thinner membrane formed will break more easily when stripped off. Third, ethanol in the solution accelerates its solidification, but a high concentration of ethanol has an antimicrobial effect.

In practice, the PVA solution is slightly irritating to human skin. Repeated or long-term application to human skin may cause a mild skin rash. While impurities on surfaces would not affect the collection of target substances, they may inhibit the detection using either PCR or plate assay. In addition, the consumables in detection, such as PCR primers and nutrient broths, should be adjusted and optimized according to the sampling targets.

In this study, we focused on introducing the new sampling approach, whereas its effect on few surfaces and species was measured because the high sampling efficiency exhibited in the measurement was considered due to a different mechanism in the newly presented approach, which could scarcely be affected by the surface/species material. The optimum m_e in sampling living microbes was only measured for *S. aureus*, although it may also be affected by sampling conditions or as the sampling approach is optimized. In addition, we did not attempt to shorten the drying time because this was outside the scope of this study.

In future studies, we will develop techniques for accelerating the solidification of the PVA solution, such as mildly heating the surface, increasing the concentration of the solution or sampling under a negative pressure or an air flow. The ingredients in the solutions may be adjusted to fit specific requirements. For example, acetone, which has a low boiling point (BP = 56.5 °C), can be used as an additive to or a substitute for ethanol (BP = 78.4 °C) and water (BP = 100 °C) in PVA solutions to accelerate its solidification; propylene glycol can be used as a substitute for glycerol; and surfactants (e.g., fatty acids) may be added to efficiently sample lipophilic substances. Polyvinyl acetate can also be used to completely or partially replace the solute, as the former has shown a higher adhesiveness [53] and might improve sampling efficiency on surfaces with complex topographies.

In this study, a swab was used as an auxiliary tool to mix the sampling target with the PVA solution and to strip off the solidified membrane. Other tools can also be used as a substitute for swabs to detach the membrane. According to our experiment, no significant differences in sampling efficiency were found between using swabs and forceps to strip off the membrane.

5. Conclusions

This study proposed a new approach for sampling on non-absorbent surfaces. To sample a surface, a PVA solution was applied to the surface. The solution engulfed target substances as it solidified to a "soluble tape", and the "tape" releases the substances when it is re-dissolved in water. As a complement of existing surface sampling methods, the new approach exhibited high sampling efficiency and consistency. The measured efficiencies exceeded 96% in collecting microspheres and DNA of different microorganisms from glass and stainless steel surfaces.

Supplementary Materials: The following supporting information can be downloaded at: https://www.mdpi.com/article/10.3390/jerph191912571/s1, Video S1: Video demonstrating the new sampling approach; Supplementary File S2: The amplification efficiencies in qPCR and the deviation in measuring sampling efficiency. Ref [54] is cited in the Supplementary file.

Author Contributions: Conceptualization, P.Z.; Formal analysis, N.Z.; Funding acquisition, Y.L.; Investigation, P.-T.C.; Methodology, P.Z.; Project administration, Y.L.; Software, P.Z.; Supervision, Y.L.; Validation, P.Z., P.-T.C. and N.Z.; Visualization, P.Z.; Writing—review and editing, P.Z., P.-T.C. and Y.L. All authors have read and agreed to the published version of the manuscript.

Funding: This research was funded by a GENERAL RESEARCH FUND project of the Hong Kong Research Grants Council, grant number 17203420.

Institutional Review Board Statement: Not applicable.

Informed Consent Statement: Not applicable.

Data Availability Statement: Not applicable.

Conflicts of Interest: The authors declare no conflict of interest.

References

1. Ismail, R.; Aviat, F.; Michel, V.; Le Bayon, I.; Gay-Perret, P.; Kutnik, M.; Federighi, M. Methods for recovering microorganisms from solid surfaces used in the food industry: A review of the literature. *Int. J. Env. Res. Public Health* **2013**, *10*, 6169–6183. [CrossRef]
2. Wood, J.P.; Adrion, A.C. Review of decontamination techniques for the inactivation of *Bacillus anthracis* and other spore-forming bacteria associated with building or outdoor materials. *Environ. Sci. Technol.* **2019**, *53*, 4045–4062. [CrossRef]
3. Weir, M.H.; Shibata, T.; Masago, Y.; Cologgi, D.L.; Rose, J.B. Effect of surface sampling and recovery of viruses and non-spore-forming bacteria on a quantitative microbial risk assessment model for fomites. *Environ. Sci. Technol.* **2016**, *50*, 5945–5952. [CrossRef]
4. Silvestri, E.E.; Yund, C.; Taft, S.; Bowling, C.Y.; Chappie, D.; Garrahan, K.; Brady-Roberts, E.; Stone, H.; Nichols, T.L. Considerations for estimating microbial environmental data concentrations collected from a field setting. *J. Expo. Sci. Environ. Epidemiol.* **2017**, *27*, 141–151. [CrossRef]
5. Marques, M.; Domingo, J.L. Contamination of inert surfaces by SARS-CoV-2: Persistence, stability and infectivity. A review. *Environ. Res.* **2021**, *193*, 110559. [CrossRef] [PubMed]
6. Turnage, N.L.; Gibson, K.E. Sampling methods for recovery of human enteric viruses from environmental surfaces. *J. Virol. Methods* **2017**, *248*, 31–38. [CrossRef] [PubMed]
7. van Loosdrecht, M.C.; Lyklema, J.; Norde, W.; Zehnder, A.J. Bacterial adhesion: A physicochemical approach. *Microb. Ecol.* **1989**, *17*, 1–15. [CrossRef] [PubMed]
8. Griffith, C. Surface sampling and the detection of contamination. In *Handbook of Hygiene Control in the Food Industry*; Elsevier: Amsterdam, The Netherlands, 2016; pp. 673–696.
9. Madsen, A.M.; Phan, H.U.T.; Laursen, M.; White, J.K.; Uhrbrand, K. Evaluation of methods for sampling of *Staphylococcus aureus* and other *Staphylococcus* species from Indoor surfaces. *Ann. Work. Expo. Health* **2020**, *64*, 1020–1034. [CrossRef] [PubMed]
10. Lyons, A.K.; Rose, L.J.; Noble-Wang, J. Recovery efficiency of two glove-sampling methods. *Infect. Control Hosp. Epidemiol.* **2021**, *43*, 390–392. [CrossRef]
11. Rabuza, U.; Šostar-Turk, S.; Fijan, S. Efficiency of four sampling methods used to detect two common nosocomial pathogens on textiles. *Text. Res. J.* **2012**, *82*, 2099–2105. [CrossRef]
12. Buttner, M.P.; Cruz, P.; Stetzenbach, L.D.; Klima-Comba, A.K.; Stevens, V.L.; Emanuel, P.A. Evaluation of the Biological Sampling Kit (BiSKit) for large-area surface sampling. *Appl. Environ. Microbiol.* **2004**, *70*, 7040–7045. [CrossRef] [PubMed]
13. Conover, D.M.; Gibson, K.E. Comparison of two plain soap types for removal of bacteria and viruses from hands with specific focus on food service environments. *Food Control* **2016**, *69*, 141–146. [CrossRef]
14. Chowdhury, D.; Tahir, S.; Legge, M.; Hu, H.; Prvan, T.; Johani, K.; Whiteley, G.S.; Glasbey, T.O.; Deva, A.K.; Vickery, K. Transfer of dry surface biofilm in the healthcare environment: The role of healthcare workers' hands as vehicles. *J. Hosp. Infect.* **2018**, *100*, e85–e90. [CrossRef] [PubMed]
15. Exum, N.G.; Kosek, M.N.; Davis, M.F.; Schwab, K.J. Surface sampling collection and culture methods for *Escherichia coli* in household environments with high fecal contamination. *Int. J. Env. Res. Public Health* **2017**, *14*, 947. [CrossRef]

16. Zhao, P.; Chan, P.-T.; Gao, Y.; Lai, H.-W.; Zhang, T.; Li, Y. Physical factors that affect microbial transfer during surface touch. *Build. Environ.* **2019**, *158*, 28–38. [CrossRef]
17. Rose, L.; Jensen, B.; Peterson, A.; Banerjee, S.N.; Arduino, M.J. Swab materials and *Bacillus anthracis* spore recovery from nonporous surfaces. *Emerg. Infect. Dis.* **2004**, *10*, 1023–1029. [CrossRef] [PubMed]
18. Cutler, N.A.; Oliver, A.E.; Viles, H.A.; Whiteley, A.S. Non-destructive sampling of rock-dwelling microbial communities using sterile adhesive tape. *J. Microbiol. Methods* **2012**, *91*, 391–398. [CrossRef] [PubMed]
19. Ogai, K.; Nagase, S.; Mukai, K.; Iuchi, T.; Mori, Y.; Matsue, M.; Sugitani, K.; Sugama, J.; Okamoto, S. A comparison of techniques for collecting skin microbiome samples: Swabbing versus tape-stripping. *Front. Microbiol.* **2018**, *9*, 2362. [CrossRef]
20. Hedman, J.; Knutsson, R.; Ansell, R.; Radstrom, P.; Rasmusson, B. Pre-PCR processing in bioterrorism preparedness: Improved diagnostic capabilities for laboratory response networks. *Biosecur. Bioterror. Biodef. Strategy Pract. Sci.* **2013**, *11* (Suppl. 1), S87–S101. [CrossRef]
21. Van Horn, K.G.; Audette, C.D.; Tucker, K.A.; Sebeck, D. Comparison of 3 swab transport systems for direct release and recovery of aerobic and anaerobic bacteria. *Diagn. Microbiol. Infect. Dis.* **2008**, *62*, 471–473. [CrossRef]
22. Hong-Geller, E.; Valdez, Y.E.; Shou, Y.; Yoshida, T.M.; Marrone, B.L.; Dunbar, J.M. Evaluation of *Bacillus anthracis* and *Yersinia pestis* sample collection from nonporous surfaces by quantitative real-time PCR. *Lett. Appl. Microbiol.* **2010**, *50*, 431–437. [CrossRef] [PubMed]
23. Whyte, W.; Carson, W.; Hambraeus, A. Methods for calculating the efficiency of bacterial surface sampling techniques. *J. Hosp. Infect.* **1989**, *13*, 33–41. [CrossRef]
24. Perkins, M.A.; Osterhues, M.A.; Farage, M.A.; Robinson, M.K. A noninvasive method to assess skin irritation and compromised skin conditions using simple tape adsorption of molecular markers of inflammation. *Ski. Res. Technol.* **2001**, *7*, 227–237. [CrossRef] [PubMed]
25. Yamaguchi, N.; Ishidoshiro, A.; Yoshida, Y.; Saika, T.; Senda, S.; Nasu, M. Development of an adhesive sheet for direct counting of bacteria on solid surfaces. *J. Microbiol. Methods* **2003**, *53*, 405–410. [CrossRef]
26. Daley, P.; Castriciano, S.; Chernesky, M.; Smieja, M. Comparison of flocked and rayon swabs for collection of respiratory epithelial cells from uninfected volunteers and symptomatic patients. *J. Clin. Microbiol.* **2006**, *44*, 2265–2267. [CrossRef] [PubMed]
27. Buttner, M.P.; Cruz, P.; Stetzenbach, L.D.; Cronin, T. Evaluation of two surface sampling methods for detection of *Erwinia herbicola* on a variety of materials by culture and quantitative PCR. *Appl. Environ. Microbiol.* **2007**, *73*, 3505–3510. [CrossRef]
28. Moore, G.; Griffith, C. Problems associated with traditional hygiene swabbing: The need for in-house standardization. *J. Appl. Microbiol.* **2007**, *103*, 1090–1103. [CrossRef]
29. Brown, G.S.; Betty, R.G.; Brockmann, J.E.; Lucero, D.A.; Souza, C.A.; Walsh, K.S.; Boucher, R.M.; Tezak, M.; Wilson, M.C.; Rudolph, T. Evaluation of a wipe surface sample method for collection of *Bacillus* spores from nonporous surfaces. *Appl. Environ. Microbiol.* **2007**, *73*, 706–710. [CrossRef]
30. Bisha, B.; Brehm-Stecher, B.F. Simple adhesive-tape-based sampling of tomato surfaces combined with rapid fluorescence in situ hybridization for Salmonella detection. *Appl. Environ. Microbiol.* **2009**, *75*, 1450–1455. [CrossRef]
31. Khamisse, E.; Firmesse, O.; Christieans, S.; Chassaing, D.; Carpentier, B. Impact of cleaning and disinfection on the non-culturable and culturable bacterial loads of food-contact surfaces at a beef processing plant. *Int. J. Food Microbiol.* **2012**, *158*, 163–168. [CrossRef]
32. Lutz, J.K.; Crawford, J.; Hoet, A.E.; Wilkins, J.R., 3rd; Lee, J. Comparative performance of contact plates, electrostatic wipes, swabs and a novel sampling device for the detection of *Staphylococcus aureus* on environmental surfaces. *J. Appl. Microbiol.* **2013**, *115*, 171–178. [CrossRef]
33. Zhao, P.; Li, Y. Modeling and experimental validation of microbial transfer via surface touch. *Environ. Sci. Technol.* **2021**, *55*, 4148–4161. [CrossRef]
34. Ahmad, A.L.; Yusuf, N.M.; Ooi, B.S. Preparation and modification of poly (vinyl) alcohol membrane: Effect of crosslinking time towards its morphology. *Desalination* **2012**, *287*, 35–40. [CrossRef]
35. Parisi, E.I.; Bonelli, N.; Carretti, E.; Giorgi, R.; Ingo, G.M.; Baglioni, P. Film forming PVA-based cleaning systems for the removal of corrosion products from historical bronzes. *Pure Appl. Chem.* **2018**, *90*, 507–522. [CrossRef]
36. Zhao, P.; Li, Y. New sequential-touch method to determine bacterial contact transfer rate from finger to surface. *J. Appl. Microbiol.* **2019**, *127*, 605–615. [CrossRef] [PubMed]
37. Wei, N.; Nakajima, F.; Tobino, T. A microcosm study of surface sediment environmental DNA: Decay observation, abundance estimation, and fragment length comparison. *Environ. Sci. Technol.* **2018**, *52*, 12428–12435. [CrossRef]
38. Gurtner, C.; Heyrman, J.; Piñar, G.; Lubitz, W.; Swings, J.; Rölleke, S. Comparative analyses of the bacterial diversity on two different biodeteriorated wall paintings by DGGE and 16S rDNA sequence analysis. *Int. Biodeterior. Biodegrad.* **2000**, *46*, 229–239. [CrossRef]
39. Op De Beeck, M.; Lievens, B.; Busschaert, P.; Declerck, S.; Vangronsveld, J.; Colpaert, J.V. Comparison and validation of some ITS primer pairs useful for fungal metabarcoding studies. *PLoS ONE* **2014**, *9*, e97629.
40. Fujita, S.I.; Senda, Y.; Nakaguchi, S.; Hashimoto, T. Multiplex PCR using internal transcribed spacer 1 and 2 regions for rapid detection and identification of yeast strains. *J. Clin. Microbiol.* **2001**, *39*, 3617–3622. [CrossRef]
41. Masago, Y.; Shibata, T.; Rose, J.B. Bacteriophage P22 and *Staphylococcus aureus* attenuation on nonporous fomites as determined by plate assay and quantitative PCR. *Appl. Environ. Microbiol.* **2008**, *74*, 5838–5840. [CrossRef]

42. Mubarak, S.M.; Al-Koofee, D.A.; Radhi, O.A.; Ismael, J.M.; Al-Zubaidi, Z.F. An optimization and common troubleshooting solving in polymerase chain reaction technique. *Syst. Rev. Pharm.* **2020**, *11*, 427–436.
43. Baker, M.I.; Walsh, S.P.; Schwartz, Z.; Boyan, B.D. A review of polyvinyl alcohol and its uses in cartilage and orthopedic applications. *J. Biomed. Mater. Research. Part B Appl. Biomater.* **2012**, *100*, 1451–1457. [CrossRef]
44. Urzı, C.; De Leo, F. Sampling with adhesive tape strips: An easy and rapid method to monitor microbial colonization on monument surfaces. *J. Microbiol. Methods* **2001**, *44*, 1–11. [CrossRef]
45. Holmes, S. An overview of current surgical instrument and other medical device decontamination practices. In *Decontamination in Hospitals and Healthcare*; Elsevier: Amsterdam, The Netherlands, 2020; pp. 443–482.
46. Barash, M.; Reshef, A.; Brauner, P. The use of adhesive tape for recovery of DNA from crime scene items. *J. Forensic Sci.* **2010**, *55*, 1058–1064. [CrossRef] [PubMed]
47. Garrett, T.R.; Bhakoo, M.; Zhang, Z. Bacterial adhesion and biofilms on surfaces. *Prog. Nat. Sci.* **2008**, *18*, 1049–1056. [CrossRef]
48. Piškur, J.; Rupprecht, A. Aggregated DNA in ethanol solution. *FEBS Lett.* **1995**, *375*, 174–178. [CrossRef]
49. Adelnia, H.; Ensandoost, R.; Shebbrin Moonshi, S.; Gavgani, J.N.; Vasafi, E.I.; Ta, H.T. Freeze/thawed polyvinyl alcohol hydrogels: Present, past and future. *Eur. Polym. J.* **2022**, *164*, 110974. [CrossRef]
50. Morrison, L.R. Glycerol. In *Kirk-Othmer Encyclopedia of Chemical Technology*; John Wiley & Sons, Inc.: Hoboken, NJ, USA, 2000.
51. Katzenberger, R.H.; Rosel, A.; Vonberg, R.P. Bacterial survival on inanimate surfaces: A field study. *BMC Res. Notes* **2021**, *14*, 97. [CrossRef]
52. McDonnell, G.; Russell, A.D. Antiseptics and disinfectants: Activity, action, and resistance. *Clin. Microbiol. Rev.* **1999**, *12*, 147–179. [CrossRef]
53. Murray, G. *Handbook of Materials Selection for Engineering Applications*; CRC Press: Boca Raton, FL, USA, 1997.
54. Bustin, S.A.; Benes, V.; Garson, J.A.; Hellemans, J.; Huggett, J.; Kubista, M.; Mueller, R.; Nolan, T.; Pfaffl, M.W.; Shipley, G.L.; et al. The MIQE guidelines: Minimum information for publication of quantitative real-time PCR experiments. *Clin. Chem.* **2009**, *55*, 611–622. [CrossRef]

toxics

MDPI

Article

Cleaning the Flue in Wood-Burning Stoves Is a Key Factor in Reducing Household Air Pollution

Mizanur Rahman [1], Hans Petersen [2], Hammad Irshad [3], Congjian Liu [1], Jacob McDonald [3], Akshay Sood [4], Paula M. Meek [4,5] and Yohannes Tesfaigzi [1,*]

1 Pulmonary and Critical Care Medicine, Department of Medicine, Brigham and Women's Hospital, Harvard Medical School, Boston, MA 02115, USA
2 Chronic Obstructive Pulmonary Disease Program, Lovelace Biomedical Research Institute, Albuquerque, NM 87108, USA
3 Applied Sciences, Lovelace Biomedical Research Institute, Albuquerque, NM 87108, USA
4 Department of Internal Medicine, University of New Mexico School of Medicine and Miners Colfax Medical Center, Raton, NM 87740, USA
5 College of Nursing, University of Utah, Salt Lake City, UT 84102, USA
* Correspondence: ytesfaigzi@bwh.harvard.edu; Tel.: +1-617-525-3480

Citation: Rahman, M.; Petersen, H.; Irshad, H.; Liu, C.; McDonald, J.; Sood, A.; Meek, P.M.; Tesfaigzi, Y. Cleaning the Flue in Wood-Burning Stoves Is a Key Factor in Reducing Household Air Pollution. *Toxics* **2022**, *10*, 615. https://doi.org/10.3390/toxics10100615

Academic Editor: Roel Vermeulen

Received: 1 September 2022
Accepted: 10 October 2022
Published: 17 October 2022

Abstract: In experimental settings, replacing old wood stoves with new wood stoves results in reduced personal exposure to household air pollution. We tested this assumption by measuring $PM_{2.5}$ and levoglucosan concentrations inside homes and correlated them with wood stove age. Methods: Thirty homes in the Albuquerque, NM area were monitored over a seven-day period using in-home particulate monitors placed in a common living area during the winter months. Real-time aerosol monitoring was performed, and filter samples were analyzed gravimetrically to calculate $PM_{2.5}$ concentrations and chemically to determine concentrations of levoglucosan. A linear regression model with backward stepwise elimination was performed to determine the factors that would predict household air pollution measures. Results: In this sample, 73.3% of the households used wood as their primary source of heating, and 60% burned daily or almost daily. The mean burn time over the test week was 50 ± 38 h, and only one household burned wood 24/day (168 h). The average $PM_{2.5}$ concentration (standard deviation) for the 30 homes during the seven-day period was 34.6 μg/m^3 (41.3 μg/m^3), and median (min, max) values were 15.5 μg/m^3 (7.3 μg/m^3, 193 μg/m^3). Average $PM_{2.5}$ concentrations in 30 homes ranged from 0–15 μg/m^3 to >100 μg/m^3. Maximum $PM_{2.5}$ concentrations ranged from 100–200 μg/m^3 to >3000 μg/m^3. The levoglucosan levels showed a linear correlation with the total $PM_{2.5}$ collected by the filters (R^2 = 0.92). However, neither mean nor peak $PM_{2.5}$ nor levoglucosan levels were correlated with the age (10.85 ± 8.54 years) of the wood stove ($R^2 \leq 0.07$, $p > 0.23$). The final adjusted linear regression model showed that average $PM_{2.5}$ was associated with reports of cleaning the flue with a beta estimate of 35.56 (3.47–67.65) and R^2 = 0.16 (p = 0.04). Discussion: Cleaning the flue and not the wood stove age was associated with household air pollution indices. Education on wood stove maintenance and safe burning practices may be more important in reducing household air pollution than the purchase of new stoves.

Keywords: wood stove; wood smoke; household air pollution; stove age; particulate matter; stove maintenance

1. Introduction

Exposure to wood smoke (WS) is increasing not only in low-income countries but also in America, Canada, Europe, the Taigas in Canada, Alaska, and Siberia [1]. Climate change and other factors contribute to the rise in prevalence of wildfire events [2], causing populations in large areas being exposed to outdoor air pollution and toxic particulate matter (PM). In addition, household air pollution is a major concern [3,4] because approximately one-third of the world's population, comprising over 2.4 billion people, still uses solid fuels, such as wood, coal, or biomass (vegetable remains and dung), for cooking and heating their homes [5,6]. During the winter months, 30% of ambient fine particles ($PM_{2.5}$) mass stems from wood burning used for heating and cooking in some areas of the United States (US) [7]. More recently, exposure to household air pollution has increased during the COVID-19 pandemic as people were confined to their homes for longer periods [8].

According to the guidelines published by the World Health Organization (WHO) in September 2021, the levels of 24 h mean particulate matter (PM) concentration is 45 $\mu g/m^3$ for PM_{10} and 15 $\mu g/m^3$ for $PM_{2.5}$ [9]. However, indoor $PM_{2.5}$ concentrations in the US often exceed health-based air quality standards, especially in homes that use stoves for cooking or heating. In Montana, mean indoor $PM_{2.5}$ concentrations of 45 $\mu g/m^3$ and 51 $\mu g/m^3$ were reported in homes with wood stoves [10,11], exceeding the WHO 24 h standard for $PM_{2.5}$ of 15 $\mu g/m^3$. However, in addition to PM, levoglucosan content is a large fraction of the emitted fine particles from wood burning [12]. Levoglucosan is a product of pyrolysis generated during the combustion of wood and a major constituent of $PM_{2.5}$, and, therefore, it has been proposed as a tracer of WS [12].

An estimated 3.8 million premature deaths are caused each year from illnesses attributable to household air pollution due to heating or cooking with inefficient stoves using either solid fuel or kerosene [5]. Exposure to household pollutants is particularly high among women and young children, and this contributes to many deaths in children under 5 years of age. These deaths are primarily due to acute lower respiratory infections such as pneumonia and in adult women due to chronic obstructive pulmonary disease (COPD) [13,14]. However, even exposure to low levels of WS PM can cause oxidative stress in lung cells and elicit airway inflammation. Thus, exposure to WS can be a major cause of respiratory illness in all susceptible individuals [15] and has been implicated in respiratory illness, including COPD exacerbations [16], lower respiratory infections [17], and cough and wheezing [18,19]. Epidemiological studies suggest that WS exposure may cause an increased risk of infection and reduced lung function [20–22]. Several controlled exposure studies also demonstrated a clear association between exposure to WS particulates and respiratory dysfunction [23,24]. A 25 $\mu g/m^3$ increase in 6-d mean indoor $PM_{2.5}$ concentrations was associated with the presence of lower respiratory tract infection in children [25]. Exposure to WS affects not only the respiratory system but also increases the risk for cancer (lung, head and neck, cervical), interstitial lung disease, cardiovascular diseases, hypertension, low birth weight, and reduces growth rate of children [26,27].

Over 90% of the total PM from biomass burning is smaller than 2.5 μm, which can enter the alveolar region and pass into circulation [28,29]. Because wood stoves are major sources of household air pollution, several intervention strategies have been implemented to reduce indoor $PM_{2.5}$. Open fires generate high concentrations of WS particulate matter of 2000–30,000 $\mu g/m^3$, and the use of improved wood stoves reduces exposures to the 1000–5000 $\mu g/m^3$ range [11]. Modern technologies in biomass combustion, such as automatic small-scale wood pellet appliances and larger domestic heating plants, are commonly more efficient and emit much lower levels of PM [30]. Changeout programs of older wood stove models with new EPA-certified wood stoves in Libby, Montana, showed a >70% reduction in indoor $PM_{2.5}$ concentrations [10,31]. These studies determined a drop of basal mean values from 51.2 to 15 $\mu g/m^3$ $PM_{2.5}$ after the changeout of stoves [10]. The follow-up study showed that the mean $PM_{2.5}$ level of 45.0 $\mu g/m^3$ before changeout was reduced to 21.0 $\mu g/m^3$ over the following three winters. However, over subsequent winters, the average concentrations across homes varied, and several homes actually showed increased concentrations [11]. Despite the overall reduction in indoor pollutants, the study suggested that not only the introduction of a new wood stove but other factors contribute to the level of pollutants. Recently, the same research group proposed that a lack of cleaning of chimneys may also contribute to indoor pollution [32]. Therefore, simply replacing old woodstoves with newer improved ones may not have a long-term benefit of reducing exposure to WS $PM_{2.5}$.

Additional studies are necessary to determine whether newer wood stoves reduce household air pollution [33]. Confirmatory studies should not only have accurate measurements of PM emission from wood stoves but also consider including chemical markers that define the source of pollutants. Therefore, the main objective of the present study was to document the age of wood stoves, in addition to important factors, such as stove maintenance and frequency of cleaning the flue, as these variables were not included in earlier studies. Further, earlier investigations did not consider the possibility that other indoor pollutants, such as $PM_{2.5}$, generated from cigarette smoking can affect indoor air quality. The present study measured the wood pyrolysis product levoglucosan to confirm that the $PM_{2.5}$ stems from wood burning specifically. Therefore, by including both $PM_{2.5}$ and levoglucosan, the present study was designed to elucidate whether the age of wood stoves is a key determinant of reducing household air pollution in "real-life settings" or whether other factors, such as flue cleaning, affect indoor air pollutant levels.

2. Materials and Methods

2.1. Study Population

The sample for this investigation was drawn from those currently enrolled in the Lovelace Smokers Cohort (LSC), who reported yes to the question, "have you been exposed to WS over the last year". Further details of the LSC have been described previously [34–36]. Most LSC participants were recruited through newspaper or television advertisements, and ongoing recruitment continues using these methods in Albuquerque, an urban, diverse, high-altitude Southwestern community.

2.2. Study Design

This analysis was part of a larger study that tested the development of a self-report questionnaire concerning exposure to household WS. Exposure to WSwas self-reported in response to a question administered at study entry as part of the general health survey. The question "Have you ever been exposed to WS for 12 months or longer" provided no additional details about the type, intensity, and duration of WS exposure. The original research design was a cross-sectional sample, monitoring the particulate matter and levoglucosan concentrations over seven days in 30 homes. The homes (n = 30) were selected from those originally contacted by a study coordinator and those contacted by word of mouth. All individuals were enrolled during the heating season when wood stoves were active in the home to obtain real-world experience. This analysis reports on these 30 homes' internal environment exposures.

Demographic information such as age, gender, ethnicity, race, smoking history, and history of respiratory disease was obtained using the American Thoracic Society (ATS)-DLD-78 questionnaire, with some questions added about in-home exposures to smoking. Demographic information concerning wood stove maintenance and burning details were also asked, including what type of fuel was burned and cleaning of the flue and stove. The thirty homes were monitored over the seven-day period using in-home particulate monitors placed in a common living area during the winter months of 2013–2014. The Teflon filter was conditioned a minimum of 24 h prior to and after sample collection at 25 °C and 40% relative humidity. Filter samples were collected with a Personal Environmental Monitor that had a 2.5-micron size selective inlet (PEM, Model 200, PEM-10-2.5, MSP Corporation, Shoreview, MN, USA) (Figure S1). Real-time aerosol monitoring was performed by using a DustTrak Aerosol Monitor (Model 8520, TSI, Inc., Shoreview, MN, USA). The aerosol sample enters through a multi-nozzle, single-stage impactor to remove particles with an aerodynamic diameter (AD) larger than 2.5-µm in diameter. Particles smaller than the impactor cut-point were collected on a 37 mm diameter filter. Two different types of filters were used during this study. For the first 12 deployments, PTFE Zefluor filters, pore size, 3.0-µm (Part No. 60230, Pall Life Sciences, Ann Arbor, MI, USA) were used. However, it was observed that due to heavy particulate loading in some residences and a long sampling time (1 week), the pressure dropped across the filters and reduced the sampling flow rate or caused the failure of the pump. Due to this consequence, Zefluor filters were replaced with PallFlex Membrane Filters (Type: Fiberfilm T60A20, Pall Life Sciences, Ann Arbor, MI, USA) for the remaining household deployments. With this change, no drops in the sampling flow rate were observed despite the high loading of the filters.

2.3. In-Home Particulate Samples and Analysis

$PM_{2.5}$ Filter Sampling System: Dust Track was set up in the room with the wood stoves, and PM was measured. The measurement of PM in Dust Track identified WS as the major constituent of the filter particulate loading. The $PM_{2.5}$ filter sampling system consisted of two major parts: the PEM and the Leland Legacy Air Sampling Pump. The PEM (Model 200, PEM-10-2.5, MSP Corporation, Shoreview, MN, USA) is a lightweight personal sampler for collecting airborne particles in the PM_{10} and $PM_{2.5}$ size range with a cut-point of 2.5 μm AD, as used in this study (Figure S2). The flow rate through the sample was maintained at 10 ± 1 L/min to maintain a 2.5 μm cut-point. The Leland Legacy Air Sampling Pump (SKC, Inc., Eighty-Four, PA, USA) was used to provide the required sampling flow rate through the PEM. The PEM sampler was installed on a vertical rod about 4–5 ft. from the ground (typical breathing zone while sitting), with the inlet holes aligned parallel with the floor to avoid gravitational settling. Additional detail on the sampling system can be found in the Supplementary Materials.

Sampling Deployment Procedure: Pre-deployment began with the Personal Environmental Monitor (PEM) being prepared by cleaning the impaction surface, and a thin film of grease was applied to the impaction surface to minimize particle bounce and re-entrainment. The filter was weighed using a microbalance (Model MX5, Mettler Toledo, Columbus, OH, USA) and installed in the PEM. Then, the PEM was connected to the sampling pump, and the sampling flow rate was adjusted to achieve a 10 ± 1 L/min flow rate. The flow rate was measured by installing a TSI flow meter (Model: 4100, TSI, Inc., Shoreview, MN, USA) between the sampling pump and the PEM. Pickup and post-deployment began with both samplers (DustTrak and PEM) being stopped with the final flow rate, and the pressure drop across the PEM sampler was measured and recorded similarly to pre-deployment. The filter was weighed and stored in a −80 °C freezer until levoglucosan analysis could be performed. Additional detail on the sampling deployment procedure is described in the Supplementary Figure S2.

Chemical Analysis: Levoglucosan (1, 6-anhydro-b-D-glucopyranose), a cellulose combustion product, is a tracer species for WS, mainly because of its high resistance to degradation. Levoglucosan levels were determined in the collected PM to determine whether the PM was primarily from wood burning, as 23% of the study participants were also current cigarette smokers. The amount of levoglucosan analysis of the filter extraction solution was determined by GC-MS. The analysis was performed by Desert Research Institute, Reno, Nevada, USA, using the following analytical method. Details of the chemical analyses are described in Supplementary Materials. Due to storage and chemical analysis failure, levoglucosan values were obtained in only 23 of 30 homes. Data were reported in ng/m^3 units.

Statistical Analysis: Summary demographic statistics for continuous variables consisted of means and standard deviation (S.D.), and categorical variables are presented as proportions. We conducted Pearson's correlations to examine associations of the woodstove age with the average of $PM_{2.5}$, peak value of $PM_{2.5}$, or levoglucosan levels of in-home particulate measures (n = 30) and whether or not the stove was maintained regularly (yes/no) or the flue cleaned (yes/no). Based on our hypothesis of associations with the subject's proximity to the stove (based on three questions: (a) Over the past week, when wood was burning in the stove/fireplace, there was some smoke in the room? (b) When wood is burning, how close to the stove/fireplace are you? (c) Usually, when wood was burning in the stove/fireplace, I was in the same room?), stove age, number of cigarettes smoked per day, cleaning the flue, cleaning the stove, income, and education, we performed a linear regression model with backward stepwise elimination to determine the factors that would predict household air pollution measures. All analyses were conducted using SAS version 9.4 (Cary, NC, USA).

3. Results

3.1. Demographic Characteristics of Subjects and the Homes

The mean age of the study participants was almost 60 years, 43% were male, and about half were Hispanic (Table 1). Approximately a quarter (23%) were current smokers, and two-thirds (66.7%) of the individuals reported some chronic conditions (Table 1). The majority of the study participants reported living in a home with 5 rooms and 1–2 individuals residing in the home. The stove was reported to be serviced in the last year in 65.5% of the households, with the flue being cleaned in over half of the sample within the past year. Only 30% reported using a humidifier, and only one household used an air filter in their home.

Table 1. Characteristics of the study participants in 30 homes studied.

Variable	Mean/Number	Percentage/SD
Age (<46 years)	3	10.0%
Gender Male	13	43.3%
Hispanic	14	46.7%
Reported Chronic Illness	20	66.7%
Currently Smoking	7	23.3%
Education (>high school)	27	90.0%
Annual Household Income (expense less than income)	3	10.0%
Number of individuals in the home	2.29	0.87
Number of rooms in the home	4.87	0.51
Stove maintained in last year (%Yes)	19	65.5%
Flue cleaned in the last year (%Yes)	16	53.3%
Used a humidifier (%Yes)	9	30%
Used an air filter (%Yes)	1	3.3%

Among the 30 households in this study, 73.3% used wood as their primary source of heating, and 60% burned wood daily or almost daily. The mean burn time over the test week was 50 ± 38 h, with only one home burning 24 h a day (Figure 1A). On average, the measured $PM_{2.5}$ ranged between 0–15 μg/m^3 and >100 μg/m^3 (Figure 1B). Over the 6 days, the average $PM_{2.5}$ was >100 μg/m^3 in only 1 home, 15 homes (50% of the sample) showed 0–15, 6 homes (20%) measured 15–35 μg/m^3, and in 8 homes, $PM_{2.5}$ ranged between 35 and 100 μg/m^3 (Figure 1B). Interestingly, the peak measurements ranged from 100–200 μg/m^3 to >3000 μg/m^3 in four homes (Figure 2).

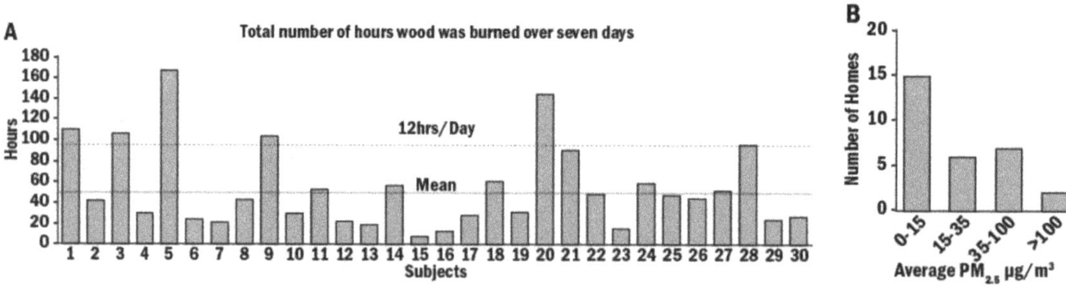

Figure 1. Total number of hours of wood burning in 30 homes. Total wood-burning time over the 7 days was calculated, and the mean wood-burning time over the test week was 50 ± 38 h with only one home burring 24 h a day (**A**). Average $PM_{2.5}$ frequency in different homes. Particulate mass was collected on filter, and total $PM_{2.5}$ emission was recorded over the test week from the 30 homes. Average $PM_{2.5}$ range was >100 μg/m^3 in 1 home, 0–15 μg/m3 in 15 homes, 15–35 μg/m^3 in 6 homes, and 35–100 μg/m^3 in 8 homes (**B**).

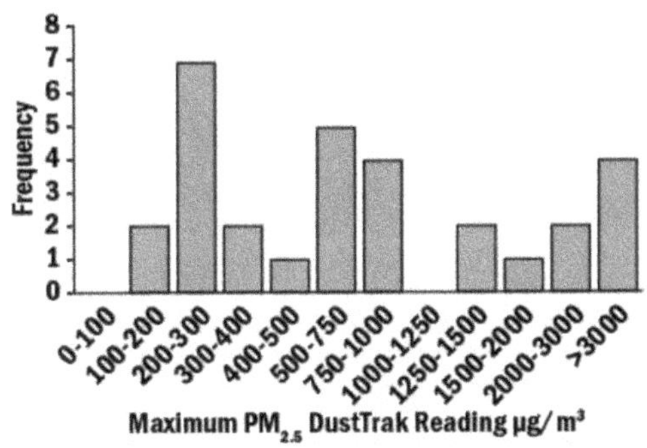

Figure 2. $PM_{2.5}$ concentration in DustTrack from 30 homes. A DustTrak Aerosol Monitor was used to monitor real time aerosol. Mean $PM_{2.5}$ ranged from 100–200 µg/m^3 to >3000 µg/m^3 in the 30 homes.

3.2. WS Was the Cause of the Measured PM

There was a linear relationship ($R^2 = 0.83$) between $PM_{2.5}$ in the filter measurement concentration and DustTrak average concentration reading. Levoglucosan collected by filters showed a positive linear relationship with ($R^2 = 0.92$, $p < 0.01$) the particulate mass collected on the filter (Figure 3).

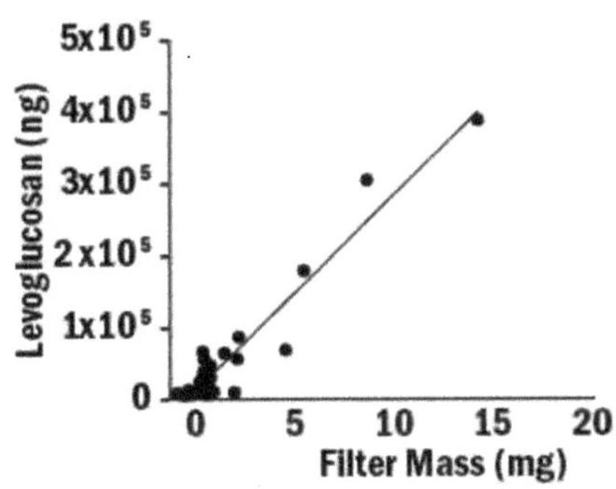

Figure 3. Linear association between levoglucosan and filter mass. Levoglucosan was extracted from the filter, and the amount of levoglucosan was determined by GC-MS analysis from the filter extraction solution. The level of levoglucosan was associated with filter mass, $R^2 = 0.92$, $p =< 0.01$.

3.3. Age of Wood Stoves Was Not Associated with the PM and Levoglucosan

The result was shown as Figure 4.

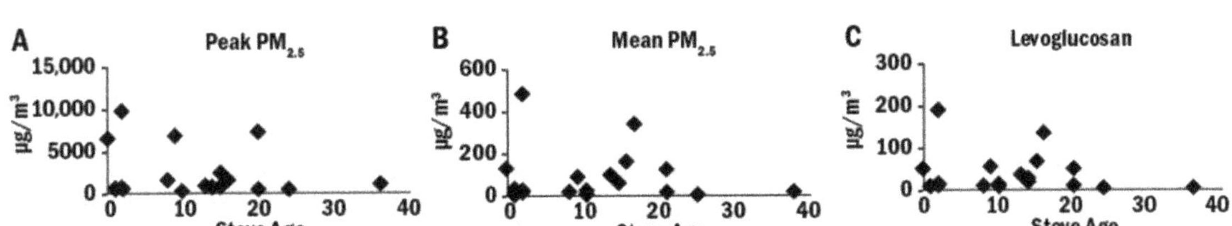

Figure 4. Age of the wood stoves was not correlated with $PM_{2.5}$ or levoglucosan. Neither Peak (**A**) nor mean (**B**) of $PM_{2.5}$ emission was associated with the age of the wood stoves. Similarly, the level of levoglucosan was not associated with the age of the stoves (**C**).

3.4. The Age of Wood Stoves Did Not Play a Role in $PM_{2.5}$ or Levoglucosan Levels

The age of wood stoves was not correlated with either $PM_{2.5}$ or levoglucosan (Figure 4). Over the seven-day measurement period, peak $PM_{2.5}$ emission ($R^2 = 0.01$, $p = 0.555$) (Figure 4A) or mean $PM_{2.5}$ values ($R^2 = 0.00$, $p = 0.893$) (Figure 4B) were not associated with the age of the stoves. Similarly, the mean concentration of the levoglucosan ($R^2 = 0.07$, $p = 0.230$) was not associated with the age of the wood stoves (Figure 4C).

3.5. Cleaning of Wood Stoves Was Associated with $PM_{2.5}$ Emission

Using a univariate linear regression model, we identified that only the variable "flue cleaning" was associated with PM emission (Table 2). When using the full model that includes all variables in multivariate linear regression, we did not find an association of any of the co-variates with PM (Table 3). After backward selection, only stove flue was left in the model, so the final model is the same as the result from the univariable table. The linear regression model showed that the average $PM_{2.5}$ was associated with reports of cleaning the flue with a beta estimate for stove flume of 35.56 (3.47–67.65) and $R^2 = 0.16$ ($p = 0.03$).

Table 2. Univariate linear regression.

Variable	Beta-Est.	95% CI	*p*-Value
Income	8.416	−8.7–25.6	0.35
Education	0.8414	−27.6–29.2	0.95
Stove_Clean	9.3082	−27.5–446.1	0.62
Stove_flue	35.5638	3.1–68.0	0.04
Cig.per.Day	−0.6049	−3.9–2.6	0.72
Proximity	0.6271	−15.5–16.8	0.94
Stove Age	−0.2954	−2.4–1.8	0.78

Table 3. Multivariate linear regression.

	Est.	95% CI	*p*-Value
Intercept	11.22	−196.1–173.7	0.91
Income	7.02	−17.1–31.2	0.58
Education	4.15	−35.9–44.3	0.84
Stove_Clean	2.7	−39.8–45.2	0.9
Stove_flue	32.84	−9.3–75.0	0.15
Cig.per.Day	0.44	−3.6–4.5	0.83
Proximity	−0.47	−17.7–16.8-	0.577
Stove Age	0.02	−2.2–2.3	0.987

4. Discussion

The current study identified that the age of wood stoves is not correlated with PM emission, but it is rather the maintenance and cleaning of the flue that is correlated with household air pollution due to wood stoves. Furthermore, the $PM_{2.5}$ emission level was positively associated with the level of levoglucosan. In the current study, the level of indoor $PM_{2.5}$ was significantly higher than the WHO's recommended levels, suggesting that people who use wood stoves are exposed to high levels of WS, even in high-income countries. Our findings are similar to the levels of indoor $PM_{2.5}$ in homes with wood stoves that were measured in earlier studies [10,11,31,37,38]. Over 24 h, $PM_{2.5}$ concentrations ranged from 24 to 60 $\mu g/m^3$, whereas sampling over a 2 h cooking period exceeded 1000 $\mu g/m^3$ [38]. The peak levels of PM emission are expected to be high during the cooking or heating periods. In another study, the measured $PM_{2.5}$ levels were higher (1910 to 6030 $\mu g/m^3$) than in our study when the measurements were taken during the cooking period [39]. However, these high concentrations in PM emissions may be the result of the types of stoves used. Mean $PM_{2.5}$ levels of 5310 $\mu g/m^3$ and maximum $PM_{2.5}$ levels of 13,800 $\mu g/m^3$ were measured in the homes that used open fire and in some homes with Plancha or Lorena wood stoves. However, all the homes in our study used stoves with chimneys, and the

$PM_{2.5}$ was measured over the 7-day period. The peak value of $PM_{2.5}$ in our study exceeded 3000 $\mu g/m^3$ during the cooking and heating period in four homes (>10% of homes studied). These levels of $PM_{2.5}$ concentrations are usually thought to be present only in low-income countries [40].

Changing older wood stoves with newer EPA-certified wood stoves was widely encouraged to help reduce household air pollution [10,31,37,41]. The wood stove changeout program in Libby reduced indoor $PM_{2.5}$ concentrations by >70%, but the first study did not provide any information on the sustainability of the effect. The follow-up study conducted multiple samplings over subsequent winters and found large variability in the average $PM_{2.5}$ levels across the homes: several homes had higher concentrations than the concentration pre-changeout [11]. In the original study [10], only one measurement was taken following the changeout. Although multiple samplings from 21 homes in the follow-up study suggested, on average, a 53% reduction in $PM_{2.5}$ in 16 homes, 7 homes demonstrated no reduction post-changeout. Interestingly, samplings from seven of the homes exhibited even higher levels of $PM_{2.5}$ than pre-changeout [10]. Another study revealed [41] that the wood stove changeout program reduced phenolics and PAH compounds on average by 64%, while the mass of $PM_{2.5}$ was reduced by only 20%. However, in that study, EPA-certified stoves were installed with efficient burning woods. The efficient combustion of wood in modern, certified stoves may also contribute to lowering PM emissions. Wood with higher moisture content produces a higher quantity of PM compared to dry wood [42,43]. High moisture content causes incomplete wood combustion resulting in high emission of PM. Therefore, indoor WS levels are likely determined by the type of wood used, moisture content, the combustion appliances, as well as the combustion phase, which affects PM generation [44–46]. These studies determined that types of wood and airflow play a role in the amount and different types of PM, including the generation of pyrolysis products. Studies with more controlled burning conditions are needed to determine the contribution of these factors to PM generation by wood burning.

Next, we measured levoglucosan, which forms from the pyrolysis of starch and cellulose of wood. Hydrolysis or biodegradation, even the combustion of fossil fuel, does not produce levoglucosan. Cellulose combustion generates levoglucosan, which is considered a tracer for biomass burning [47–50]. Depending on the air supply, the relative range of levoglucosan to total particle emission from wood burning was reported to be 3–17%. Although, as a fine particle, levoglucosan constitutes a large fraction of total emitted particles from wood burning. Studies with levoglucosan measurements as a marker of wood smoke from wood stoves are sparse. The measurements of levoglucosan in our study were carried out over 7 days, and levoglucosan is stable over 10 days [51]. The levoglucosan level detected in the homes of the current study was similar to the levels previously reported (mean 300 ng/m^3) in a different study [52]. The $PM_{2.5}$ level was associated with levoglucosan, but no association was found between levoglucosan and the age of the stoves, which is in line with the finding on the association between $PM_{2.5}$ and the age of the stoves. However, reports of cleaning the flue were able to explain 16% of the variance in the $PM_{2.5}$ level [32]. In addition to best burning practices, operating and maintenance, including flue cleaning, may have a significant influence on the emission of PM. A previous study [32] showed higher $PM_{2.5}$ in the homes that reported cleaning their chimney more than 12 months before the sampling period compared to those that cleaned chimneys within 6 months of the sampling period. Regular cleaning of the flue is crucial to reduce indoor PM emissions, as the ashes can accumulate and clog the passage of the WS outdoors. Further, air intake vents can improve wood-burning efficiency [47]. Cleaning of the flue may increase wood-burning efficiency by air intakes and generate less PM, thus reducing PM emissions. To reduce indoor household pollution, education on wood stove maintenance and safe burning practices are more important than replacing old stoves with new stoves.

The limitations of this study include the relatively small number of homes studied, and extensive characterization of the WS in all 30 homes could provide detailed information on exposures experienced by all study participants. Future studies should inquire about the experience and skill of the individual who operates and maintains the wood stove and include homes without wood stoves for comparison. Furthermore, determining the effect of different types of wood in smoke generation and identifying the wood type that may be less likely to clog the flue would be the most efficient path to reduce indoor PM emissions. Although all the homes were in the same area, the effect of ambient air pollution on the indoor air should be considered. Further, 23% of the household participants currently smoked, and although smoking did not seem to be a factor associated with PM, it is possible that it could have affected the measurement. The strengths of our studies are the positive association of $PM_{2.5}$ with levoglucosan levels, confirming that the PM indeed represents WS.

5. Comparative Section

In line with earlier studies (45–58), our study further determined that levoglucosan is a useful wood-burning marker to study indoor air pollution in connection with WS. Furthermore, the homes included in this study were selected at random, but all the homes were with similar architecture and ventilation, and the type of wood stoves are representative of the stoves used in the area. Indoor air pollution varies in different geographical regions or areas. We are not aware of an earlier study that was conducted in New Mexico that investigated the age of wood stoves and household air pollution.

6. Conclusions and Future Perspectives

Although changeout with EPA-certified woodstoves was suggested as a strategy to minimize exposure to $PM_{2.5}$, our study suggests that the age of wood stoves is not associated with the level of indoor PM but rather with flue cleaning. In addition, better characterization of $PM_{2.5}$ is recommended to ensure the origin is warranted. Additionally, further studies with bigger sample sizes are needed to elucidate whether moisture content in the wood and cleaning the flue improve indoor air quality. Installation of proper stoves, operation following best burning practices with proper wood selection with maintenance, and regularly cleaning flue may reduce the level of PM emissions from wood stoves. Whether smoke from different types of wood or moisture content may result in the rapid clogging of the flue should also be investigated. Finally, whether flue cleaning will reduce household air pollution to an extent that translates into health benefits will need further investigation.

Supplementary Materials: The following supporting information can be downloaded at: https://www.mdpi.com/article/10.3390/toxics10100615/s1, Figure S1: PEM-10-2.5 in disassembled form. Multiple jets of single stage impactor can be seen on the red inlet part. The impaction surface on the left shows collected particles greater than 2.5-μm. The center part shows the filter (mostly black with a white boundary).; Figure S2: PEM installed on a rod attached to an electric cooler enclosure

Author Contributions: Data acquisition was by P.M.M., A.S., H.I., J.M. and Y.T.; data analysis by P.M.M., H.P., A.S. and Y.T.; manuscript writing by M.R., P.M.M., A.S., J.M., H.I. and Y.T.; detailed review of final version by M.R., H.P., H.I., A.S., P.M.M. and Y.T.; formal analysis by C.L. All authors have read and agreed to the published version of the manuscript.

Funding: This work was funded by NIH R15 HL115544 and RO1 HL140839 and RO1 HL068111.

Institutional Review Board Statement: All procedures performed for the study were in accordance with the approval of the Ethical Review Authority.

Informed Consent Statement: All the subjects gave informed consent for the study.

Data Availability Statement: Not applicable.

Acknowledgments: We thank Elise Bennett for preparing the figures.

Conflicts of Interest: The authors do not have conflict to report.

References

1. Kang, J.-H.; Hwang, H.; Lee, S.-J.; Choi, S.-D.; Kim, J.-S.; Hong, S.; Hur, S.D.; Baek, J.-H. Record of North American boreal forest fires in northwest Greenland snow. *Chemosphere* **2021**, *276*, 130187. [CrossRef] [PubMed]
2. Black, C.; Tesfaigzi, Y.; Bassein, J.A.; Miller, L.A. Wildfire smoke exposure and human health: Significant gaps in research for a growing public health issue. *Environ. Toxicol. Pharmacol.* **2017**, *55*, 186–195. [CrossRef] [PubMed]
3. Leung, D.Y.C. Outdoor-indoor air pollution in urban environment: Challenges and opportunity. *Front. Environ. Sci.* **2015**, *2*, 69. [CrossRef]
4. Copiello, S.; Grillenzoni, C. The spread of 2019-nCoV in China was primarily driven by population density. Comment on "Association between short-term exposure to air pollution and COVID-19 infection: Evidence from China" by Zhu et al. *Sci. Total Environ.* **2020**, *744*, 141028. [CrossRef]
5. World Health Organization. Household Air Pollution and Health. Available online: https://www.who.int/news-room/fact-sheets/detail/household-air-pollution-and-health (accessed on 27 July 2022).
6. Perrone, M.G.; Larsen, B.R.; Ferrero, L.; Sangiorgi, G.; De Gennaro, G.; Udisti, R.; Zangrando, R.; Gambaro, A.; Bolzacchini, E. Sources of high PM2.5 concentrations in Milan, Northern Italy: Molecular marker data and CMB modelling. *Sci. Total Environ.* **2012**, *414*, 343–355. [CrossRef]
7. Fullerton, D.G.; Bruce, N.; Gordon, S. Indoor air pollution from biomass fuel smoke is a major health concern in the developing world. *Trans. R. Soc. Trop. Med. Hyg.* **2008**, *102*, 843–851. [CrossRef]
8. Zhang, Y.; Zhao, B.; Jiang, Y.; Xing, J.; Sahu, S.K.; Zheng, H.; Ding, D.; Cao, S.; Han, L.; Yan, C.; et al. Non-negligible contributions to human health from increased household air pollution exposure during the COVID-19 lockdown in China. *Environ. Int.* **2022**, *158*, 106918. [CrossRef]
9. Kan, H. World Health Organization air quality guidelines 2021: Implication for air pollution control and climate goal in China. *Chin. Med. J.* **2022**, *135*, 513–515. [CrossRef]
10. Ward, T.; Palmer, C.; Bergauff, M.; Hooper, K.; Noonan, C. Results of a residential indoor $PM_{2.5}$ sampling program before and after a woodstove changeout. *Indoor Air* **2008**, *18*, 408–415. [CrossRef]
11. Noonan, C.W.; Navidi, W.; Sheppard, L.; Palmer, C.P.; Bergauff, M.; Hooper, K.; Ward, T.J. Residential indoor $PM_{2.5}$ in wood stove homes: Follow-up of the Libby changeout program. *Indoor Air* **2012**, *22*, 492–500. [CrossRef]
12. Hedberg, E.; Johansson, C.; Johansson, L.; Swietlicki, E.; Brorström-Lundén, E. Is Levoglucosan a Suitable Quantitative Tracer for Wood Burning? Comparison with Receptor Modeling on Trace Elements in Lycksele, Sweden. *J. Air Waste Manag. Assoc.* **2006**, *56*, 1669–1678. [CrossRef]
13. Balmes, J.R. When Smoke Gets in Your Lungs. *Proc. Am. Thorac. Soc.* **2010**, *7*, 98–101. [CrossRef]
14. Gordon, S.B.; Bruce, N.G.; Grigg, J.; Hibberd, P.L.; Kurmi, O.P.; Lam, K.B.H.; Mortimer, K.; Asante, K.P.; Balakrishnan, K.; Balmes, J.; et al. Respiratory risks from household air pollution in low and middle income countries. *Lancet Respir. Med.* **2014**, *2*, 823–860. [CrossRef]
15. Camp, P.G.; Ramirez-Venegas, A.; Sansores, R.H.; Alva, L.F.; McDougall, J.E.; Sin, D.D.; Paré, P.D.; Müller, N.L.; Silva, C.I.S.; Rojas, C.E.; et al. COPD phenotypes in biomass smoke- versus tobacco smoke-exposed Mexican women. *Eur. Respir. J.* **2014**, *43*, 725–734. [CrossRef]
16. Viniol, C.; Vogelmeier, C.F. Exacerbations of COPD. *Eur. Respir. Rev.* **2018**, *27*, 170103. [CrossRef]
17. Ezzati, M.; Kammen, D.M. Quantifying the effects of exposure to indoor air pollution from biomass combustion on acute respiratory infections in developing countries. *Environ. Health Perspect.* **2001**, *109*, 481–488. [CrossRef]
18. Honicky, R.E.; Osborne, J.S., 3rd; Akpom, C.A. Symptoms of Respiratory Illness in Young Children and the Use of Wood-Burning Stoves for Indoor Heating. *Pediatrics* **1985**, *75*, 587–593. [CrossRef]
19. Koenig, J.Q.; Larson, T.V.; Hanley, Q.; Rebolledo, V.; Dumler, K.; Checkoway, H.; Wang, S.Z.; Lin, D.Y.; Pierson, W.E. Pulmonary Function Changes in Children Associated with Fine Particulate Matter. *Environ. Res.* **1993**, *63*, 26–38. [CrossRef]
20. Reed, M.D.; Campen, M.J.; Gigliotti, A.P.; Harrod, K.S.; McDonald, J.D.; Seagrave, J.C.; Mauderly, J.L.; Seilkop, S.K. Health Effects of Subchronic Exposure to Environmental Levels of Hardwood Smoke. *Inhal. Toxicol.* **2006**, *18*, 523–539. [CrossRef]
21. Ghio, A.J.; Soukup, J.M.; Case, M.; Dailey, L.A.; Richards, J.; Berntsen, J.; Devlin, R.B.; Stone, S.; Rappold, A. Exposure to wood smoke particles produces inflammation in healthy volunteers. *Occup. Environ. Med.* **2011**, *69*, 170–175. [CrossRef]
22. Kurmi, O.P.; Semple, S.; Simkhada, P.; Smith, W.C.S.; Ayres, J.G. COPD and chronic bronchitis risk of indoor air pollution from solid fuel: A systematic review and meta-analysis. *Thorax* **2010**, *65*, 221–228. [CrossRef]
23. Sällsten, G.; Gustafson, P.; Johansson, L.; Johannesson, S.; Molnar, P.; Strandberg, B.; Tullin, C.; Barregard, L. Experimental Wood Smoke Exposure in Humans. *Inhal. Toxicol.* **2006**, *18*, 855–864. [CrossRef]
24. Barregard, L.; Sällsten, G.; Gustafson, P.; Andersson, L.; Johansson, L.; Basu, S.; Stigendal, L. Experimental Exposure to Wood-Smoke Particles in Healthy Humans: Effects on Markers of Inflammation, Coagulation, and Lipid Peroxidation. *Inhal. Toxicol.* **2006**, *18*, 845–853. [CrossRef]
25. Walker, E.S.; Semmens, E.O.; Belcourt, A.; Boyer, B.B.; Erdei, E.; Graham, J.; Hopkins, S.E.; Lewis, J.L.; Smith, P.G.; Ware, D.; et al. Efficacy of Air Filtration and Education Interventions on Indoor Fine Particulate Matter and Child Lower Respiratory Tract Infections among Rural U.S. Homes Heated with Wood Stoves: Results from the KidsAIR Randomized Trial. *Environ. Health Perspect.* **2022**, *130*, 47002. [CrossRef]

26. Steenland, K.; Pillarisetti, A.; Kirby, M.; Peel, J.; Clark, M.; Checkley, W.; Chang, H.H.; Clasen, T. Modeling the potential health benefits of lower household air pollution after a hypothetical liquified petroleum gas (LPG) cookstove intervention. *Environ. Int.* **2018**, *111*, 71–79. [CrossRef]
27. Sood, A.; Petersen, H.; Blanchette, C.M.; Meek, P.; Picchi, M.A.; Belinsky, S.A.; Tesfaigzi, Y. Wood Smoke Exposure and Gene Promoter Methylation Are Associated with Increased Risk for COPD in Smokers. *Am. J. Respir. Crit. Care Med.* **2010**, *182*, 1098–1104. [CrossRef]
28. Linak, W.P.; Miller, C.A.; Wendt, J.O. Comparison of Particle Size Distributions and Elemental Partitioning from the Combustion of Pulverized Coal and Residual Fuel Oil. *J. Air Waste Manag. Assoc.* **2000**, *50*, 1532–1544. [CrossRef]
29. Goodarzi, F. Morphology and chemistry of fine particles emitted from a Canadian coal-fired power plant. *Fuel* **2006**, *85*, 273–280. [CrossRef]
30. Sigsgaard, T.; Forsberg, B.; Annesi-Maesano, I.; Blomberg, A.; Bølling, A.; Boman, C.; Bønløkke, J.; Brauer, M.; Bruce, N.; Héroux, M.-E.; et al. Health impacts of anthropogenic biomass burning in the developed world. *Eur. Respir. J.* **2015**, *46*, 1577–1588. [CrossRef]
31. Ward, T.J.; Palmer, C.P.; Noonan, C.W. Fine Particulate Matter Source Apportionment Following a Large Woodstove Changeout Program in Libby, Montana. *J. Air Waste Manag. Assoc.* **2010**, *60*, 688–693. [CrossRef]
32. Walker, E.S.; Noonan, C.W.; Semmens, E.O.; Ware, D.; Smith, P.; Boyer, B.B.; Erdei, E.; Hopkins, S.E.; Lewis, J.; Belcourt, A.; et al. Indoor fine particulate matter and demographic, household, and wood stove characteristics among rural US homes heated with wood fuel. *Indoor Air* **2021**, *31*, 1109–1124. [CrossRef] [PubMed]
33. Gall, E.T.; Carter, E.M.; Earnest, C.M.; Stephens, B. Indoor Air Pollution in Developing Countries: Research and Implementation Needs for Improvements in Global Public Health. *Am. J. Public Health* **2013**, *103*, e67–e72. [CrossRef] [PubMed]
34. Lange, P.; Celli, B.R.; Agustí, A.; Jensen, G.B.; Divo, M.; Faner, R.; Guerra, S.; Marott, J.L.; Martinez, F.D.; Martinez-Camblor, P.; et al. Lung-Function Trajectories Leading to Chronic Obstructive Pulmonary Disease. *New Engl. J. Med.* **2015**, *373*, 111–122. [CrossRef] [PubMed]
35. Hunninghake, G.M.; Cho, M.H.; Tesfaigzi, Y.; Soto-Quiros, M.E.; Avila, L.; Lasky-Su, J.; Stidley, C.; Melén, E.; Söderhäll, C.; Hallberg, J.; et al. *MMP12*, Lung Function, and COPD in High-Risk Populations. *New Engl. J. Med.* **2009**, *361*, 2599–2608. [CrossRef] [PubMed]
36. Silverman, E.K.; Vestbo, J.; Agusti, A.; Anderson, W.; Bakke, P.S.; Barnes, K.C.; Barr, R.G.; Bleecker, E.R.; Boezen, H.M.; Burkart, K.M.; et al. Opportunities and Challenges in the Genetics of COPD 2010: An International COPD Genetics Conference Report. *COPD J. Chronic Obstr. Pulm. Dis.* **2011**, *8*, 121–135. [CrossRef] [PubMed]
37. Fleisch, A.F.; Rokoff, L.B.; Garshick, E.; Grady, S.T.; Chipman, J.; Baker, E.R.; Koutrakis, P.; Karagas, M.R. Residential wood stove use and indoor exposure to PM2.5 and its components in Northern New England. *J. Expo. Sci. Environ. Epidemiol.* **2020**, *30*, 350–361. [CrossRef]
38. Ranathunga, N.; Perera, P.; Nandasena, S.; Sathiakumar, N.; Kasturiratne, A.; Wickremasinghe, R. Effect of household air pollution due to solid fuel combustion on childhood respiratory diseases in a semi urban population in Sri Lanka. *BMC Pediatr.* **2019**, *19*, 306. [CrossRef]
39. Naeher, L.P.; Smith, K.R.; Leaderer, B.P.; Mage, D.; Grajeda, R. Indoor and outdoor PM2.5 and CO in high- and low-density Guatemalan villages. *J. Expo. Anal. Environ. Epidemiol.* **2000**, *10*, 544–551. [CrossRef]
40. Mannucci, P.M.; Franchini, M. Health Effects of Ambient Air Pollution in Developing Countries. *Int. J. Environ. Res. Public Health* **2017**, *14*, 1048. [CrossRef]
41. Ward, T.J.; Palmer, C.P.; Houck, J.E.; Navidi, W.C.; Geinitz, S.; Noonan, C.W. Community Woodstove Changeout and Impact on Ambient Concentrations of Polycyclic Aromatic Hydrocarbons and Phenolics. *Environ. Sci. Technol.* **2009**, *43*, 5345–5350. [CrossRef]
42. Singh, D.; Tassew, D.D.; Nelson, J.; Chalbot, M.-C.G.; Kavouras, I.G.; Demokritou, P.; Tesfaigzi, Y. Development of an Integrated Platform to Assess the Physicochemical and Toxicological Properties of Wood Combustion Particulate Matter. *Chem. Res. Toxicol.* **2022**, *35*, 1541–1557. [CrossRef]
43. Singh, D.; Tassew, D.D.; Nelson, J.; Chalbot, M.-C.G.; Kavouras, I.G.; Tesfaigzi, Y.; Demokritou, P. Physicochemical and Toxicological Properties of wood Smoke Particulate Matter as a Function of Wood Species and Combustion Condition. *J. Hazard. Mater.* **2022**, *441*, 129874. [CrossRef]
44. Johansson, L.S.; Leckner, B.; Gustavsson, L.; Cooper, D.; Tullin, C.; Potter, A. Emission characteristics of modern and old-type residential boilers fired with wood logs and wood pellets. *Atmospheric Environ.* **2004**, *38*, 4183–4195. [CrossRef]
45. Bølling, A.K.; Pagels, J.; Yttri, K.E.; Barregard, L.; Sallsten, G.; Schwarze, P.E.; Boman, C. Health effects of residential wood smoke particles: The importance of combustion conditions and physicochemical particle properties. *Part. Fibre Toxicol.* **2009**, *6*, 29. [CrossRef]
46. Fitzpatrick, E.M.; Ross, A.B.; Bates, J.; Andrews, G.; Jones, J.; Phylaktou, H.; Pourkashanian, M.; Williams, A. Emission of Oxygenated Species from the Combustion of Pine Wood and its Relation to Soot Formation. *Process Saf. Environ. Prot.* **2007**, *85*, 430–440. [CrossRef]
47. Rokoff, L.B.; Koutrakis, P.; Garshick, E.; Karagas, M.R.; Oken, E.; Gold, D.R.; Fleisch, A.F. Wood Stove Pollution in the Developed World: A Case to Raise Awareness Among Pediatricians. *Curr. Probl. Pediatr. Adolesc. Health Care* **2017**, *47*, 123–141. [CrossRef]

48. Elias, V.O.; Simoneit, B.R.; Cordeiro, R.C.; Turcq, B. Evaluating levoglucosan as an indicator of biomass burning in Carajás, amazônia: A comparison to the charcoal record. *Geochim. Cosmochim. Acta* **2001**, *65*, 267–272. [CrossRef]
49. Simoneit, B.R.T.; Elias, V.O. Detecting Organic Tracers from Biomass Burning in the Atmosphere. *Mar. Pollut. Bull.* **2001**, *42*, 805–810. [CrossRef]
50. Simoneit, B.R.T.; Schauer, J.J.; Nolte, C.; Oros, D.R.; Elias, V.O.; Fraser, M.P.; Rogge, W.F.; Cass, G.R. Levoglucosan, a tracer for cellulose in biomass burning and atmospheric particles. *Atmospheric Environ.* **1999**, *33*, 173–182. [CrossRef]
51. Fraser, M.P.; Lakshmanan, K. Using Levoglucosan as a Molecular Marker for the Long-Range Transport of Biomass Combustion Aerosols. *Environ. Sci. Technol.* **2000**, *34*, 4560–4564. [CrossRef]
52. Weichenthal, S.; Kulka, R.; Lavigne, E.; van Rijswijk, D.; Brauer, M.; Villeneuve, P.J.; Stieb, D.; Joseph, L.; Burnett, R.T. Biomass Burning as a Source of Ambient Fine Particulate Air Pollution and Acute Myocardial Infarction. *Epidemiology* **2017**, *28*, 329–337. [CrossRef] [PubMed]

International Journal of
Environmental Research and Public Health

MDPI

Article

Vulnerable Factors Affecting Urinary N-Methylformamide Concentration among Migrant Workers in Manufacturing Industries in Comparison with Native Workers in the Republic of Korea (2012–2019)

Kyung-Eun Lee *, Kayoung Park, Dong Joon Park and Sungkyun Park

Korea Occupational Safety and Health Agency, Ulsan 44429, Korea
* Correspondence: kyeong85@kosha.or.kr

Abstract: Background: Occupational studies on N-N-dimethylformamide (DMF) exposure among migrant workers in Korea are scarce. We determined the urine concentration of N-methylformamide (NMF) among migrant workers with DMF exposure and compared the data with those of native workers. Methods: Data were collected from Workers' Special Health Examination and Work Environment Monitoring databases during 2014–2019. Workers aged ≥20 years were eligible to participate in the special health examination for DMF exposure. Urine concentrations of NMF were determined and compared between migrant and native workers. We also evaluated the factors affecting the difference in the urine concentration of NMF between the migrant and native workers. Multiple logistic regression was performed by adding confounders step by step. Results: Among 9259 subjects, 504 (5.2%) were migrant workers. The mean urinary concentration of NMF was 6.73 mg/L in migrant workers, which was significantly higher than that in native workers (2.06 mg/L, $p < 0.001$). The odds of a urine concentration of NMF > 30 mg/L were significantly higher in migrant workers than in native workers after adjusting for sex and age (odds ratio [OR] = 7.31, 95% confidence interval [CI] = 4.66–11.45). However, the odds between the native and migrant workers were not significantly different when fully adjusted for confounders (OR = 1.12, 95% CI = 0.65–1.94). Conclusions: The excessive exposure to DMF among migrant workers was attributed not to differences in biological characteristics but to their work environment. Workers must have awareness of the use of protective equipment and knowledge of hazardous chemicals that they may be exposed to, especially at the workplace.

Keywords: migrant workers; SMEs; N,N-methylformamide; threshold limit values

Citation: Lee, K.-E.; Park, K.; Park, D.J.; Park, S. Vulnerable Factors Affecting Urinary N-Methylformamide Concentration among Migrant Workers in Manufacturing Industries in Comparison with Native Workers in the Republic of Korea (2012–2019). *Int. J. Environ. Res. Public Health* **2022**, *19*, 13450. https://doi.org/10.3390/ijerph192013450

Academic Editors: Roel Vermeulen, Lauren Petrick and Maaike van Gerwen

Received: 2 September 2022
Accepted: 12 October 2022
Published: 18 October 2022

1. Introduction

Nearly 4.4% of the total workers globally are immigrants, who leave their country of origin and move to another country in search of occupational opportunities [1,2]. The proportion of migrant workers has rapidly increased both globally and in the Republic of Korea (hereafter, Korea) to resolve manpower shortage [3]. According to the Survey on Immigrants' Living Conditions and Labor Force conducted by the Ministry of Justice and Statistics Korea, the number of migrant workers increased by 6.0% from the previous year to 884,000 as of 2018, of which 45.6% were engaged in manufacturing [4]. Migrant workers were mainly recruited in small and high-risk companies wherein native workers avoided work; this caused frequent occupational injuries among migrant workers than among native workers [5–7]. It is assumed that the incidence of occupational diseases such as sub-acute intoxication, which shows inconspicuous or latent progression compared to mechanical injuries, is further underestimated among migrant workers [5,8,9].

The Korea Occupational Safety and Health Agency (KOSHA) reported that 25% of the intoxication-related accidents due to 10 chemicals at workplace occurred in migrant workers during 2005–2016 [10]. N-N-dimethylformamide (DMF), which could induce liver toxicity, was the most common chemical, accounting for 27.3% of the intoxication-related accidents (9 accidents affecting 24 employees), of which 33% involved migrant workers in the workplace [10]. Workers might be exposed to DMF when using chemicals, resins, fibers, coatings, inks, and adhesives, mostly in synthetic leather and textile industries. Many migrant workers who are engaged in manufacturing processes might have high exposure to DMFs at workplace because immigrants without language skills are relegated to dangerous and high-risk jobs [2,7]. Occupational health studies of migrant workers who are at risk of exposure to DMF are rare despite their social vulnerabilities because of the lack of information owing to deficiencies in the national monitoring system [2,11,12].

The present study aimed to determine the urinary concentration of N-methylformamide (NMF), a biological monitoring index (BEI) of DMF, among migrant workers who are at risk of exposure to DMF and compared with the data of native workers in the manufacturing industry. This study also assessed occupational factors associated with the risk of DMF intoxication, using data from occupational health examinations during 2014–2019.

2. Materials and Methods

2.1. Data Source

Data were obtained from the Workers' Special Health Examination (WSHE) and Work Environment Monitoring (WEM) databases during 2014–2019, managed by the KOSHA. In accordance with Articles 125, 129, and 130 of the Occupational Safety and Health Act in Korea, the WSHE and WEM are mandatory health screening and exposure monitoring programs performed periodically among workers who are regularly exposed to any of the listed hazardous substances and various physical environments specified by the Ministry of Occupational and Employ (MOE). All companies in Korea have a unique identification number managed by the MOE. Using the companies' own identification numbers granted by the MOE, the 2 databases of WSHE and WEM were linked by the exposure agent and the year of examination. The highest record of TWA at the same company and year was attached without weights to WSHE in the workers at risk of exposure. Among two or more records collected in a subject throughout 3 years, the oldest record was used in the analysis.

2.2. Study Design

This cross-sectional study was designed to enroll employees aged ≥20 years who were eligible to participate in the special health examination for DMF exposure during 2014–2019. However, participants (a) who had no records of the urine concentration of NMF and (b) who were employed in places where the airborne concentration of DMF had never been measured were excluded from the study. The enrolled population was classified into 2 groups based on whether the participants were migrant workers or native workers. If several records were found from the WSHE and WEM for a worker, then the earliest record was included in the analysis.

2.3. Measurements

We assessed the characteristics known to affect DMF exposure and its metabolism in the human body, including demographics (sex and age), companies where the subjects are employed (number of workers, types of industries), and DMF exposure factor (time weighted average (TWA), years of employment). According to the definition used by the European Union, companies with less than 5 and between 5 and 49 employees were classified as micro- and small-sized enterprises (MSEs), and companies with 50 or more employees were classified as medium-and large-sized enterprises [13]. For industry types in manufacturing, the 5 dominant industries where the migrant workers were the most hired as employees were chosen and classified as follows: rubber and plastic products, apparel, textiles, chemicals and chemical products, leather and related products, and others, defined as other manufacturing except the 5 industries. As for the workplace exposure indicators of DMF, the employees were classified into the "Low," "Moderate," and "High" groups based on the recorded urine concentration of NMF and TWA of DMF in the workplace.

Employees with records of "0" or "ND (not detected)" or "Trace" were considered in the "Low" group. Others were classified into 2 groups, namely, "Moderate" and "High" based on the threshold limit values (TLVs, 5 ppm) and biological exposure indices (BEIs, 30 mg/L) adopted by the American Conference of Governmental Industrial Hygienists (ACGIH).

2.4. Statistical Analysis

Student's *t*-test and Pearson's chi-square test were used to compare the baseline characteristics between native and migrant workers. To compare the urine concentration of NMF by the types of manufacture industries, the generalized linear models were used in estimating the marginal means with 95% confidence intervals. The risk of exceeding a urine NMF concentration of 30 mg/L among manufacturing workers was assessed using multiple logistic regression. To evaluate the factors affecting the difference between migrant and native workers in the urinary concentration of NMF, 3 models were analyzed by adding demographics (sex, age), companies where the subjects were employed (number of workers, types of industries), and the DMF exposure factor (TWA, years of employment) as covariates. Statistical analysis was performed using SAS, version 9.4 (SAS Institute, Cary, NC, USA).

2.5. Ethical Considerations

The study protocol was approved by the ethical committee of the Institutional Review Board of the Occupational Safety and Health Research Institute (approval no. OSHRI-2021-HR-005). The need for informed consent was waived because of the retrospective nature of the study and the anonymity of the data after de-identification.

3. Results

Among the employees aged ≥20 years who were eligible to participate in the special health examination for DMF exposure during 2014–2019, 9763 subjects were included in this study. The included subjects had records of urine concentrations of NMF and TWA of DMF at their workplace. Among the included subjects, 5.2% were classified as migrant workers (n = 504) and the others as native workers (n = 9259).

Table 1 shows the various characteristics of the native and migrant workers, despite the fact that the proportion of male workers was higher than that of female workers in both groups ($p = 0.794$). The proportion of younger workers was higher in the migrant worker group. Subjects aged 40–49 and ≥50 years accounted for 22.0% and 18% among native workers, and migrant workers accounted for only 12.7% and 5.8%, respectively ($p < 0.001$). In addition, 60.6% of the migrant workers were employed in small-or medium-sized companies with less than 50 employees (n = 305), while 19.6% of the native workers (n = 1814) were employed in such a company ($p < 0.001$, Table 1). In terms of manufacturing type, more than half of the migrant workers were employed in the manufacture of rubber and plastics (24.8%), apparel (14.1%), and textile (11.5%) industries. On the contrary, native workers at risk of DMF exposure were employed the most (38.3%) in the manufacturing of chemicals and chemical products, where 11.5% of the migrant workers were employed. The rate of exposure to DMF exceeding the threshold limit values (TLVs) was significantly higher in migrant workers (21.6%, n = 109) than in native workers (6.3%, $p < 0.001$). The rate of urine concentration of NMF exceeding the recommended limit by ACGIH (30 mg/L) was significantly higher in migrant workers (5.6%, n = 28) than in native workers (2.2%, $p < 0.001$). The mean duration (in years) of employment among migrant workers was 1.61 years, which was significantly shorter than that among native workers (7.78 years). The mean concentration of NMF in the urine was 6.73 mg/L among migrant workers, which was also significantly higher than that among native workers (2.06 mg/L, $p < 0.001$, Figure 1). Both in the native and the migrant workers in the manufacture of leather and related products, the mean concentration of urine NMF were high compared with those in other manufacture industries. However, the mean difference of urine concentration of

NMF between the native and the migrant workers in the manufacture of leather and related products was not significant (Figure 1).

Table 1. Characteristics of native and migrant workers exposed to DMF in manufacturing industries (2012–2019).

Characteristics	Native Workers (N = 9259)	Migrant Workers (N = 504)	p *
Sex, n (%)			0.794
Male	7629 (82.4)	413 (81.9)	
Female	1630 (17.6)	91 (18.1)	
Age group, n (%)			<0.001
20–29 years	2868 (31.0)	214 (42.5)	
30–39 years	2681 (29.0)	197 (39.1)	
40–49 years	2041(22.0)	64 (12.7)	
≥50 years	1669 (18.0)	29 (5.8)	
Number of workers, n (%)			<0.001
<5	112 (1.2)	23 (4.6)	
5–50	1702 (18.4)	282 (56.0)	
≥50	7447 (80.4)	199 (39.5)	
Types of industries/Manufacturers of, n (%)			<0.001
Rubber and plastic products	1034 (11.2)	125 (24.8)	
Wearing apparel	230 (2.5)	71 (14.1)	
Textiles	406 (4.4)	58 (11.5)	
Chemicals and chemical products	3549 (38.3)	58 (11.5)	
Leather and related products	236 (2.5)	46 (9.1)	
Other manufacturing	3804 (41.1)	146 (29.0)	
TWA of DMF, n (%) [II]			<0.001
Low	5439 (58.8)	134 (26.6)	
Moderate	3233 (34.9)	261 (51.8)	
High	580 (6.3)	109 (21.6)	
Urine concentration of NMF, n (%) [II]			<0.001
Low	3013 (32.5)	90 (17.9)	
Moderate	6151 (66.4)	386 (76.6)	
High	95 (2.2)	28 (5.6)	
Years of employment, years (SD)	7.78 (9.02)	1.61 (2.06)	<0.001

* p values were calculated using Pearson's chi-square and Student's t-tests. [II] Employees with records of "0" or "ND (not detected)" or "Trace" were considered in the "Low" group. Others were classified into 2 groups, namely, "Moderate" and "High" based on the threshold limit values (TLVs, 5 ppm) and biological exposure indices (BEIs, 30 mg/L) adopted by the ACGIH. DMF = Dimethylformamide; SD = Standard deviation; TWA = Time-weighted average; ND = Not detected; TLVs = Threshold limit values; ACGIH = American Conference of Governmental Industrial Hygienists.

Table 2 shows the factors that cause differences in the body absorption of DMF between the native and migrant workers. The odds of urine concentration of NMF exceeding 30 mg/L were significantly higher in the migrant worker group than in the native worker group when adjusting for sex and age (odds ratio [OR] = 7.31, 95% confidence interval [CI] = 4.66–11.45; Model 1). However, the risk among migrant workers dramatically decreased after adjusting additionally for the characteristics of companies (number of workers, type of manufacturing); however, the risk among migrant workers was significantly higher than that among native workers (OR = 1.72, 95% CI = 1.01–2.94; Model 2). The ORs were 8.50 (95% CI = 3.77–19.15) and 4.52 (95% CI = 2.87–7.13) in the MSEs compared with those in the medium- and large-sized enterprises (Model 2). The types of manufacturing products were also significantly associated with increasing the risks of excessive exposure to DMF, which was particularly high in manufacturing leather and related products (OR = 18.10, 95% CI = 9.87–33.21), apparel (OR = 3.64, 95% CI = 1.57–8.44), and textiles (OR = 3.51, 95% CI = 1.81–6.82) when compared with that in other manufacturing (Model 2). Finally, when fully adjusted for confounders including the ambient concentration of the DMF level at workplace and years of employment, the odds between the groups of the native and migrant workers were not significantly different (OR = 1.12, 95% CI = 0.65–1.94; Model 3).

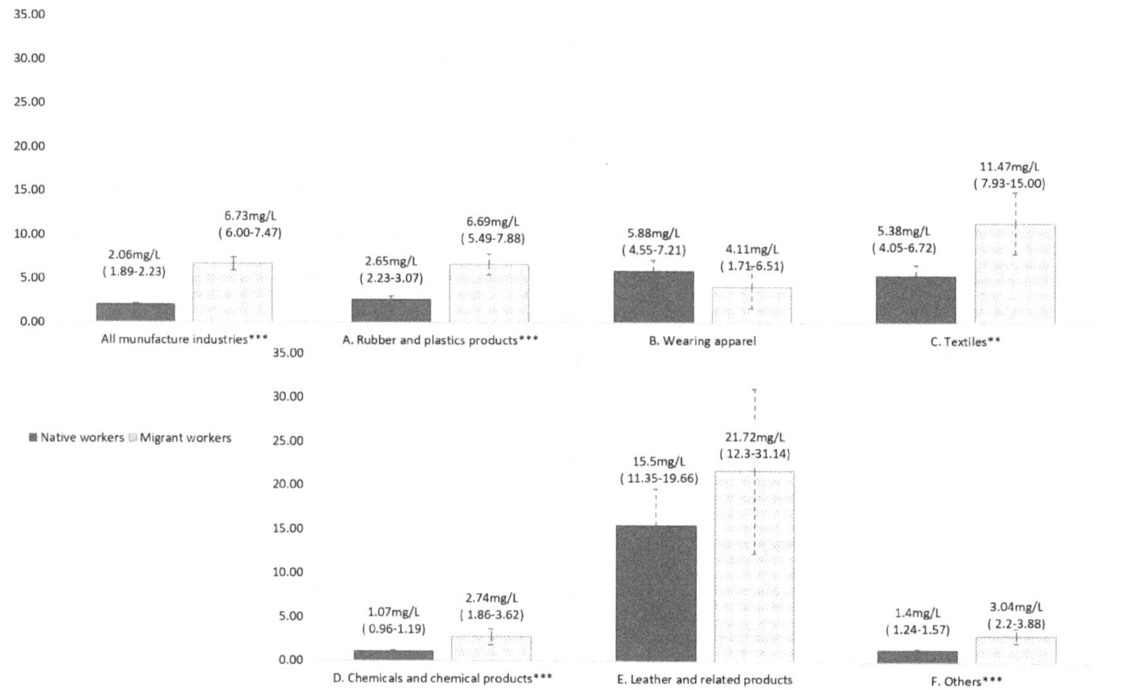

Figure 1. Urine NMF concentration in the native and migrant workers by the types of manufacture industries. (**A–F**) show the urine NMF concentrations by the native and migrant workers in the manufactures of rubber and plastics products, wearing apparels, textiles, chemicals and chemical products, leather and related products and other manufacturing industries, respectively. Estimated marginal means (95% confidence intervals) were calculated using generalized linear model. Non-detected values under LOD or those recorded as "0" were calculated by substitution with half of the minimally detected value (=0.0005 mg/dL). **, $p < 0.01$; ***, $p < 0.001$; NMF = N′-methylformamide; LOD = limit of detection.

Table 2. Risk of urine concentration of N′-methylformamide > 30 mg/L among the workers.

Characteristics	Model 1		Model 2		Model 3	
	OR	(95% CI)	OR	(95% CI)	OR	(95% CI)
Group						
Native workers		Reference		Reference		Reference
Migrant workers	7.31	(4.66–11.45)	1.72	(1.01–2.94)	1.12	(0.65–1.94)
Sex						
Male	0.94	(0.59–1.50)	1.52	(0.92–2.50)	1.64	(0.97–2.76)
Female		Reference		Reference		Reference
Age group						
20–29 years		Reference		Reference		Reference
30–39 years	0.69	(0.4–1.2)	0.53	(0.30–0.94)	0.54	(0.30–0.97)
40–49 years	1.46	(0.88–2.45)	0.80	(0.45–1.42)	0.73	(0.39–1.34)
≥50 years	2.74	(1.70–4.42)	.89	(0.50–1.60)	1.06	(0.57–1.96)

Table 2. *Cont.*

Characteristics	Model 1		Model 2		Model 3	
	OR	(95% CI)	OR	(95% CI)	OR	(95% CI)
Number of workers						
<5			8.50	(3.77–19.15)	11.44	(5.00–26.18)
5–50			4.52	(2.87–7.13)	3.39	(2.13–5.38)
≥50			Reference		Reference	
Types of industries/Manufacturer of						
Rubber and plastic products			2.06	(1.07–3.99)	0.93	(0.48–1.83)
Wearing apparel			3.64	(1.57–8.44)	1.77	(0.76–4.11)
Textiles			3.51	(1.81–6.82)	1.71	(0.87–3.38)
Chemicals and chemical products			0.58	(0.27–1.23)	0.88	(0.41–1.89)
Leather and related products			18.10	(9.87–33.21)	7.32	(3.95–13.56)
Other manufacturing			Reference		Reference	
TWA of DMF						
Not detected or 0					Reference	
≤5 ppm					7.65	(3.19–18.35)
>5 ppm					36.03	(14.61–88.83)
Years of employment					0.92	(0.88–0.96)

Model 1 was adjusted for sex and age. Model 2 was adjusted for Model 1 + Characteristics of companies (the number of total workers and the type of manufacturers). Model 3 was adjusted for Model 2 + DMF exposure (TWA of DMF and years of employment). DMF = dimethylformamide; TWA = time-weighted average; OR = odds ratio; CI = confidence interval.

4. Discussion

To the best of our knowledge, this study is the first to assess the exposure to DMF in migrant workers by comparing the data with those of the native workers using the TWA of DMF at the workplace and evaluate the urine NMF concentration at the national level. We found that the mean urinary concentration of NMF was significantly high among migrant workers compared with that among native workers despite the fact that the time of exposure and age of the migrant workers are much shorter and lower than those of the native workers. This finding supports the notion that migrant workers who are at risk of DMF exposure in manufacturing industries are highly vulnerable to occupational exposure to toxic materials. The difference in the urinary NMF concentration between the migrant and native workers could be attributed to the difference in biological metabolism according to ethnic specificity [14]. However, the fact that the significant difference in urinary NMF concentration was remarkably decreased after adjusting for several work environmental factors demonstrated that the main causes for the increase in NMF concentration in the urine were mostly attributed to the occupational environment in migrant workers (Table 2).

One of the attributing factors exceeding the BEIs in migrant workers is the size of the companies based on the number of employees. In this study, the number of employees working in MSEs was 3 times higher among migrant workers (60.6%) than among native workers (19.6%). The occupational environment in the MSEs tends to be unsafe and unregulated compared with that in medium- and large-sized enterprises because of limited resources, little knowledge about risk management processes, and deficiencies in organizational processes [15–17]. The International Labor Organization also reported that the MSEs attracted the majority of vulnerable workers at risk of exploitation, and migrant workers who face specific risks linked to cultural and language issues and access to social protection, as well as barriers to healthcare, were included in these MSEs. Therefore, national programs on capacity-building must be developed and adapted in the sector of occupational safety and health targeting MSEs, considering their heterogenic characteristics, including those for migrant workers [17].

The type of products manufactured at the workplace is another main factor affecting the level of exposure to DMF, leading to the difference between the migrant workers and the native workers. We found that a higher proportion of migrant workers was involved in the manufacturing industries known to lead to a high risk of exposure to DMF. The types of manufacturing products highly associated with increasing DMF exposure were rubber and plastics, leather and related products, wearing apparel, and textiles, which is consistent with the results of previous studies [18–20].

In this study, the risk of exposure to DMF exceeding the BEIs was significantly high even after adjusting for the air concentration of DMF in the industries that manufacture leather and related products compared with that in other manufacturing industries. Previous studies have also reported the discrepancy between BEIs and the air concentration of DMF because of the co-exposure to other chemicals or dermal absorption [18,19,21]. Kim (2014) explained that the greater the proportion of processes requiring manual handling of organic solvents, such as in the leather manufacturing industry, the larger may be difference between the biological exposure index of DMF and the measured air concentration due to the amount of additional exposure through dermal absorption [18]. On the other hand, the discrepancy between TWA and BEIs could also be attributed from the different levels of data unit. The highest measures of TWAs in the workplace were equally assigned to the workers, in spite that the workers exposed differently according to their types of working. Considering that the TWA of DMF was measured in two or more workers who were most exposed in the workplace at the same time, the TWA of DMF could be overestimated. However, the main result that the migrant workers were in vulnerable to DMF exposure in workplace would not be changed in terms that the ORs in the migrant workers was dramatically decreased after adjusting with the characteristics of companies in Table 2.

This study has several limitations that need to be considered when interpreting the findings. First, the WSHE and WEM databases used in this study contained secondary data collected from nationally designated institutions; residual confounding cannot be ruled out because of the lack of variables, especially information about specific jobs as well as co-exposures that affect the BEIs. Second, because the data used in this study were collected from heterogeneous institutions located across the country, the homogeneity of data might not be consistent despite KOSHA's efforts to control quality in the WSHE and WEM. Therefore, we compared urinary NMF concentrations between the native and migrant workers twice after removing the limit of detection (LOD) value and after imputation to and the half of the minimum in the observed values (=0.0005 mg/L) for sensitivity analysis. Although the urine concentration of NMF in workers could be underestimated due to the substitution to the half of the minimally detected value, the overall results would not be changed. Because assessing whether the urine concentration exceeded 30 mg/L was rarely affected by the heterogeneity of LODs among various devices. Third, we modified the measured values to categorical scales in the adjusted regression to minimize the effect of heterogeneity. Finally, the small number in subgroup could have affected a wide range of confidence intervals and the risk of overestimation when producing ORs with fully adjustment with multiple variables (Table S1).

5. Conclusions

In conclusion, the results of this study support the vulnerability of migrant workers in terms of their working environment and DMF exposure by analyzing data from a national database [8,22]. The results also showed that the excessive exposure to DMF among migrant workers was attributed not to their biological characteristics but to their work environment, including working in small-sized companies, having poor ventilation systems leading to high ambient levels of DMF, and using inappropriate personal protective equipment among highly manipulating processing workers [9]. It has been reported that DMF exposure is affected mainly by the effectiveness of local exhaust ventilation and appropriate wearing of personal protective equipment [23]. Therefore, it is necessary to develop an educational program, especially in small-sized enterprises, that considers the diversity of languages of migrant workers so that they can be advised on the proper use of personal protective equipment and provided with knowledge of the dangers of chemicals they are exposed to in their workplace.

Supplementary Materials: The following supporting information can be downloaded at: https://www.mdpi.com/article/10.3390/jerph192013450/s1, Table S1: Subgroup analysis between the native and the migrant workers based on the level of urine concentration of N′-methylformamide

Author Contributions: Conceptualization, K.-E.L.; Formal analysis, D.J.P.; Investigation, K.P.; Methodology, K.-E.L.; Project administration, K.P.; Software, S.P.; Supervision, K.-E.L.; Writing—original draft, K.-E.L.; Writing—review & editing, D.J.P. and S.P. All authors have read and agreed to the published version of the manuscript.

Funding: This research received no external funding.

Institutional Review Board Statement: The study was conducted in accordance with the Declaration of Helsinki, and approved by the ethical committee of the Institutional Review Board of the Occupational Safety and Health Research Institute (approval no. OSHRI-2021-HR-005, 25 February 2021).

Informed Consent Statement: The need for informed consent was waived due to the retrospective nature of the study using anonymization.

Data Availability Statement: Data was obtained from Occupational Safety and Health Institute (OSHRI), Korea Occupational Safety and Health Agency (KOSHA) and are available with the permission of OSHRI KOSHA.

Conflicts of Interest: The authors declare no conflict of interest.

References

1. International Labour Office. *ILO Global Estimates of Migrant Workers and Migrant Domestic Workers: Results and Methodology*; International Labour Office: Geneva, Switzerland, 2015.
2. Moyce, S.C.; Schenker, M. Migrant workers and their occupational health and safety. *Annu. Rev. Public Health* **2018**, *39*, 351–365. [CrossRef] [PubMed]
3. Shin, J.J. A transnational approach to the integration of migrant workers with focus on the Koreas employment permit system. *Omnes J. Multicult. Soc.* **2017**, *7*, 128–153. [CrossRef]
4. Ministry of Justice, Statistics Korea. Survey on Immigrants' Living Conditions and Labor Force. Available online: http://kostat.go.kr/portal/korea/kor_nw/1/3/4/index.board?bmode=read&bSeq=&aSeq=372125&pageNo=1&rowNum=10&navCount=10&currPg=&searchInfo=&sTarget=title&sTxt=;2018 (accessed on 30 December 2020).
5. Lee, S.W.; Kim, K.S.; Kim, T.W. The status and characteristics of Industrial accidents for migrant workers in Korea compared with native workers. *Korean J. Occup. Environ. Med.* **2008**, *20*, 351–361. [CrossRef]
6. Jeong, B.Y.; Park, M.H. Analysis of accident characteristics of foreign workers in domestic chemical industry. *J. Ergon. Soc. Korea* **2018**, *37*, 169–182.
7. Yi, K.-H.; Cho, H.-H.; You, K.-H. The comparative study on the occupational injury rate and mortality rate of the total workers and foreign workers. *J. Korean Soc. Saf.* **2012**, *27*, 96–104.
8. Pérez, E.R.; Benavides, F.G.; Levecque, K.; Love, J.G.; Felt, E.; Van Rossem, R. Differences in working conditions and employment arrangements among migrant and non-migrant workers in Europe. *Ethn. Health* **2012**, *17*, 563–577. [CrossRef] [PubMed]
9. Hargreaves, S.; Rustage, K.; Nellums, L.B.; McAlpine, A.; Pocock, N.; Devakumar, D.; Waldridge, R.; Abubakar, R.; Kristensen, K.L.; Himmels, J.W.; et al. Occupational health outcomes among international migrant workers: A systematic review and meta-analysis. *Lancet Glob. Health* **2019**, *7*, E872–E882. [CrossRef]
10. Kim, H.-A.; Kim, S.K.; Eoh, W.-S.; Lee, C.-K.; Kim, S.-B.; Heo, Y. *A Study on the Current Status of Distribution Management of Acute Intoxication to Chemicals at Workplaces*; 2016-OSHRI-779; Occupational Safety and Health Research Institute, Korea Occupational Safety and Health Agency: Ulsan, Korea, 2016.
11. Song, Y.H.; Kim, K.S.; Lee, S.W.; Choi, S.H. The health status including biologic exposure indices of migrant workers: Based on 2005 special health examination data. *Korean J. Occup. Environ. Med.* **2008**, *20*, 63–73. [CrossRef]
12. Yi, K.-H.; Cho, H.-H. The survey study on working conditions and industrial safety and health of foreign workers. *J. Korea Saf. Manag. Sci.* **2012**, *14*, 53–63.
13. Berisha, G.; Pula, J.S. Defining small and medium enterprises: A critical review. *Acad. J. Bus. Admin. Law Soc. Sci.* **2015**, *1*, 17–27.
14. Berardesca, E.; Maibach, H. Ethnic skin: Overview of structure and function. *J. Am. Acad. Dermatol.* **2003**, *48*, S139–S142. [CrossRef] [PubMed]
15. Barbosa, C.; Azevedo, R.; Rodrigues, M.A. Occupational safety and health performance indicators in SMEs: A literature review. *Work* **2019**, *64*, 217–227. [CrossRef] [PubMed]

16. Rodrigues, M.A.; Arezes, P.; Leão, C.P. Risk assessment: Getting the 'big picture'. In *Safety Management in Small and Medium Sized Enterprises (SMEs)*; CRC Press: Boca Raton, FL, USA, 2017.
17. Park, J.; Park, J.S.; Han, B.; Kim, Y. Vulnerability of employees in businesses with fewer than five workers (micro-enterprises) to occupational safety and health problems. *Am. J. Ind. Med.* **2017**, *60*, 1056–1065. [CrossRef] [PubMed]
18. Kim, D.; Byun, K.; Park, J.-O.; Lee, M.-Y. Analysis on urinary N-methylformamide of Korean workers according to industrial classification and countermeasure for exposure control of N,N-dimethylformamide. *J. Korean Soc. Occup. Environ. Hyg.* **2014**, *24*, 345–352. [CrossRef]
19. Kuo, H.W.; Huang, Y.S.; Lo, J.C.; Cheng, T.J.; Wu, M.J. Exposure to solvents in a synthetic leather manufacturing plant. *Int. Arch. Occup. Environ. Health* **2000**, *73*, 275–280. [CrossRef] [PubMed]
20. Lee, J.; Hahm, M.; Huh, D.A.; Byeon, S.H. Prioritizing type of industry through health risk assessment of occupational exposure to dimethylformamide in the workplace. *Int. J. Environ. Res. Public Health* **2018**, *15*, 503. [CrossRef] [PubMed]
21. Chang, H.Y.; Yun, Y.D.; Yu, Y.C.; Shih, T.S.; Lin, M.S.; Kuo, H.W.; Chen, K.M. The effects of simultaneous exposure to methyl ethyl ketone and toluene on urinary biomarkers of occupational N,N-dimethylformamide exposure. *Toxicol. Lett.* **2005**, *155*, 385–395. [CrossRef] [PubMed]
22. Ronda, E.; Agudelo-Suárez, A.A.; García, A.M.; López-Jacob, M.J.; Ruiz-Frutos, C.; Benavides, F.G. Differences in exposure to occupational health risks in Spanish and foreign-born workers in Spain (Itsal Project). *J. Immigr. Minor. Health* **2013**, *15*, 164–171. [CrossRef] [PubMed]
23. Hur, S.-J.; Suh, C.-H.; Lee, C.-K.; Kim, J.-H.; Kim, D.-H.; Son, B.-C.; Lee, C.-H.; Chang, G.-R.; Lee, J.-T. Evaluation of work environment, health care management and exposure to chemicals in the workplaces using dimethylformamide (DMF). *J. Korean Soc. Occup. Environ. Hyg.* **2010**, *20*, 225–235.

toxics

Article

Effects of Different Concentrations of Oil Mist Particulate Matter on Pulmonary Fibrosis In Vivo and In Vitro

Huipeng Nie, Huanliang Liu, Yue Shi, Wenqing Lai, Xuan Liu, Zhuge Xi * and Bencheng Lin *

Tianjin Institute of Environmental and Operational Medicine, Tianjin 300050, China
* Correspondence: zhugexi2003@sina.com (Z.X.); linbencheng123@126.com (B.L.)

Highlights:

What are the main findings?

- OMPM exposure induced lung lesions and PF in rats.

What is the implication of the main finding?

- OMPM exposure induced PF via TGF-β1/Smad3 and TGF-β1/p38 MAPK signaling pathways.
- OMPM exposure induced fibrotic phenotypes in the BEAS-2B cell line.
- OMPM exposure activated the differentiation of the HFL-1 cell line.
- OMPM exposure led to the BEAS-2B cell line being able to induce the differentiation of the HFL-1 cell line.

Abstract: Oil-mist particulate matter (OMPM) refers to oily particles with a small aerodynamic equivalent diameter in ambient air. Since the pathogenesis of pulmonary fibrosis (PF) has not been fully elucidated, this study aims to explore the potential molecular mechanisms of the adverse effects of exposure to OMPM at different concentrations in vivo and in vitro on PF. In this study, rats and cell lines were treated with different concentrations of OMPM in vivo and in vitro. Sirius Red staining analysis shows that OMPM exposure could cause pulmonary lesions and fibrosis symptoms. The expression of TGF-β1, α-SMA, and collagen I was increased in the lung tissue of rats. The activities of MMP2 and TIMP1 were unbalanced, and increased N-Cadherin and decreased E-Cadherin upon OMPM exposure in a dose-dependent manner. In addition, OMPM exposure could activate the TGF-β1/Smad3 and TGF-β1/MAPK p38 signaling pathways, and the differentiation of human lung fibroblast HFL-1 cells. Therefore, OMPM exposure could induce PF by targeting the lung epithelium and fibroblasts, and activating the TGF-β1/Smad3 and TGF-β1/MAPK p38 signaling pathways.

Keywords: OMPM; TGF-β1/Smad3; TGF-β1/MAPK p38; PF

Citation: Nie, H.; Liu, H.; Shi, Y.; Lai, W.; Liu, X.; Xi, Z.; Lin, B. Effects of Different Concentrations of Oil Mist Particulate Matter on Pulmonary Fibrosis In Vivo and In Vitro. *Toxics* **2022**, *10*, 647. https://doi.org/10.3390/toxics10110647

Academic Editors: Roel Vermeulen and Swapna Upadhyay

Received: 25 August 2022
Accepted: 25 October 2022
Published: 28 October 2022

1. Introduction

Pulmonary fibrosis (PF) is a complex pathophysiological process that is common to most interstitial lung diseases. The pathological characteristics of pulmonary fibrosis are the diffuse chronic inflammation of alveolar walls and interstitial fibrosis; eventually, the pulmonary parenchyma is gradually replaced by fibrous scar tissue and the loss of respiratory function [1,2]. As the disease progresses, patients with pulmonary fibrosis eventually die due to respiratory failure. In recent years, the incidence of pulmonary fibrosis has increased significantly in China and abroad, and the incidence of pulmonary fibrosis in China is 3–5/100,000 per year [3,4]. The early diagnosis of pulmonary fibrosis is difficult due to its insidious onset and no specific clinical manifestations [5]. In addition, there is no special treatment for PF, resulting in high mortality rates of patients with PF. Epidemiological studies confirmed that environmental factors may induce PF, and participate in the occurrence and development of pulmonary interstitial fibrosis [6,7]. For example, the exposure to metal dust and wood dust can lead to increased risk of pulmonary interstitial fibrosis [8,9].

Oil-mist particulate matter (OMPM) refers to oily particles with a small aerodynamic equivalent diameter in ambient air, which is one of the main pollution factors in the urban atmosphere [10]. Due to the small particles, OMPM can enter the deep respiratory tract and deposit in all levels of the bronchus and alveoli [11]. At the same time, OMPM have a larger surface-area-to-volume ratio, which can adsorb more pathogenic microorganisms, acid oxides, transition metal components, and polycyclic aromatic hydrocarbons, resulting in increased biological toxicity [4,12]. A large number of studies confirmed that exposure to particulate matter in ambient air is closely related to the increased morbidity and mortality of respiratory diseases [13]. OMPM exposure can be involved in the occurrence and development of pulmonary infection, COPD, and other respiratory diseases through pulmonary inflammation and oxidative stress [13,14]. However, it is not clear whether oil-mist exposure is involved in the occurrence and development of PF [15]. In recent decades, with the development of the global industry and the process of globalization, marine resources have become the commanding heights of human exploitation and utilization. With the increase in large-scale operations conducted by humans at sea, the OMPM pollution emitted from these operations is becoming increasingly serious. As a result, the local marine environment has undergone great changes and continues to expand. At the same time, the entry of OMPM into the air and marine environment also poses a serious threat to marine life and human health. Furthermore, China's regulations for offshore operations are as follows: the instantaneous discharge of washing water for offshore operations must not exceed 30 L/kN, and the total discharge must not exceed 1/30,000th of the cargo capacity of the last voyage [11,16]. A large amount of oil pollution discharge adversely affects the marine ecological environment and the health of offshore workers [17]. In recent years, the relationship between OMPM and ecological health has become a hot global issue.

The pathological basis of early pulmonary fibrosis is acute alveolitis, which involves inflammatory cells, immune effector cells, and cytokines in the formation of alveolitis [7,18]. Recent studies found that a variety of cytokines play an important role in the formation of fibrosis [2,5]. These cytokines include promoting and inhibiting fibrosis. They coordinate and resist each other, and determine the outcome of lung-tissue injury and repair. Pulmonary fibrosis is a disease in which the functions of the two major types of factors lose balance during the repair of a lung injury, resulting in the damage of lung function and tissue structure [19]. TGF-β1 is a key cytokine causing fibrosis that can promote excessive proliferation and differentiation of lung fibroblasts, and then promote the excessive accumulation of extracellular matrices such as collagen in the interstitium and lung alveoli, leading to the occurrence and development of pulmonary fibrosis [20]. Studies showed that increased TGF-β1 levels are found in multifactorial pulmonary fibrosis, including bleomycin, cyclophosphamide, and radiation [2,21]. TGF-β1 content was closely correlated with pulmonary fibrosis. Numerous studies found that TGF-β1 can induce the transformation of fibroblasts into myofibroblasts by promoting the expression of Smad3 and p38MAPK, can stimulate the mitosis and synthesis and deposition of pulmonary fibroblasts, and induce the occurrence of pulmonary fibrosis [22,23]. Therefore, the TGF-β1/Smad3 and TGF-β1/p38 MAPK signaling pathways are the most important pathways promoting the formation of PF [24].

In our study, rats and cell lines were treated with different concentrations of OMPM to establish PF models in vivo and in vitro, and the molecular mechanism of TGF-β1/Smad3 and TGF-β1/p38 MAPK signaling pathways of the occurrence and development of PF induced by OMPM exposure was studied indepth using histopathology, molecular biology, and cell biology.

2. Materials and Methods

2.1. OMPM Collection

This study was carried out in a large offshore operation. A large amount of OMPM was collected during the working hours of the offshore operation. As described in previous studies (Nie et al., 2022 [25]), a media flow sampler (100 L/min) and a quartz filter (Whatman, UK, φ90 mm) were used to collect the OMPM. Before sampling, a blank quartz-filter membrane was heated in a muffle furnace at 550 °C for 4 h to remove the residual organic matter and other organic impurities. After balancing in a dryer for 24 h, we weighed the film on 1/10,000 scales. After sampling, the filter membrane was placed in a dryer for 24 h and weighed.

2.2. Animal Model and Grouping

SPF Wistar rats (6–8 weeks old, 200 ± 20 g) were purchased from Beijing Vital River Laboratory Animal Technology Co., Ltd. (Beijing, China) The animals were fed in the Experimental Animal Center of Tianjin Institute of Environmental and Occupational Medicine. We followed the regulations of the Experimental Animal Ethics Committee of the Tianjin Institute of Environmental and Occupational Medicine (IACUC of AMMS-04-2020-016). The rats drank water and ate freely. The breeding condition of the rats was room temperature at 24 ± 2 °C and humidity at 50% ± 10%.

To establish the animal model, the rats were directly exposed to different concentrations of OMPM. According to the dose per unit body weight, the equivalent dose of OMPM in rats was equivalent to 6.3 times that of humans. Therefore, the dose of OMPM in rats was 16 mg/m^3 multiplied by 6.3, which was about 100 mg/m^3. After about a week's acclimation, the rats were randomly divided into 3 groups: the high concentration of OMPM (HC group, the concentration of OMPM was 100 mg/m^3), low concentration of OMPM (LC group, the concentration of OMPM was 50 mg/m^3 for half treatment of high concentration) and control (the concentration of OMPM was 0 mg/m^3) groups, with 18 rats from each sex in each group, and OMPM exposure was 6.5 h a day for 42 consecutive days.

2.3. Cell Culture and Treatment

Human lung epithelial cell line BEAS-2B and human lung fibroblast cell line HFL-1 were purchased from the Institute of Biochemistry and Cell Biology, Chinese Academy of Sciences (Shanghai, China). The cell lines were cultured in a DMEM medium (Solarbio, Beijing, China) supplemented with 10% fetal bovine serum (FBS, Gibco, China) in a 37 °C incubator. A single-concentration cell exposure system (HRH-CES1332, Beijing Huironghe Technology Co., Ltd., Beijing, China) was used to expose the cells of BEAS-2B and HFL1 to OMPM for 1 h at different concentrations of 0, 5, 10, 20, 30, 40 mg/m^3.

2.4. Histopathology

The rat lung tissues were fixed in 4% paraformaldehyde (PFA) for 24 h, and the tissue paraffin slides were prepared. The paraffinized lung slides were deparaffinized and stained with Sirius Red to observe the tissue morphology. The images were obtained using a bioelectron microscope (OLYMPUS, Tokyo, Japan, BX51).

2.5. Immunofluorescence

The different groups of BEAS-2B and HFL-1 cell lines were used to prepare the paraffin slides. After dewaxing and dehydrating, antigen retrieval was performed at 95 °C for 10 min. The paraffin slices were blocked with 5% BSA at room temperature for 1 h. For immunofluorescence, FITC-labeled fluorescent primary antibody collagen I (Affinity, Jiangsu, China, dilution: 1:200) and α-SMA (CST, Boston, USA, dilution: 1:200) were incubated overnight at 4 °C. On the next day, sections were rewarmed, washed with PBS, and antirabbit IgG (H + L) Alexa Fluor® 488 conjugate secondary antibody (CST, Boston, USA, dilution: 1:500) was incubated in the dark for 1 h. After restaining with DAPI, the slices were observed with a fluorescence microscope (OLYMPUS, Japan, BX51) and photographed. The percentage of fluorescently labeled cells in rat lung tissues was counted with different fluorescently labeled cell-counting methods and the fluorescence intensity statistics of Image J. In addition, we anonymously shuffled the image order statistics to prevent statistical preference. Three fields of view were also analyzed. The magnification was 100×.

2.6. Western Blotting

The rat lung tissues and cell samples were scraped and resuspended in a 0.5 mL RIPA buffer (CWBIO, Jiangsu, China). Tissue or cell lysates were placed on ice for 1 h. After centrifugation at 10,000× *g* for 20 min, the protein concentration was determined using the BCA kit (Beyotime, Shanghai, China). Then, 20 μg samples were electrophoresed via 10% SDS-PAGE (Yeasen Biotechnology, Shanghai, China) and transferred onto a PVDF membrane. Samples were blocked with 5% nonfat milk in TBST solution at room temperature for 2 h and then probed with primary antibodies p38 (CST, Boston, USA, dilution: 1:1000), p-p38 (CST, Boston, USA, dilution: 1:1000), Smad3 (CST, USA, dilution: 1:1000), p-Smad3 (CST, Boston, USA, dilution: 1:1000), GAPDH (Utibody, Tianjin, China, dilution: 1:2000) at 4 °C overnight. Secondary antibodies used for detection included HRP-conjugated antirabbit IgG and antimouse IgG (Bioss, Beijing, China, dilution: 1:3000), and the protein expression was detected using an ECL chemiluminescence detection kit (Boster, Wuhan, China).

2.7. ELISA Assay

The rat serum or lung tissues and cell culture medium supernatants were obtained, and the samples were stored at −80 °C. α-SMA, COL1, N-cadherin, E-cadherin, MMP-2, and TIMP1 activities were determined using ELISA kits (Cloud-Clone, Beijing, China).

2.8. Statistical Analysis

Classical scientific software SPSS 25.0 was used for statistical data analysis. The data are expressed as mean ± standard deviation. The t-test was used for comparison between the two groups, and one-way ANOVA was used among the multiple groups. and comparisons between groups were performed using LSD (homogeneity of variance) or Dunnett's T3 (non-homogeneity of variance). $p < 0.05$ was considered to be statistically significant. The analyzed data were plotted using GraphPad Prism 5.0 software.

3. Results and Discussion

3.1. OMPM Exposure Induced Lung Lesions and PF in Rats

To investigate the potential effect of OMPM exposure on PF induction, the rats exposed to OMPM were examined histologically. According to our previous study (Nie et al., 2022 [25]), the OMPM was composed of tiny particles with various shapes and structures, with very irregular shapes, an uneven surface, uneven distribution, and different particle sizes. Otherwise, OMPM came in granular, blocky, strip, column, needle, and lamellar shapes, most of which clustered together, and some particles even overlapped. In addition, according to previous EDS data analysis (Nie et al., 2022 [25]), OMPM was mainly composed of aluminosilicate, inorganic salt crystals, organic carbon, random carbon, and other substances. The structure of a substance determines the properties; therefore, we found via HE staining that the long-term exposure to OMPM might trigger PF (Nie et al., 2022 [25]). Sirius Red staining is a classical pathological method to observe the distribution of collagen in tissues. Compared with the control group, the expression of collagens I and III in the lung tissues of rats in the HC and LC groups was significantly upregulated (Figure 1), indicating that the lung tissues of the rats in these two groups presented interstitial lung lesions and pulmonary inflammation with PF-like symptoms. In addition, in the HC group, the degree of lung-tissue lesions and fibrosis was more serious, which may have been due to the long-term exposure of high concentrations of OMPM to aggravate the lung injury of rats. There are few pathological studies on the effect of PM2.5 exposure on pulmonary fibrosis in animals. Hirofumi Kamata et al. confirmed that 14 nm carbon black nanoparticles could aggravate lung inflammation and fibrosis in rats, which is consistent with our results [26,27].

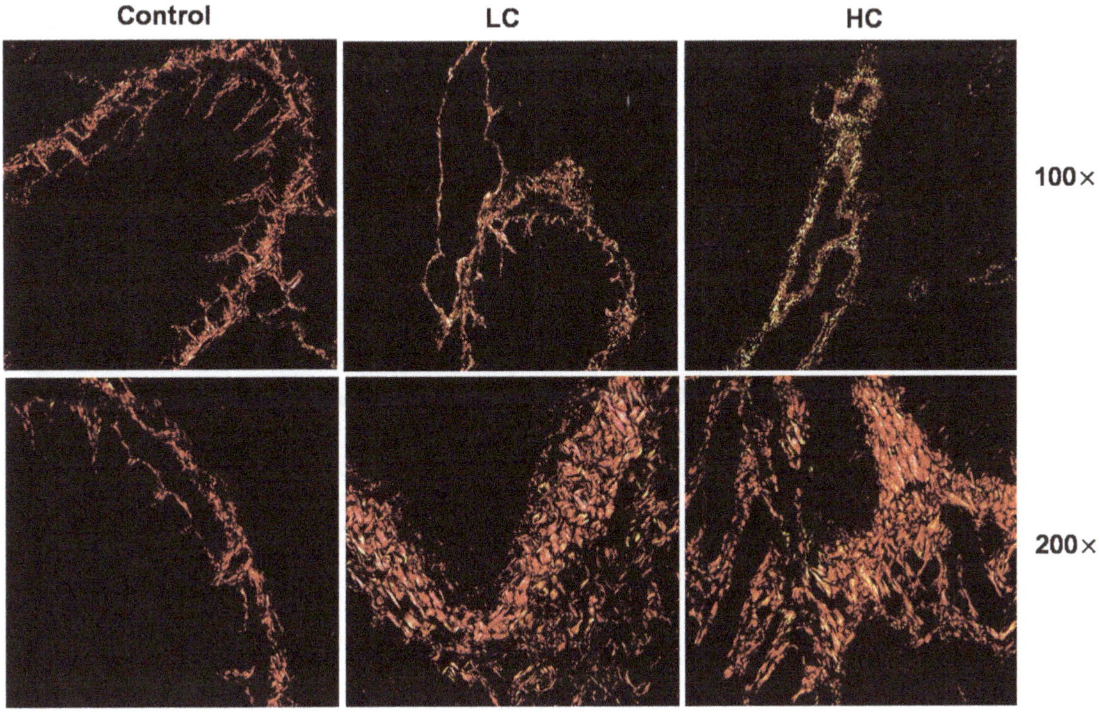

Figure 1. OMPM-induced PF in Wistar rat lung tissues with Sirius Red staining; 100×, bar: 100 μm; 200×, Bar: 50 μm.

Furthermore, Figure 2A–C show the classical PF biomarkers via an ELISA assay. We observed significant accumulation of COL1, α-SMA, and TGF-β1 in rat lung tissues after OMPM treatment, suggesting that OMPM may induce the extensive activation of lung fibroblasts and the accumulation of an extracellular matrix, thereby triggering the occurrence of PF. Meanwhile, compared with the LC group, the PF lesions were more serious in the HC group, which is consistent with the Sirius Red staining results, and also indicates that OMPM induced PF in a dose-dependent manner. In addition, Matrix metalloproteinase (MMP) and the tissue inhibitor of metalloproteinase (TIMP) are two important factors in the synthesis and degradation regulation of the extracellular matrix (ECM) [28]. MMP-2 mainly degrades collagens I and III, the main components of the extracellular matrix, and TIMP-1 can inhibit the activity of MMP-2 [29]. In the process of pulmonary fibrosis, MMP-2 activity decreases, while TIMP-1 activity increases continuously. Previous studies showed that the imbalance of the two activities is an important factor leading to the progression of PF [22]. In our study, compared with the control group, a significant imbalance of MMP-2 and TIMP-1 activity occurred (Figure 2D,E), indicating the occurrence of PF upon OMPM exposure. Moreover, with increased N-cadherin and decreased E-cadherin compared to the control rats (Figure 2F,G), two EMT and PF biomarkers further validated the OMPM-induced lung lesions and PF in rats. Taken together, these results indicate that OMPM might cause significant lung lesions and PF symptoms in a dose-dependent manner in vivo. In the rat model of lung injury and fibrosis constructed in this study, TGF-β1 was significantly increased in the peripheral blood, which was consistent with the changes in COL1 and α-SMA. TGF-β1 could promote intercellular adhesion and cross-cell interaction, and amplify the inflammatory cascade. Previous studies showed that TGF-β1 knockdown attenuates bleomycin-induced inflammatory cell infiltration and pulmonary fibrosis in rats, and attenuates fibrosis-related mediators [30,31].

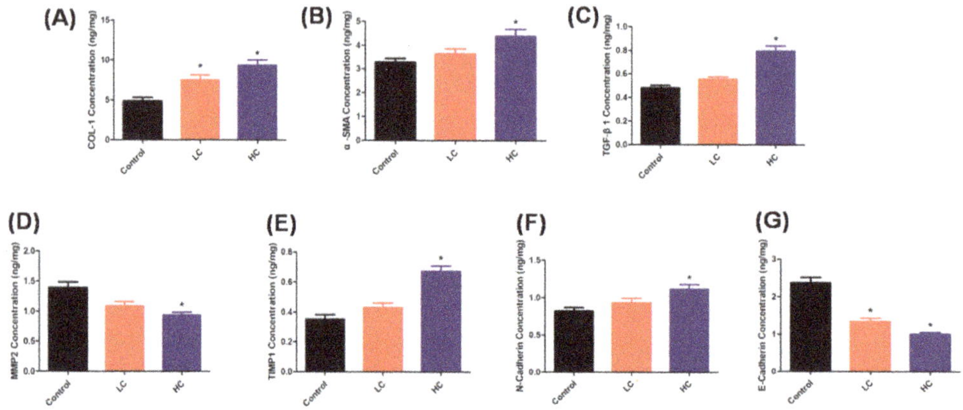

Figure 2. PF biomarkers were determined via ELISA assay after OMPM exposure at different concentrations in rat serum and lung tissues. (**A**) COL-1 concentration; (**B**) α-SMA concentration; (**C**) TGF-β1 concentration; (**D**) MMP2 concentration; (**E**) TIMP1 concentration; (**F**) N-cadherin concentration; (**G**) E-cadherin concentration. Data represent mean ± SD. * $p < 0.05$ vs. the control group.

3.2. OMPM Exposure Induced PF via the TGF-β1/Smad3 and TGF-β1/p38 MAPK Signaling Pathways

TGF-β1 is mainly composed of eosinophils, macrophages, fibroblasts, and secreted muscle fibroblasts. The TGF-β1/Smad3 and TGF-β1/p38 MAPK signaling pathways are keys to normal wound repair and repair-regulating mechanisms [15,32]. The signaling pathways are characteristic of many fibrosis diseases, and are fibroblast recruiters, activating and differentiating the central adjustment, thus playing a core role in the pathogenesis of PF [14,33]. A previous study found that the upregulation of the TGF-β1/Smad3 signaling pathway in the PM2.5 exposure group was significantly higher than that in the control group at each time point, suggesting that PM2.5 exposure aggravated PF by increasing the TGF-β1 [23,29]. However, it is not clear whether OMPM could induce PF via TGF-β1/Smad3 and TGF-β1/p38 MAPK signaling pathways. To solve this mystery, TGF-β1/Smad3 and TGF-β1/p38 MAPK signaling pathways were observed with Western blotting. Consistent with what we suspected, compared with the control group, there was a significant increase in the p-Smad3/Smad3 and p-p38/p38 ratios in the HC and LC groups (Figure 3A), which indicated that the TGF-β1/Smad3 and TGF-β1/p38 MAPK signaling pathways were activated, and the degree of activation of the TGF-β1/Smad3 and TGF-β1/p38 MAPK signaling pathways in the HC group was higher than that in the LC group (Figure 3B,C), which is consistent with the previous experimental data. Therefore, OMPM exposure may have induced PF via the TGF-β1/Smad3 and TGF-β1/p38 MAPK signaling pathways. In recent years, with the deepening of research on the structure and biological function of Smad3 and p38 MAPK, the role of Smad3 and p38 MAPK in organ fibrosis has gradually attracted attention [34]. Studies confirmed that the formation of the Smad3 heterodimer and the nuclear translocation of p38 MAPK are key links that are the biomarkers of the activation of the TGF-β1/Smad3 and TGF-β1/p38 MAPK signaling pathways [35]. Therefore, TGF-β1/Smad3 and TGF-β1/p38 MAPK signaling pathways play an important role in the PF and are the most closely related to the process of PF.

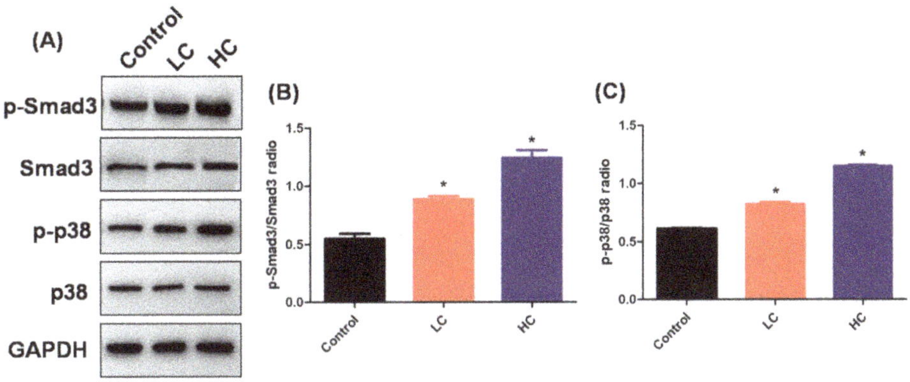

Figure 3. OMPM exposure induced PF via the TGF-β1/Smad3 and TGF-β1/p38 MAPK signaling pathways in rat lung tissues. (**A**) The phosphorylation of Smad3 and p38 after OMPM exposure detected with Western blotting; (**B**) ratio of p-Smad3/Smad3; (**C**) ratio of p-p38/p38. Data represent mean ± SD. * $p < 0.05$ vs. the control group.

3.3. OMPM Exposure Induced Fibrotic Phenotypes in the BEAS-2B Cell Line

Inhalable particulate matter could come into direct contact with and injure the bronchi after being inhaled into the human body, resulting in the inflammation and lesions of lung epithelial tissue, which is one of the main causes of PF [17,19]. Therefore, in the in vitro experiments, human bronchial epithelial BEAS-2B cell line was used to study the relationship between an OMPM-exposed lung epithelial cell line and fibrotic phenotypes. After the exposure of BEAS-2B cells to different types of OMPM, compared with the control group, ELISA and imunofluorescence showed that α-SMA, collagen I, and TGF-β1 levels in OMPM-treated cells were significantly increased and dose-dependent (Figure 4A–C,I,J), indicating that the high concentration of OMPM was more direct to damage to the BEAS-2B cell line. We further measured the levels of MMP2, TIMP1, and E-cadherin in the BEAS-2B cell line with ELISA. Alveolar macrophages can release some cytokines and proinflammatory factors under the imbalance of MMP2 and TIMP1, which can, in turn, further stimulate fibroblast and lung epithelial cells to secrete a variety of cytokines, such as IL-8, IL-6, and adhesion factors. At the same time, the imbalance of MMP2 and TIMP1 can cause varying inflammatory cell aggregation, and may have a profound effect on the pulmonary inflammatory response and fibrous hyperplasia. As shown in Figure 4D–G, in OMPM-exposed cells, the level of MMP2 was largely inhibited, and the level of TIMP1 was largely improved, suggesting that the imbalance of MMP2 and TIMP1 activities might induce the occurrence of PF upon OMPM exposure. Moreover, in previous studies, N-cadherin significantly increased in the peripheral blood, consistent with TGF-β1, and procollagens I and III, which promoted intercellular adhesion and intercellular interaction, and amplifying the inflammatory cascade. In our study, increased N-cadherin and decreased E-cadherin further validated that OMPM exposure could induce PF in the BEAS-2B cell line. Additionally, there was a significant increase in p-Smad3/Smad3 and p-p38/p38 ratios upon OMPM exposure in the BEAS-2B cell line, with a dose-dependent tendency (Figure 4H). Taken together, these data suggest that OMPM exposure could induce fibrotic phenotypes in the BEAS-2B cell line, which is in perfect agreement with the results of the animal experiments.

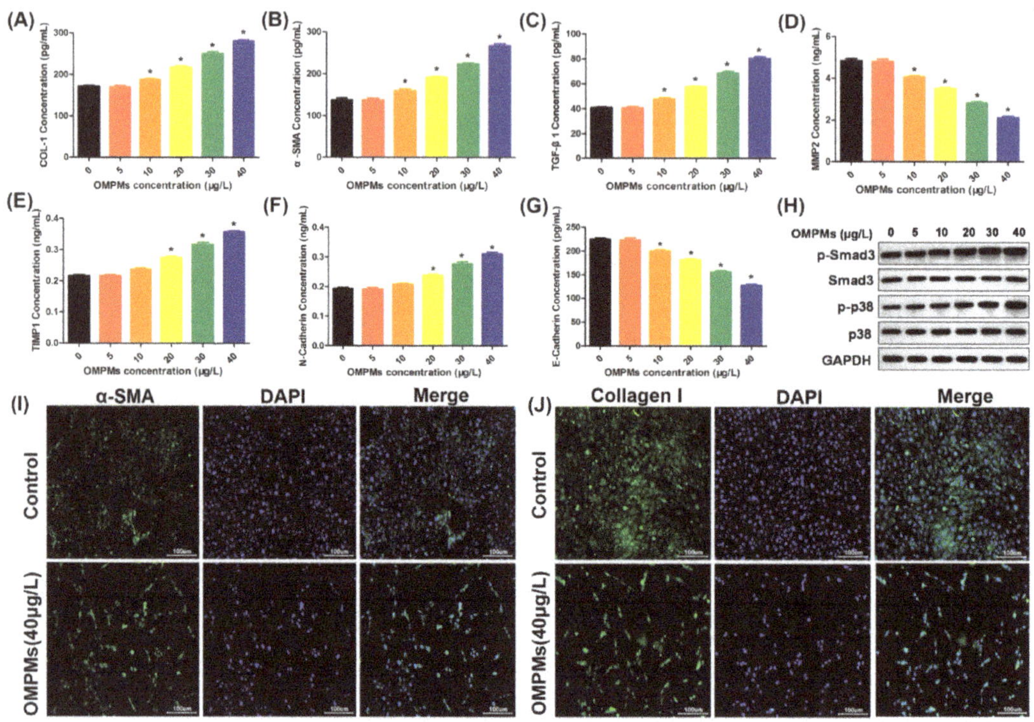

Figure 4. OMPM exposure induced fibrotic phenotypes in the BEAS-2B cell line. (**A**) COL-1 concentration; (**B**) α-SMA concentration; (**C**) TGF-β1 concentration; (**D**) MMP2 concentration; (**E**) TIMP1 concentration; (**F**) N-cadherin concentration; (**G**) E-cadherin concentration. (**H**) Phosphorylation of Smad3 and p38 after OMPM exposure detected with Western blotting. (**I**) α-SMA expression after OMPM exposure was detected using immunofluorescence. (**J**) Collagen I expression after OMPM exposure was detected using immunofluorescence; 100×, bar: 100 μm. Data represent mean ± SD. * $p < 0.05$ vs. the control group.

3.4. OMPM Exposure Activated the Differentiation of the HFL-1 Cell Line

The massive differentiation of lung fibroblasts is the main cause of lung-tissue inflammation and pulmonary fibrosis [1,36]. Therefore, we chose the HFL-1 cell line as the research object to explore whether OMPM could induce the differentiation of lung fibroblasts. As shown in Figure 5A–C, different concentrations of OMPM exposure triggered a dose-dependent increase in the expression of TGF-β1, collagen I, and α-SMA in the HFL-1 cell line. Moreover, the imbalance of MMP2 and TIMP1, and increased N-cadherin and decreased E-cadherin further validated that OMPM exposure could activate the differentiation of the HFL-1 cell line (Figure 5D–G), suggesting that OMPM exposure might exacerbate the ECM deposition and EMT process in the HFL-1 cell line. Furthermore, immunofluorescence assay verified that a treatment with a high concentration of OMPM (40 mg/m^3) caused a significant increase in the expression of myofibroblast markers α-SMA and collagen I (Figure 5I,J). A significant increase in p-Smad3/Smad3 and p-p38/p38 ratios was observed after OMPM treatment (Figure 5H), which might have contributed to the rapid profibrotic activation to OMPM in the HFL-1 cell line. Taken together, these data indicate that OMPM promoted the activation of the TGF-β1/Smad3 and TGF-β1/p38 MAPK signaling pathways, and lung fibrotic differentiation phenotypes in the HFL-1 cell line.

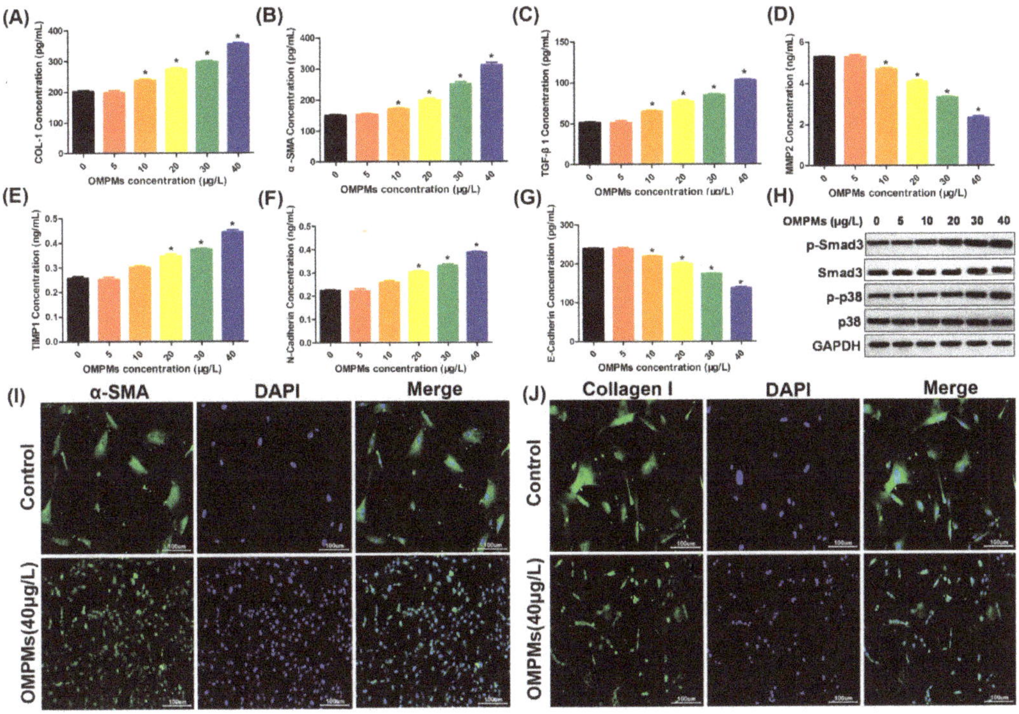

Figure 5. OMPM exposure activated the differentiation of the HFL-1 cell line. (**A**) COL-1 concentration; (**B**) α-SMA concentration; (**C**) TGF-β1 concentration; (**D**) MMP2 concentration; (**E**) TIMP1 concentration; (**F**) N-Cadherin concentration; (**G**) E-cadherin concentration. (**H**) Phosphorylation of Smad3 and p38 after OMPM exposure detected with Western blotting. (**I**) α-SMA expression after OMPM exposure was detected using immunofluorescence. (**J**) Collagen I expression after OMPM exposure was detected using immunofluorescence; 100×, bar: 100 μm. Data represent mean ± SD. * $p < 0.05$ vs. the control group.

3.5. OMPM Exposed BEAS-2B Cell Line Could Induce Differentiation of HFL-1 Cell Line

Dysfunctional lung epithelial cells often cause pulmonary fibrosis by interacting with lung fibroblasts [37]. Therefore, we hypothesized that OMPM-induced pulmonary epithelial dysfunction might also lead to pulmonary fibrosis phenotypes by interacting with pulmonary fibroblasts. To assess this hypothesis, a culture medium of non-OMPM-M and OMPM-M, cell supernatants of non-OMPM-E and OMPM-E groups of BEAS-2B cell lines were used to incubate the HFL-1 cell line. As shown in Figure 6A, B, compared with the non-OMPM-M, OMPM-M, and non-OMPM-E groups, the medium supernatants of the OMPM-E groups of the BEAS-2B cell line triggered an increase in the expression of collagen I and α-SMA in the HFL-1 cell line. Once the TGF-β1 inhibitor had been added, however, the levels of collagen I and α-SMA returned to normal. Similar to previous results, an imbalance of MMP2 and TIMP1, increased N-cadherin, and decreased E-cadherin further validated that the medium supernatants of the OMPM-E groups of the BEAS-2B cell line could activate the differentiation of the HFL-1 cell line, and the TGF-β1 inhibitor had the same effect (Figure 6C–F). Furthermore, immunofluorescence staining showed that, compared with the non-OMPM-M, OMPM-M, non-OMPM-E, and TGF-β1 inhibitor groups, the expression of α-SMA and collagen I was increased in the cell supernatants of the OMPM-E group-treated HFL-1 cell line (Figure 6G,H). Consistent with immunofluorescence staining, Western blotting showed that a significant increase in the p-Smad3/Smad3 and p-p38/p38 ratios was observed after the treatment of the medium supernatants of the

OMPM-E group of the BEAS-2B cell line, indicating that the OMPM-exposed BEAS-2B cell line could activate the TGF-β1/Smad3 and TGF-β1/p38 MAPK signaling pathways (Figure 6I). The results confirm our hypothesis that the OMPM-exposed BEAS-2B cell line could promote the activation of the TGF-β1/ Smad3 and TGF-β1/p38 MAPK signaling pathways and lung fibrotic differentiation phenotypes in the HFL-1 cell line. The activation of the TGF-β1/ Smad3 and TGF-β1/p38 MAPK signaling pathways, and lung fibrotic differentiation phenotypes in the HFL-1 cell line can be regarded as the initial event of PF in vitro [38,39].

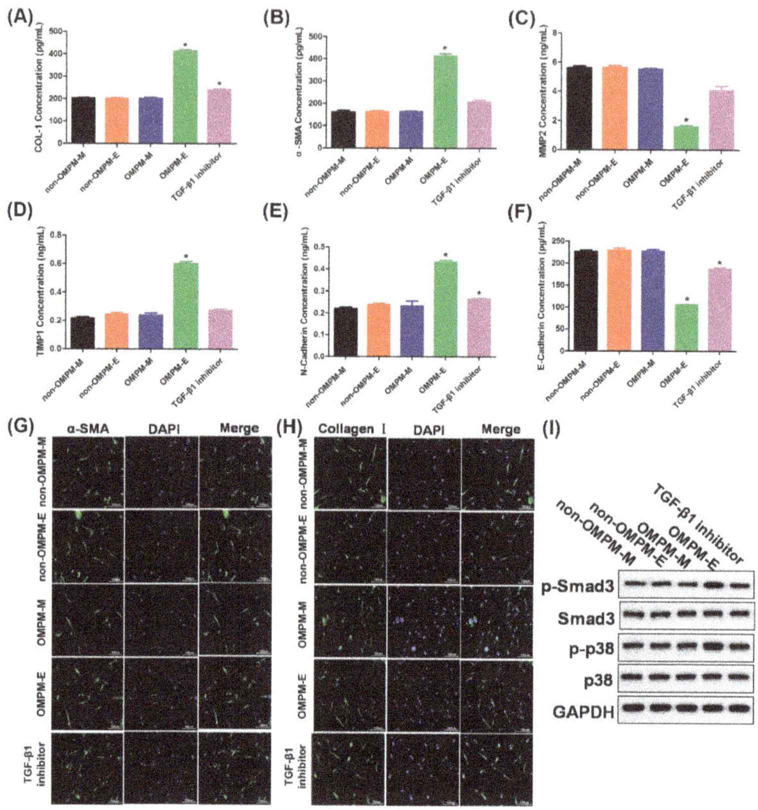

Figure 6. Different ways with which OMPM-exposed BEAS-2B cell lines could induce the differentiation of the HFL-1 cell line. (**A**) COL-1 concentration; (**B**) α-SMA concentration; (**C**) MMP2 concentration; (**D**) TIMP1 concentration; (**E**) N-cadherin concentration; (**F**) E-cadherin concentration. (**G**) α-SMA expression after the treatment of the medium of non-OMPM-M and OMPM-M cell supernatants of non-OMPM-E and OMPM-E groups of BEAS-2B cell lines was detected using immunofluorescence. (**H**) Collagen I expression after the treatment of the medium of non-OMPM-M and OMPM-M cell supernatants of non-OMPM-E and OMPM-E groups of BEAS-2B cell lines was detected using immunofluorescence; 100×, bar: 100 μm. (**I**) Phosphorylation of Smad3 and p38 after the treatment of the medium of non-OMPM-M and OMPM-M cell supernatants of non-OMPM-E and OMPM-E groups of BEAS-2B cell lines was detected with Western blotting. Non-OMPM-M, medium for BEAS-2B without OMPM treatment before it was used for culture; OMPM-M, medium for BEAS-2B with OMPM treatment before it was used for culture; non-OMPM-E, BEAS-2B that was not exposed to OMPM; OMPM-M, BEARS-2B that was exposed to OMPM. Data represent mean ± SD. * $p < 0.05$ vs. the non-OMPM-M, OMPM-M, and non-OMPM-E groups.

4. Conclusions

Our work systematically demonstrated that OMPM plays an important role in inducing PF in vivo and in vitro. First, OMPM could cause lung tissue lesions and PF in rats. In addition, in the in vitro cell experiments, we further confirmed the molecular mechanism of the profibrotic effect of OMPM, and OMPM exposure could trigger the EMT development and fibroblast differentiation of pulmonary bronchial epithelial cells. Therefore, our work provides a theoretical and practical basis for filling in the blank of whether there is a health risk of OMPM in offshore operations, and has constructive significance for the sustainable development of offshore operations.

Author Contributions: H.N.: conceptualization, methodology, validation, project administration, writing—original draft, writing—review and editing. H.L.: methodology, validation, data curation, formal analysis. Y.S.: methodology, validation, data curation, formal analysis. W.L.: methodology, validation, data curation, formal analysis. X.L.: methodology, validation, data curation. Z.X.: conceptualization, supervision, writing—review and editing, funding acquisition. B.L.: conceptualization, writing—review and editing, funding acquisition, project administration. All authors have read and agreed to the published version of the manuscript.

Funding: This research was funded by Zhuge Xi and Bencheng Lin, grant number 21WQ046.

Institutional Review Board Statement: The study was conducted according to the guidelines of the Declaration of Helsinki, and approved by Institutional Review Board of the Experimental Animal Ethics Committee of the Tianjin Institute of Environmental and Occupational Medicine. Approval Code: IACUC of AMMS-04-2020-016. Approval Date: 2020-04-01.

Data Availability Statement: Not applicable.

Conflicts of Interest: This manuscript has not been published in whole or in part and is not being considered for publication elsewhere. This manuscript does not violate or infringe upon any existing copyright/s license/s from any third party. All authors contributed significantly to work and agree with the manuscript's content. There are no conflict of interest.

References

1. Chen, K.F.; Tsai, Y.P.; Lai, C.H.; Xiang, Y.K.; Chuang, K.Y.; Zhu, Z.H. Human health-risk assessment based on chronic exposure to the carbonyl compounds and metals emitted by burning incense at temples. *Environ. Sci. Pollut. Res. Int.* **2021**, *28*, 40640–40652. [CrossRef] [PubMed]
2. Fois, A.G.; Paliogiannis, P.; Sotgia, S.; Mangoni, A.A.; Zinellu, E.; Pirina, P.; Carru, C.; Zinellu, A. Evaluation of oxidative stress biomarkers in idiopathic pulmonary fibrosis and therapeutic applications: A systematic review. *Respir. Res.* **2018**, *19*, 51. [CrossRef] [PubMed]
3. Frey, A.; Lunding, L.P.; Ehlers, J.C.; Weckmann, M.; Zissler, U.M.; Wegmann, M. More Than Just a Barrier: The Immune Functions of the Airway Epithelium in Asthma Pathogenesis. *Front. Immunol.* **2020**, *11*, 761. [CrossRef]
4. Weiskirchen, R.; Weiskirchen, S.; Tacke, F. Organ and tissue fibrosis: Molecular signals, cellular mechanisms and translational implications. *Mol. Asp. Med.* **2019**, *65*, 2–15. [CrossRef] [PubMed]
5. Lai, C.H.; Chen, Y.C.; Lin, K.A.; Lin, Y.X.; Lee, T.H.; Lin, C.H. Adverse pulmonary impacts of environmental concentrations of oil mist particulate matter in normal human bronchial epithelial cell. *Sci. Total Environ.* **2022**, *809*, 151119. [CrossRef] [PubMed]
6. Liu, N.; Feng, J.; Lu, X.; Yao, Z.; Liu, Q.; Lv, Y.; Han, Y.; Deng, J.; Zhou, Y. Isorhamnetin Inhibits Liver Fibrosis by Reducing Autophagy and Inhibiting Extracellular Matrix Formation via the TGF-beta1/Smad3 and TGF-beta1/p38 MAPK Pathways. *Mediat. Inflamm.* **2019**, *2019*, 6175091. [CrossRef] [PubMed]
7. Zhang, L.; Ren, L.; Shan, K.; Guo, X.; Wang, J.; Cui, B.; An, J. Serum Inflammatory Cytokines Comparison in Gastric Cancer Therapy. *Open Med.-Warsaw.* **2019**, *14*, 300–306. [CrossRef] [PubMed]
8. Liu, S.; Zhang, W.; Zhang, F.; Roepstorff, P.; Yang, F.; Lu, Z.; Ding, W. TMT-Based Quantitative Proteomics Analysis Reveals Airborne PM2.5-Induced Pulmonary Fibrosis. *Int. J. Environ. Res. Public Health* **2018**, *16*, 98. [CrossRef]
9. Murphy, M.P.; McEnery, T.; McQuillan, K.; McElvaney, O.F.; McElvaney, O.J.; Landers, S.; Coleman, O.; Bussayajirapong, A.; Hawkins, P.; Henry, M.; et al. alpha1 Antitrypsin therapy modulates the neutrophil membrane proteome and secretome. *Eur. Respir. J.* **2020**, *55*, 1901678. [CrossRef]
10. Deng, X.; Zhang, F.; Wang, L.; Rui, W.; Long, F.; Zhao, Y.; Chen, D.; Ding, W. Airborne fine particulate matter induces multiple cell death pathways in human lung epithelial cells. *Apoptosis* **2014**, *19*, 1099–1112. [CrossRef] [PubMed]
11. Cakmak, S.; Dales, R.; Kauri, L.M.; Mahmud, M.; Van Ryswyk, K.; Vanos, J.; Liu, L.; Kumarathasan, P.; Thomson, E.; Vincent, R.; et al. Metal composition of fine particulate air pollution and acute changes in cardiorespiratory physiology. *Environ. Pollut.* **2014**, *189*, 208–214. [CrossRef]

12. Xu, Z.; Li, Z.; Liao, Z.; Gao, S.; Hua, L.; Ye, X.; Wang, Y.; Jiang, S.; Wang, N.; Zhou, D.; et al. PM2.5 induced pulmonary fibrosis in vivo and in vitro. *Ecotoxicol. Environ. Saf.* **2019**, *171*, 112–121. [CrossRef] [PubMed]
13. Zhang, J.; Pan, W.; Long, Z.; Wang, C.; Feng, Z. Study of the Oil Mist Filtration Performance: Pressure Drop Characteristics and Filter Efficiency Model. *Aerosol Air Qual. Res.* **2017**, *17*, 1063–1072. [CrossRef]
14. Yin, P.; Guo, J.; Wang, L.; Fan, W.; Lu, F.; Guo, M.; Moreno, S.B.R.; Wang, Y.; Wang, H.; Zhou, M.; et al. Higher Risk of Cardiovascular Disease Associated with Smaller Size-Fractioned Particulate Matter. *Environ. Sci. Technol. Lett.* **2020**, *7*, 95–101. [CrossRef]
15. Chu, H.; Xin, J.; Yuan, Q.; Zhang, X.; Pan, W.; Zeng, X.; Chen, Y.; Ma, G.; Ge, Y.; Du, M.; et al. Evaluation of vulnerable PM2.5-exposure individuals: A repeated-measure study in an elderly population. *Environ. Sci. Pollut. Res.* **2018**, *25*, 11833–11840. [CrossRef]
16. Bergstra, A.D.; Brunekreef, B.; Burdorf, A. The effect of industry-related air pollution on lung function and respiratory symptoms in school children. *Environ. Health* **2018**, *17*, 30. [CrossRef] [PubMed]
17. Camelo, A.; Dunmore, R.; Sleeman, M.A.; Clarke, D.L. The epithelium in idiopathic pulmonary fibrosis: Breaking the barrier. *Front. Pharmacol.* **2014**, *4*, 173. [CrossRef] [PubMed]
18. Xu, M.; Wang, X.; Xu, L.; Zhang, H.; Li, C.; Liu, Q.; Chen, Y.; Chung, K.F.; Adcock, I.M.; Li, F. Chronic lung inflammation and pulmonary fibrosis after multiple intranasal instillation of PM2.5 in mice. *Environ. Toxicol.* **2021**, *36*, 1434–1446. [CrossRef] [PubMed]
19. Gaston, B.; May, W.J.; Sullivan, S.; Yemen, S.; Marozkina, N.V.; Palmer, L.A.; Bates, J.N.; Lewis, S.J. Essential role of hemoglobin beta-93-cysteine in posthypoxia facilitation of breathing in conscious mice. *J. Appl. Physiol.* **2014**, *116*, 1290–1299. [CrossRef] [PubMed]
20. Ghatak, S.; Hascall, V.C.; Markwald, R.R.; Feghali-Bostwick, C.; Artlett, C.M.; Gooz, M.; Bogatkevich, G.S.; Atanelishvili, I.; Silver, R.M.; Wood, J.; et al. Transforming growth factor beta1 (TGFbeta1)-induced CD44V6-NOX4 signaling in pathogenesis of idiopathic pulmonary fibrosis. *J. Biol. Chem.* **2017**, *292*, 10490–10519. [CrossRef] [PubMed]
21. Guo, W.; Saito, S.; Sanchez, C.G.; Zhuang, Y.; Gongora Rosero, R.E.; Shan, B.; Luo, F.; Lasky, J.A. Translational Research in Acute Lung Injury and Pulmonary Fibrosis: TGF-beta1 stimulates HDAC4 nucleus-to-cytoplasm translocation and NADPH oxidase 4-derived reactive oxygen species in normal human lung fibroblasts. *Am. J. Physiol. Lung Cell Mol. Physiol.* **2017**, *312*, L936–L944. [CrossRef]
22. de Brito, J.M.; Mauad, T.; Cavalheiro, G.F.; Yoshizaki, K.; de André, P.A.; Lichtenfels, A.J.F.C.; Guimarães, E.T.; Rivero, D.H.R.F.; Antonangelo, L.; Oliveira, L.B.; et al. Acute exposure to diesel and sewage biodiesel exhaust causes pulmonary and systemic inflammation in mice. *Sci. Total Environ.* **2018**, *628–629*, 1223–1233. [CrossRef] [PubMed]
23. Hewlett, J.C.; Kropski, J.A.; Blackwell, T.S. Idiopathic pulmonary fibrosis: Epithelial-mesenchymal interactions and emerging therapeutic targets. *Matrix Biol.* **2018**, *71–72*, 112–127. [CrossRef] [PubMed]
24. Asker, S.; Asker, M.; Yeltekin, A.C.; Aslan, M.; Ozbay, B.; Demir, H.; Turan, H. Serum levels of trace minerals and heavy metals in severe COPD patients with and without pulmonary hypertension. *Int. J. Chronic Obs. Pulm. Dis.* **2018**, *13*, 1803–1808. [CrossRef] [PubMed]
25. Nie, H.; Liu, H.; Shi, Y.; Lai, W.; Liu, X.; Xi, Z.; Lin, B. Combined multi-omics analysis reveals oil mist particulate matter-induced lung injury in rats: Pathological damage, proteomics, metabolic disturbances, and lung dysbiosis. *Ecotoxicol. Environ. Saf.* **2022**, *241*, 113759. [CrossRef] [PubMed]
26. Allen, S.; Allen, D.; Phoenix, V.R.; Le Roux, G.; Durántez Jiménez, P.; Simonneau, A.; Binet, S.; Galop, D. Atmospheric transport and deposition of microplastics in a remote mountain catchment. *Nat. Geosci.* **2019**, *12*, 339–344. [CrossRef]
27. Almeida, M.; Martins, M.A.; Soares, A.M.V.; Cuesta, A.; Oliveira, M. Polystyrene nanoplastics alter the cytotoxicity of human pharmaceuticals on marine fish cell lines. *Environ. Toxicol. Pharmacol.* **2019**, *69*, 57–65. [CrossRef]
28. Ornatowski, W.; Lu, Q.; Yegambaram, M.; Garcia, A.E.; Zemskov, E.A.; Maltepe, E.; Fineman, J.R.; Wang, T.; Black, S.M. Complex interplay between autophagy and oxidative stress in the development of pulmonary disease. *Redox Biol.* **2020**, *36*, 101679. [CrossRef] [PubMed]
29. Saito, S.; Zhuang, Y.; Shan, B.; Danchuk, S.; Luo, F.; Korfei, M.; Guenther, A.; Lasky, J.A. Tubastatin ameliorates pulmonary fibrosis by targeting the TGFbeta-PI3K-Akt pathway. *PLoS ONE* **2017**, *12*, e0186615. [CrossRef] [PubMed]
30. Cole, M.; Coppock, R.; Lindeque, P.K.; Altin, D.; Reed, S.; Pond, D.W.; Sørensen, L.; Galloway, T.S.; Booth, A.M. Effects of Nylon Microplastic on Feeding, Lipid Accumulation, and Moulting in a Coldwater Copepod. *Environ. Sci. Technol.* **2019**, *53*, 7075–7082. [CrossRef] [PubMed]
31. Dawson, A.L.; Kawaguchi, S.; King, C.K.; Townsend, K.A.; King, R.; Huston, W.M.; Bengtson Nash, S.M. Turning microplastics into nanoplastics through digestive fragmentation by Antarctic krill. *Nat. Commun.* **2018**, *9*, 1001. [CrossRef]
32. Zychowski, K.E.; Kodali, V.; Harmon, M.; Tyler, C.R.; Sanchez, B.; Ordonez Suarez, Y.; Herbert, G.; Wheeler, A.; Avasarala, S.; Cerrato, J.M.; et al. Respirable Uranyl-Vanadate-Containing Particulate Matter Derived From a Legacy Uranium Mine Site Exhibits Potentiated Cardiopulmonary Toxicity. *Toxicol. Sci.* **2018**, *164*, 101–114. [CrossRef]
33. Zhang, Z.Q.; Zhang, C.Z.; Shao, B.; Pang, D.H.; Han, G.Z.; Lin, L. Effects of abnormal expression of fusion and fission genes on the morphology and function of lung macrophage mitochondria in SiO2-induced silicosis fibrosis in rats in vivo. *Toxicol. Lett.* **2019**, *312*, 181–187. [CrossRef]

34. Hahladakis, J.N.; Velis, C.A.; Weber, R.; Iacovidou, E.; Purnell, P. An overview of chemical additives present in plastics: Migration, release, fate and environmental impact during their use, disposal and recycling. *J. Hazard. Mater.* **2018**, *344*, 179–199. [CrossRef] [PubMed]
35. Keller, A.S.; Jimenez-Martinez, J.; Mitrano, D.M. Transport of Nano- and Microplastic through Unsaturated Porous Media from Sewage Sludge Application. *Environ. Sci. Technol.* **2020**, *54*, 911–920. [CrossRef] [PubMed]
36. Chi, Y.; Huang, Q.; Lin, Y.; Ye, G.; Zhu, H.; Dong, S. Epithelial-mesenchymal transition effect of fine particulate matter from the Yangtze River Delta region in China on human bronchial epithelial cells. *J. Environ. Sci.* **2018**, *66*, 155–164. [CrossRef]
37. Dysart, M.M.; Galvis, B.R.; Russell, A.G.; Barker, T.H. Environmental particulate (PM2.5) augments stiffness-induced alveolar epithelial cell mechanoactivation of transforming growth factor beta. *PLoS ONE* **2014**, *9*, e106821. [CrossRef]
38. Leppänen, I.; Lappalainen, T.; Lohtander, T.; Jonkergouw, C.; Arola, S.; Tammelin, T. Capturing colloidal nano- and microplastics with plant-based nanocellulose networks. *Nat. Commun.* **2022**, *13*, 1814. [CrossRef]
39. Malafaia, G.; de Souza, A.M.; Pereira, A.C.; Gonçalves, S.; da Costa Araújo, A.P.; Ribeiro, R.X.; Rocha, T.L. Developmental toxicity in zebrafish exposed to polyethylene microplastics under static and semi-static aquatic systems. *Sci. Total Environ.* **2020**, *700*, 134867. [CrossRef]

toxics

MDPI

Article

Genotoxic Biomonitoring in Children Living near the El Fraile Mine Tailings in Northern Guerrero State, Mexico

María Elena Calderon-Segura [1,*], Alejandro Ramírez-Guzmán [2], Oscar Talavera-Mendoza [2], Yolanda Carbajal-López [3], María del Carmen Martínez-Valenzuela [4], Martha Elena Mora-Herrera [5], Liliana Salinas-Alcántara [1] and Patricia Hurtado-Brito [1]

1 Laboratorio de Toxicología Ambiental, Departamento de Ciencias Ambientales, Instituto de Ciencias de la Atmósfera y Cambio Climático, Universidad Nacional Autónoma de México, Ciudad Universitaria Coyoacán, Ciudad de México 04510, Mexico

2 Escuela Superior de Ciencias de la Tierra, Universidad Autónoma de Guerrero, Ex Hacienda de San Juan Bautista s/n, Taxco el Viejo 40323, Mexico

3 Facultad de Ciencias Químico Biológicas, Universidad Autónoma de Guerrero, Av. Lázaro Cárdenas s/n, Chilpancingo de los Bravo 39087, Mexico

4 Instituto de Investigaciones en Ambiente y Salud, Universidad Autónoma de Occidente, Boulevar Macario Gaxiola, Carretera Internacional, Los Mochis 81200, Mexico

5 Laboratorio de Fisiología y Biotecnología Vegetal, Centro Universitario Tenancingo, Universidad Autónoma, Tenancingo 52400, Mexico

* Correspondence: mcalderon@atmosfera.unam.mx

Citation: Calderon-Segura, M.E.; Ramírez-Guzmán, A.; Talavera-Mendoza, O.; Carbajal-López, Y.; Martínez-Valenzuela, M.d.C.; Mora-Herrera, M.E.; Salinas-Alcántara, L.; Hurtado-Brito, P. Genotoxic Biomonitoring in Children Living near the El Fraile Mine Tailings in Northern Guerrero State, Mexico. *Toxics* **2022**, *10*, 674. https://doi.org/10.3390/toxics10110674

Academic Editors: Maaike van Gerwen, Roel Vermeulen and Lauren Petrick

Received: 29 September 2022
Accepted: 3 November 2022
Published: 8 November 2022

Abstract: A genotoxic study was conducted with 101 elementary school children (56 girls and 45 boys) in the 6–7, 8–9, and 10–12 age ranges from El Fraile rural community, which is located beside the El Fraile mine tailings in Taxco of Alarcon City, in northern Guerrero State, Mexico. For this, we used the alkaline comet assay in exfoliated buccal mucosa cells, scoring three genotoxic parameters: tail intensity, tail moment, and tail length. Additionally, we detected oxidative DNA damage through urinary 8-OHdG levels by enzyme-linked immunosorbent assay. We also evaluated a control group consisting of 101 children in the same age ranges from Chilpancingo City, Guerrero, who had never lived near mining zones. Genotoxic results showed that there was a significant increase in three genotoxic parameters and urinary 8-OHdG levels in the exposed children group compared with the control group. Analysis of MANOVA revealed that boys aged 8 and 9 years had higher DNA damage than girls from the same exposure group, and Spearman's analysis identified a positive correlation between DNA damage and sex and age. This study provides the first valuable genotoxic data in children living in areas with environmental pollution.

Keywords: DNA damage; exfoliated buccal mucosa cells; urinary 8-OHdG adduct; children; El Fraile Taxco mine tailings; México

1. Introduction

The Taxco Mining District in southern Mexico was one of the richest producers of precious metals (Ag and Au) in the Americas during colonial (1523–1810) times. Throughout much of the 20th Century, Taxco mines were renowned for their production of base metals (Cu, Pb, and Zn) and Ag. More than 450 years of intensive mining activity has resulted in the discharge of large quantities of tailings and mining wastes into the surrounding areas, and some of these tailings are located within communities (Santa Rosa, Dolores, and El Fraile) (Figure 1) [1,2]. The El Fraile rural community, which is located beside the El Fraile mine tailings and near the abandoned El Fraile mine, is 15 km from Taxco of Alarcon City, in northern Guerrero State, Mexico (Figure 1). In this community, people occasionally use leachates as an alternative source of domestic water during the dry (winter/spring) seasons. Furthermore, edible plants such as tomatoes, peppers, and corn for human consumption are grown in the local soil, and animals also consume plants that grow near El Fraile tailings, increasing the health risk in the trophic chain [1,2] (Figure 1). Many studies in this area have indicated that the water, soil, and plants contain heavy metals such as Ni, Hg, Pb, As, and Cd, and other elements such as Fe, Mn, and Cu [1,2]. High heavy metal levels in human urine [3,4], vegetal tissues, soil [1], and water [5] clearly indicate high exposure of the general population. However, developing children are the most sensitive population to the adverse effects of these heavy metals [6]. Children living near or on mining residue are routinely exposed to a large number of genotoxic heavy metals by inhalation, dermal absorption, drinking water, and contaminated food [7].

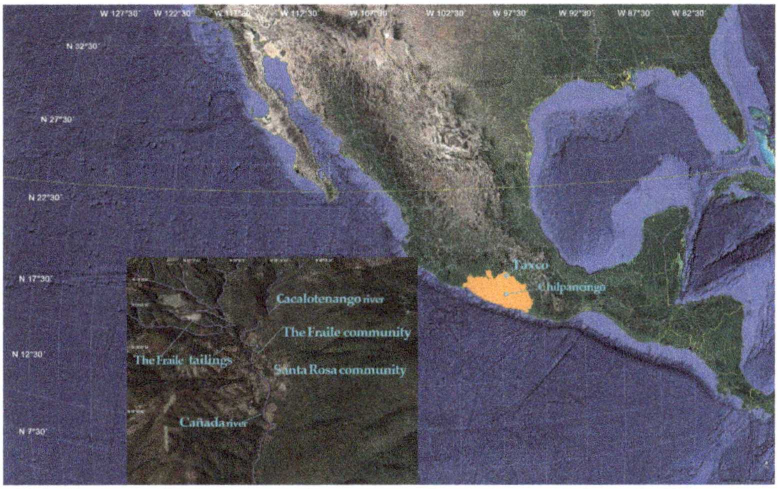

Figure 1. Study areas. The El Fraile community is located beside the El Fraile mine tailings in Taxco of Alarcon City in Northern, and Chilpancingo City in South, Guerrero State, México.

Different scientific studies have suggested that certain modifications in the central nervous system may also originate during early development by exposure to heavy metals [8–10]. Neural disorders in adults have been linked to environmental factors acting in the preconceptual, embryonic, fetal, and infantile phases of life [9,10]. The impact of such influences early in life may be partially mediated through oxidative stress that involves DNA damage and other biochemistry and molecular mechanisms [11,12]. Other studies point to a relationship between exposure to heavy metals and an increased risk of cancer [13] in the lungs [14] and in the liver [15], among others [11].

Heavy metals directly or indirectly exert their toxicity by producing a large quantity of reactive nitrogen species (RNS) or reactive oxygen species (ROS), such as hydroxyl and superoxide radicals, as well as nonradical species such as H_2O_2. [16,17]. Among these, hydroxyl radicals are the most reactive species; thus, they can chemically attack DNA, causing strand breaks, and oxidation of their bases gives rise to a range of oxidative DNA damage products, including 8-hydroxyadenine, 5,6-dihydroxyuracil, 5-hydroxycytosine, and several others [18].

One lesion of this type is 8-hydroxy-deoxyguanosine (8-OHdG), which is formed by the reaction of the hydroxyl radical at the C-8 position of the guanine on DNA; 8-OHdG is excised by constitutive enzymatic repair systems, appears in the blood, and is subsequently excreted in the urine [19]. The 8-OHdG level in the blood or urine is measured as a marker of oxidative DNA damage induced by environmental pollution [20,21]. It is associated with the development of various diseases, tumors, and cancers. Urinary 8-OHdG, in particular, has been the measurement most frequently used to indicate the extent of oxidative DNA damage and is an excellent molecular biomarker for mutagenic and/or carcinogenic agents. Furthermore, 8-OHdG can pair with both adenine and cytosine, but if an adenine and guanine (A:G) mismatch is not repaired, there will be a transversion of regular pairs adenine and timidine (A:T), and cytosine and guanine (C:G) [22,23].

Monitoring of biological effects as a measure of the internally effective dose is relevant for the assessment of ultimate health risks, such as cancer [24]. Many biomarkers, such as micronuclei [4], sister chromatid exchanges [25], and alkaline comet assays, have been developed to estimate exposure and to assess the risk of adverse health effects in an early phase by environmental genotoxins [26,27]. In human populations, alkaline comet assays can be easily assessed in diverse cellular lines and exfoliated epithelial cells (e.g., oral, urothelial, nasal) to obtain a measure of genome damage induced in vivo [28,29]. Primary DNA damage is considered to be an important initial event in carcinogenesis and in the development of diseases [30].

The alkaline comet assay (single-cell gel electrophoresis) has become the preferred test for the qualitative and quantitative assessment of DNA damage in single cells, and is capable of detecting DNA single- and double-strand breaks, alkali-labile sites, incomplete excision repair sites, and genomic structural discontinuities [31]. Given the recognized effects of heavy metal exposure on children's health, this research evaluated DNA damage in a group of 101 children (6 to 12 years old) living beside the El Fraile mine tailings, and a nonexposure group consisting of 101 nonexposed children of the same ages from Chilpancingo City, Guerrero, Mexico, who had never lived near mine zones or been exposed to heavy metals, through the two biomarkers of exposure, urinary 8-OHdG adduct and comet alkaline assay in exfoliated buccal mucosa cells.

2. Materials and Methods

2.1. Study Areas

El Fraile mining tailings are located between the parallels 18°30′ and 18°35′ and meridians 99°32′30″ and 99°40′ in the southeastern part of Taxco de Alarcón City. The El Fraile community is located beside the El Fraile mine tailings, 100 m from the abandoned El Fraile mine; 15 km from Taxco of Alarcon City in Northern Guerrero State, México (Figure 1).

2.2. Study Populations

The exposure group consisted of 101 elementary school children living in the El Fraile rural community among mining tailings, 100 m from the abandoned El Fraile mine. The control group also consisted of 101 elementary school children from Chilpancingo City in South Guerrero State, México, who had never lived near or in mining zones (Figure 1). The characteristics of both groups are described in Table 1. The Ethics Committee of the Faculty of Medicine from Universidad Nacional Autonoma de Mexico approved the research procedures used in this study, and a clinical questionnaire was administered to parents aiming to obtain information on food consumption habits, leisure activities, and current and former occupations of the parents. All donors were otherwise healthy. The parents were provided with information about the genetic research and informed that access to exfoliated buccal mucosa cells and urine samples are noninvasive methods for biological specimen collection and are relatively safe. Written informed consent was obtained from the parents of each participant in accordance with the Declaration of Helsinki Ethical Principles (2008). The age range of the exposure and control groups was 6 to 12 years old. The exposure group consisted of 56 girls and 45 boys, and the control group included 52 girls and 49 boys (Table 1). Before biological sample collection, exclusion criteria were applied for both exposure and control groups, and parents were asked for detailed information regarding age, diet, habits, present or previous history of any disease or exposure to diagnostic X-rays, medical examinations, and years of living in their current home. None of the healthy children used any medications during the study period.

Table 1. Sociodemographic characteristics in the (El Fraile, Taxco de Alarcón)-exposure and (Chilpancingo)-control groups.

Variable	Control Group [a]	Exposure Group [a]
Age ranges	6–12	6–12
Gender	52 girls 49 boys	56 girls 45 boys
Living area	Urban	Rural
Diet	Not special	Not special
Lifestyle Children play in garden	80%	85%
Type of drinking water and zone of residence (close to or far from mine tailings)	Bottled 60%	Bottled 100%
Frequency of children's contact with soil Every day	10%	85%
Frequency of consuming fruits and edible vegetables grown on site	No	100%
Frequency of Cacalotenango water river use	None	100%
Frequency of rain water use Every day 2–3 times/week Once a week	None	70%
Children consuming meat from animals grown on site	None	85%
Children consuming milk from cows grown on site	None	80%
Socioeconomic status (parental education level)	Elementary, high school	Elementary school

[a] n = 101.

2.3. *Oral Mucosa Samples Collection*

Exfoliated buccal mucosa samples were collected from the exposure and control children groups with a sterile spoon for each child and kept in sterile microtubes with phosphate-buffered saline (PBS; pH 7.4; GIBCO™ Thermo Fisher Scientific, Grand Island, New York, NY, USA). All samples were kept on ice and immediately transferred to the Geosciences Laboratory (Escuela Superior de Ciencias de la Tierra, Universidad Autónoma de Guerrero), and centrifuged at 1500 rpm for 10 min. The cellular layer was diluted with PBS and immediately assessed in the alkaline comet assay [29,31].

2.4. *Urine Collection*

Urine samples of all children (exposure and control groups) were collected from the first morning void in polypropylene containers (25–50 mL), kept on ice, transferred to the Environmental Toxicology laboratory, and centrifuged at 800 g for 10 min. The supernatant was stored at −80 °C until analysis of the urinary 8-OHdG levels by enzyme-linked immunosorbent assay [26].

2.5. *Alkaline Comet Assay*

Exfoliated buccal mucosa cells were kept in a sterile microtube with 180 µL low melting point agarose (0.5%, 37 °C) (Sigma-Aldrich, Darmstadt, Germany), mixed gently, and 90 µL were placed on a slide coated with a thin layer of normal melting point agarose (1 %, 37 °C) (Sigma-Aldrich, Darmstadt, Germany) and covered with a coverslip. Two slides were utilized for each child's sample. The slides were maintained at 4 °C for 3–5 min to solidify the agarose. The coverslip was then carefully removed, and the slides were immersed in a staining jar containing a freshly prepared cold lysis solution (2.5 M NaCl, 100 mM EDTA, 10 mM Tris, 1% Triton X − 100 and 10% DMSO, pH 10) (Amresco, Solon, OH, USA) at 4 °C for 1 h. The slides were placed in a horizontal electrophoresis chamber (Thermo Fisher Scientific, USA) containing freshly prepared cold electrophoresis alkaline buffer (300 mM NaOH and 1 mM EDTA, pH 13) (Amresco, Solon, OH, USA) for 20 min to unwind the DNA. Electrophoresis was performed at 25 V (0.7 V/cm) and 300 mA for 20 min in darkness to prevent additional DNA damage. The slides were washed three times with freshly prepared neutralization buffer (0.4 M Tris, pH 7.5) for 5 min, fixed with cold absolute methanol for 5 min, and air-dried at room temperature. Then, 60 µL GelRed (4 µg/mL; Biotium, Inc., Fremont, CA, USA) was added to each slide to stain the DNA. The slides were labeled with a code that was unfamiliar to the viewer and examined using a Nikon Eclipse fluorescence microscope equipped with an excitation filter (515–560 nm) and a barrier filter (590 nm). To visualize the DNA damage, the slides were observed at 40× magnification using Comet IV Software (Perceptive Instruments, Bury St Edmunds, UK). Three parameters were used to determine genotoxicity: (a) tail length (the distance between the first and last DNA fragments); (b) tail moment (the tail length weighted by the percentage of tail DNA); and tail intensity (the percentage of DNA in the tail) in 50 randomly selected nuclei on each slide (two slides per child for a total of 100 nuclei) (Figure 2). Comets with completely fragmented DNA (hedgehog-like figures with no apparent head) were excluded from the evaluation [32].

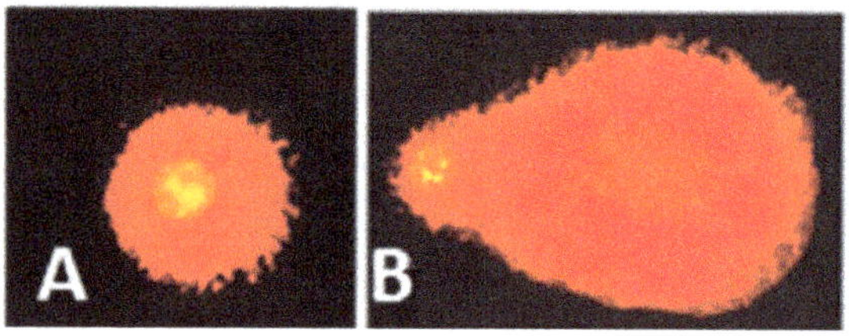

Figure 2. Nuclei without comet (without DNA damage) (**A**) and with comet (with DNA damage) (**B**) from exfoliated buccal mucosa cells.

2.6. Detection of Urinary 8-OHdG Levels by Enzyme-Linked Immunosorbent Assay (ELISA)

The 8-OHdG adduct levels were measured in a 100 µL urine sample from each child in triplicate using an 8-OHdG EIA kit (Cayman Chemicals, Ellsworth Rd, Ann Arbor, MI, USA) for ELISA, following the procedure suggested by the manufacturers. First, 50 µL EIA buffer, 50 µL 8-hydroxy-2′-deoxy-guanosine tracer AChE, and 50 µL 8-OHdG monoclonal antibody were added to each well, except the blank. The plate was covered with plastic film and incubated for 2 h at 4 °C in an orbital shaker (Rocker 25 Labnet). Then, the wells were washed five times with wash buffer, and 200 µL Ellman's Reagent was added. The plate was covered with plastic film and incubated for 2 h at room temperature in an orbital shaker (Rocker 25 Labnet) in darkness. The plates were read at a wavelength of 450 nm in an ELISA Reader (Elx800 BioTec) using Gen 5 ELISA software. The data obtained were processed in a calculus sheet (provided by Cayman Chemicals), and the 8-OHdG concentrations were obtained with an 8-OHdG standard curve (10.3–3000 ng/mL) and expressed in ng/mL [26].

2.7. Statistical Analysis

The results of the tail length (µm), tail moment (%), tail intensity (%), and urinary 8-OHdG levels were reported as the mean ± SEM of 101 children, for range of age and gender of the exposure and control groups. The data were analyzed using Student's *t*-test analysis to determine statistical significance between the exposure and control groups at $p < 0.05$. The relationship between the three genotoxic parameters and the level of 8-OHdG with age and gender were determined using a MANOVA test, $p < 0.01$ (Spearman's rank correlation analysis), with SPSS Statistics.

3. Results

3.1. Characteristics of the Participant Children

The sociodemographic and dietary data from the exposure and control groups are shown in Table 1. The data suggest that food, air, soil, and water are major sources of heavy metal exposure in children who live near the El Fraile tailings that contribute to increased DNA damage. School outdoor areas in which schoolchildren play also constitute important outdoor sources of heavy metals at schools. Parents also reported using water from mine tailings to wash clothing, and the Cacalotenango River in the dry season was sometimes used as drinking water or to wash clothing (Figure 1). Additionally, in the indoor air of classrooms and in the respective outdoor areas, the reported heavy metal levels were strongly influenced by their proximity to the mine tailings. Outdoor air penetration impacts the indoor air of classrooms, mainly if natural ventilation is used throughout the day (including during class hours). Ambient air infiltration is strongly dependent on the ventilation rates of the classrooms in the dry season.

3.2. DNA Damage in Buccal Mucosa Cells in the (El Fraile)-Exposed and (Chilpancingo)-Control Groups

The results of the genotoxic analysis of the buccal mucosa cells from the exposure and control groups are shown in Table 2. There was a significant increase in the mean of the three genotoxic parameters of tail intensity (TI), tail moment (TM), and tail length (TL) in the exposed child group compared with the control group (Student's *t*-test, $p < 0.05$). Comparing genotoxic parameters between exposure and control groups by age and gender, we observed that there were significant differences in the three genotoxic parameters (Tables 2 and 3). However, when we compared three genotoxic parameter means from the exposure group by age and gender, we found that there were no significant differences in the three genotoxic parameters in the children in the 6–7 years group when compared with the children in the 10–12 years group (MANOVA, $p < 0.01$) (Table 2). However, the children aged 8–9 years had a greater increase in the mean of the three genotoxic parameters than the children aged 6–7 and 10–12 years (MANOVA, $p < 0.01$). Comparing the genotoxic effect with gender, we found that the boys' exposure group had a higher increase in all three genotoxic parameters than the girls ($p < 0.05$) (Table 3). There was a positive correlation between sex and age, and a significant increase in DNA damage by Spearman´s rank correlation analysis (r = 0.65) (Table 3).

Table 2. Mean of the genotoxic parameters in buccal mucosa cells in the (El Fraile, Taxco de Alarcón)-exposure and (Chilpancingo)-control groups in Guerrero State, Mexico.

Children Groups (Age Ranges)	n	Tail Intensity (%)	Tail Moment (%)	Tail Length (μm)
		Mean ± SEM	Mean ± SEM	Mean ± SEM
Exposure	101	37.68 ± 4.59 *	19.79 ± 3.50 *	60.89 ± 6.18 *
6–7	29	45.41 ± 5.40 *	17.86 ± 5.07 *	56.32 ± 4.36 *
8–9	39	49.94 ± 4.51 **	24.54 ± 6.64 **	71.14 ± 4.98 **
10–12	33	39.94 ± 4.31 *	17.25 ± 4.95 *	55.81 ± 6.10 *
Control	101	18.38 ± 1.52	4.24 ± 1.54	36.59 ± 1.86
6–7	30	16.76 ± 1.44	3.91 ± 2.30	46.13 ± 4.84
8–9	39	18.81 ± 1.38	4.58 ± 2.77	34.27 ± 4.08
10–11	32	14.53 ± 1.66	4.16 ± 2.65	44.01 ± 2.03

* Significant differences between exposure and control groups were obtained by a Student's *t*-test, $p < 0.05$) and a MANOVA test ** $p < 0.01$.

Table 3. Mean of the genotoxic parameters by gender in buccal mucosa cells in the (El Fraile, Taxco de Alarcón)-exposure and (Chilpancingo)-control groups in Guerrero State, Mexico.

Gender	Children's Groups	n	Tail Intensity (%)	Tail Moment (%)	Tail Length (μm)
			Mean ± SEM	Mean ± SEM	Mean ± SEM
	Exposure				
Girls		56	44.33 ± 5.26 *	22.01 ± 5.49 *	58.73 ± 5.71 *
Boys		45	58.70 ± 6.44 **	28.93 ± 3.01 **	68.13 ± 4.99 **
	Control				
Girls		52	14.33 ± 5.04	5.01 ± 4.49	28.73 ± 4.71
boys		49	17.70 ± 2.88	6.29 ± 3.01	24.13 ± 6.11

* Significant differences among exposure and control groups were obtained by an analysis with Student´s *t*-test $p < 0.05$ and a MANOVA test ** $p < 0.01$.

3.3. Oxidative DNA Damage in the Exposure and Control Groups

The mean urinary 8-OHdG levels from the exposure and control groups are presented in Table 4. There was a significant increase in the mean urinary 8-OHdG level (5.65 ± 0.33 ng/mL) in the exposure group compared to the control group value (1.64 ± 0.88 ng/mL) (Student's *t*-test, $p < 0.05$). There was a statistically significant increase in the mean urinary 8-OHdG levels in children aged 8–9 years (6.95 ± 0.24 ng/mL) compared with the urinary 8-OHdG adduct means from children aged 6–7 (5.15 ± 0.2 ng/mL) and 10–12 years (4.35 ± 0.54 ng/mL) from the exposure group (MANOVA, $p < 0.01$) (Table 4). Comparing the 8-OHdG adduct mean level of the girls with the boys from the exposure group, we observed that oxidative DNA damage was significantly greater in the boys (6.88 ± 0.12 ng/mL) than in the girls (5.18 ± 0.24 ng/mL) (Table 5). There was a positive correlation between sex and age, and a significant increase in oxidative DNA damage with Spearman´s rank correlation analysis (r = 0.72) (Table 5).

Table 4. Mean of the urinary 8-OHdG levels in the (El Fraile, Taxco de Alarcón)-exposure and (Chilpancingo)-control groups in Guerrero State, Mexico.

Age Ranges	Children Groups	(n)	8-OHdG Level (ng/mL) Mean ± SEM
	Exposure	101	5.65 ± 0.33 *
6–7		29	5.15 ± 0.67 *
8–9		39	6.95 ± 0.24 **
10–12		33	4.35 ± 0.54 *
	Control	101	1.64 ± 0.88
6–7		30	1.13 ± 0.24
8–9		39	1.88 ± 0.12
10–12		32	1.74 ± 0.30

* Significant differences between exposure and control groups were obtained by an analysis with Student's *t*-test, $p < 0.05$, and a MANOVA test, ** $p < 0.01$.

Table 5. Means of urinary 8OHdG (ng/mL) levels by gender in the (El Fraile, Taxco de Alarcón)-exposure and (Chilpancingo)-control groups, in Guerrero State, Mexico.

Gender	Children Groups	n	8-OHdG Level (ng/mL) Mean ± SEM
	Exposure		
Girls		56	5.18 ± 0.24 *
Boys		45	6.88 ± 0.12 **
	Control		
Girls		52	1.73 ± 0.24
Boys		49	1.38 ± 0.08

* Significant differences between exposure and control groups were obtained by an analysis with Student´s *t* test, $p < 0.05$, and a MANOVA test ** $p < 0.01$.

4. Discussion

Previous studies have shown that the metal concentrations in the El Fraile mine tailings are very high because of geochemical, physicochemical, and environmental factors [2,5]. Heavy metals such as As (140–3627 mg/kg), Cd (0.5–434 mg/Kg), Pb (148–0900 mg/kg), Cu (0.002–55%), Zn (0.021–3.86%), and Fe (2.4–35.7%) have been identified in this zone [1,2]. El Fraile leachates, which are used as domestic water, typically exceed the Mexican Drinking Water Guidelines for sulfate, hardness, Fe, Mn, Pb, and As, while acidic leachates exceed the Mexican Guidelines for Industrial Discharge Waters for pH, As (<0.01–12.0 mg/L), Fe (0.025–2352 mg/L), Mn (0.1–732 mg/L), Zn (<0.025–1465 mg/L)1), and Pb (<0.01–0.351 mg/L) [1,2]. Moreover, a metal analysis of maize plants reported that Pb was in the range of 25.2 to 300.9 mg/kg, and Zn was in the range of 88.9 to 504.8 mg/ kg; the aerial part concentrations were in the range of 15.5 to 555.6 mg/kg for Zn and 2.2 to 10.8 mg/kg for Pb, which represent other risk factors since corn is the food base of the Mexican diet. High heavy metal levels in vegetal tissues, air, soil, and drinking water in the El Fraile area clearly indicate that the general population is highly exposed to heavy metals [33].

Developing children are the most sensitive population with regard to adverse effects from these environmental genotoxins [7,10]. In this regard, this study is the first to evaluate DNA damage using the alkaline comet assay in exfoliated buccal mucosa cells from 101 children (56 girls and 45 boys) aged 6–7, 8–9, and 10–12 years living beside the El Fraile mine tailings. In addition, oxidative DNA damage was quantified with the detection of urinary 8-OHdG adduct levels by ELISA [20,34]. We also evaluated a control group consisting of 101 children (52 girls and 49 boys) from Chilpancingo City, Guerrero, Mexico, within the same age ranges, who had never lived near or in mining zones. The results of the genotoxic analysis evidenced a higher increase in DNA damage in exfoliated buccal mucosa cells of the exposure children group compared with the control group ($p < 0.05$), demonstrated by a high increase in the three genotoxic parameter means (TI, TM, and TL) ($p < 0.05$). We applied a MANOVA to the data on the three genotoxic parameters in children aged 6–7, 8–9, and 10–12 years from the exposure group with respect to sex and age. We found that boys aged 8–9 years had higher DNA damage than girls ($p < 0.01$). Moreover, we found a positive Spearman's rank correlation between increased DNA damage and sex (boys) and age (8–10 years) in the exposure group ($r = 0.65$; $p < 0.01$).

Our results agree with other genotoxic studies in children living near mining zones. Children exposed to Pb and As from the Villa de la Paz mine, Mexico, had significant increases in the tail length (67.6 μm) and the tail moment (6.08%) in human peripheral blood lymphocytes [35]. Another study also reported a significant increase in DNA damage (49.1–61.8%), with tail lengths ranging from 25.3 to 29.2 μm in peripheral blood lymphocytes from children aged 6 to 11 years exposed to soil and water contaminated with Pb and AS in the region of Lagunera, Mexico [36]. Children aged 4 to 11 years living 200 km away from a mining zone and exposed to Pb and As in Luis Potosi, Mexico, evidenced a significant increase in the tail moment (5.2 μm) and tail length (3.5 μm) in peripheral lymphocytes [37]. In children aged 6 to 11 years who drank water contaminated with Pb and As, a significant increase was observed in the tail moment (>4.8) of lymphocytes [38]. A reduction in telomere length and mitochondrial DNA were observed in children (6 to 15 years old) exposed to As and Pb that lived in an industrial area in Salamanca, Mexico [39].

This study evidenced that children of all ages from the exposure group showed a significant increment in DNA damage in relation to the control group, but the children aged 8–9 years had a higher increment in their genotoxic parameters ($p < 0.05$) than children aged 6–7 and 10–12 years from the same exposure group ($p < 0.01$). In peripheral blood cells from 9-year-old children exposed to Pb in Poland, a significant increase was found in the frequency of micronuclei [40], as well as an increase in DNA strand breaks in children living around a waste incinerator [41]. These results possibly indicate that children, especially boys aged 8 to 9 years old, are more vulnerable to heavy metal exposure than children aged 6–7 and 10–12 years, due to the childrens' physiological conditions and exposure time;

they consume more food and have high gastrointestinal uptake and other physiological factors (detoxification system immaturity) [42,43]. Furthermore, boys in these age ranges spend more time outside, and they have more environmental exposure to heavy metals by air inhalation, soil dermal contact or absorption, or by ingestion of contaminated water and food than girls [10,44]. Previous studies have demonstrated that children's unique behaviors and metabolic rate often place them at risk for absorption of higher doses of heavy metals in contaminated environments in comparison to adults [9,45]. Many studies have also indicated that direct soil exposure and hand-to-mouth activity are the principal routes that contribute to heavy metal exposure in children [43,46]. Unintentional ingestion of heavy metals may lead to a considerably higher dose than an adult because of the greater intake of food or fluids per pound of body weight [42]. Heavy metals are introduced into the atmosphere by air particles, into the food chain by deposition on crop plants, and into drinking water, as well as by soil dust inhalation [1–5,7,33]. Regarding heavy metals in body tissues, the brain, kidneys, liver, heart, bones, and blood are the main target organs for heavy metal accumulation in children [10,11,15]. Most likely, children living in the El Fraile rural community are exposed to a high heavy metal mix that produces ROS or NOS that directly or indirectly interact with their DNA, which has caused its fragmentation, as evidenced by greater comets in buccal mucosa cell nuclei.

In parallel, we analyzed oxidative DNA damage through urinary 8-OHdG adduct levels in the children from the exposure and control groups, and our results showed that the urinary 8-OHdG level was three times higher in the exposure group than in the control group ($p < 0.01$). When we applied a MANOVA test to urinary 8-OHdG mean levels in the children from the 6–7, 8–9, and 10–12 age ranges and by sex from the exposure group, we observed that the boys aged 8–10 years had five times more 8-OHdG adduct than boys from the nonexposure group (MANOVA, $p < 0.01$). Furthermore, we found a positive Spearman's rank correlation between increased oxidative DNA damage and sex (boys) and age (8–9 years) in the exposure group ($r = 0.72$; $p < 0.01$). Our results are consistent with various published studies, in which children exposed to chromium and a heavy metal mix showed a significant increase in urinary 8-OHdG levels [47]. Chronic exposure to low levels of Cd produces oxidative stress in children [48]. Children within the age range of 3 to 6 years with high Pb exposure from an electronic waste recycling center showed a significant increase in urinary 8-OHdG levels [41]. In adolescents aged 12–14 years old, exposure to heavy metals in Milazzo-Valle del Mela (Sicily, Italy) led to higher levels of urinary 8OHdG adduct [49]. A significant increase in urinary 8-OHdG levels in children exposed to As [50] and salivary 8-OHdG levels in children 5–7 years old exposed to water contaminated with As [51] was reported. An increase was observed in the 8-OHdG adduct in children 5–8 years old with a high blood lead (Pb) level, and in children 6–8 years old with high Cd levels from Montevideo, Uruguay [52,53]. In children living near the petrochemical industry in Central Taiwan, elevated 8-OHdG levels were associated with copper, arsenic, strontium, cadmium, and mercury exposure [12].

In our study, urinary 8-OHdG levels detected by ELISA in both the exposure and control groups agreed with what has been reported in other studies [34,52]. Our results were supported by high levels of heavy metals and metalloids reported in blood and urine samples in children from the El Fraile community within the same age ranges that had been included in our study [3]. Our results clearly indicate that the bodies of the children had experienced an increase in oxidative DNA damage due to high environmental exposure to heavy metals, which produced a greater increase in ROS, such as OH•, which are capable of oxidizing mainly guanine to produce an 8-OHdG adduct, which is excreted in urine, as demonstrated by higher urinary 8-OHdG levels.

8-Hydroxy-deoxyguanosine (8-OHdG) is one of the most widely used biomarkers for detecting oxidative DNA damage [22,23]. The formation of DNA adducts, one of the earliest molecular changes caused by environmental chemicals, may promote the initiation and progression of carcinogenesis [54–56]. Furthermore, these data supported the hypothesis that children who live in the El Fraile community have higher oxidative stress in their body, which consequently reduces their genome repair mechanisms, and a low capacity of endogenous antioxidants in the child's body makes them more vulnerable to genome oxidation. Several studies have shown that populations that are chronically exposed to heavy metals, such as Cd, Ni, Pb, and Hg, have a significant increase in oxidative stress [11,12,16,56] and DNA damage, such as strand breaks, mutations, chromosomal aberrations, oxidation of their bases, and decreased endogenous antioxidant levels [15,50,53–55]. Moreover, an increase in ROS and NOS can modify cellular signaling pathways, transcription factors, and gene regulation, which are mechanisms that play a crucial role in carcinogenesis and immunosuppression in childhood [57,58]. Chronic exposure to drinking water containing high levels of metals and metalloids is associated with various gastrointestinal diseases [46], diabetes, and cardiovascular diseases [15,30].

It is worth mentioning that our study had a few limitations. First, we could not detect heavy metal concentrations in urine samples or in blood in either the exposure or control groups to associate them with DNA strand breaks and urinary 8-OHdG levels, or with sociodemographic variables. However, it is important to mention that this study is the first to evaluate DNA damage with two early genetic biomarker exposures to environmental pollution using noninvasive methods, such as buccal mucosa cells and urine samples, to demonstrate that children from the El Fraile rural community with high heavy metal pollution are at a high health risk; the results can support the prevention of developmental diseases.

5. Conclusions

The results of this study demonstrate that children (boys and girls) living in the El Fraile rural community near the El Fraile mine tailings are exposure to an area heavily polluted by toxic heavy metal mixes and are at a high health risk, since the genotoxic results demonstrated an increase in DNA damage (genome strand breaks and oxidation) in the exposure group relative to the control group through the two genetic biomarkers of alkaline comet assay and urinary 8-OHdG adduct. However, the boys were shown to have more DNA damage than the girls. Nonetheless, diverse factors can contribute to genotoxic actions due to the exposure of children to heavy metals, such as lifestyle, water, soil, and food contamination, and the synergistic or additive actions of heavy metal mixes.

Author Contributions: M.E.C.-S., O.T.-M., A.R.-G., Y.C.-L., M.d.C.M.-V., M.E.M.-H., L.S.-A. and P.H.-B. participated in the realization of the research article. All authors have read and agreed to the published version of the manuscript.

Funding: This study was financially supported through the Programa de Apoyo a Proyectos de Investigacion e Innovacion Tecnologica, Direccion General de Apoyo al Personal Academico [DGAPAUNAM-IN215911] of the Universidad Nacional Autonoma de Mexico (UNAM).

Institutional Review Board Statement: The Ethics Committee of the Faculty of Medicine from Universidad Nacional Autonoma de Mexico approved the research procedures used in this study, and a clinical questionnaire that was administered to parents.

Informed Consent Statement: All parents signed the informed consent letter for each participant in the genotoxic study, in accordance with the Declaration of Helsinki Ethical Principles (2008).

Data Availability Statement: Not applicable.

Acknowledgments: The authors thank the parents and children who participated in this study. As well as Rocio Vaca Paulin for statistical analysis data, and Ricardo Olvera Calderon and Marco Ku-Centurion for their technical assistance during the development of this project.

Conflicts of Interest: The authors declare that there is no conflict of interest.

References

1. Talavera-Mendoza, O.; Armienta-Hernández, M.A.; García-Abundis, J.; Flores-Mundo, N. Geochemistry of leachates from the El Fraile sulfide tailings piles in Taxco, Guerrero, southern Mexico. *Environ. Geochem. Health* **2006**, *28*, 243–255. [CrossRef] [PubMed]
2. Romero, F.M.; Nuñéz, M.E.; Gutiérrez, M.A.; Armienta, A.E.; Cineros-Gómez, A.E. Evaluation of the potential of indigenous calcareous shale for neutralization and removal of arsenic and heavy metals from acid mine drainage in the Taxco mining area, Mexico. *Arch. Environ. Contam. Toxicol.* **2011**, *60*, 191–203. [CrossRef] [PubMed]
3. Moreno, M.E.; Acosta-Saavedra, L.C.; Meza-Figueroa, D.; Vera, E.; Cebrian, M.E.; Ostrosky-Wegman, P.; Calderón-Aranda, E.S. Biomonitoring of metal in children living in a mine tailings zone in Southern Mexico: A pilot study. *Int. J. Hyg. Environ. Health* **2010**, *213*, 252–258. [CrossRef] [PubMed]
4. Soto-Ríos, M.L.; Rothenberg, S.; Gonsebatt, M.E.; Talavera-Mendoza, O. Cytogenotoxicity in uroepithelial cells of women exposed to mercury in a mining area. *Occup. Environ. Med.* **2010**, *67*, 620–624. [CrossRef]
5. Armienta, M.A.; Talavera, O.; Morton, O.; Barrera, M. Geochemistry of Metals from Mine Tailings in Taxco, Mexico. *Bull. Environ. Contam. Toxicol.* **2003**, *71*, 387–393. [CrossRef]
6. de Burbure, C.; Buchet, J.P.; Bernard, A.; Leroyer, A.; Nisse, C.; Haguenoer, J.M.; Bergamaschi, E.; Mutti, A. Biomarkers of renal effects in children and adults with low environmental exposure to heavy metals. *J. Toxicol. Environ. Health A* **2003**, *66*, 783–798. [CrossRef]
7. Al Osman, M.; Yang, F.; Massey, Y.I. Exposure routes and health effects of heavy metals on children. *Biometals* **2019**, *32*, 563–573. [CrossRef]
8. Bellinger, D.C. Very low lead exposures and children's neurodevelopment. *Curr. Opin. Pediatr.* **2008**, *20*, 172–177. [CrossRef]
9. Dunn, A.M.; Burns, C.; Satiler, B. Environmental Health children. *J. Pediatr. Health Care* **2003**, *17*, 223–231. [CrossRef]
10. Wright, R.O. Environment, susceptibility windows, development, and child health. *Curr. Opin. Pediatr.* **2017**, *29*, 211–217. [CrossRef]
11. Granot, E.; Kohen, R. Oxidative stress in childhood in health and disease states. *Clin. Nutr.* **2004**, *23*, 3–11. [CrossRef]
12. Killian, B.; Yuan, T.H.; Tsai, C.H.; Chiu, T.H.T.; Chen, Y.H.; Chan, C.C. Emission-Related Heavy Metal Associated with Oxidative Stress in Children: Effect of Antioxidant Intake. *Int. J. Environ. Res. Public Health* **2020**, *17*, 3920. [CrossRef] [PubMed]
13. Romaniuk, A.; Lyndin, M.; Sikora, V.; Lyndina, Y.; Romaniuk, S.; Sikora, K. Heavy metals effect on breast cancer progression. *J. Occup. Med. Toxicol.* **2017**, *12*, 32. [CrossRef]
14. Loft, S.; Svoboda, P.; Kasai, H.; Tjønneland, A.; Vogel, U.; Møller, P.; Overvad, K.; Raaschou-Nielsen, O. Prospective study of 8-oxo-7,8-dihydro-2′-deoxyguanosine excretion and the risk of lung cancer. *Carcinogenesis* **2006**, *27*, 1245–1250. [CrossRef] [PubMed]
15. Korashy, H.M.; Attafi, I.M.; Famulski, K.S.; Bakheet, S.A.; Hafez, M.M.; Alsaad, A.M.S.; Al-Ghadeer, A.R.M. Gene expression profiling to identify the toxicities and potentially relevant human disease outcomes associated with environmental heavy metal exposure. *Environ. Pollut.* **2017**, *221*, 64–74. [CrossRef] [PubMed]
16. Ercal, N.; Gurer-Orhan, H.; Aykin-Burns, N. Toxic metals and oxidative stress part I: Mechanisms involved in metal induced oxidative damage. *Curr. Top. Med. Chem.* **2001**, *1*, 529–539. [CrossRef]
17. Marnett, L.J. Oxyradicals and DNA damage. *Carcinogenesis* **2000**, *21*, 361370. [CrossRef]
18. Cooke, M.S.; Olinski, R.; Loft, S. Measurement and meaning of oxidatively modified DNA lesion in Urine. *Cancer Epidemiol. Biomark. Prev.* **2008**, *17*, 13–14. [CrossRef]
19. Hartwig, A.; Schwerdtle, T. Interactions by carcinogenic metal compounds with DNA repair processes: Toxicological implications. *Toxicol. Lett.* **2002**, *127*, 47–54. [CrossRef]
20. Drury, J.A.; Jeffers, G.; Cooke, R.W. Urinary 8-hydroxydeoxyguanosine in infants and children. *Free Radic. Res.* **1998**, *28*, 423–438. [CrossRef]
21. Poulsen, H.E.; Nadal, L.L.; Broedbaek, K.; Nielsen, P.E.; Weimann, A. Detection and interpretation of 8-oxodG and 8-oxoGua in urine, plasma and cerebrospinal fluid. *Biochim. Biophys. Acta* **2014**, *1840*, 801–808. [CrossRef] [PubMed]
22. Cheng, K.C.; Cahill, D.S.; Kasai, H.; Nishimura, S.; Loeb, L.A. 8-Hydroxyguanine, an abundant form of oxidative DNA damage causes G-t and A-C substitutions. *J. Biol. Chem.* **1992**, *267*, 166–172. [CrossRef]
23. Chiou, C.C.; Chang, P.Y.; Chan, E.C.; Wu, T.L.; Tsao, K.C.; Wu, J.T. Urinary 8-hydroxydeoxyguanosine and its analogues as DNA marker of oxidative stress: Development of an ELISA and measurement in both bladder and prostate cancers. *Clin. Chim. Acta* **2003**, *334*, 87–94. [CrossRef]
24. Bernard, A. Biomarkers of metal toxicity in population studies: Research potential and interpretation issues. *J. Toxicol. Environ. Health. Part A* **2008**, *71*, 1259–1265. [CrossRef] [PubMed]
25. Calderón-Segura, M.E.; Gómez-Arroyo, S.; Villalobos-Pietrini, R.; Espinosa-Ramírez, M. In vivo and in vitro promutagen activation by *Vicia faba* of thiocarbamate herbicides molinate and butylate to products inducing sister chromatid exchanges in human lymphocyte cultures. *Mutat. Res.* **1999**, *438*, 81–88. [CrossRef]
26. Rodríguez-Romero, M.I.; Gómez-Arroyo, S.; Villalobos-Pietrini, R.; Martínez-Valenzuela, C.; Calderón-Ezquerro, M.C.; Cortés-Eslava, J.; Arenas-Huertero, F.; Calderón-Segura, M.E. Evaluation of 8-hidroxy-2′-deoxiguanosine (8-OHdG) adduct levels and DNA strand breaks in human peripheral lymphocytes in vitro exposed to polycyclic aromatic hydrocarbons with or without animal metabolic activation. *Toxicol. Mech. Methods* **2012**, *22*, 170–183. [CrossRef]
27. Tsukahara, H. Biomarkers for oxidative stress: Clinical application in pediatric medicine. *Curr. Med. Chem.* **2007**, *14*, 339–351. [CrossRef]
28. Martínez-Valenzuela, C.; Waliszewski, S.M.; Amador-Muñoz, O.; Meza, E.; Calderón-Segura, M.E.; Zenteno, E.; Huichapan-Martínez, J.; Caba, M.; Félix-Gastélum, R.; Longoria-Espinoza, R. Aerial pesticide application causes DNA damage in pilots from Sinaloa, Mexico. *Environ. Sci. Pollut. Res.* **2016**, *24*, 2412–2420. [CrossRef]

29. Carbajal-López, Y.; Gómez-Arroyo, S.; Villalobos-Pietrini, R.; Calderón-Segura, M.E.; Martínez-Arroyo, A. Biomonitoring of agricultural workers exposed to pesticide mixtures in Guerrero state, Mexico, with comet assay and micronucleus test. *Environ. Sci. Pollut. Res.* **2016**, *23*, 2513–2520. [CrossRef]
30. Zhushan, F.; Shuhua, X. The effects of heavy metals on human metabolism. *Toxicol. Mech. Methods* **2020**, *30*, 167–176. [CrossRef]
31. Singh, N.P.; McCoy, M.T.; Tice, R.R.; Schneider, E.L. A simple technique for quantification of low levels of DNA damage in individual cells. *Exp. Cell Res.* **1988**, *175*, 184–191. [CrossRef]
32. Calderón-Segura, M.E.; Gómez-Arroyo, S.; Cortés-Eslava, J.; Martínez-Valenzuela, C.; Mojica-Vázquez, L.H.; Sosa-López, M.; Flores-Ramírez, D.; Romero-Velázquez, Z.E. In Vitro cytotoxicity and genotoxicity of Furia ®180 SC (Zeta-cypermethrin) and Bulldock 125®SC (β-Cyfluthrin) pyrethroid insecticides in human peripheral blood lymphocytes. *Toxicol. Mech. Methods* **2018**, *28*, 268–278. [CrossRef] [PubMed]
33. Díaz-Villaseñor, E. Transferencia de Metales Entre Suelo y Plantas de Maíz (*Zea mays* L.), Sembradas en Terrenos Impactados Por Jales Mineros en la Región de Taxco, Guerrero. Master's Thesis, Universidad Autónoma de Guerrero, Iguala, Mexico, November 2006.
34. Breton, J.; Sichel, F.; Bianchini, F.; Prevost, V. Measurement of 8-Hydroxy-2′deoxyguanosine by a commercially available ELISA test: Comparison with HPLC/electrochemical detection in calf thymus DNA and determination in human serum. *J. Anal. Lett.* **2003**, *36*, 123–134. [CrossRef]
35. Yáñez, L.; García-Nieto, E.; Rojas, E.; Carrizales, L.; Mejía, J.; Calderón, J.; Razo, I.; Díaz-Barriga, F. DNA damage in blood cells from children exposed to arsenic and lead in a mining area. *Environ. Res.* **2003**, *93*, 231–240. [CrossRef]
36. Méndez-Gómez, J.; García-Vargas, G.G.; López-Carrillo, L.; Calderón-Aranda, E.S.; Gómez, A.; Vera, E.; Valverde, M.; Cebrián, M.E.; Rojas, E. Genotoxic Effects of Environmental Exposure to Arsenic and Lead on Children in Region Lagunera, Mexico. *Ann. N. Y. Acad. Sci.* **2008**, *1140*, 358–367. [CrossRef]
37. Jasso-Pineda, Y.; Díaz-Barriga, F.; Calderón, J.; Yáñez, L.; Carrizales, L.; Pérez-Maldonado, I.N. DNA damage and decreased DNA repair in peripheral blood mononuclear cells in individuals exposed to arsenic and lead in a mining site. *Biol. Trace Elem. Res.* **2012**, *146*, 141–149. [CrossRef]
38. Jasso-Pineda, Y.; Díaz-Barriga, F.; Yáñez-Estrada, L.; Pérez-Vázquez, F.J.; Pérez-Maldonado, I.N. DNA damage in Mexican children living in high-risk contaminated scenarios. *Sci. Total Environ.* **2015**, *518–519*, 38–48. [CrossRef]
39. Alegría-Torres, J.A.; Pérez-Rodríguez, R.Y.; García-Torres, L.; Costilla-Salazar, R.; Rocha-Amador, D. Exposure to arsenic and lead in children from Salamanca México, effects on telomeric lengthening and mitochondrial DNA. *Environ. Sci. Pollut. Res. Int.* **2020**, *27*, 6420–6428. [CrossRef]
40. Kapka, L.; Baumgartner, A.; Siwińska, E.; Knudsen, L.E.; Anderson, D.; Mielzyńska, D. Environmental lead exposure increases micronuclei in children. *Mutagenesis* **2007**, *22*, 201–207. [CrossRef]
41. Xu, X.; Liao, W.; Lin, Y.; Dai, Y.; Shi, Z.; Huo, X. Blood concentrations of lead, cadmium, mercury and their association with biomarkers of DNA oxidative damage in preschool children living in an e-waste recycling area. *Environ. Geochem. Health* **2018**, *40*, 1481–1494. [CrossRef]
42. Moya, J.; Bearer, C.F.; Etzel, R.A. Children's behavior and physiology and how it affects exposure to environmental contaminants. *Pediatrics* **2004**, *113*, 996–1006. [CrossRef] [PubMed]
43. Sly, J.L.; Carpenter, D.O. Special vulnerability of children to environmental exposures. *Rev. Environ. Health* **2012**, *27*, 151–157. [CrossRef] [PubMed]
44. Neri, M.; Bonassi, S.; Knudsen, L.E.; Sram, R.J.; Holland, N.; Ugolini, D.; Merlo, D.F. Children's exposure to environmental pollutants and biomarkers of genetic damage I. Overview and critical issues. *Mutat. Res.* **2006**, *612*, 1–13. [CrossRef] [PubMed]
45. Engström, K.S.; Vahter, M.; Lindh, C.; Teichert, F.; Singh, R.; Concha, G.; Nermell, B.; Farmer, P.B.; Strömberg, U.; Broberg, K. Low 8-oxo-7,8-dihydro-2-oxi-deoxyguanosine levels and influence of genetic background in an Andean population exposed to high levels of arsenic. *Mutat. Res.* **2010**, *683*, 98–105. [CrossRef] [PubMed]
46. Kim, J.J. Ambient air pollution: Health hazards to children. *Pediatrics* **2004**, *114*, 1699–1707. [CrossRef]
47. Sughis, M.; Nawrot, T.S.; Haufroid, V.; Nemery, B. Adverse Health Effects of Child Labor: High Exposure to Chromium and Oxidative DNA Damage in Children Manufacturing Surgical Instruments. *Environ. Health Perspect.* **2012**, *120*, 1469–1474. [CrossRef]
48. Kippler, M.; Hossain, M.B.; Lindh, C.; Moore, S.E.; Kabir, I.; Vahter, M.; Broberg, K. Early life low-level cadmium exposure is positively associated with increased oxidative stress. *Environ. Res.* **2012**, *112*, 164–170. [CrossRef]
49. Pizzino, G.; Bitto, A.; Interdonato, M.; Galfo, F.; Irrera, N.; Mecchio, A.; Pallio, G.; Ramistella, V.; De Luca, F.; Minutoli, L.; et al. Oxidative stress and DNA repair and detoxification gene expression in adolescents exposed to heavy metals living in the Milazzo-Valledel Mela area (Sicily, Italy). *Redox Biol.* **2014**, *2*, 686–693. [CrossRef]
50. Xu, Y.; Wang, Y.; Zheng, Q.; Li, X.; Li, B.; Jin, Y.; Sun, X.; Sun, G. Association of oxidative stress with arsenic methylation in chronic arsenic-exposed children and adults. *Toxicol. Appl. Pharmacol.* **2008**, *232*, 142–149. [CrossRef]
51. Hinhumpatch, P.; Navasumrit, P.; Chaisatra, K.; Promvijit, J.; Mahidol, C.; Ruchirawat, M. Oxidative DNA damage and repair in children exposed to low levels of arsenic in utero and during early childhood: Application of salivary and urinary biomarkers. *Toxicol. Appl. Pharmacol.* **2013**, *273*, 569–579. [CrossRef]
52. Roy, A.; Queirolo, E.; Peregalli, F.; Mañay, N.; Martínez, G.; Kordas, K. Association of blood lead levels with urinary F2-8α Isoprostane and 8-hydroxy-2-deoxy-Guanosine concentrations in first-grade Uruguayan children. *Environ. Res.* **2015**, *140*, 127–135. [CrossRef] [PubMed]

53. Kordas, K.; Roy, A.; Vahter, M.; Ravenscroft, J.; Mañay, N.; Peregalli, F.; Martínez, G.; Queirolo, E.I. Multiple-metal exposure, diet, and oxidative stress in Uruguayan school children. *Environ. Res.* **2018**, *166*, 507–515. [CrossRef] [PubMed]
54. Fukuda, M.; Yamauchi, H.; Yamamoto, H.; Aminaka, M.; Murakami, H.; Kamiyama, N.; Miyamoto, Y.; Koitabashi, Y. The evaluation of oxidative DNA damage in children with brain damage using 8-hydroxydeoxyguanosine levels. *Brain Dev.* **2008**, *30*, 131–136. [CrossRef] [PubMed]
55. Wong, R.H.; Kuo, C.H.; Hsu, M.L.; Wang, T.Y.; Chang, P.I.; Wu, T.H.; Huang, S. Increased Levels of 8-Hydroxy-2′-Deoxyguanosine Attributable to Carcinogenic Metal Exposure among School children. *Environ. Health Perspect.* **2005**, *113*, 1386–1390. [CrossRef] [PubMed]
56. Pineda-Zavaleta, A.P.; García-Vargas, G.; Borja-Aburto, V.H.; Acosta-Saavedra, L.C.; Vera-Aguilar, E.; Gómez-Muñoz, A.; Cebrián, M.E.; Calderón-Aranda, E.S. Nitric oxide and superoxide anion production in monocytes from children exposed to arsenic and lead in region Lagunera, Mexico. *Toxicol. Appl. Pharmacol.* **2004**, *198*, 283–290. [CrossRef]
57. Leonard, S.S.; Harris, G.K.; Shi, X. Metal-induced oxidative stress and signal transduction. *Free Radic. Biol. Med.* **2004**, *37*, 1921–1942. [CrossRef] [PubMed]
58. Endo, K.; Miyashita, Y.; Sasaki, H.; Ebisuno, M.; Ohira, M.; Saiki, A.; Koide, N.; Oyama, T.; Takeyoshi, M.; Shirai, K. Probucol and atorvastatin decrease urinary 8-hydroxy-2′-deoxyguanosine in patients with diabetes and hypercholesterolemia. *J. Atheroscler. Thromb.* **2006**, *13*, 68–75. [CrossRef]

toxics

MDPI

Article

Determination of 19 Steroid Hormones in Human Serum and Urine Using Liquid Chromatography-Tandem Mass Spectrometry

Zhong-Min Li [1,2] and Kurunthachalam Kannan [1,2,*]

[1] Department of Pediatrics, New York University Grossman School of Medicine, New York, NY 10016, USA
[2] Department of Environmental Medicine, New York University Grossman School of Medicine, New York, NY 10016, USA
* Correspondence: kurunthachalam.kannan@nyulangone.org; Tel.: +1-(212)-263-1546

Abstract: This paper describes a methodology for simultaneous determination of 19 steroid hormones, *viz.* estrone, estradiol, estriol, testosterone, 5α-dihydrotestosterone, androstenedione, androstenediol, dehydroepiandrosterone, progesterone, pregnenolone, 17α-OH-progesterone, 17α-OH-pregnenolone, cortisone, cortisol, 11-deoxycortisol, 11-deoxycorticosterone, 11-dehydrocorticosterone, aldosterone, and corticosterone, in 500-μL of urine or serum/plasma. The method was optimized using isotopically labeled internal standards and liquid-liquid extraction followed by detection using liquid chromatography-electrospray ionization-tandem mass spectrometry (LC-MS/MS). Dansylation of estrogens significantly improved their sensitivities (~11- to 23-fold) and chromatographic separation. The respective limit of detection (LOD) and limit of quantification (LOQ) of all analytes were 0.04–0.28 and 0.14–0.92 ng/mL in human urine, and 0.11–0.35 and 0.38–1.18 ng/mL in human serum/plasma. Recoveries of all analytes (except for progesterone) fortified at 10, 20, and 200 ng/mL in urine and serum were 80–120%, with standard deviations ranging from 0 to 17.3%. Repeated analysis of similarly fortified urine and serum samples yielded intra-day and inter-day variations of 0–21.7% and 0.16–11.5%, respectively. All analytes except cortisone exhibited weak matrix effects in urine and serum (−13.9–18.2%). The method was further validated through the analysis of the National Institute of Standards and Technology (NIST) plasma Standard Reference Material (SRM1950) with certified concentrations for cortisol, progesterone, and testosterone (coefficient of variation: 3–11%). The developed method was applied in the analysis of urine samples from 20 volunteers, which revealed the occurrence of 16 analytes with detection frequencies (DFs) > 80%. Furthermore, 15 analytes were found in plasma SRM1950, indicating the feasibility of our method in the analysis of steroid hormones in urine and serum/plasma. This method will facilitate analysis of steroid hormones in population-based biomonitoring studies.

Keywords: steroid hormones; estrogens; androgens; progestogens; corticosteroids; urine; serum

Citation: Li, Z.-M.; Kannan, K. Determination of 19 Steroid Hormones in Human Serum and Urine Using Liquid Chromatography-Tandem Mass Spectrometry. *Toxics* **2022**, *10*, 687. https://doi.org/10.3390/toxics10110687

Academic Editors: Roel Vermeulen, Lauren Petrick, Maaike van Gerwen and Kelvin S. Y. Leung

Received: 10 October 2022
Accepted: 9 November 2022
Published: 12 November 2022

1. Introduction

Steroid hormones are cholesterol derivatives that play critical roles in regulating water and salt homeostasis, metabolism, stress response, and in initiating and maintaining sexual differentiation and reproduction [1]. Steroid hormones are synthesized through a cascade-like pathway in the adrenal cortex, the gonads and the placenta and are released into the blood stream to act in both peripheral target tissues and the central nervous system. The homeostasis of steroid hormones is regulated through hypothalamus-pituitary-gonadal (HPG) and hypothalamus-pituitary-adrenal (HPA) axes [2,3], which consist stimulatory hormones and feedback loops. Steroid hormones can be classified as estrogens, androgens, progestogens, and corticosteroids (Figure 1), according to their structures and genomic receptors to which they bind [4]. In general, estrogens (18 carbons with an aromatic ring) are female reproductive hormones, androgens (19 carbons) are male reproductive hormones, progestogens (21 carbons) are pregnancy hormones, and corticosteroids (21 carbons) are stress hormones [5].

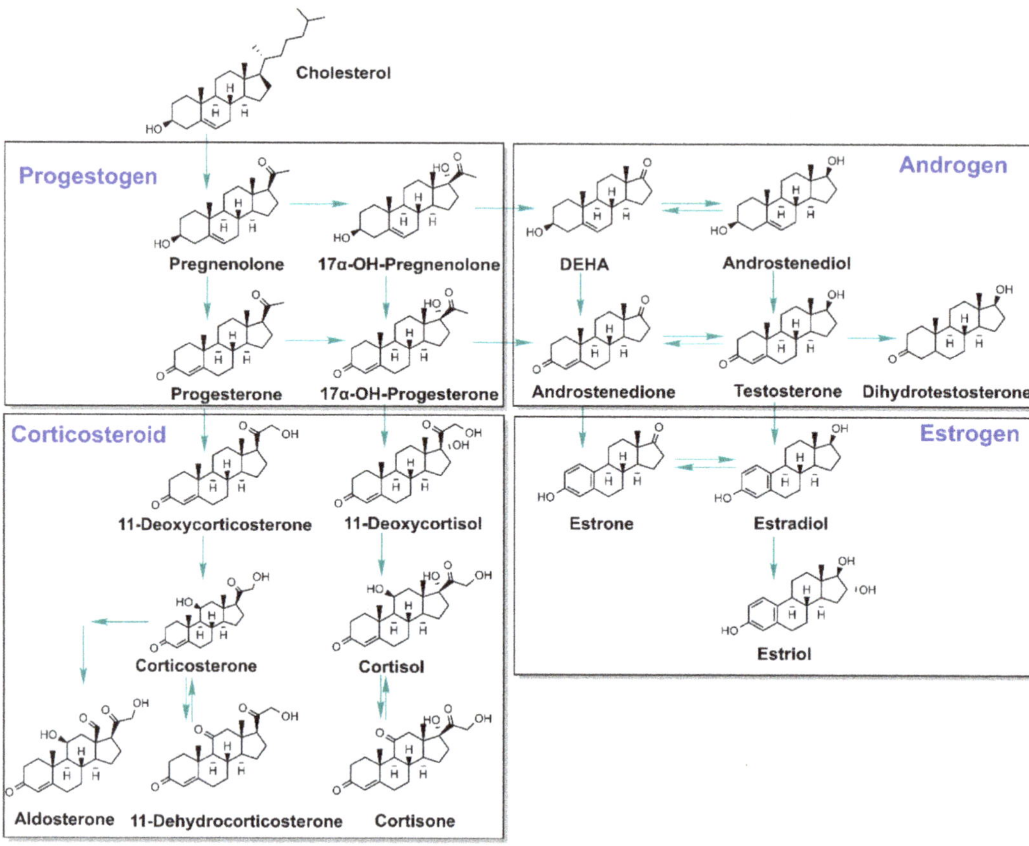

Figure 1. Molecular structures and metabolic pathway of the 19 steroid hormones investigated in this study.

The lipophilic steroid hormones undergo phase I and phase II metabolism, and are excreted mainly through urine as glucuronides, sulfates, diglucuronides, disulfates, and sulfoglucuronides [1,6]. Perturbation in steroid hormone homeostasis in blood, urine, saliva, and hair has been measured in investigations focused on cancer [7–11], stress [12], and endocrine-disruption by environmental chemicals such as bisphenols [13], phthalates [14] and triclosan [13].

Traditional methods of analysis of steroid hormones in human specimens (such as serum) are radioimmunoassays (RIAs) and enzyme-linked immunosorbent assays (ELISAs). These methods are highly sensitive, but lack selectivity due to nonspecific antigen-antibody interactions [15]. Besides, RIAs and ELISAs usually target one analyte (i.e., hormone) per assay. Methods based on mass spectrometry offer improved robustness, specificity and accuracy, and have been routinely used for steroid hormone analysis in recent years. For instance, gas chromatography-mass spectrometry (GC-MS) methods provide high sensitivity, selectivity and accuracy. However, due to the low volatilities of steroid hormones, analysis by GC-MS requires derivatization, a laborious and time consuming step [16,17]. To overcome this problem, studies have applied liquid chromatography coupled with tandem mass spectrometry (LC-MS/MS) [18–22]. Nevertheless, challenges exist in simultaneous measurement of ultra-trace levels of several steroid hormones in serum or urine. Estrogens exhibit low sensitivities due to their low ionization efficiencies [23], leading to difficulties in analysis at trace levels of these hormones [24]. Besides, many steroid hormones have similar chemical structures and molecular fragmentation patterns, which make their mass spectrometric identification difficult. The loss of one or two water moiety from the molecule is a common mass fragmentation feature observed for steroid hormones. Several reported earlier studies entailed long time (e.g., 40 min) for chromatographic separation [5,19], which can hamper high throughput analysis in large-scale population-based studies.

We describe a method for sensitive and selective determination of four classes of steroid hormones simultaneously in human urine and serum/plasma. The goals of this study were: (1) to develop a LC-MS/MS method for the determination of physiologically relevant concentrations of 19 steroid hormones in a single extraction and injection; (2) to evaluate and validate the method for sensitivity, accuracy and precision in human urine and serum/plasma; and (3) to assess the feasibility of the method in measuring steroid hormones in urine samples from the general population and in pooled human serum/plasma. Nineteen steroid hormones, comprising 3 estrogens (estrone, estradiol and estriol), 5 androgens (testosterone, 5α-dihydrotestosterone, androstenedione, androstenediol, and dehydroepiandrosterone (DEHA)), 4 progestogens (progesterone, pregnenolone, 17α-OH-progesterone, and 17α-OH-pregnenolone), and 7 corticosteroids (cortisone, cortisol, 11-deoxycortisol, aldosterone, 11-deoxycorticosterone, 11-dehydrocorticosterone, and corticosterone) were the target analytes.

2. Materials and Methods

2.1. Chemicals and Reagents

The chemical structures of the 19 steroid hormones investigated in this study are shown in Figure 1. Individual certified stock solutions of cortisone, 11-deoxycortisol, aldosterone, corticosterone, progesterone, 17α-OH-progesterone, pregnenolone, estrone, estradiol, testosterone, dehydroepiandrosterone (DEHA), 5α-dihydrotestosterone, 17α-OH-pregnenolone, 11-dehydrocorticosterone, androstenediol, $^{13}C_3$-cortisone, $^{13}C_3$-cortisol, $^{13}C_3$-11-deoxycortisol, D_4-aldosterone, D_4-corticosterone, $^{13}C_3$-progesterone, $^{13}C_3$-17α-OH-progesterone, $^{13}C_2$-D_2-pregnenolone, $^{13}C_3$-estrone, $^{13}C_2$-estradiol, $^{13}C_3$-estriol, $^{13}C_3$-testosterone, $^{13}C_3$-DEHA, $^{13}C_3$-5α-OH-dihydrotestosterone, and $^{13}C_2$-D_2-17α-OH-pregnenolone (100 μg/mL) with purities of 95–98% were purchased from Cambridge Isotope Laboratories (Andover, MA, USA). Neat cortisol and estriol (purity: ≥95%) were obtained from Cambridge Isotope Laboratories (Andover, MA, USA). Individual stock solutions of cortisol and estriol (100 μg/mL) were prepared in methanol (MeOH). Individual certified stock solutions of 11-deoxycorticosterone, androstenedione and $^{13}C_3$-11-deoxycorticosterone (100 μg/mL) with purities of ≥98% were purchased from Sigma-Aldrich (St. Louis, MO, USA). All stock solutions were stored at −20 °C. Working solutions of analytical standards were diluted from stock solutions using acetonitrile (ACN).

Synthetic urine was purchased from Cerilliant (Round Rock, TX, USA). Pooled human serum was from Sigma-Aldrich (St. Louis, MO, USA). Plasma Standard Reference Material (SRM1950) was purchased from the National Institute for Standards and Technology (NIST; Gaithersburg, MD, USA) through Sigma-Aldrich (St. Louis, MO, USA). β-glucuronidase/arylsulfatase enzyme (ALS; β-glucuronidase: ~100,000 units/mL; arylsulfatase: ~47,500 units/mL) from *Helix pomatia* was obtained from Roche Life Science through Sigma-Aldrich (St. Louis, MO, USA). L-Ascorbic acid, dansyl chloride (DC), formic acid (88%), and ammonium acetate (NH_4Ac) of analytical grade were from Sigma-Aldrich (St. Louis, MO, USA). Water, ACN, methyl *tert*-butyl ether (MTBE), and ethyl acetate (EtAc) of LC-MS grade were from Fisher Scientific (Waltham, MA, USA). Bond Elut C18 (60 mg/3 mL), Bond Elut Plexa (60 mg/3 mL), and Bond Elut NEXUS (60 mg/3 mL) solid phase extraction (SPE) cartridges were obtained from Agilent Technologies (Santa Clara, CA, USA).

2.2. Sample Collection

Spot urine samples were collected randomly in 50-mL polypropylene (PP) tubes from 11 male (age: 18–64 years, average 38.0 years) and 9 female (age: 15–59 years, average 33.3 years) volunteers from New York City during May–June 2022. Immediately after collection, samples were stored at −20 °C until analysis. The urine samples were deidentified and therefore fell under the 'exempt' category of the New York University Institutional Review Board.

2.3. Preparation of Urine and Serum Samples

A 500-µL aliquot of urine was pipetted into a 15-mL borosilicate glass culture tube. Synthetic urine and pooled human urine fortified with target analytes (at concentrations of 10, 20, and 200 ng/mL) were used for method development and validation. Then, 2.5–12.5 ng of isotopically labelled internal standards were spiked into each sample, and 500 µL of 1 M NH_4Ac buffer (pH 5.5) containing 2.5 mg of L-ascorbic acid and 20 µL of ALS enzyme (2000 units) were added. After gentle mixing, the samples were incubated overnight (~15 h) at 37 °C by shaking at 100 rpm (Jeio Tech Co., Seoul, South Korea). Thereafter, 1 mL of HPLC-grade water was added to each sample, followed by the addition of 4 mL MTBE/EtAc (5:1, *v/v*). After ultra-sonication for 15 min and shaking for 30 min in a reciprocating shaker, the samples were subjected to centrifugation at 3000 rpm for 5 min, and the supernatant was transferred into a new glass tube. The liquid-liquid extraction (LLE) was repeated twice with 3 mL of MTBE/EtAc (5:1, *v/v*). The combined extracts were subsequently evaporated to dryness under N_2 at 25 °C. Estrogens were then selectively derivatized by the addition of 125 µL of sodium bicarbonate buffer (0.1 M; pH 9.0) and 125 µL of dansyl chloride (1 mg/mL in acetone), vortexed vigorously (~30 s), and the sample tubes were immediately kept at 60 °C (on a hot plate) for 5 min. The extracts were evaporated to dryness under N_2 at 25 °C, reconstituted in 250 µL MeOH, and transferred into amber glass vials for LC-MS/MS analysis.

Serum samples (500 µL) were processed by following the procedure described above, without enzymatic hydrolysis. Pooled human serum fortified with target analytes (at concentrations of 10, 20, and 200 ng/mL) was used for method development and validation.

2.4. LC-MS/MS Analysis

The quantification of target analytes was performed on an ABSciex 5500+ Q-trap mass spectrometer (Framingham, MA, USA) coupled to an ExionLC HPLC (SCIEX, Redwood City, CA, USA). Analytes were separated on an Eclipse Plus C18 RRHD column (150 × 2.1 mm, 1.8 µm; Agilent Technologies, Santa Clara, CA, USA) coupled to a BetaSil C18 guard column (20 × 2.1 mm, 5 µm; Thermo Fisher Scientific, Waltham, MA, USA). The mobile phase, maintained at a flow rate of 0.2 mL/min, consisted water (A) and ACN (B) each containing 0.1% formic acid (*v/v*). The following gradient program was applied: hold at 10% B for 0.5 min, linear ramp to 40% B over 0.5 min, linear ramp to 70% B over 9 min, then linear ramp to 95% B over 0.5 min, hold at 95% B for 4.5 min, then return to initial conditions over 0.5 min, and equilibrate for 1.5 min prior to the next injection. The mobile phase flow was diverted to waste during the first 3.5 min after sample injection. The column was maintained at room temperature; the autosampler was kept at 15 °C; and the injection volume was 10 µL.

All analytes were measured using electrospray ionization (ESI) multiple reaction monitoring (MRM) in positive-ion mode. The optimized MRM parameters including collision energy (CE), declustering potential (DP), collision cell exit potential (CXP), entrance potential (EP), and dwell time are given in Table 1 and Table S1 (Supplementary Material). The IonSpray voltage was 5.5 kV; the ion source temperature was 500 °C; the curtain gas and collision gas flow rates were set at 20 and 9 psi, respectively; and the ion source gases 1 and 2 were set at 70 and 60 psi, respectively. Analyst software v1.7.2 (ABSciex, Framingham, MA, USA) was used for data analysis.

Table 1. Target analytes (19 steroid hormones), isotopically labeled internal standards, and their MRM parameters. MRM parameters include precursor ion (Q1), product ion (Q3), declustering potential (DP), entrance potential (EP), collision energy (CE), collision cell exit potential (CXP), and dwell time.

Name	CAS	Q1 (*m/z*)	Q3 (*m/z*)	DP (V)	CE (V)	EP (V)	CXP (V)	Dwell (ms)
			Target analytes					
DC-Estrone		504	171	125	35	10	15	30
DC-Estradiol		506	171	125	35	10	15	30
DC-Estriol		522	171	125	35	10	15	30
Testosterone	58-22-0	289	97	100	29	10	12	30
5α-Dihydrotestosterone	521-18-6	291	255	100	22	10	14	30
Androstenedione	63-05-8	287	97	172	25	10	14	30
Androstenediol	521-17-5	273	255	93	17	10	8	30
DEHA	53-43-0	271	253	80	23	10	12	30
Progesterone	57-83-0	315	97	130	23	10	10	30
Pregnenolone	145-13-1	317	299	98	13	10	16	30
17α-OH-Progesterone	68-96-2	331	109	114	39	10	6	30
17α-OH-Pregnenolone	387-79-1	333	297	80	23	10	18	30
Cortisone	53-06-5	361	163	152	31	10	20	30
Cortisol	50-23-7	363	121	144	29	10	12	30
11-Deoxycortisol	152-58-9	347	97	161	27	10	10	30
11-Deoxycorticosterone	64-85-7	331	97	160	27	10	16	30
11-Dehydrocorticosterone	72-23-1	331	97	168	26	10	11	30
Corticosterone	50-22-6	347	329	100	23	10	18	30
Aldosterone	52-39-1	361	343	120	24	10	12	30
			Internal standards					
DC-$^{13}C_3$-Estrone		507	171	125	35	10	15	30
DC-$^{13}C_2$-Estradiol		508	171	125	35	10	15	30
DC-$^{13}C_3$-Estriol		525	171	125	35	10	15	30
$^{13}C_3$-Testosterone		292	100	120	27	10	14	30
$^{13}C_3$-5α-Dihydrotestosterone		294	258	130	21	10	12	30
$^{13}C_3$-Dehydroepiandrosterone		274	256	125	14	10	14	30
$^{13}C_3$-Progesterone		318	100	170	30	10	6	30
$^{13}C_2$-D_2-Pregnenolone		321	303	110	11	10	22	30
$^{13}C_3$-17α-OH-Progesterone		334	100	138	27	10	12	30
$^{13}C_2$-D_2-17α-OH-Pregnenolone		319	301	90	15	10	16	30
$^{13}C_3$-Cortisol		366	124	152	30	10	22	30
$^{13}C_3$-11-Deoxycortisol		350	112	132	32	10	17	30
$^{13}C_3$-11-Deoxycorticosterone		334	100	154	27	10	18	30
D_4-Corticosterone		351	333	126	17	10	18	30
D_4-Aldosterone		365	347	100	26	10	24	30

Abbreviations: DC-Estrone, dansylated estrogen; DC-Estradiol, dansylated estradiol; DC-Estriol, dansylated estriol; DEHA, dehydroepiandrosterone.

2.5. Method Validation

Calibration curves of all analytes were constructed both in solvent and in fortified urine and serum matrices. Calibration standards were prepared in ACN at concentrations ranging from 0.05 to 2000 ng/mL, with 10–50 ng/mL of isotopically labelled internal standards, and were derivatized by following the procedure described above. The matrix-matched calibration curves were obtained by spiking all analytes at different concentrations (0.05–2000 ng/mL) in urine and serum extracts.

Matrix effects were evaluated by calculating the percentage of signal enhancement or suppression, as shown in Equation (1):

$$\text{Matrix effect} = \left(\frac{\text{A}}{\text{B}} - 1\right) \times 100\% \quad (1)$$

where A and B are the slopes of the matrix-matched calibration curve, and calibration curve from pure solvent, respectively.

Accuracy of the method was determined as the recoveries of analytes spiked at three different concentrations (10, 20 and 200 ng/mL) in both pooled urine and serum. Precision was assessed as intra-day and inter-day variations in measured concentrations in three pooled urine or serum samples fortified at 10, 20 and 200 ng/mL and was calculated as the coefficient of variation (CV%). The inter-day CV was measured by repeated injection of fortified samples on three different days. Accuracy of the method was further assessed through the analysis of NIST SRM1950 plasma, which has certified reference values for cortisol, progesterone and testosterone.

To calculate the method limit of detection (LOD) and limit of quantification (LOQ) of steroid hormones in urine, six replicates of synthetic urine were fortified with each target analyte at 1 ng/mL, a concentration that yielded signal-to-noise (*S/N*) ratios of 6.4–62.4. LOD and LOQ of target analytes in urine were calculated as 3 and 10 times the standard deviation (SD), respectively. Similarly, LOD and LOQ in serum were calculated as 3 and 10 times the SD of concentrations, respectively, measured in pooled human serum spiked at 1 ng/mL.

2.6. Quality Assurance and Quality Control

Quality control (QC) samples include procedural blanks (LC-MS grade water instead of urine or serum), matrix blanks (pooled urine or serum), and matrix spiked samples (analytes fortified in pooled urine or serum at three different concentrations: 10, 20 and 200 ng/mL). The NIST SRM sample was also included in the analysis of hormones in plasma.

3. Results and Discussion

3.1. Chromatography and Mass Spectrometry

The development of the method started with optimization of MS/MS parameters for 19 target analytes, through the infusion of a standard solution (200 ng/mL in MeOH containing 0.2% formic acid) directly into the mass spectrometer. Addition of formic acid to the standard solution (as an additive) over acetic acid yielded better signal intensity. ACN, as the solvent, yielded lower backpressure in comparison to MeOH.

Chromatographic separation of steroid hormones in urine or serum is critical because several of these analytes exhibit the same mass spectrometric MRM transitions (Table 1). Few earlier studies employed lengthy chromatographic separation time to isolate and resolve individual analytes [5,19]. We compared the performance of several reversed-phase LC columns and found that Eclipse Plus C18 RRHD column (150 × 2.1 mm, 1.8 μm; Agilent Technologies, Santa Clara, CA, USA) provided good peak shape and chromatographic separation for most of the analytes within a run time of 10 min (Figure S1). Nevertheless, several important biomarkers were not resolved chromatographically. Estriol and aldosterone, as well as estrone and DEHA co-eluted at the same retention times. Furthermore, estrogens exhibited low signal intensities (Figure 2) due to their low ionization efficiencies. To improve selectivity and sensitivity of estrogens, we applied an estrogen-specific derivatization step using dansyl chloride, as reported earlier [18,19,25]. Following dansylation, the signal intensities of estrone, estradiol, and estriol improved significantly with an enhancement in *S/N* by 23, 16 and 11 times, respectively, in comparison to intensities obtained prior to derivatization. These results are consistent with those reported earlier [26]. Furthermore, the retention times of dansyl-estrone (DC-estrone), DC-estradiol, and DC-estriol were different from all other analytes, indicating improved selectivity (Figures 2 and 3). Nevertheless, androstenedione and 17α-OH-progesterone co-eluted, which, however, was deemed acceptable because of their different MRM transitions (Table 1). Our method, with a total run time of 17 min, was faster than the previous methods [5,19], a feature favorable for large-scale human biomonitoring studies.

						Intra-Day	Inter-Day		Day	Day		Day	Day
						Estrogen							
Estrone	0.9998	0.11	0.38	1.09	112 ± 5	4.08	2.18	103 ± 3	2.97	6.06	97.3 ± 3.7	3.79	0.78
Estradiol	0.9989	0.14	0.48	0.92	88.5 ± 3.8	4.33	3.17	92.3 ± 8.9	9.63	1.14	93.3 ± 1.8	3.29	2.68
Estriol	0.9982	0.17	0.55	−6.49	110 ± 2	1.89	2.46	114 ± 5	3.97	2.39	99.0 ± 3.1	3.15	1.13
						Androgen							
Testosterone	0.9958	0.22	0.75	−0.12	118 ± 8	6.37	2.16	116 ± 6	5.19	1.26	119 ± 1	0.64	2.63
5α-Dihydrotestosterone	0.9998	0.18	0.60	0.00	104 ± 7	6.28	4.78	114 ± 3	2.67	0.78	108 ± 1	0.80	3.02
Androstenedione	0.9985	0.16	0.52	18.2	114 ± 3	2.68	4.09	117 ± 3	2.58	1.48	110 ± 3	2.73	2.08
Androstenediol	0.9985	0.24	0.81	5.34	96.1 ± 1.2	1.20	3.88	104 ± 6	5.30	3.00	99.7 ± 1.4	1.45	2.20
DEHA	0.9990	0.17	0.56	−10.2	102 ± 5	5.05	6.43	102 ± 8	7.62	2.40	107 ± 3	2.84	3.89
						Progestogen							
Progesterone	0.9997	0.17	0.58	−7.08	110 ± 5	4.48	4.40	115 ± 5	4.24	6.72	110 ± 3	2.62	5.82
Pregnenolone	0.9992	0.35	1.18	−0.62	96.6 ± 12.1	12.5	11.5	91.5 ± 1.3	1.45	8.55	100 ± 2	2.18	1.75
17α-OH-Progesterone	0.9994	0.13	0.42	2.99	109 ± 6	5.36	2.95	108 ± 2	1.84	1.42	100 ± 1	1.00	1.00
17α-OH-Pregnenolone	0.9974	0.24	0.80	−13.9	101 ± 3	3.04	5.80	100 ± 7	6.60	4.14	92.8 ± 3.5	3.73	2.51
						Corticosteroid							
Cortisone	0.9990	0.25	0.84	38.7	119 ± 9	7.33	4.33	118 ± 8	6.58	1.04	119 ± 5	4.21	2.61
Cortisol	0.9981	0.24	0.79	−3.48	90.0 ± 14.1	15.7	1.24	80.0 ± 17.3	21.7	0.96	94.0 ± 8.8	9.32	2.06
11-Deoxycortisol	0.9997	0.17	0.58	7.22	107 ± 3	2.46	2.55	111 ± 2	1.88	1.99	105 ± 3	2.72	1.91
11-Deoxycorticosterone	0.9989	0.21	0.70	0.64	97.2 ± 1.6	1.68	1.01	103 ± 4	3.67	2.22	96.0 ± 3.5	3.61	2.04
11-Dehydrocorticosterone	0.9981	0.23	0.78	6.76	106 ± 9	8.04	7.61	107 ± 1	1.08	4.73	103 ± 2	2.03	4.22
Corticosterone	0.9995	0.17	0.56	−11.0	94.7 ± 6.4	6.71	0.16	100 ± 4	4.36	2.66	88.7 ± 1.3	1.49	1.83
Aldosterone	0.9997	0.22	0.73	9.73	102 ± 6	5.59	1.41	106 ± 5	4.34	4.74	98.8 ± 4.7	4.70	1.11

Abbreviation: DEHA, dehydroepiandrosterone; ME, matrix effect; LOD, limit of detection; LOQ, limit of quantification; CV, coefficient of variation. [a] R value in serum matrix.

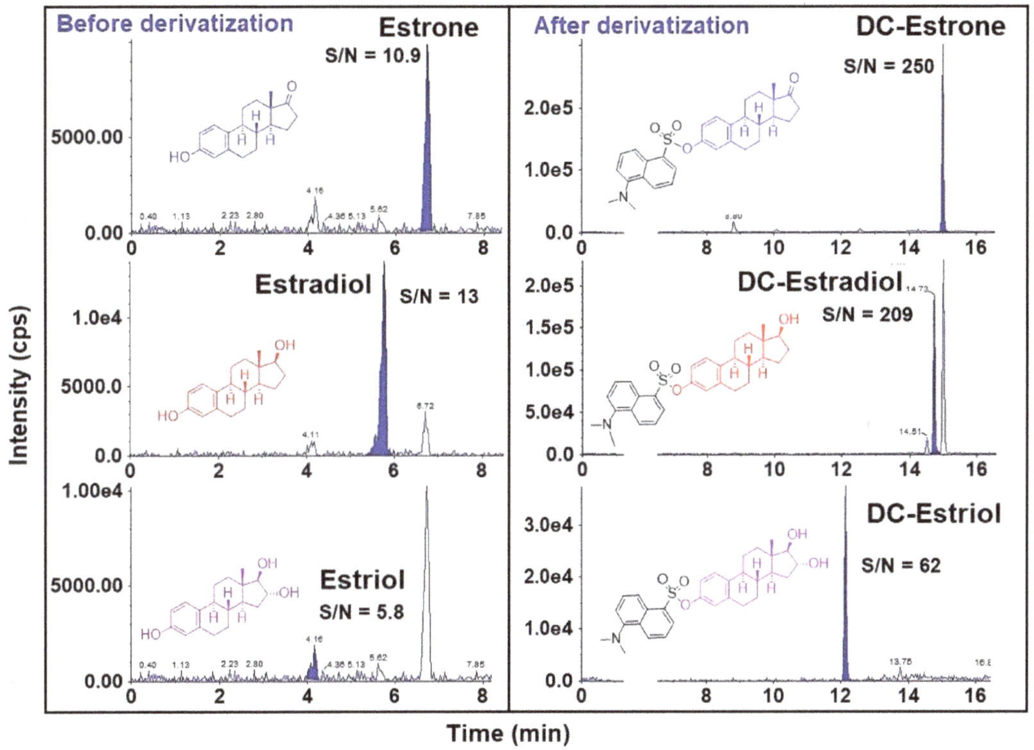

Figure 2. Comparison of LC-MS/MS responses of estrone, estradiol, and estriol before and after dansylation. Concentrations of the analytes were 1 ng/mL; Injection volume was 10 μL.

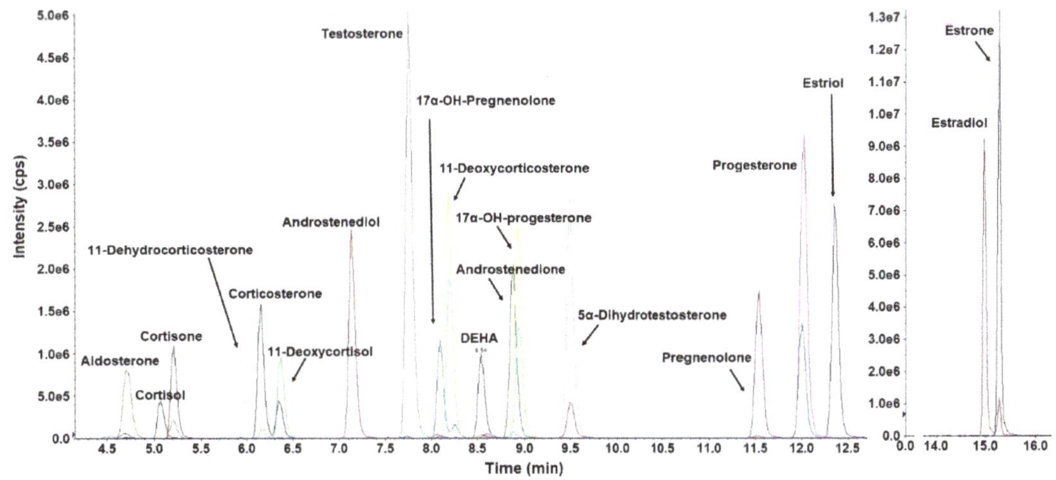

Figure 3. Representative LC-MS/MS chromatograms of 19 steroid hormones in solvent. Concentrations of the target analytes were 100 ng/mL; Injection volume was 10 μL; estrone, estradiol, and estriol were derivatized using dansyl chloride. Estrone and estradiol were shown separately because of their significantly higher intensities than the other analytes.

3.2. Optimization of Sample Purification

Given the lipophilic nature of steroid hormones, liquid-liquid extraction (LLE) with non-polar solvents (e.g., MTBE, EtAc and dichloromethane) and SPE with reversed-phase sorbents (e.g., C18) were considered appropriate to efficiently extract the analytes from urine and serum/plasma [19,20,26]. We compared the performance of LLE using MTBE/EtAc (5:1, *v/v*) [20] and different types of SPE cartridges including Bond Elut C18, Bond Elut Plexa, and Bond Elut NEXUS (Supplementary Material). The recoveries of most of the target analytes through the extraction (LLE) and purification (SPE) steps were acceptable (Table S2). However, strong matrix effect following SPE (Figure S2) resulted in poor resolution of pregnenolone and $^{13}C_2$-D_2-pregnenolone. Resolution of this compound was excellent in serum samples after LLE. Therefore, samples were analysed using LC-MS/MS after LLE with MTBE/EtAc (5:1, *v/v*).

In human urine, steroid hormones are present almost exclusively as phase II metabolites. Thus, deconjugation is an essential step to measure the total concentrations of steroid hormones. ALS enzyme was reported to be efficient for the deconjugation of phase II metabolites of various endogenous and exogenous compounds [20,27]. Therefore, urine samples were incubated with ALS enzyme prior to extraction.

3.3. Method Validation

A 16-point calibration curve was prepared at concentrations ranging from 0.05 to 2000 ng/mL, with 10–50 ng/mL of isotopically labelled internal standards. Calibration curves prepared in pure solvent yielded R-values in the range of 0.9973–0.9999 for all steroid hormones (Table S3). Matrix-matched calibration curves at fortification levels of 0.05–2000 ng/mL also exhibited excellent R-values in both synthetic urine (0.9974–0.9998) and pooled human serum (0.9958–0.9998) (Tables 2 and 3).

Accuracy of the method was assessed by spike-recovery tests conducted in triplicate in pooled human urine and pooled human serum. Average recoveries of each analyte at different fortification levels (10, 20, and 200 ng/mL) were measured. In human urine, the recoveries of all hormones (except for progesterone) at three fortification levels were in the range of 81.7–117%, 85.0–120%, and 82.7–119%, respectively, and the relative standard deviation (SD) of triplicate measurements was 1.00–15.5%, 0–13.0%, and 0.30–11.0%, respectively. These results suggest acceptable accuracies for most of the analytes. Nevertheless, recoveries of progesterone were in the range of 117–141% (Table 2). This may indicate that certain untargeted hormone metabolites (or precursors) exhibit the same retention and MRM transition with progesterone, considering their similar chemical structures [1]. In human serum, the recoveries of all hormones at three fortification levels were in the range of 88.5–119%, 80.0–118%, and 88.7–119%, respectively, and the SD of triplicate analyses was in the range of 1.2–14.1%, 1–17.3%, and 1.0–8.8%, respectively (Table 3). The determined accuracies of all analytes were similar to those reported in an earlier study (84–122%) [20].

Table 2. Method validation parameters for the analysis of 19 steroid hormones in human urine.

Analytes	R [a]	LOD (ng/mL)	LOQ (ng/mL)	ME%	10 ng/mL			20 ng/mL			200 ng/mL		
					Recovery%	CV%		Recovery%	CV%		Recovery%	CV%	
						Intra-Day	Inter-Day		Intra-Day	Inter-Day		Intra-Day	Inter-Day
Estrogen													
Estrone	0.9990	0.17	0.56	−6.56	111 ± 9	8.39	3.24	114 ± 3	2.78	2.39	105 ± 5	4.61	0.55
Estradiol	0.9992	0.04	0.14	0.92	93.3 ± 11.5	12.4	2.84	120 ± 10	8.67	3.10	118 ± 11	9.05	4.08
Estriol	0.9981	0.13	0.42	−2.16	110 ± 5	4.18	9.68	109 ± 3	3.12	6.83	99.6 ± 5.3	5.31	3.65
Androgen													
Testosterone	0.9997	0.14	0.46	−0.81	113 ± 5	4.20	0.71	119 ± 1	0.60	1.34	115 ± 3	2.10	1.71
5α-Dihydrotestosterone	0.9974	0.15	0.49	−2.51	82.5 ± 1.3	1.50	4.17	85.0 ± 1.9	2.20	3.21	82.7 ± 2.0	2.40	0.74
Androstenedione	0.9994	0.22	0.73	7.30	109 ± 13	11.7	2.39	109 ± 11	10.4	2.17	119 ± 7	6.20	3.68
Androstenediol	0.9994	0.23	0.76	0.49	102 ± 12	11.7	2.56	119 ± 8	6.40	0.56	99.8 ± 4.8	4.83	0.65
DEHA	0.9996	0.24	0.81	−1.67	117 ± 6	4.95	3.88	91.7 ± 0	0.00	4.27	105 ± 8	7.45	2.58
Progestogen													
Progesterone	0.9990	0.15	0.50	−2.65	141 ± 11	7.50	3.93	132 ± 8	6.30	4.45	117 ± 8	6.60	4.77
Pregnenolone	0.9998	0.23	0.75	−2.48	106 ± 6	6.00	1.77	111 ± 1	0.90	2.70	110 ± 4	3.72	2.29
17α-OH-Progesterone	0.9997	0.21	0.70	0.75	110 ± 1	0.60	1.49	115 ± 13	11.2	2.31	113 ± 4	3.30	1.69
17α-OH-Pregnenolone	0.9998	0.12	0.41	−2.45	112 ± 4	3.10	3.86	112 ± 1	1.10	1.00	96.8 ± 4.0	4.20	2.43
Corticosteroid													
Cortisone	0.9985	0.13	0.44	29.6	101 ± 12	11.5	4.53	115 ± 8	7.08	3.41	117 ± 4	3.47	4.25
Cortisol	0.9998	0.20	0.65	−1.74	81.7 ± 15.5	19.0	1.34	100 ± 9	9.35	1.20	96.5 ± 4.3	4.49	3.12
11-Deoxycortisol	0.9995	0.19	0.64	5.84	112 ± 3	2.37	2.63	116 ± 4	3.80	4.37	106 ± 3	2.76	1.17
11-Deoxycorticosterone	0.9998	0.25	0.83	−2.47	97.1 ± 9.9	10.2	1.29	99.2 ± 1.4	1.45	2.42	99.9 ± 2.0	2.02	1.58
11-Dehydrocorticosterone	0.9990	0.28	0.92	5.41	86.7 ± 5.5	6.35	2.56	90.2 ± 3.2	3.56	1.82	84.1 ± 2.5	2.93	3.20
Corticosterone	0.9993	0.26	0.85	5.52	107 ± 4	3.37	2.20	101 ± 2	1.72	1.39	101 ± 1.3	1.31	1.13
Aldosterone	0.9997	0.19	0.63	4.42	95.5 ± 2.6	2.77	7.16	101 ± 10	9.47	7.59	90.0 ± 0.3	0.32	4.59

Abbreviation: DEHA, dehydroepiandrosterone; ME, matrix effect; LOD, limit of detection; LOQ, limit of quantification; CV, coefficient of variation. [a] R value in urine matrix.

Table 3. Method validation parameters for the analysis of 19 steroid hormones in human serum.

We further assessed and validated the accuracy of the method through the analysis of NIST human plasma SRM1950. The concentrations of testosterone, progesterone, and cortisol determined in our method agreed well with the certified reference values, with variations within 11% (Table 4), further confirming validity of the developed method.

Table 4. Steroid hormones measured in plasma Standard Reference Material (SRM1950) using the method developed in this study and compared to the certified reference values. Reference values were only available for cortisol, progesterone, and testosterone.

Analytes	Our Results (ng/mL)	Reference Value (ng/mL)	Variation%
	Estrogen		
Estrone	0.14	n.a.	–
Estradiol	<LOD	n.a.	–
Estriol	<LOD	n.a.	–
	Androgen		
Testosterone	2.47	2.214	11
5α-Dihydrotestosterone	0.20	n.a.	–
Androstenedione	0.74	n.a.	–
Androstenediol	0.58	n.a.	–
DEHA	1.99	n.a.	–
	Progestogen		
Progesterone	1.36	1.482	8.6
Pregnenolone	0.41	n.a.	–
17α-OH-Progesterone	0.58	n.a.	–
17α-OH-Pregnenolone	0.94	n.a.	–
	Corticosteroid		
Cortisol	81.4	83.9	3
Cortisone	20.9	n.a.	–
11-Deoxycortisol	0.17	n.a.	–
11-Deoxycorticosterone	<LOD	n.a.	–
11-Dehydrocorticosterone	0.67	n.a.	–
Corticosterone	1.57	n.a.	–
Aldosterone	<LOD	n.a.	–

Abbreviation: n.a., not available; DEHA, dehydroepiandrosterone; LOD, limit of detection.

We assessed precision of the method as intra-day and inter-day variations from repeated analysis of fortified human urine and serum samples. In human urine, the intra-day CVs of target analytes fortified at 10, 20, and 200 ng/mL were 0.60–19.0%, 0–11.2%, and 0.32–9.05%, respectively, whereas the inter-day CVs were 0.71–9.68%, 0.56–7.59%, and 0.55–4.77%, respectively (Table 2). In human serum, the intra-day CVs were 1.20–15.7%, 1.08–21.7%, and 0.64–9.32%, respectively, whereas the inter-day CVs were 0.16–11.5%, 0.78–8.55%, and 0.78–5.82%, respectively (Table 3). These results suggested acceptable precision (CVs < 20%) for all analytes except cortisol in serum (9.32–21.7%). The relatively low precision of cortisol was likely due to its high background concentration in urine and serum (Tables 4 and S4). The method precision calculated in this study was similar to that reported in an earlier study (intra-day CV: 2–11%; inter-day CV: 5–24%) [20].

Matrix effect is a common phenomenon in ESI-LC-MS analysis, produced by ionization enhancement or suppression by matrix components. Typically, values in the range of −20 to +20% are considered as weak suppression/enhancement. Use of isotopically labelled internal standards of analytes (and isotopic dilution method of quantification) is an acceptable practice to compensate for matrix effects. In this study, all analytes except cortisone exhibited weak matrix effects in both urine (−6.56 to 7.30%) and serum (−13.9 to 18.2%), indicating that the optimized method adequately removed interfering matrix components (Tables 2 and 3). However, a moderate ion enhancement was found for cortisone in urine (29.6%) and serum (38.7%), which could be explained by the lack of isotope labelled cortisone available in this study ($^{13}C_3$-cortisol was used as the internal standard for cortisone). Use of isotopically labelled cortisone would reduce the uncertainty in measurement associated with matrix-effect for this analyte.

The sensitivity of the method was determined as LOD and LOQ in both human urine and serum. In human urine, the LODs ranged from 0.04 ng/mL (estradiol) to 0.28 ng/mL (11-

dehydrocorticosterone), and the LOQs were between 0.14 ng/mL (estradiol) and 0.92 ng/mL (11-dehydrocorticosterone) (Table 2). In human serum, the LODs ranged from 0.11 ng/mL (estrone) to 0.35 ng/mL (pregnenolone), and the LOQs were between 0.38 ng/mL (estrone) and 1.18 ng/mL (pregnenolone) (Table 3). Thus, the method provides adequate sensitivity for the determination of 19 steroid hormones in human urine and serum/plasma. The method sensitivities were similar to those reported in the literature (e.g., LODs: 0.20–1.00 ng/mL in urine [20], and 0.04–1.35 in fetal bovine serum [5]) although the earlier studies used different methods to calculate LOD. In particular, the LODs for estrogens in urine in our study (0.04–0.17 ng/mL) were significantly lower than those reported in an earlier study (0.20–0.83 ng/mL) [20].

3.4. Method Application

The validated method was applied for the analysis of 20 spot urine samples collected from healthy volunteers (11 males aged 18–64 y, and 9 females aged 15–59 y) during May–June 2022 from New York City. Sixteen analytes were found with detection frequencies (DFs) $\geq$ 80% in urine samples. The average concentrations of all analytes ranged from 0.34 to 284 ng/mL. In both males and females, DEHA was the most abundant (accounting for 39% and 32% of the total concentrations in males and females, respectively), followed by estradiol (20% and 23%, respectively), cortisol (13% and 14%, respectively), cortisone (8% and 9%, respectively), and androstenediol (8% and 9%, respectively) (Figure 4 and Table S4). Representative chromatograms of all analytes in urine samples are shown in Figure S3.

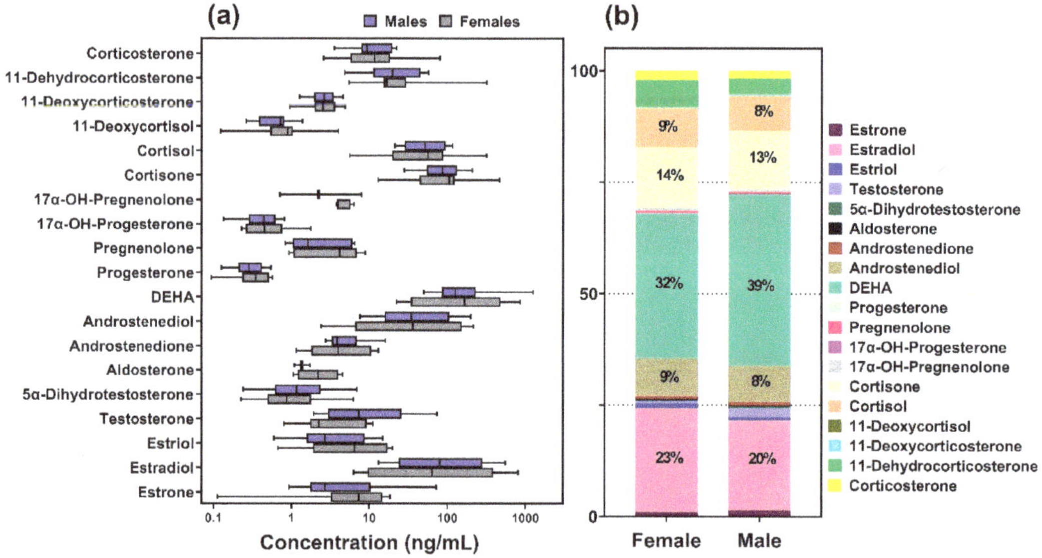

Figure 4. Concentrations (**a**) and relative distribution (**b**) of 19 steroid hormones measured in urine from 11 males and 9 females from New York, United States. Mean concentrations were used in the calculation of the percent contribution.

The method was applied in the analysis of human plasma by measuring all analytes in SRM1950. Among the 19 steroid hormones, 15 analytes were quantified at concentrations ranging from 0.14 (estrone) and 81.4 ng/mL (cortisol). Aldosterone, 11-deoxycorticosterone, estradiol, and estriol were below the respective LODs in the sample. In pooled human serum purchased from Sigma-Aldrich, 12 out of the 19 analytes were found at concentrations in the range of 0.23 (11-deoxycortisol) to 72.5 ng/mL (cortisol) (Table S5). Representative chromatograms of all analytes in serum samples are shown in Figure S4. Overall, the method is suitable for the determination of 19 steroid hormones in human urine and serum/plasma of the general population.

4. Conclusions

We developed and validated a HPLC-MS/MS method for simultaneous determination of 19 steroid hormones in human urine and serum/plasma. Liquid-liquid extraction with MTBE/EtAc (5:1, *v/v*) yielded acceptable recoveries of all analytes from these matrices. Chromatographic and mass spectrometric conditions were optimized to separate and detect the hormones at physiologically relevant concentrations. Dansylation of estrone, estradiol, and estriol improved their chromatographic resolution and sensitivities.

The method was validated for accuracy, precision, sensitivity, linearity, and through the analysis of certified plasma standard reference material. The method was applied in the analysis of 19 hormones in real human urine and serum/plasma. The method can also be used in the analysis of steroid hormones in other matrices (e.g., hair and saliva) with slight modifications. The method can be applied in the analysis of steroid hormones in human urine and serum/plasma in population-based biomonitoring studies.

Supplementary Materials: The following supporting information can be downloaded at: https://www.mdpi.com/article/10.3390/toxics10110687/s1. Supplementary method: sample cleanup procedure using SPE. Figure S1. Representative LC-MS/MS chromatograms of 19 steroid hormones without derivatization. Concentrations of the target analytes were 100 ng/mL; Injection volume was 10 µL. Figure S2. Comparison of pregnenolone in serum following extraction with LLE, Bond Elut C18, Bond Elut Plexa, and Bond Elut NEXUS. Analyte concentration was 100 ng/mL, and the injection volume was 10 µL. Figure S3. Representative chromatograms of 19 steroid hormones found in human urine (injection volume: 10 µL). Figure S4. Representative chromatograms of 19 steroid hormones found in human serum (injection volume: 10 µL). Table S1. Optimized MRM parameters for the estrogens. Table S2. Comparison of the spike-recoveries of all analytes in serum following extraction with LLE, Bond Elut C18 (60 mg/3 mL), Bond Elut Plexa (60 mg/3 mL), and Bond Elut NEXUS (60 mg/3 mL). Analytes were spiked at 100 ng/mL, injection volume was 10 µL. Table S3. R value of the calibration curve of all analytes in neat solution. Table S4. Steroid hormone concentrations measured in twenty human urine samples from the general population in New York, USA, in 2022. Table S5. Steroid hormone concentrations measured in the pooled human serum purchased from Sigma-Aldrich.

Author Contributions: Z.-M.L.: Methodology, Data curation, Formal analysis, Writing—original draft. K.K.: Conceptualization, Funding acquisition, Supervision, Writing—review and editing. All authors have read and agreed to the published version of the manuscript.

Funding: The research reported here was supported, in part, by the US National Institute of Environmental Health Sciences (NIEHS) under award number U2CES026542 (KK). The content is solely the responsibility of the authors and does not necessarily represent the official views of the NIEHS.

Institutional Review Board Statement: The study was conducted in accordance with the Declaration of Helsinki, and approved by the Institutional Review Board of the New York University for the analysis of de-identified urine samples (under exempt category) to demonstrate application of the method developed in this study.

Informed Consent Statement: Informed consent was waived as the study was deemed exempt human study and only de-identified specimens were analyzed.

Data Availability Statement: Not applicable.

Conflicts of Interest: The authors declare that there are no conflict of interest.

References

1. Schiffer, L.; Barnard, L.; Baranowski, E.S.; Gilligan, L.C.; Taylor, A.E.; Arlt, W.; Shackleton, C.H.L.; Storbeck, K.H. Human steroid biosynthesis, metabolism and excretion are differentially reflected by serum and urine steroid metabolomes: A comprehensive review. *J. Steroid. Biochem. Mol. Biol.* **2019**, *194*, 105439. [CrossRef] [PubMed]
2. Papadopoulos, A.S.; Cleare, A.J. Hypothalamic-pituitary-adrenal axis dysfunction in chronic fatigue syndrome. *Nat. Rev. Endocrinol.* **2011**, *8*, 22–32. [CrossRef] [PubMed]
3. Acevedo-Rodriguez, A.; Kauffman, A.S.; Cherrington, B.D.; Borges, C.S.; Roepke, T.A.; Laconi, M. Emerging insights into hypothalamic-pituitary-gonadal axis regulation and interaction with stress signalling. *J. Neuroendocrinol.* **2018**, *30*, e12590. [CrossRef] [PubMed]
4. Falkenstein, E.; Tillmann, H.-C.; Christ, M.; Feuring, M.; Wehling, M. Multiple actions of steroid hormones—A focus on rapid, nongenomic effects. *Pharmacol. Rev.* **2000**, *52*, 513–556.
5. Boggs, A.S.; Bowden, J.A.; Galligan, T.M.; Guillette, L.J., Jr.; Kucklick, J.R. Development of a multi-class steroid hormone screening method using Liquid Chromatography/Tandem Mass Spectrometry (LC-MS/MS). *Anal. Bioanal. Chem.* **2016**, *408*, 4179–4190. [CrossRef]
6. Taylor, J.I.; Grace, P.B.; Bingham, S.A. Optimization of conditions for the enzymatic hydrolysis of phytoestrogen conjugates in urine and plasma. *Anal. Biochem.* **2005**, *341*, 220–229. [CrossRef]
7. Baglietto, L.; Severi, G.; English, D.R.; Krishnan, K.; Hopper, J.L.; McLean, C.; Morris, H.A.; Tilley, W.D.; Giles, G.G. Circulating steroid hormone levels and risk of breast cancer for postmenopausal women. *Cancer Epidemiol. Biomark. Prev.* **2010**, *19*, 492–502. [CrossRef]
8. Peck, J.D.; Hulka, B.S.; Poole, C.; Savitz, D.A.; Baird, D.; Richardson, B.E. Steroid hormone levels during pregancy and incidence of maternal breast cancer. *Cancer Epidemiol. Biomark. Prev.* **2002**, *11*, 361–368.
9. Eliassen, A.H.; Missmer, S.A.; Tworoger, S.S.; Spiegelman, D.; Barbieri, R.L.; Dowsett, M.; Hankinson, S.E. Endogenous steroid hormone concentrations and risk of breast cancer among premenopausal women. *J. Natl. Cancer Inst.* **2006**, *98*, 1406–1415. [CrossRef]
10. Lukanova, A.; Lundin, E.; Micheli, A.; Arslan, A.; Ferrari, P.; Rinaldi, S.; Krogh, V.; Lenner, P.; Shore, R.E.; Biessy, C.; et al. Circulating levels of sex steroid hormones and risk of endometrial cancer in postmenopausal women. *Int. J. Cancer* **2004**, *108*, 425–432. [CrossRef]
11. Lukanova, A.; Lundin, E.; Akhmedkhanov, A.; Micheli, A.; Rinaldi, S.; Zeleniuch-Jacquotte, A.; Lenner, P.; Muti, P.; Biessy, C.; Krogh, V.; et al. Circulating levels of sex steroid hormones and risk of ovarian cancer. *Int. J. Cancer* **2003**, *104*, 636–642. [CrossRef] [PubMed]
12. Herane Vives, A.; De Angel, V.; Papadopoulos, A.; Strawbridge, R.; Wise, T.; Young, A.H.; Arnone, D.; Cleare, A.J. The relationship between cortisol, stress and psychiatric illness: New insights using hair analysis. *J. Psychiatr. Res.* **2015**, *70*, 38–49. [CrossRef] [PubMed]
13. Pollock, T.; Arbuckle, T.E.; Guth, M.; Bouchard, M.F.; St-Amand, A. Associations among urinary triclosan and bisphenol A concentrations and serum sex steroid hormone measures in the Canadian and U.S. Populations. *Environ. Int.* **2021**, *146*, 106229. [CrossRef] [PubMed]
14. Aimuzi, R.; Wang, Y.; Luo, K.; Jiang, Y. Exposure to phthalates, phenols, and parabens mixture and alterations in sex steroid hormones among adolescents. *Chemosphere* **2022**, *302*, 134834. [CrossRef] [PubMed]
15. Soldin, S.J.; Soldin, O.P. Steroid hormone analysis by tandem mass spectrometry. *Clin. Chem.* **2009**, *55*, 1061–1066. [CrossRef] [PubMed]
16. Adlercreutz, H.; Kiuru, P.; Rasku, S.; Wahala, K.; Fotsis, T. An isotope dilution gas chromatographic-mass spectrometric method for the simultaneous assay of estrogens and phytoestrogens in urine. *J. Steroid. Biochem. Mol. Biol.* **2004**, *92*, 399–411. [CrossRef] [PubMed]
17. Xu, X.; Duncan, A.M.; Merz-Demlow, B.E.; Phipps, W.R.; Kurzer, M.S. Menstrual cycle effects on urinary estrogen metabolites. *J. Clin. Endocrinol. Metab.* **1999**, *84*, 3914–3918. [CrossRef]
18. Xu, X.; Veenstra, T.D.; Fox, S.D.; Roman, J.M.; Issaq, H.J.; Falk, R.; Saavedra, J.E.; Keefer, L.K.; Ziegler, R.G. Measuring fifteen endogenous estrogens simultaneously in human urine by high-performance liquid chromatography-mass spectrometry. *Anal. Chem.* **2005**, *77*, 6646–6654. [CrossRef]
19. Xu, X.; Keefer, L.K.; Ziegler, R.G.; Veenstra, T.D. A liquid chromatography-mass spectrometry method for the quantitative analysis of urinary endogenous estrogen metabolites. *Nat. Protoc.* **2007**, *2*, 1350–1355. [CrossRef]
20. Zhou, Y.; Cai, Z. Determination of hormones in human urine by ultra-high-performance liquid chromatography/triple-quadrupole mass spectrometry. *Rapid Commun. Mass Spectrom.* **2020**, *34* (Suppl. 1), e8583. [CrossRef]
21. Gao, W.; Kirschbaum, C.; Grass, J.; Stalder, T. LC-MS based analysis of endogenous steroid hormones in human hair. *J. Steroid. Biochem. Mol. Biol.* **2016**, *162*, 92–99. [CrossRef] [PubMed]
22. Gao, W.; Stalder, T.; Kirschbaum, C. Quantitative analysis of estradiol and six other steroid hormones in human saliva using a high throughput liquid chromatography-tandem mass spectrometry assay. *Talanta* **2015**, *143*, 353–358. [CrossRef] [PubMed]
23. Wang, Q.; Mesaros, C.; Blair, I.A. Ultra-high sensitivity analysis of estrogens for special populations in serum and plasma by liquid chromatography-mass spectrometry: Assay considerations and suggested practices. *J. Steroid. Biochem. Mol. Biol.* **2016**, *162*, 70–79. [CrossRef] [PubMed]

24. Stanczyk, F.Z.; Clarke, N.J. Measurement of estradiol—Challenges ahead. *J. Clin. Endocrinol. Metab.* **2014**, *99*, 56–58. [CrossRef]
25. Nelson, R.E.; Grebe, S.K.; DJ, O.K.; Singh, R.J. Liquid chromatography-tandem mass spectrometry assay for simultaneous measurement of estradiol and estrone in human plasma. *Clin. Chem.* **2004**, *50*, 373–384. [CrossRef]
26. Honda, L.; Becerra-Herrera, M.; Richter, P. Liquid chromatography-time-of-flight high-resolution mass spectrometry study and determination of the dansylated products of estrogens and their hydroxylated metabolites in water and wastewater. *Anal. Bioanal. Chem.* **2018**, *410*, 7909–7919. [CrossRef]
27. Zhu, H.; Chinthakindi, S.; Kannan, K. A method for the analysis of 121 multi-class environmental chemicals in urine by high-performance liquid chromatography-tandem mass spectrometry. *J. Chromatogr. A* **2021**, *1646*, 462146. [CrossRef]

International Journal of
Environmental Research and Public Health

MDPI

Article

Diurnal Nonlinear Recurrence Metrics of Skin Temperature and Their Association with Metabolic Hormones in Contrasting Climate Settings: A Randomized Cross-Over Trial

Konstantinos C. Makris [1,*], Pantelis Charisiadis [1], Thibaut Delplancke [1], Nikolaos Efthymiou [1] and Alessandro Giuliani [2]

1 Cyprus International Institute for Environmental and Public Health, Cyprus University of Technology, 3041 Limassol, Cyprus
2 Environment and Health Department, Istituto Superiore di Sanità, Viale Regina Elena 299, 00161 Rome, Italy
* Correspondence: konstantinos.makris@cut.ac.cy; Tel.: +357-25-002-398

Abstract: The urban overheating phenomenon in Mediterranean cities is a societal challenge with vast implications for the protection of public health. An additional analysis of the pilot TEMP randomized 2 × 2 cross-over trial was set up, using wearable sensor-based skin temperature measurements (n = 14). The study objectives were to: (i) assess the recurrence patterns of skin temperature measurements in individuals spending time in two climatologically contrasting settings (urban versus mountainous), and (ii) evaluate the association between the diurnal nonlinear recurrence quantification analysis (RQA) metrics and metabolic hormone levels. The intervention was a short-term stay (5–7 days) in a mountainous, climate-cooler setting (range 600–900 m altitude), which is about a 1 h drive from the main urban centres of Cyprus. The RQA analysis showed a blunting phenomenon on the nonlinear temporal dynamics of skin temperature time series observed in the urban setting. Compared with the mountainous setting, a more stable (and thus less adaptive) profile of skin temperature dynamics in the urban setting appeared, being less deterministic and with a smaller degree of complexity. No significant ($p > 0.05$) associations were observed between the leptin or cortisol and any of the skin temperature dynamical descriptors. However, there were marginal associations between the adiponectin and laminarity (beta = 0.24, 95%CI: −0.02, 0.50, $p = 0.07$) and with determinism (beta = 0.23, 95%CI: −0.037, 0.50, $p = 0.09$). We found dysregulations in skin temperature temporal dynamics of the study population while residing in the urban setting when compared with the cooler mountainous setting; these dysregulations took the form of reduced cycle duration and complexity, while skin temperature dynamics became less responsive to perturbations and less regular in magnitude. More research is needed to better understand heat stress temporal dynamics and their influence on human health. Trial registration: This trial is registered with ClinicalTrials.gov; number: NCT03625817.

Keywords: climate; sensors; recurrence; temporal; metabolism; synchronization

Citation: Makris, K.C.; Charisiadis, P.; Delplancke, T.; Efthymiou, N.; Giuliani, A. Diurnal Nonlinear Recurrence Metrics of Skin Temperature and Their Association with Metabolic Hormones in Contrasting Climate Settings: A Randomized Cross-Over Trial. *Int. J. Environ. Res. Public Health* **2022**, *19*, 15195. https://doi.org/10.3390/ijerph192215195

Academic Editors: Maaike van Gerwen, Roel Vermeulen and Lauren Petrick

Received: 10 September 2022
Accepted: 8 November 2022
Published: 17 November 2022

Publisher's Note: MDPI stays neutral with regard to jurisdictional claims in published maps and institutional affiliations.

1. Introduction

The climate and health nexus is continuously gaining momentum in a changing climate that adversely impacts planetary health [1]. A recent report by the World Health Organization (WHO) calls for the immediate adoption of interventions to reduce urban overheating [2]. Most scientific evidence has been collected on ambient outdoor temperature effects on human health [3,4]. Evidence on associations between indoor air temperatures and health outcomes is scarce. The literature suggests linkages between indoor air temperatures >26 °C and an increased proportion of respiratory distress calls ($p = 0.056$) during summer in New York City [5].

Several studies have noted differences in environmental stressor exposure (e.g., particulate matter) between actual personal exposure measurements and estimates based on central monitoring station values [6]. This represents a historic issue in environmental epidemiology as the specific impact of exposure errors on effect estimates is not straightforward and depends on several factors, including study design, types of error, and the relationships between the outcome and the independent variables [7–9]. The growing availability and lower cost of wearable sensors allows for improving exposure measurement errors by providing individual-level measurements that are closer to the true personal exposure value. The choice of wearable sensors finds increasing use in environmental epidemiology studies, especially in those applying the human exposome framework and its exposomic tools, utilizing sensor data for multiple exposure stressors [10–12].

The deployment of such wearable sensors allows for capturing the exposure variance in 24-h skin temperature measurements as a result of human mobility and activity across different indoor and outdoor (micro)environments [10]. A diurnally varying temperature profiling was observed indoors for the majority of participants with a higher number of temperature peaks recorded indoors than outdoors [10]. Skin temperature is known to respond quite quickly to changes in ambient temperatures (within seconds) [13]. These time–activity varying patterns are often characterized by nonlinear temporal trends that are hard, if not impossible, to capture with standard biostatistical algorithms, such as linear regression models.

The body's core temperature system is well controlled by the circadian clock and peripheral systems such as the thyroid gland. Skin temperature may be more responsive to episodic or short-term temperature gradient changes during the day in an ambient environment than the body core temperature system. As such, recurrence patterns of body or skin temperature as the body attempts to thermoregulate its core temperature and maintain homeostasis become highly important and they are characterized by dynamically evolving cyclic processes. It is important to stress that ensuring an interfacial synchronization of interacting systems will likely maintain a resilient biological state, whereas, on the contrary, systematic disruptions in synchronization (desynchronization) will likely lead to differential toxicity [14]. The global dimension of chronos (time in Greek) entails a dynamic equilibrium brought about by transient and temporally dynamic input systems or processes.

Synchronization or *synchronie* (in French), a term invented by the Swiss glossologist F. de Saussure, is defined as the totality of elements that operate during the same time period within a system, including the network of interactions between the system's elements. Besides their entrainment with external cycles (e.g., the day/night circadian rhythm), organisms host a multiplicity of endogenous oscillatory temporal structures. This is an obliged consequence of the fact that any biological process encompasses a relation network among its players by being concurrently confined into a finite space [15]. From single protein molecules whose internal oscillatory dynamics stems from the mutual relationship among amino acid residues solicited by thermal noise [16] to ecological systems whose relative abundance of species oscillation is driven by their relative position in the food web [17], everything oscillates. However, such characteristic periods and amplitudes of natural oscillations span many orders of magnitude, and they can be modified by external environmental stressors. These modifications can alter the harmonic character of the oscillatory behaviour and disrupt their stationary character [18]. This requires the adoption of non-linear time series analysis tools [19] in order for the temporal dynamics of disease processes to be captured and appreciated at the right scale.

Typical examples of such dynamically oscillating proportional and reversible disease models [20] would be the effects of proportional dosing effects of pesticides on their rapid binding to cholinesterase enzyme, blocking the enzyme and causing a series of respiratory, gastrointestinal or central nervous system symptoms due to accumulation of acetylcholine in red blood cells; another example would be the severity of respiratory symptoms (cough, phlegm) being proportional to use of cigarettes packs per day, or the irritant respiratory effects associated with proportional ground level ozone exposures. For example, exposures to a chemical that do not saturate the enzymatic capacity either by being low in concentration or because exposures occur outside a specific chrono-window of a day with higher enzymatic activity would result in larger effect size per unit of exposure than other cases.

The above-mentioned examples of environment–health associations have never been studied before under the proposed temporal dynamics mode of finer temporal resolution for monitoring and analysing time series for both exposures and effects [14,18]. It has been shown earlier that studying the temporally dynamic network of systems would facilitate the detection/capture of critical transitions of early-warning signals of the pre-disease state so as to prevent drastic deterioration [15]. The transition from a resilient to a perturbed state cannot be captured with the conventional quantitative tools of environmental epidemiology for the aforementioned reasons. The temporal dynamics towards a perturbed state involves a myriad of steps that cannot be known in detail, calling for a global search for 'network biomarkers' that are sensitive and responsive to changes in correlation degree more than to actual values of single players [15].

The need for a deeper analysis of the daily patterns of skin temperature fluctuations emerged from the findings of the main analysis of the TEMP trial [10,11]. The main TEMP analysis showed alterations in metabolic hormones (e.g., leptin) associated with a short-term stay (5–7 days) in climatologically cooler areas than those observed in geographically adjacent urban settings (~1-h driving distance) during a typical Mediterranean summer [10,11]. Towards this, an additional analysis of the pilot TEMP study dataset was set up, using the finely resolved time series data that were available, based on the wearable sensor-based skin temperature measurements. Thus, the study objectives were to: (i) assess the recurrence patterns of skin temperature measurements in individuals spending time in two climatologically contrasting settings (urban versus mountainous), and (ii) evaluate the association between the recurrence quantification analysis (RQA) metrics and metabolic hormone levels.

2. Methods

2.1. Trial Oversight

This study involved participants from the TEMP 2 × 2 cross-over pilot trial of healthy, non-obese adults (n = 41) (NCT03625817). Participants were included in this analysis, if they had complete skin (excluded if ≥80% skin temperature values were below 30 °C during the 24-h) or personal air temperature measurements in both settings (n =16). Detailed information about the study methodology can be found in the original publication [11]. In brief, the trial studied the influence of setting and skin temperature or personal air temperature gradient on metabolic hormone profiling. Data and sample collection took place in Cyprus from July until the end of September 2018.

No participants were involved in setting the research question or the outcome measures, nor were they involved in developing plans for the design or implementation of the study.

The trial protocol was approved by the Cyprus National Bioethics Committee (EEBK/ΕΠ/2018/33) and written informed consent was signed by the participants. The trial was performed in accordance with the principles of the Declaration of Helsinki. The authors assume responsibility for the accuracy and completeness of the data and analyses, as well as for the fidelity of the trial.

2.2. Trial Population and Intervention Characteristics

Eligible participants were healthy adults between 20–60 years old, having their permanent residence in one of the two largest cities of Cyprus (Nicosia and Limassol), and who intended on staying in the Troodos area (a mountainous rural area in Cyprus) for a period of at least 5 to 7 consecutive days between July–September 2018. Recruitment was made possible via the dissemination of flyers, radio ads and telephone communication with governmental and other non-profit organizations (Troodos Development Company, Troodos Tourist Board, etc.) that contacted community leaders of the mountainous area and helped us to collect and create a list of potential participants. Using this list, potential participants were informed about the study, screened for eligibility criteria and agreed to participate via telephone communication. Exclusion criteria were applied to those who were pregnant, obese, suffering from a chronic condition (hypertension, diabetes, metabolic syndrome, cancer), receiving pharmaceutical treatment for impaired glucose levels, hypertension, antidepressants or thyroxin medication for thyroid disorders, or those who had travelled to another country in the week prior to the study's initiation. The intervention was a short-term (5–7 days) stay in the cooler-climate rural areas of the Troodos mountains.

2.3. Randomization and Masking

Random allocation of the intervention was carried out in blocks to two groups that differed in the sequence of the treatments. Block randomisation was performed using the function RAND() in Excel with a randomly generated list by a research investigator with no trial involvement. Group 1 included volunteers initially staying in their permanent house in the urban area and then moving to the rural/mountainous area for at least 5–7 days; Group 2 included individuals who initially stayed in the mountainous area for at least 5–7 days and then returned to their permanent residence in the urban area.

The blinding of the participants to group assignment was not possible with the present study design. The blinding of the researchers to the subjects' identity was achieved by coding all study materials (urine containers, questionnaires, and diaries). The study personnel were not informed of the group assignment.

2.4. Trial Procedures

Participants spent their short-term stay in the rural communities of the mountainous Troodos area in Cyprus with an altitude range of 600–900 m above sea level.

The researcher visited the participant one day before the first sampling day (T after being in the allocated setting). Upon giving informed consent, the participant completed the baseline questionnaire and the Munich ChronoType Questionnaire. Study materials were provided. Participants were asked to collect all urine voids separately, to note their day's activities in the activity diary and to wear the temperature sensors continuously until mid-day (noon) of the day following the sampling day. A washout period of at least 9 days between settings was observed (average: 22 days, range: 9–41 days).

2.5. Sensors

Wearable sensors of skin and personal air temperature, including activity tracking (as a surrogate marker of physical activity) (e-TACT, BodyCAP Medical, Hérouville-Saint-Clair, France) were used. The skin sensor was attached to the armpit, while the personal air temperature sensor was attached as a hanging tag on the individual's chest. For the skin temperature sensor, only data collected with temperatures $\geq$ 30 °C were included in the final data analysis (n = 14), while for personal air sensor data we had 15 eligible participants. The following input parameters were set for the sensors: temperature measurement period (1 min), sampling frequency (50 Hz), the actimetry period (1 min), the accelerometer sensitivity threshold to consider an activity as important or not (0.1 g) and the measurement range (2G). The activity sensor tracking mode used the classic TAT method (time above threshold).

2.6. Urine Sample Collection

Two 24 h periods of urine sampling per participant were obtained, a 24 h sampling period at the end of the 5–7 days period in the urban setting and another 24 h sampling period at the end of the mountainous/rural setting period. On the sampling day, the first morning void urine sample was discarded and the first morning void of the following day was obtained. All bladder content was emptied into separate vials whenever participants urinated. The time stamp of each urination and its volume were recorded in a diary. All samples were stored in the freezer until handed to the laboratory, where aliquoting was performed and samples were stored at −80 °C until analysis.

2.7. RQA Analysis

The notion of recurrence is simple: for any ordered time series, a recurrence is a value which repeats itself within an assigned tolerance (=radius). Thus, given a reference point, X0, and a sphere of radius r centred on it, a point X is said to recur if:

$$Br(X0) = \{X: ||X - X0|| \leq r\} \tag{1}$$

Equation (1) can be generalized in terms of the distance between any two i, j points considered in a Euclidean space: a recurrence is scored anytime the distance between two i, j points is lower than a pre-determined threshold r.

In the case of a time series of length n, the $n \times (n - 1)/2$ distinct pairwise distances between the n points are computed in terms of Euclidean distances in a phase space having as axes p lagged copies of the series, with p = embedding dimension.

If we let $p = 3$, out of an original series S = 10, 11, 21, 32, 41, 35, 40, 19, then by shifting the series by a fixed lag, we obtain the corresponding 3-dimensional phase space:

10 11 21
11 21 32
21 32 41
32 41 35
41 35 40
35 40 19
40 19 .
19 . .

The above embedding matrix (EM) has as rows subsequent epochs of the series of length p. The matrix of the Euclidean distances between all the rows (epochs) of EM is converted into a binary matrix, recording a value of 1 (scoring of a recurrence) every time the distance is lower than r, and 0 otherwise.

Such a binary distance matrix is called a Recurrence Plot (RP) and is graphically depicted by a black dot at every recurrent pair. Given the symmetric character of distance, RP aspect is symmetrical along the main diagonal. We report here an example of a recurrence plot (top panel), together with the original signal from which it was derived (Figure 1).

The RPs generate quantitative descriptors of the time series by different indexes describing the number and pattern of recurrent point; the simplest descriptor is recurrence rate (REC):

REC = $1/N^2$ SUM (Rij), being R = 1 if i, j recur, and 0 otherwise; N = time series length.

Determinism (DET) corresponds to the proportion of recurrences (Rij) appearing in sequence, with respect to the total number of recurrences:

DET = SUM(Drij)/SUM(Rij), where Drij are those recurrences forming diagonal lines.

Laminarity (LAM) is the analogue of DET as for horizontal (vertical) lines, namely is the proportion of recurrences appearing in horizontal (vertical) lines:

LAM = SUM(Lrij)/SUM(Rij) where Lrij are those recurrences forming horizontal(vertical) lines.

The above three RQA metrics can be considered as the basic ones that in turn generate secondary indexes, namely:

Entropy (ENT) corresponds to the Shannon Entropy of the length distribution of deterministic lines. Given the definition of deterministic lines as recurrences appearing in sequential order, they can form lines with a length going from 2 to N; the relative proportion of these length classes gives rise to ENT expressed in terms of the Shannon entropy index as:

ENT = SUM (p(l)log(p(l)), where the SUM is extended from 2 to N and p(l) = proportion of deterministic lines of length = l.

Divergence (DIV) corresponds to the inverse of MAXLINE (the maximum length of deterministic lines):

DIV = 1/MAXLINE.

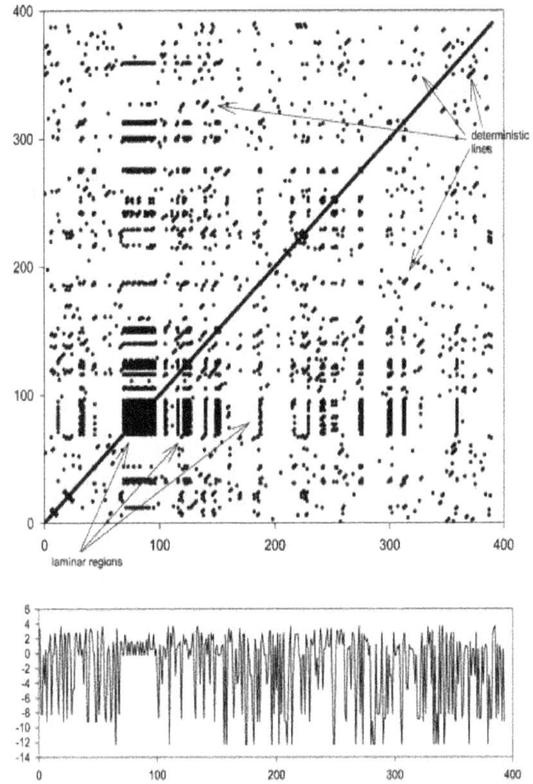

Figure 1. The black dots correspond to recurrences of pairs of points having their time coordinates reported on the x and y axes of the graph. Adjacent-in-time recurrences define deterministic lines (parallel to the main diagonal), while laminar regions are horizontal (vertical) lines of recurrences in the plot. Both deterministic and laminar lines are indicated by arrows in the figure. In this case, embedding dimension $p = 4$, and the radius was set to 1/10 of mean distance (modified from [21]).

The name divergence stems from the mathematical demonstration of the strong link between DIV and the first Lyapunov exponent of a dynamical system [22] that corresponds to the 'instability' of the dynamics. Analogously, Lmean is the average length of deterministic line and Vmean the average length value of vertical (laminar) lines.

The slope of the linear relation linking the number of recurrences (Rij) and the distance from main diagonal of RP is called TREND (TRE). This is a measure of non-stationarity of the series: keeping in mind the meaning of recurrences, an RP having recurrences only in the vicinity of main diagonal implies a drift affecting the series that changes its statistics along the time dimension (i.e., progressively going away from the main diagonal of RP).

The brief discussion of the meaning of TRE allows us to understand the huge success RQA enjoys in fields as diverse as theoretical physics and psychology to ecology, molecular biology and engineering [21–24]. RQA does not impose any constraint of length, stationarity, functional form, type of dynamics, while allowing for an immediate appreciation of the essential features of the studied dynamics. Thus, an increase in determinism points to the approaching of a phase transition (e.g., an epileptic seizure in the case of Electroencephalography, the onset of fatigue in electromyography) or a more 'constrained' and less 'environmentally sensible' dynamics [25–28]. On the other hand, an increase in ENT implies a richer deterministic dynamics profile with the activation of a higher number of drivers.

In summary, the following recurrence quantification analysis (RQA) metrics were calculated, serving as relevant nonlinear markers of changes in dynamical systems and physiological states [29–31]: (i) recurrence rate (REC) denoting the density of recurrence points in a recurrence plot; (ii) determinism (DET), which is the percentage of recurrence points that form diagonal lines in the recurrence plot of minimal length; (iii) laminarity (LAM), which is the number of recurrence points forming vertical lines (intermittency); (iv) divergence (DIV), which equals the inverse of Lmax, where Lmax is the maximal diagonal line length; (v) entropy (ENT) denoting the complexity of the deterministic structure in the system; vi) trend (TRE), which is the regression coefficient of a linear relationship between the density of recurrence points in a line parallel to the LOI (line of identity) and its distance to the LOI, (vii) the mean diagonal line length (Lmean); and (viii) the mean vertical line length (Vmean).

The input parameters of the RQA analysis were: the embedding dimension (ED = 2), the number of lagged (time lag = 1) replicas of the original sampled series of temperature consecutive values, used to construct the Takens' vectors and the maximum distance between two phase-space points (vector points having a number of components equal to ED) to be considered a recurrence, which was set as radius = 0.1 times the mean distance between distinct time point pairs. This threshold was selected to keep low REC and have a consistently high DET not affected by adjacent recurrences due to chance. This way, the saturation of recurrence is avoided (too many recurrent points can give rise to false deterministic lines). The time series of the skin temperature fluctuating dynamics were taken for the same subject sampled in both urban and mountainous environments considered as proxies of 'business as usual' (urban) and 'more relaxed' (mountainous) settings. The fact that RQA catches the features of the single-subject physiological dynamics [21] asks for a paired statistical analysis (paired *t*-test) for RQA descriptors comparing the two conditions on a single subject basis. This means that the observation of largely sign-invariant changes of RQA descriptors going from urban to mountain conditions is much more relevant than the average change over the entire population.

The minimal length of vertical/diagonal lines to be considered in the RQA plot were lmin = 2, while distance-to-border points near the border of the recurrence matrix were ignored when computing the RQA metrics (distance to border = 2). The non-linear RQA modelling and its metrics were constructed with the aid of the *nonlinearTseries* R package and the *ggplot* R package for visualization of RP using our custom code.

2.8. Outcomes

The primary outcomes of this study were the metabolic (leptin and adiponectin) and stress hormones (cortisol). Metabolic hormones (leptin and adiponectin) were quantified using validated kits for human urine (immunoassays, Elabscience, Houston, TX, USA). Cortisol was measured using gas chromatography coupled with mass spectrometry [32]. Urinary creatinine was determined using the picric acid-based spectrophotometric method (Jaffe method) [33].

2.9. Statistical Analysis

The baseline characteristics were presented overall and by study group. Categorical variables were described with sample size and percentages and compared by the chi-square test. Approximately normally distributed continuous variables were described with means and standard deviations (SD) and compared using the *t*-test. Non-normal continuous variables were described with medians and interquartile ranges (IQR) and compared using the Wilcoxon test. Variables with skewed distributions were log transformed to approximate normality. Biomarker values below the LOD were imputed to LOD/2. All outcome variables were adjusted for creatinine levels to account for urine dilution. The percentiles of each RQA metric were calculated overall and by setting, while log transformed RQA metrics by setting were compared using a paired t test comparing the subject recurrence dynamics in the urban and mountain settings. A principal component analysis (PCA) of the RQA metrics allowed for the calculation of the eigen values and associated variance for each principal component; we retained principal component 1 (PC1) for inclusion in the below mixed effect models.

For the analysis of the primary outcomes, mixed effect models were fitted for the RQA metrics (log-transformed) as a function of the setting (mountainous vs. urban) adjusted for age and sex and accounting for both between and within subject variability. Linear mixed effect models were also fitted for the adipokines and cortisol hormones (creatinine-adjusted and log-transformed) as a function of the RQA metric and the time (first morning vs. night) of the diurnal biomarker measurements (sample type), accounting for both between and within subject variability. All mixed effect models included participant-level (repeated measures within person) random intercepts with an unstructured covariance matrix.

One set of sensitivity analysis was undertaken using a radius of 0.01 to produce the recurrence plots. Statistical tests and 95% confidence intervals were two-sided with the statistical significance level set at 5%. All analyses were performed in R (v.5.2) with Rstudio (v.1.1.463).

3. Results

3.1. Participant Characteristics

From the initial pool of 41 volunteers who agreed to participate, a total of 31 volunteers were included in the original analysis [11], while a total of 16 volunteers were retained for this study, because these were the participants with fully available skin temperature measurements in both settings, minus two participants; one was excluded because ≥80% skin temperature values were below 30 °C, and another was excluded, because of abnormally high leptin levels (mean 3926 ± 209 ng/L) (n = 14). Most participants were females (78.6%) with a mean age of 42 years and normal BMI (Table 1). A high level of education was reported for the participants, with the majority holding at least a university/college degree (64%). At baseline, most participants were non-smokers (79%) with intermediate chronotype (50%), while about half of them reported that they never/rarely consume alcohol (57%) and reported spending about 5.4 h per day on screens (Table 1).

3.2. Nonlinear Temporal Temperature Dynamics and Their Metrics

The individual-level recurrence plots (RP) based on the continuous 24-h measurement period using 1-min intervals for the skin temperature gradient are shown in Figure 2, from which all RQA metrics were calculated.

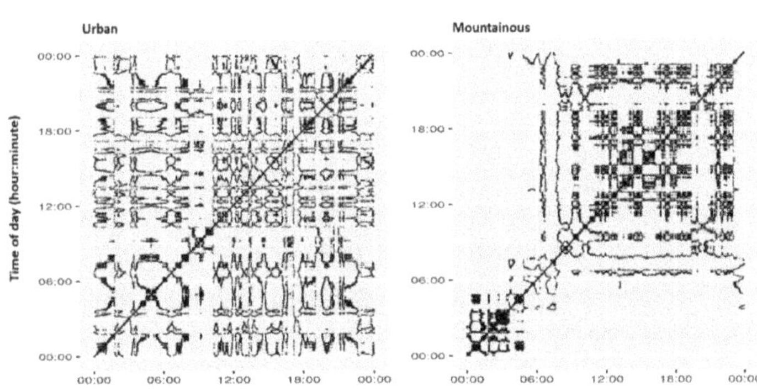

Figure 2. Indicative individual-level recurrence plots (RP) (24-h, 1-min intervals) for skin temperature gradients of the TEMP RCT with subjects spending time in an urban (left) setting and crossing to a cooler mountainous setting (right) during summer. The axes of the RPs correspond to subsequent time points, while darkened pixels mark recurrences, i.e., the dynamics returning to an already-visited state corresponding to the quasi-equivalence of the temperature value of the X and Y coordinates.

Table 1. Demographics and baseline characteristics of the study population (overall, and by group).

	Overall	First Urban	First Mountainous	*p* Value *
n	14	8	6	
Age (mean (SD))	42.00 (9.11)	40.86 (8.01)	43.33 (10.88)	NA
Sex = Male (%)	3 (21.4)	2 (25.0)	1 (16.7)	1.000
BMI (mean (SD))	24.94 (3.22)	25.89 (3.67)	23.68 (2.19)	NA
BMI category = overweight (%)	5 (35.7)	4 (50.0)	1 (16.7)	0.469
Educational level (%)				0.161
Secondary	5 (35.7)	3 (37.5)	2 (33.3)	
University/college	3 (21.4)	3 (37.5)	0 (0.0)	
Master/PhD	6 (42.9)	2 (25.0)	4 (66.7)	
Chronotype ^ (%)				0.117
early	4 (40.0)	4 (57.1)	0 (0.0)	
intermediate	5 (50.0)	2 (28.6)	3 (100.0)	
late	1 (10.0)	1 (14.3)	0 (0.0)	
Smoking status (%)				0.239
Smoker	2 (14.3)	2 (25.0)	0 (0.0)	
Non-smoker	11 (78.6)	5 (62.5)	6 (100.0)	
Former smoker	1 (7.1)	1 (12.5)	0 (0.0)	
Alcohol frequency (%)				0.823
Weekly	3 (21.4)	2 (25.0)	1 (16.7)	
Monthly	3 (21.4)	2 (25.0)	1 (16.7)	
Rarely/never	8 (57.1)	4 (50.0)	4 (66.7)	
Physical exercise = No (%)	7 (50.0)	7 (87.5)	0 (0.0)	0.007
Screen time (hr/day) (mean (SD))	5.36 (3.25)	5.88 (3.48)	4.67 (3.08)	NA
Days in mountainous setting (mean (SD))	7.71 (3.10)	7.00 (1.20)	8.67 (4.59)	NA
Washout days (mean (SD))	11.88 (7.28)	11.88 (7.28)	NA	NA

* the above variables were tested for differences between the two groups by chi-square tests for categorical variables and *t*-tests for normally distributed continuous variables. ^ the chronotype category cut-offs are: early (<3:00), intermediate (3:00–5:00) and late (>5:00). [29].

The average max daily air temperatures recorded using governmental station measurements followed the pattern of urban > mountainous settings, e.g., Nicosia (35 °C) > Limassol (33 °C) > Troodos (25 °C). Median values by setting for all RQA metrics can be found in the Supplementary File. The skin temperature profiling showed a diurnal pattern, with the highest values during the later afternoon towards the night and the lowest in the morning, which is consistent with earlier findings on distal and proximal skin temperature profiling regardless of the season [34]. Because of the diurnal character of the skin temperature measurements, the nonlinear temporal dynamics RQA algorithm was applied. Keeping in mind that DET corresponds to the proportion of recurrent pairs that occur along diagonal lines (i.e., appearing one after the other) out of total recurrent pairs, the low value of REC (overall mean REC = 0.06; number of recurrent pairs/total number of pairs) ensured that the observed deterministic line of high repetitiveness (mean DET = 0.91) did not occur by chance. The high DET value was indicative of a time series with a highly ordered structure (low system randomness). The high Lmean indicator (overall mean = 6.7) suggested the system's stability in its attractor for a long time. The Lmax descriptor corresponds to the length of the longest deterministic line, with the DIV value (inverse of Lmax) being quite low (DIV = 0.005); this metric has been correlated with the so called 'Lyapunov exponent', i.e., the speed of change of a system's dynamics [25]. The indicator of ENT denotes the degree of 'richness' of a deterministic structure. This parameter corresponds to the Shannon entropy of the length of the various deterministic lines in the RP. Therefore, if all diagonal lines have the same length, then ENT goes to zero; if there are many different lengths with almost equal frequency, ENT goes to the maximum.

The presence of several laminar states is indicative of intermittent patterns, suggesting the classical route to chaos; this was demonstrated in this study with the 'ellipsoid' figures ('butterfly' like patterns, Figure 1), pointing to 'Brownian motion' trajectories, i.e., to trajectories in which the system moves along nearby states and then returns to its original state. The Vmean descriptor corresponds to the length of the mean vertical line (Vmean = 7.92); the higher this is, the more unstable the system, with states changing more frequently. Long vertical lines (Vmean) suggest high laminarity (LAM = 0.94), i.e., intermittency-based patterns in the temporal dynamics of the skin temperature system. TRE was almost close to zero (TRE = −0.00005), suggesting the stationarity of the time series, i.e., the underlying system works with the same dynamics all along the observed time window; the more negative the TRE, the less stationary the system is.

The RQA metrics, when analysed by setting (urban vs. mountainous) showed interesting patterns. In effect, most of the monitored RQA metric values, i.e., Lmean, Vmean, DET, PC1, LAM, ENT and REC, were consistently higher ($p < 0.05$) in the mountainous setting than those observed in the urban setting (Table 2).

Table 2. Linear mixed-effect models of log-transformed RQA metrics as a function of mountainous setting (in comparison to the urban setting) accounting for the repeated measurements.

RQA	Term	Estimate	Std.Error	*p* Value	Lower Bound	Upper Bound
Lmean	Mountainous	0.841	0.232	0.0006	0.384	1.297
Vmean	Mountainous	0.83	0.231	0.00069	0.380	1.286
DET	Mountainous	0.780	0.230	0.00139	0.325	1.228
PC1	Mountainous	0.745	0.224	0.00160	0.306	1.185
DIV	Mountainous	−0.652	0.211	0.00319	−1.106	−0.238
LAM	Mountainous	0.764	0.231	0.00170	0.311	1.217
ENT	Mountainous	0.707	0.243	0.00500	0.232	1.182
TRE	Mountainous	−0.322	0.236	0.17777	−0.785	0.140
REC	Mountainous	0.181	0.253	0.477567	−0.314	0.676

Thus, a blunting phenomenon on the nonlinear temporal dynamics of skin temperature time series was observed in the urban setting. Compared with the mountainous setting, a more stable (and thus less adaptive) profile of skin temperature dynamics in the urban setting appeared, being less deterministic and with a smaller degree of complexity. For the personal air temperature dataset, only the TRE, ENT, Vmean and Lmean metrics was significantly higher in the mountainous setting, while the rest of them were not ($p > 0.05$) (SI Section).

The mixed-effect models of the three hormones and the skin temperature-based RQA metrics are shown in Table 3. No significant ($p > 0.05$) associations were observed between the leptin or cortisol and any of the RQA metrics as calculated from either the skin or the personal air sensor datasets. However, there were marginal associations in the skin temperature dataset between adiponectin and LAM (beta = 0.24, 95%CI: −0.02, 0.50, $p = 0.065$) and with DET (beta = 0.23, 95%CI: −0.035, 0.50, $p = 0.087$) (Table 3). For the personal air dataset, there were marginal associations between adiponectin and Lmean (beta = −0.31, 95%CI: −0.597, -0.031, $p = 0.030$) and with DIV (beta = −0.22, 95%CI: −0.456, −0.016, $p = 0.067$) (SI Section).

Table 3. Linear mixed-effect models of log-transformed biomarkers (leptin, adiponectin and cortisol) as a function of first morning sample (in comparison to the last-before-sleep sample) accounting for the repeated measurements and for age.

RQA Metric	Estimate	Std.Error	*p*-Value	Lower Bound	Upper Bound
Leptin					
TREND	0.214	0.141	0.1358313	−0.069	0.497
Lmean	−0.129	0.128	0.3171442	−0.387	0.128
LAM	0.121	0.128	0.3483536	−0.136	0.380
Vmean	−0.114	0.128	0.3812657	−0.372	0.144
DET	0.102	0.131	0.4397261	−0.161	0.365
DIV	−0.083	0.142	0.5616446	−0.368	0.202
PC1	0.073	0.131	0.5808591	−0.191	0.337
REC	0.039	0.129	0.7627006	−0.220	0.299
ENT	−0.013	0.129	0.9180662	−0.273	0.246
Adiponectin					
RQA Metric	**Estimate**	**Std.Error**	***p*-Value**	**Lower Bound**	**Upper Bound**
LAM	0.246	0.130	0.0647188	-0.015	0.507
DET	0.233	0.133	0.0871753	−0.035	0.501
REC	0.191	0.133	0.1562231	−0.075	0.459
PC1	0.175	0.136	0.2050513	−0.098	0.449
ENT	0.122	0.134	0.3678370	−0.147	0.391
DIV	−0.132	0.148	0.3747916	−0.430	0.164
TREND	0.111	0.148	0.4552794	−0.186	0.409
Lmean	−0.030	0.135	0.8276175	−0.302	0.242
Vmean	0.006	0.136	0.9606054	−0.266	0.279
Cortisol					
RQA Metric	**Estimate**	**Std.Error**	***p*-Value**	**Lower Bound**	**Upper Bound**
Vmean	0.101	0.128	0.4343851	−0.157	0.361
ENT	0.091	0.127	0.4763856	−0.165	0.348
Lmean	0.080	0.128	0.5329737	−0.178	0.339
LAM	0.068	0.129	0.5995378	−0.192	0.329
DET	0.066	0.130	0.6145308	−0.196	0.328
PC1	0.067	0.134	0.6177130	−0.201	0.336
TREND	−0.028	0.137	0.8357969	−0.307	0.249
DIV	0.025	0.144	0.8621418	−0.264	0.315
REC	0.017	0.131	0.8923230	−0.246	0.281

Model details: Leptin, adiponectin and cortisol are creatinine adjusted and log transformed. Random intercepts for the repeated samples within participants with unstructured covariance matrix.

4. Discussion

This was an additional analysis of the original TEMP trial [10,11] that used wearable sensors to measure diurnal (24-h long, one-minute intervals) skin temperature temporal profiles of healthy, non-obese individuals crossing from one setting to another (urban to mountainous and vice versa) after a 7-day long stay in each setting during the Mediterranean summer. A notable difference of about 8–10 °C in daily max air temperature between the urban and mountainous settings during the study period was observed using governmental station measurement data. Skin temperature sensors generated data between 12,240 and 27,000 time point counts. We found dysregulations in skin temperature temporal dynamics of the study population while residing in the urban setting when compared with the cooler mountainous setting in Cyprus. These dysregulations took the form of reduced cycle duration and complexity, while skin temperature dynamics became less responsive to perturbations and less regular in magnitude. Simply by looking at the respective median skin temperatures in the two settings, it was obvious they were quite similar

(35.1 °C [34.6, 35.6] vs. 35.05 °C [34.55, 35.7]) [10]. However, the RQA analysis offered an alternative perspective in inferring about the temporal dynamics of the skin temperature datasets by showing that the short-term (7-day) stay in an urban setting during the hot Mediterranean summer tended to blunt the cyclical properties of skin temperature diurnal dynamics observed for the same individuals in the cooler climate of the mountainous setting.

Skin temperature is an interesting "interface-variable" between the core body and ambient temperature regimes [34], being a faster responder to changes in ambient air temperature than that of core body temperature. As such, it may be the preferred marker of truly experienced temperature gradients at the individual level, in the absence of core body temperature measurements. There were indeed differences in the ambient air temperature levels between the two selected urban and mountainous settings in the summer, where the median [interquartile range, IQR] personal air temperature in the mountainous settings was 1.5 °C lower than that in the urban settings (27.1 °C [25.4, 29.2] vs. 28.6 °C [27.1, 30.5], $p < 0.001$), being consistent with the Mediterranean climate [10]. The cooler ambient air temperature profile of the mountainous setting vs. that of the urban setting during the summer period was confirmed with data obtained from the Cyprus Meteorological Service; data showed that Nicosia city (35 °C) > Limassol city (33 °C) > Troodos mountainous area (25 °C). Skin temperature seems to integrate the observed ambient air temperature gradient together with the time activity patterns of the study participants during the monitoring period in the respective setting.

Although heart rate and skin temperature temporal dynamics in a 24 h period are inversely related, there are similarities in the observed skin temperature sensitivity to human mobility and changes in environmental setting with those perturbations documented for heartbeat regulation in response to environmental stimuli as age progresses, where heart rate dynamics become increasingly predictable (constrained) on a beat-to-beat basis [21]. Moving beyond the traditional analyses of temperature time series with daily max, daily mean and regression models, this nonlinear RQA approach offers a new perspective on the characterization of the temperature regime experienced by individuals in different settings during a 24 h period. Skin temperature temporal dynamics in the urban setting were also characterized by lower entropy, i.e., lower temperature network complexity when compared with the more complex network cycles of skin temperature in the mountainous setting. Cycle regularity and response to perturbation as expressed by DET were also evidently reduced in the urban setting than those observed in the mountainous setting. This may be characteristic of the time activity patterns of individuals in the urban setting, where changes in different microenvironments of variable ambient temperatures and different activities during the day than those experienced in the more relaxing environment of the mountainous setting may be observed. Indeed, a higher number of temperature peaks was found in indoor (micro)environments than outdoors, suggesting higher fluctuations in personal air temperature levels across different indoor (micro)environments [10]. In this respect, it is worth noting that a more efficiently regulated system in physiology corresponds to a more sensitive system (and thus, more rapid adaptation) to environmental changes; this is, for example, the case with heartbeat dynamics [21] showing a marked increase in stability (higher recurrence) and decrease in complexity with increasing age. This situation mirrors what was observed in this study, with a marked decrease in complexity in urban settings with respect to mountainous settings.

Circadian and metabolic systems operate in synchronization in healthy individuals, where entraining agents of circadian rhythm, such as light, feeding, temperature and circadian markers (e.g., melatonin, cortisol, etc.) [35] are vital components of synchronized circadian systems [36]. Capturing the temporally dynamic process of dysregulation in biological systems is the key to better understanding the changes and processes undermining the state(s) of synchronization and resilience in biological systems. For example, the monitoring of perinatal temporal dynamics in the metabolism of heavy metals in children was able to predict well the emergence of autism spectrum disorder cases using the RQA algorithm and its metrics [28].

Observational studies often utilize centrally sited air temperature data from meteorological stations, where exposure misclassification bias is a concern. To accurately depict reliable individual-based dynamics of exposure/effect associations, a sufficiently large number of time-repeated samples per subject in the disease-relevant etiologic time interval, confined in a relatively long time series relative to the variation of the outcomes of interest, is warranted. Missing the right scale to look for biological rhythms is by far the most common methodological error in chrono-observations. This error stems from both epistemological faults and practical limitations, the two most relevant sources of error being: (i) the idea that the external *zeitgeiber* (e.g., day/night cycle) is the only source of biological rhythms without any consideration of endogenous sources. This consideration makes largely irrelevant the detection of rhythmic activity at the individual level and pushes towards aggregate-level population studies in which the variable of interest (e.g., a biomarker) is measured at few external conditions (e.g., morning/evening) considering only two time-points per subject; and (ii) the inherent difficulty of obtaining a sufficiently rich time series at the individual level. The most common bias is the time-fluctuating amplitude of the cyclic process, where in most cases, the most relevant point to address is the modification of the temporal correlation structure (e.g., increase/decrease in determinism or harmonic character of the time series). While this consciousness is widespread in physiology (e.g., heartbeat time series) [21], this is only in its infancy for environmental epidemiology studies. Recurrence analysis may detect state changes in complex biological systems, with this being practiced for more than three decades in a vast array of scientific disciplines (physiology, mathematics, physics, biology, medicine, neurology, pathology), but never applied before in the environmental health sciences [29–31].

A strength of this study was the multidisciplinary approach taken in coupling knowledge and information obtained from the RQA novel application in environmental health sciences and further coupling them with classical metabolic and stress hormone measurements. The randomized design is another strength of the study, together with the use of a wearable sensor, which is also novel method in behavioural change trials. Sensor data collection was continuous at the individual level, overcoming the exposure measurement error associated with temperature measurements obtained from a few specific discrete time points or measuring stations. Other strengths include the repeated measures and the multivariable adjustment. This study has few limitations. The sample size was small due to its proof-of-concept mode. A wide time window of several hours during the collection of the first morning void sample introduced additional variability in the magnitude of metabolic hormone diurnal patterns that was not accounted for.

5. Conclusions

The urban overheating phenomenon in Mediterranean cities is a societal challenge with vast implications for the protection of public health. The study documented the applicability of the RQA algorithms in delineating the temporal dynamic patterns of skin temperature as a surrogate of body core temperature, though in closer communication with alterations in external environmental stimuli that could affect skin temperature. The monitoring of personal air temperature fluctuations for urban dwellers may be key in preventing health-associated illnesses, by focusing on early-stage disease process that open up avenues for breakthroughs in precision health applications. The influence of desynchronization on an individual's temporal health dynamics will help to adapt everyday lifestyles for effective disease prevention and wellbeing in the future. Accounting for individual circadian rhythms and personalized knowledge on temperature profiling during the warmer months are the next steps forward for precision exposome and its tools (e.g., sensors) that could be used to offer individualized heat protection and control advice. Differential sensitivities to heat stress by different age groups and differential critical life stage windows may pose a health risk to sensitive individuals; thus, stratified risk assessment approaches can take these susceptible groups into consideration and improve public health.

Supplementary Materials: The following supporting information can be downloaded at: https://www.mdpi.com/article/10.3390/ijerph192215195/s1.

Author Contributions: K.C.M. conceived, designed and coordinated the study. P.C. and T.D. conducted the lab biomarker measurements. N.E. executed the primary data analysis with guidance from K.C.M. and A.G. All authors interpreted the results and drafted the manuscript. K.C.M. obtained internal funding. All authors revised the article for important intellectual content. K.C.M. is the principal investigator and guarantor of the study. All authors have read and agreed to the published version of the manuscript.

Funding: Internal funds of the principal investigator (KC Makris) were used for this study. No external funding was obtained for this study.

Institutional Review Board Statement: The study (2018) was conducted according to the guidelines of the Declaration of Helsinki, and approved by the Institutional Review Board (or Ethics Committee) of ClinicalTrials.gov (NCT03625817).

Informed Consent Statement: Informed consent was obtained from all subjects involved in the study.

Data Availability Statement: The authors commit to making the relevant anonymized individual-level data available via open access here: https://doi.org/10.5281/zenodo.7329878.

Acknowledgments: We sincerely thank all TEMP trial participants and their families for their willingness to participate in the study.

Conflicts of Interest: No competing interests: all authors have completed the ICMJE uniform disclosure form at www.icmje.org/coi_disclosure.pdf (accessed on 9 September 2022) and declare: no support from any organization for the submitted work; no financial relationships with any organization that might have an interest in the submitted work in the previous three years; and no other relationships or activities that could appear to have influenced the submitted work.

References

1. Myers, S.S. Planetary health: Protecting human health on a rapidly changing planet. *Lancet* **2017**, *390*, 2860–2868. [CrossRef]
2. Sanchez, M.G.; de' Donato, F.K.V. *Heat and Health in the WHO European Region: Updated Evidence for Effective Prevention*; World Health Organization. Regional Office for Europe: Copenhagen, Denmark, 2021.
3. World Health Organization. *Report of the systematic review on the effect of indoor heat on health. WHO Housing and Health Guidelines*; World Health Organization: Geneva, Switzerland, 2018.
4. World Health Organization. Climate Change and Health—Fact Sheet. 2017. Available online: www.who.int/mediacentre/factsheets/fs266/en (accessed on 1 February 2018).
5. Uejio, C.K.; Tamerius, J.D.; Vredenburg, J.; Asaeda, G.; Isaacs, D.A.; Braun, J.; Quinn, A.; Freese, J.P. Summer indoor heat exposure and respiratory and cardiovascular distress calls in New York City, NY, U.S. *Indoor Air* **2016**, *26*, 594–604. [CrossRef] [PubMed]
6. Wallace, L. Indoor Particles: A Review. *J. Air Waste Manag. Assoc.* **1996**, *46*, 98–126. [CrossRef] [PubMed]
7. Armstrong, B.K.; Saracci, R.; White, E. *Principles of Exposure Measurement in Epidemiology*; Oxford University Press: New York, NY, USA, 1992.
8. Zeger, S.L.; Thomas, D.; Dominici, F.; Samet, J.M.; Schwartz, J.; Dockery, D.; Cohen, A. Exposure Measurement Error in Time-Series Studies of Air Pollution: Concepts and Consequences. *Env. Health Persp.* **2000**, *108*, 419426. [CrossRef]
9. Thomas, D.; Stram, D.; Dwyer, J. Exposure measurement error: Influence on exposure-disease relationships and methods of correction. *Annu. Rev. Public Health* **1993**, *14*, 6993. [CrossRef]
10. Constantinou, A.; Oikonomou, S.; Konstantinou, C.; Makris, K.C. A randomized cross-over trial investigating differences in 24-h personal air and skin temperatures using wearable sensors between two climatologically contrasting settings. *Sci. Rep.* **2021**, *11*, 22020. [CrossRef]
11. Makris, K.C.; Konstantinou, C.; Perikkou, A.; Zdravic, A.B.; Christophi, C.A. Contrasting short-term temperature effects on the profiling of metabolic and stress hormones in non-obese healthy adults: A randomized cross-over trial. *Environ. Res.* **2020**, *182*, 109065. [CrossRef]
12. Haddad, N.; Andrianou, X.D.; Makris, K.C. A Scoping Review on the Characteristics of Human Exposome Studies. *Curr. Pollut. Rep.* **2019**, *5*, 378–393. [CrossRef]

13. Klepeis, N.E.; Nelson, W.C.; Ott, W.R.; Robinson, J.P.; Tsang, A.M.; Switzer, P.; Behar, J.V.; Hern, S.C.; Engelmann, W.H. The National Human Activity Pattern Survey (NHAPS): A resource for assessing exposure to environmental pollutants. *J. Expo. Sci. Environ. Epidemiol.* **2001**, *11*, 231–252. [CrossRef]
14. Makris, K.C. Desynchronized circadian clock and exposures to xenobiotics are associated with differentiated disease phenotypes. *BioEssays* **2021**, *43*, 2100159. [CrossRef]
15. Zheng, Z. Collective Sustained Oscillations in Complex Systems. In *Recent Trends in Chaotic, Nonlinear and Complex Dynamics*; World Scientific: Singapore, 2022; pp. 398–428. [CrossRef]
16. Tiwari, S.P.; Reuter, N. Conservation of intrinsic dynamics in proteins—What have computational models taught us? *Curr. Opin. Struct. Biol.* **2018**, *50*, 75–81. [CrossRef] [PubMed]
17. Tzuk, O.; Ujjwal, S.R.; Fernandez-Oto, C.; Seifan, M.; Meron, E. Interplay between exogenous and endogenous factors in seasonal vegetation oscillations. *Sci. Rep.* **2019**, *9*, 354. [CrossRef] [PubMed]
18. Arora, M.; Curtin, P.; Curtin, A.; Austin, C.; Giuliani, A. *Environmental Biodynamics: A New Science of How the Environment Interacts with Human Health*; Oxford University Press: New York, NY, USA, 2022.
19. Shamsan, A.; Wu, X.; Liu, P.; Cheng, C. Intrinsic recurrence quantification analysis of nonlinear and nonstationary short-term time series. *Chaos: Interdiscip. J. Nonlinear Sci.* **2020**, *30*, 093104. [CrossRef] [PubMed]
20. Smith, T.J.; Kriebel, D. *A Biologic Approach to Environmental Assessment and Epidemiology*; Oxford University Press: New York, NY, USA, 2010. [CrossRef]
21. Giuliani, A.; Piccirillo, G.; Marigliano, V.; Colosimo, A. A nonlinear explanation of aging-induced changes in heartbeat dynamics. *Am. J. Physiol.-Heart Circ. Physiol.* **1998**, *275*, H1455–H1461. [CrossRef]
22. Rampichini, S.; Vieira, T.M.; Castiglioni, P.; Merati, G. Complexity Analysis of Surface Electromyography for Assessing the Myoelectric Manifestation of Muscle Fatigue: A Review. *Entropy* **2020**, *22*, 529. [CrossRef]
23. Ouyang, G.; Li, X.; Dang, C.; Richards, D.A. Using recurrence plot for determinism analysis of EEG recordings in genetic absence epilepsy rats. *Clin. Neurophysiol.* **2008**, *119*, 1747–1755. [CrossRef]
24. Giuliani, A.; Benigni, R.; Zbilut, J.P.; Webber, C.L.; Sirabella, P.; Colosimo, A. Nonlinear Signal Analysis Methods in the Elucidation of Protein Sequence−Structure Relationships. *Chem. Rev.* **2002**, *102*, 1471–1492. [CrossRef]
25. Trulla, L.L.; Giuliani, A.; Zbilut, J.P.; Webber, C.L., Jr. Recurrence quantification analysis of the logistic equation with transients. *Phys. Lett. A* **1996**, *223*, 255–260. [CrossRef]
26. Wallot, S.; Leonardi, G. Analyzing Multivariate Dynamics Using Cross-Recurrence Quantification Analysis (CRQA), Diagonal-Cross-Recurrence Profiles (DCRP), and Multidimensional Recurrence Quantification Analysis (MdRQA)—A Tutorial in R. *Front. Psychol.* **2018**, *9*, 2232. [CrossRef]
27. Facchini, A.; Mocenni, C.; Marwan, N.; Vicino, A.; Tiezzi, E. Nonlinear time series analysis of dissolved oxygen in the Orbetello Lagoon (Italy). *Ecol. Model.* **2007**, *203*, 339–348. [CrossRef]
28. Curtin, P.; Austin, C.; Curtin, A.; Gennings, C.; Arora, M.; Tammimies, K.; Willfors, C.; Berggren, S.; Siper, P.; Rai, D.; et al. Dynamical features in fetal and postnatal zinc-copper metabolic cycles predict the emergence of autism spectrum disorder. *Sci. Adv.* **2018**, *4*, eaat1293. [CrossRef] [PubMed]
29. Marwan, N.; Riley, M.; Giuliani, A.; Webber, C.L., Jr. (Eds.). Translational Recurrences: From Mathematical Theory to Real-World Applications. In Proceedings of the 5th International Symposium on Recurrence Plots, Chicago, IL, USA, 14–16 August 2013; Springer Proceedings in Mathematics & Statistics; Springer International Publishing: Cham, Switzerland, 2013; Volume 103. [CrossRef]
30. Marwan, N.; Webber, C.L., Jr. Mathematical and Computational Foundations of Recurrence Quantifications. In *Recurrence Quantification Analysis: Theory and Best Practices*; Webber, C., Marwan, N., Eds.; Springer Series in Complexity; Springer: Berlin/Heidelberg, Germany, 2015; pp. 1–41.
31. Webber, C.L.; Marwan, N.; Facchini, A.; Giuliani, A. Simpler methods do it better: Success of Recurrence Quantification Analysis as a general-purpose data analysis tool. *Phys. Lett. A* **2009**, *373*, 3753–3756. [CrossRef]
32. Moon, J.-Y.; Jung, H.-J.; Moon, M.H.; Chung, B.C.; Choi, M.H. Heat-map visualization of gas chromatography-mass spectrometry based quantitative signatures on steroid metabolism. *J. Am. Soc. Mass Spectrom.* **2009**, *20*, 1626–1637. [CrossRef]
33. *The MAK-Collection for Occupational Health and Safety, Part IV: Biomonitoring Methods*; Wiley-VCH, Verlag, kGaA: Weinheim, Germany, 2010; Volume 12, pp. 169–184.
34. Martinez-Nicolas, A.; Meyer, M.; Hunkler, S.; Madrid, J.A.; Rol, M.A.; Meyer, A.H.; Schötzau, A.; Orgül, S.; Kräuchi, K. Daytime variation in ambient temperature affects skin temperatures and blood pressure: Ambulatory winter/summer comparison in healthy young women. *Physiol. Behav.* **2015**, *149*, 203–211. [CrossRef]
35. Dallmann, R.; Okyar, A.; Lévi, F. Dosing-Time Makes the Poison: Circadian Regulation and Pharmacotherapy. *Trends Mol. Med.* **2016**, *22*, 430–445. [CrossRef] [PubMed]
36. Phillips, M.L. Circadian rhythms: Of owls, larks and alarm clocks. *Nature* **2009**, *458*, 142–144. [CrossRef]

Journal of *Xenobiotics*

Review

Xeno-Estrogenic Pesticides and the Risk of Related Human Cancers

Vivek Kumar [1], Chandra Shekhar Yadav [2] and Basu Dev Banerjee [3,4,*]

1 Department of Biotechnology, IMS Engineering College, Dr. A.P.J. Abdul Kalam Technical University, Lucknow 226031, India
2 School of Forensic Science, National Forensic Sciences University, Gandhinagar 382010, India
3 Environmental Biochemistry & Molecular Biology Laboratory, Department of Biochemistry, University College of Medical Sciences & GTB Hospital, University of Delhi, Delhi 110095, India
4 Department of Medical Elementology & Toxicology, School of Chemical & Life Sciences, Hamdard University, New Delhi 110062, India
* Correspondence: banerjeebd@hotmail.com

Abstract: In recent decades, "environmental xenobiotic-mediated endocrine disruption", especially by xeno-estrogens, has gained a lot of interest from toxicologists and environmental researchers. These estrogen-mimicking chemicals are known to cause various human disorders. Pesticides are the most heavily used harmful xenobiotic chemicals around the world. The estrogen-mimicking potential of the most widely used organochlorine pesticides is well established. However, their effect is not as clearly understood among the plethora of effects these persistent xenobiotics are known to pose on our physiological system. Estrogens are one of the principal risk modifiers of various disorders, including cancer, not only in women but in men as well. Despite the ban on these xenobiotics in some parts of the world, humans are still at apparent risk of exposure to these harmful chemicals as they are still widely persistent and likely to stay in our environment for a long time owing to their high chemical stability. The present work intends to understand how these harmful chemicals may affect the risk of the development of estrogen-mediated human cancer.

Keywords: endocrine disruption; persistence; xenobiotics; disorders

Citation: Kumar, V.; Yadav, C.S.; Banerjee, B.D. Xeno-Estrogenic Pesticides and the Risk of Related Human Cancers. *J. Xenobiot.* **2022**, *12*, 344–355. https://doi.org/10.3390/jox12040024

Academic Editors: Roel Vermeulen, Lauren Petrick and Maaike van Gerwen

Received: 24 September 2022
Accepted: 16 November 2022
Published: 21 November 2022

Publisher's Note: MDPI stays neutral with regard to jurisdictional claims in published maps and institutional affiliations.

1. Introduction

Xenobiotics (xeno: foreign; biotic: life form) are chemicals of synthetic origin that are foreign to living systems. These chemicals are not part of the normal metabolic activity of living organisms and therefore may interfere with the functioning of the physiological system. These unwanted chemicals inside the body are known to have harmful effects on functioning and are responsible for diseases in humans. One such detrimental effect caused by some of these chemicals is xeno-estrogenicity [1].

'Xeno-estrogenicity' (xeno: foreign; estrogenicity: resembling natural estrogens) denotes the property of foreign chemicals being able to produce responses in physiological systems comparable to estrogens produced by the body by different mechanisms. Xeno-estrogens are a group of chemicals that disrupt the endocrine system by mimicking the activity of natural estrogens or producing a similar response in the body by other mechanisms. Majorly known persistent organochlorine pesticides (OCPs) and other classes of chemicals present in the environment are known to bind estrogen receptors and produce a physiological response equivalent to the hormone in various organs and tissues [2]. Therefore, these chemicals can produce adverse effects such as increasing the risk of estrogen-mediated cancers [3]. These chemicals can easily enter the human body due to their physiochemical characteristics. They not only widely persist in our environment, including water, soil, and food, but also exhibit high volatility at environmental temperatures [4]. Moreover, a much greater risk is rendered by such chemicals due to their biomagnification through to the top of the food chain to humans. Additionally, their lipophilic nature makes them more harmful, as these compounds easily bioaccumulate in the adipose tissue of humans [5]. Therefore, a large portion of the world's population, if not all, is apparently exposed to these chemicals. The health concerns due to these endocrine-disrupting chemicals are inevitable as exposure to these chemicals is continuous throughout the lifetime [6–8]. Studies have shown evidence of the continuous exposure to these compounds as their metabolites have been detected in different types of human samples. Studies reported a statistically significant correlation between cancer risk in individuals and their adipose tissue concentration [9,10].

Xenobiotics, including from pesticide exposure, are significant causes of cancer incidence worldwide [11,12]. These compounds are known to increase the risk of various human cancers amongst humans by different mechanisms such as DNA mutations, inflammation, upregulation of tumor suppressor genes, and/or downregulation of oncogenes [13]. Amongst the harmful effects of these contaminants, xeno-estrogenicity remains one of the most enigmatic and widely reported ones to contribute towards higher morbidity and mortality [14]. These estrogen-mimicking chemicals are known to increase the transcriptional activity factors associated with the natural estrogens, thus enhancing the risk of various estrogen-mediated human disorders including cancer [15–18].

Other natural and man-made compounds are known to exist with xeno-estrogenic potential in the environment, such as butylated hydroxyanisole, phthalate esters, nonylphenol, benzyl butyl phthalate, and bisphenol-A, etc. [19]. However, pesticides are the most potent xeno-estrogens amongst all the chemicals because the estrogenic potency and half-life in the environment of these compounds exceed those of any other compound [3–5].

2. Carcinogenic Effects of Organochlorine Pesticides

Pesticides are the most well-known carcinogenic chemicals to humans. Studies from both humans and animals have unambiguously confirmed the development of pesticide exposure-mediated human cancer [20]. All known OCPs possessing xeno-estrogenic potential are listed in the 'first dirty dozen' in the Stockholm Convention as they were widely reported to cause various adverse health outcomes in humans around the world [21]. Human studies support the association of various OCPs (aldrin, dieldrin, endosulfan, HCH, DDT, 2,4,5-trichlorophenoxyacetic acid, phenoxy acid herbicides, and methoxychlor). The exposure (both occupational and non-occupational) has been strongly linked to increased incidence of non-Hodgkin's lymphoma [22]; multiple myeloma [23]; soft tissue sarcomas [24]; lung cancers [25]; and cancers of the pancreas, stomach, liver, and kidney as well as urinary and gall bladder cancer [26–30].

The International Agency for Research on Cancer (IARC) has listed OCPs as potent carcinogens in animal studies [31,32]. Cancer development also depends on genetic susceptibility as well. Xenobiotic metabolic enzymes are metabolic susceptibility enzymes whose metabolization leads to the elimination of a diverse range of xenobiotics, including OCPs. These predominantly hepatic enzymes are responsible for xenobiotic detoxification in two phases: 'phase I' enzymes (functionalization reactions), mainly cytochrome P450, and 'phase II' enzymes (conjugation reactions), including glutathione S transferases (GSTs). Cytochrome P450 (abbreviated as CYP P450 and infrequently as CYP450) is a superfamily of monooxygenase enzymes that catalyze metabolism by oxidation of their substrates. On the other hand, GSTs, important phase II enzymes of the xenobiotic metabolizing enzyme family, eliminate phase I CYP450 enzymes' catalyzed intermediates, leading to their excretion out of the body [33,34].

Individual susceptibility to environmental chemical exposure leading to diseases, including cancer, is modified by genetic susceptibility, and polymorphism in such gene families is a major risk. Genetic variations in phase I and/or phase II genes may alter the activity of the corresponding enzyme responsible for the bioactivation of xenobiotics; individuals with genetically impaired xenobiotic elimination functions will be at higher risk of disease susceptibility over their lifetime. Identification of the genes responsible for the elimination of xenobiotics and their variability is key for the identification of individuals at higher risk [12,33–35].

3. Evidence of Xeno-Estrogenicity

Since the advent of modern agriculture, pesticide use has become a primary means to enhance crop productivity. Amongst all the pesticides, OCPs became the first choice of farmers around the world after their introduction, attributed to their low cost and versatile effectiveness against a wide range of agricultural pests. Table 1 presents the current usage status, commercial name, and other relevant details related to pesticides. Organochlorine pesticides were banned in developed countries after their listing in the Stockholm Convention on Persistent Organic Pollutants (an international agreement aimed at the banning of harmful environmental pollutants), which is still used for the development and control of vector-borne diseases such as malaria. The overuse of OCPs has been rising alarmingly, and it is raising serious threats to both human health and the environment [36]. Every year, billions of pounds of OCPs are used for boosting agriculture productivity to fulfill the dietary needs of the world population. It is not surprising that high levels of these harmful chemicals are also detected in water and soil [37,38].

Table 1. Current use status and year listing of commercial pesticide brands around the world.

Name of Pesticide	Commercial Name	Ban Status	Current Status	Year of Listing in Stockholm Convention on Persistent Organic Pollutants	Used for the Crops	References
Dieldrin	Hortico, Dieldrin Dust Mustex 25%, Shell Dieldrex, Yates Garden Dust	Complete	No manufacturing	2001	Corn, cotton, citrus cabbage, legumes	[39]
DDT	Cezarex, Anofex, Clorophenothane, Dicophane, Dinocide, Gesarol, Guesapon, Guesarol, Gyron, Ixodex, Neocid, Neocidol, Zerdane	Partial	For malaria control programs (countries of Africa and Asia)	2001	Amaranth, cabbage, lettuce, pumpkin, spinach	[40]
Endosulfan	Afidan, Beosit, Endocel, Hildan Cyclodan, Devisulfan, Endocide, Endosol, FMC 5462, Hex-asulfan, Hoe 2671, Insectophene, Malix, Thiodan, Thimul, Thifor, Thionex	Regulated	For certain crop–pest complexes	2011	Spinach, cauliflower, potato, brinjal, tomato, okra	[41]
Aldrin	Aldrec, Aldrex, Drinox, Octalene, Seedrin, Compound 118	Complete	No manufacturing	2007	Corn, cotton, citrus, cabbage, legumes	[39]
HCH	Forlin, Etan 3G, Gamaphex, Isotox, GermatePlus, Gamma-Mean400, Lindagram	Complete	No manufacturing	2008	Amaranth, cabbage, lettuce, pumpkin, spinach	[40]

The hormone-like activity of these man-made chemicals was realized long after they were released into our environment. Krishnan et al. (1993) accidentally found unknown chemicals to be estrogenic because they disrupted experiments conducted in laboratories while studying the effects of natural estrogens. Later experiments confirmed that the disruption of the experiments was due to the plastic tubes used in the experiments possessing estrogenic activity [42]. In humans, indirect positive evidence of "pesticide exposure-related estrogenicity" was reported among farm workers. In 1949, workers occupationally exposed to dichlorodiphenyltrichloroethane (DDT) were found to have significantly lowered sperm counts and loss of fertility [43].

These chemicals are known to stay in the environment for a long period of time, which could be from a few years to several decades in different media; for instance, the most well-known pesticide, DDT, has a half-life of up to 30 years in soil and 3 to 6 years in the human body (up to 150 years in water) [44]. Therefore, long-term exposure may lead to considerable accumulation in adipose tissue due to the lipophilic nature of agricultural chemicals [45]. High levels of pesticides, including OCPs, have been reported in different samples from all over the world, including blood, adipose tissue, placental tissue, umbilical cord, colostrum, and even cord blood, etc. [6–8]. In a recent study, groundwater polluted with pesticides was found to increase susceptibility to estrogen-mediated cancers, in spite of the ban on these chemicals for decades [46]. Likewise, alarmingly high levels of these chemicals have also been reported from other regions of the world in recent studies [47–49]. Additionally, a Brazilian study from the year 2022 linked pesticide residue with an increased risk of cancer in one province [50].

These xenobiotics contaminate water resources and animals living in pesticide-contami nated water, leading to their accumulation in the animals' bodies. The quantity of OCPs in their bodies can be several folds higher than that in the surrounding water. If these animals are consumed by other organisms, the OCPs will enter and accumulate in those animals' bodies (bioaccumulation) [51]. Studies on the OCP levels in edible fishes found alarmingly high levels of these compounds [52–54], suggesting that these compounds are bioaccumulating in our bodies at an alarming rate, mainly due to the biomagnification of pesticides through seafood. These compounds are still found in our food and increasing the risk of human diseases despite the ban.

4. Confirmation of the Pesticide's Xeno-Estrogenicity

Data from the observation of occupational exposure were not sufficient to understand the adverse effects of these OCPs. Nevertheless, animal- and cell-line-based studies unambiguously confirmed the xeno-estrogenic potential of heavily used and well-known pesticides [3,55,56].

Animal-based experiments first confirmed the biological response to DDT comparable to that for estrogens, such as increased uterine weight and vaginal epithelial cornification in lab animals, known as uterotrophic assay [57]. Later, uterotrophic assays of aldrin, dieldrin, endosulfan, and HCH confirmed that all these compounds were xeno-estrogenic in nature [55,56]. Furthermore, a cell-line-based study found evidence of the pesticides modulating estrogen-dependent genes as natural estrogens do [58].

The highest isoform of dichlorodiphenyltrichloroethane, the most well-known pesticide in the world, *p,p′*-DDT (more than 85% of w/w), was found to be estrogenic in animal and cell lines experiments [59]. Additionally, experiments proved that DDT metabolite binds to the estrogen receptor (ER), further consolidating its xeno-estrogenic activity [60].

Dieldrin was reported to cause various forms of cancer in at least seven strains of mice upon oral administration [61]. In one study, dieldrin was found to decrease the binding of natural estrogens to its receptor in female rats' uterine tissue extracts. Additionally, in another animal-based study, the dieldrin treatment was found to competitively inhibit the binding of 17β-estradiol (the most potent form of estrogen) to the receptor in the rat uterus, indicating the strong similarities between these two compounds [62]. Dieldrin treatment at a concentration of 10 μM in MCF-7 cells produced a significant increase in proliferation. The proliferation induced due to dieldrin at this concentration was 54.89% that of estradiol, indicating that it is a strong xeno-estrogen [63]. Both isoforms of the pesticide endosulfan, α- and β-endosulfan, were reported to possess xeno-estrogenic potential [64]. Furthermore, atrophy was noticed in different testicular tissues of male rats fed meal containing 10 ppm concentration of endosulfan [65]. In another study, it was noted that endosulfan-induced atrophy relates to biochemical changes in testicular tissue that translate into a significant fall in sperm counts in the testicular epididymis region and reduced spermatid counts in intratesticular tissue [66]. Endosulfan-induced testicular atrophy has also been reported to induce infertility in males due to a dramatic decrease in sperm count after exposure [67].

Chemicals produced by humans and nature possess estrogenic activity. However, the molecular details or structure rendering a chemical/compound its xeno-estrogenic activity are not well known. However, the majority of environmental xeno-estrogens share a common structural motif either of phenol or structurally similar to phenol as observed in all the OCPs [68].

The understanding of mechanisms of endocrine disruptors exerted by such chemicals is growing with time. Scientists proposed that by following these mechanisms, these environmental chemicals (including OCPs) are reported to lead to endocrine disruption (including xeno-estrogenicity) [69]. Figure 1 presents the different mechanisms of action of these compounds:

1. Mimicking the effect of endogenous steroidal hormones (androgens and estrogens).
2. Antagonizing steroidal hormones.
3. Altering the synthesis and metabolism of endogenous steroidal hormones.
4. Modifying hormone receptor expression in different tissues.

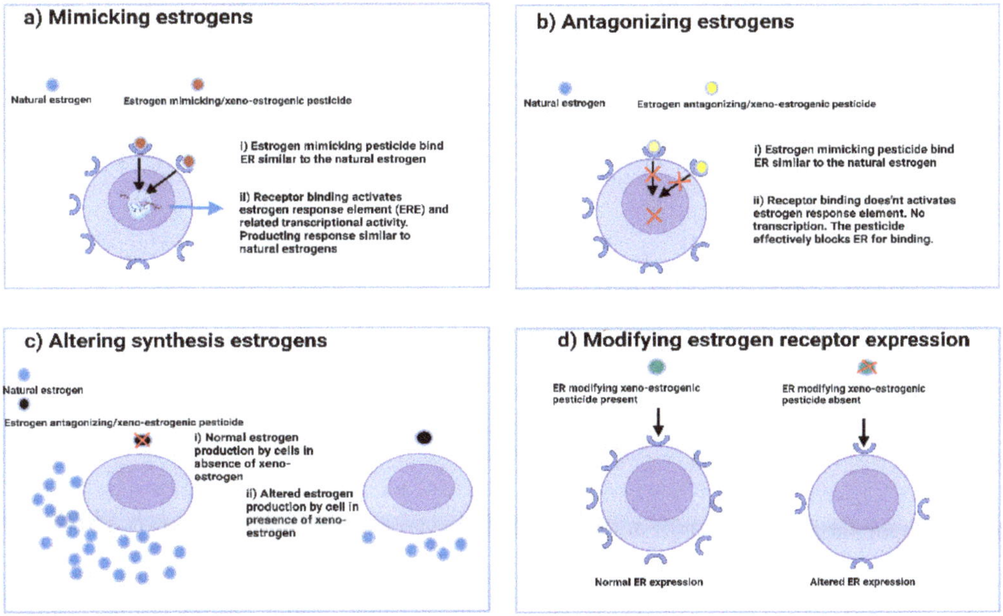

Figure 1. Different mechanisms of endocrine disruption by xeno-estrogenic pesticides: (**a**) mimicking estrogens, (**b**) antagonizing estrogens, (**c**) altering estrogen synthesis, and (**d**) modifying estrogen receptor expression.

5. Association of Xeno-Estrogenic Pesticides with Endocrine-Related Cancer

Through human studies, the role of estrogens in female as well as male cancers has been well established. Breast and endometrial cancers in females are primarily estrogen-mediated [70]. In males, declining testosterone with age and rising estrogens are known to elevate prostate cancer risk with age [71].

Therefore, the risk of estrogen-mediated cancer has been analyzed in relation to exposure to OCPs by researchers. Risk of breast, endometrium, and ovary cancers in females and risk of prostate and testis cancers in males are estrogen-dependent to some extent [71]. Studies indicate a strong relation between exposure to estrogens as the principal risk factor in the development of these estrogen-mediated cancers [15,16,18,19,32,38,57,60,62,70].

5.1. Xeno-Estrogenic Pesticides and Female Cancer

The role of these xeno-estrogenic pesticides in elevated breast cancer risk has been most extensively studied compared to any other cancer in humans [11,72–83]. Over the past few decades, the incidence and mortality of breast cancer have increased worldwide by more than 33% [72,73]. Numerous studies have linked the contaminants we humans are unavoidably exposed to as the prime cause of higher breast cancer risk. Sufficient evidence points to the fact that breast cancer is strongly related to exposure to such contaminants, including DDT and other pesticides that act as estrogenic stimulants [40]. Significantly higher levels of pesticides such as DDT along with its persistent metabolites (especially DDE and DDD), heptachlor, dieldrin, and hexacyclohexane were detected in samples of patients with carcinoma of the breast, irrespective of any other factor when compared to controls [74].

Incidence of breast carcinoma is the world's second highest, and it has the fifth highest mortality rate, while 36.8% of all incidences directly link to lifestyle and/or environmental factors in females over the age of 30 years. A significant association was found between the breast cancer epidemic and environmental contaminants exposure—especially DDT [11,75]. In one of the most significant studies linking OCPs with breast cancer, the decline in mortality rates observed in a decade (1976 to 1986) related significantly to decreases in specific OCPs (DDT and lindane) in colostrum [38]. Studies also suggest that removal of xeno-estrogens can prevent the incidence of and mortality due to breast cancer in females [76]. In prospective studies, exposure to OCPs was measured over a long time before diagnosis, and a statistically significant link was reported between breast cancer risk and DDT exposure. Similarly, two studies consistently linked the pesticide DDT with higher breast cancer risk when significant exposure occurred before teenage years (below the age of 14 years) [77,78]. These observations were further consolidated by a cohort study where it was found that above a certain cut-off concentration of dieldrin (>57.6 ng/g), morbidity and mortality amongst breast cancer patients significantly increased [78]. Dieldrin weakly induced both ER expression and cell proliferation in MCF-7 cell lines in a study, indicating dieldrin to be weakly estrogenic in females [82].

It is not clear how the estrogenic mechanism increases breast cancer risk. In a toxico-proteomic study amongst breast cancer patients, these OCPs were found to downregulate the expression of ER (a common event in a large number of breast cancer cases) by disrupting the relevant pathways [83]. In vitro studies also confirmed estrogen deregulation and increases in the concentration of cellular metabolites that activate a number of oncogenes [84,85]. In a recent study from China, OCPs were not only associated with breast cancer risk but were also found to elevate oxidative stress biomarkers in both serum ad urine [86].

Studies analyzing the role of xeno-estrogenic pesticides in endometrial cancer risk are few and rather inconclusive. Previous studies did not find any correlation between xeno-estrogenic OCP levels and endometrial cancer risk [87,88]. More human and animal experiment-based studies are needed to establish a link between these contaminants and endometrial cancer risk.

5.2. *Xeno-Estrogenic Pesticides and Male Cancer*

The role of xeno-estrogens in modifying the risk of male cancer and other related disorders is heavily debated and not well understood. In utero and early postnatal estrogen exposure is a significant contributory factor for testicular cancer risk in young men [89,90]. Studies from our lab have found a high level of these pesticides in mothers' milk and umbilical cord, indicating continuous transfer of these pesticides during the early development of children [6–8].

Endocrine-disrupting chemicals were found to contribute to increased testicular cancer risk in some studies [91–93]. Later, this hypothesis was further consolidated when it was found that exposure to different xeno-estrogens in utero increases testicular cancer risk by fourfold in men (between the ages of 16 and 59) [64]. In an occupational study, xeno-estrogenic OCPs were found to increase testicular cancer risk among youngsters born to Norwegian farmers from 1952 to 1991, reporting a higher-than-expected testicular cancer incidence compared to controls [94]. In a nested case–control study, it was found that mothers of patients with testicular cancer had significantly higher DDT metabolites in their body during the lactational period [95]. A pooled study found 1.29 times higher testicular dysfunction risk, the main cause of testicular cancer, amongst the population living in regions having higher occupational and non-occupational exposure to OCPs compared to those in regions with limited or lower chances of exposure [96].

Due to the hormonal basis of prostate development, researchers have valid reasons to understand the potential relationship of xeno-estrogenic pesticides with prostate cancer risk [97]. The human prostate gland is the organ most affected by malignant neoplasm in elderly men. With the increase in life expectancy, the incidence of cancer has increased, affecting almost 90% of males over 80 years of age. The basis for this high incidence is not clear despite decades of exhaustive research [98]. In contrast to testicular cancer, direct connections between xeno-estrogenic pesticides and prostate cancer have not been established [90,99]. Animal- and cell-line-based studies are a better indicator of the effect of xeno-estrogenic pesticides on the functioning and transformation of prostate cancer. One study reported that endocrine-disrupting chemicals can transform and/or reprogram prostate stem cells and potentially elevate prostate cancer risk in experimental animals [100]. In an animal-based study, it was found that feeding regular small doses of xeno-estrogens to pregnant females led to a significantly increased prostate weight in adulthood in their pups; increased prostate weight is one of the hallmarks of prostatic disorders, including both prostate cancer and benign prostatic hyperplasia (BPH) [101]. The most compelling proof for a link between xeno-estrogenic OCP exposure and prostate cancer comes from studies from our laboratory. We found that the levels of some of these pesticides were significantly higher amongst patients with higher-stage and more aggressive prostate cancer, indicating the development of an aggressive form of cancer. In addition, these pesticides influence prostate weight as well amongst BPH patients [102,103]. Occupational pesticide exposure is regarded as an established factor elevating prostate cancer risk [104–106]. Furthermore, numerous epidemiological studies reported a positive relation between (non-occupational) pesticide exposure and the risk of prostate cancer [106–108]. In relatively recent Asian studies, DDT and endosulfan were found to increase prostate cancer risk. This hints that in spite of the partial ban on these two compounds, they may still be increasing the risk [109,110]. However, more studies are needed to understand if there is a relation before drawing any conclusion.

6. Conclusions

The association of pesticides, including OCPs, with cancer risk is well established. However, some of the mechanisms, such as xeno-estrogenicity, that increase the risk of carcinogenic progression are not clearly understood. Accumulating evidence is consolidating the role of xeno-estrogenic OCPs in the risk of not only cancers but also a plethora of other human disorders. These chemicals, despite their apparent adverse health effects, are still used in some countries. Although the risk of some cancers such as breast cancer is clearly modified with exposure to the discussed chemicals, the link with the risk of other estrogen-mediated cancers, especially in males, is not clearly understood and needs to be studied more. Figure 2 represents the risk of different cancers reported to be associated with exposure to xeno-estrogenic pesticides. In recent decades, breast cancer in females and prostate cancer in males have become the top causes of morbidity and mortality, and both are estrogen-mediated cancers. More studies are needed to find out how much this increased incidence can be attributed to such harmful environmental factors. To the best of our knowledge, this is the only study that focuses on all-gender estrogen-mediated cancer risk modification by xeno-estrogenic OCPs. These chemicals must be completely phased out and replaced with less toxic and affordable alternatives that have negligible adverse health effects on mammalian systems.

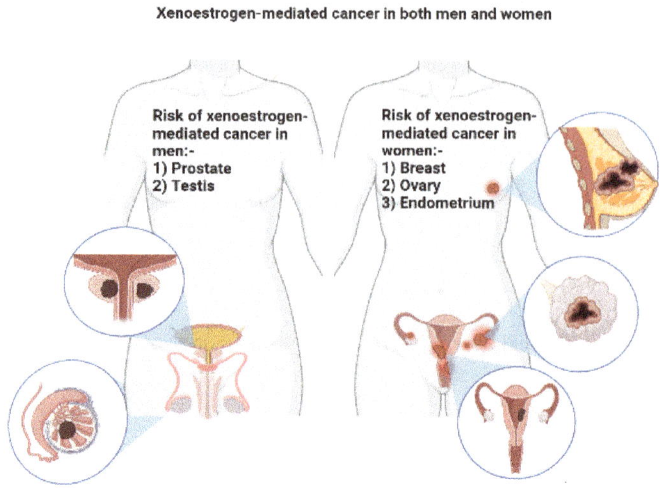

Figure 2. Risk of various estrogen-mediated cancers with xeno-estrogenic pesticide exposure.

Author Contributions: V.K., C.S.Y. and B.D.B. contributed to writing the manuscript. All authors have read and agreed to the published version of the manuscript.

Funding: This research received no external funding.

Institutional Review Board Statement: Not applicable.

Informed Consent Statement: Not applicable.

Data Availability Statement: The data used to support the findings of the study are included within the article.

Conflicts of Interest: The authors declare no conflict of interest.

References

1. Rieger, P.G.; Meier, H.M.; Gerle, M.; Vogt, U.; Groth, T.; Knackmuss, H.J. Xenobiotics in the environment: Present and future strategies to obviate the problem of biological persistence. *J. Biotechnol.* **2002**, *94*, 101–123. [CrossRef]
2. Combarnous, Y.; Nguyen, T.M.D. Membrane Hormone Receptors and Their Signaling Pathways as Targets for Endocrine Disruptors. *J. Xenobiot.* **2022**, *12*, 64–73. [CrossRef]
3. Wang, L.H.; Chen, L.R.; Chen, K.H. In vitro and vivo identification, metabolism and action of xenoestrogens: An overview. *Int. J. Mol. Sci.* **2021**, *22*, 4013. [CrossRef]
4. Jiao, L.; Lao, Q.; Chen, L.; Chen, F.; Sun, X.; Zhao, M. Concentration and influence factors of organochlorine pesticides in atmospheric particles in a coastal island in Fujian, Southeast China. *Aerosol Air Qual. Res.* **2018**, *18*, 2982–2996. [CrossRef]
5. Kumar, B.; Mishra, M.; Verma, V.K.; Kumar, S.; Sharma, C.S. Distribution of dichlorodiphenyltrichloroethane and hexachlorocyclohexane in urban soils and risk assessment. *J. Xenobiot.* **2013**, *3*, e1. [CrossRef]
6. Mustafa, M.D.; Pathak, R.; Tripathi, A.K.; Ahmed, R.S.; Guleria, K.; Banerjee, B.D. Maternal and cord blood levels of aldrin and dieldrin in Delhi population. *Environ. Monit. Assess.* **2010**, *171*, 633–638. [CrossRef]
7. Pathak, R.; Suke, S.G.; Ahmed, R.S.; Tripathi, A.K.; Guleria, K.; Sharma, C.S.; Ma-khijani, S.D.; Mishra, M.; Banerjee, B.D. Endosulfan and other organochlorine pesticide residues in maternal and cord blood in North Indian population. *Bull. Environ. Contam. Toxicol.* **2008**, *81*, 216–219. [CrossRef]
8. Pathak, R.; Ahmed, R.S.; Tripathi, A.K.; Guleria, K.; Sharma, C.S.; Makhijani, S.D.; Banerjee, B.D. Maternal and cord blood levels of organochlorine pesticides: Association with preterm labor. *Clin. Biochem.* **2009**, *42*, 746–749. [CrossRef]
9. Bagga, D.; Anders, K.H.; Wang, H.J.; Roberts, E.; Glaspy, J.A. Organochlorine pesticide content of breast adipose tissue from women with breast cancer and control subjects. *J. Natl. Cancer Inst.* **2000**, *92*, 750–753. [CrossRef]
10. Falck, F., Jr.; Ricci, A., Jr.; Wolf, M.S.; Godbold, J.; Deckers, P. Pesticides and polychlorin-ated biphenyl residues in human breast lipids and their relation to breast cancer. *Arch. Environ. Health* **1992**, *47*, 143–146.
11. Wu, S.; Zhu, W.; Thompson, P.; Hannun, Y.A. Evaluating intrinsic and non-intrinsic cancer risk factors. *Nat. Commun.* **2018**, *9*, 3490. [CrossRef] [PubMed]

12. Tamási, V.; Monostory, K.; Prough, R.A.; Falus, A. Role of xenobiotic metabolism in cancer: Involvement of transcriptional and miRNA regulation of P450s. *Cell. Mol. Life Sci.* **2011**, *68*, 1131–1146. [CrossRef] [PubMed]
13. Shah, H.K.; Sharma, T.; Banerjee, B.D. Organochlorine pesticides induce inflammation; ROS production; and DNA damage in human epithelial ovary cells: An in vitro study. *Chemosphere* **2020**, *246*, 125691. [CrossRef] [PubMed]
14. Gore, A.C.; Chappell, V.A.; Fenton, S.E.; Flaws, J.A.; Nadal, A.; Prins, G.S.; Toppari, J.; Zoeller, R.T. EDC-2: The Endocrine Society's Second Scientific Statement on Endocrine-Disrupting Chemicals. *Endocr. Rev.* **2015**, *36*, E1–E150. [CrossRef] [PubMed]
15. Eve, L.; Fervers, B.; Le Romancer, M.; Etienne-Selloum, N. Exposure to endocrine disrupting chemicals and risk of breast cancer. *Int. J. Mol. Sci.* **2020**, *21*, 9139. [CrossRef] [PubMed]
16. Badamasi, I.; Odong, R.; Masembe, C. Threats posed by xenoestrogenic chemicals to the aquatic ecosystem, fish reproduction and humans: A review. *Afr. J. Aquat. Sci.* **2020**, *45*, 243–258. [CrossRef]
17. Nishi, K.; Fu, W.; Kiyama, R. Novel estrogen-responsive genes (ERGs) for the evaluation of estrogenic activity. *PLoS ONE*. **2022**, *17*, e0273164. [CrossRef]
18. Ventura, C.; Zappia, C.D.; Lasagna, M.; Pavicic, W.; Richard, S.; Bolzan, A.D.; Monczor, F.; Núñez, M.; Cocca, C. Effects of the pesticide chlorpyrifos on breast cancer disease. Implication of epigenetic mechanisms. *J. Steroid Biochem. Mol. Biol.* **2019**, *186*, 96–104. [CrossRef]
19. Wang, X.; Ha, D.; Yoshitake, R.; Chan, Y.S.; Sadava, D.; Chen, S. Exploring the biological activity and mechanism of xenoestrogens and phytoestrogens in cancers: Emerging methods and concepts. *Int. J. Mol. Sci.* **2021**, *22*, 8798. [CrossRef]
20. Bernardes, M.F.F.; Pazin, M.; Pereira, L.C.; Dorta, D.J. Impact of pesticides on environmental and human health. In *Toxicology Studies-Cells, Drugs and Environment*; Intech: Rijeka, Croatia, 2015; pp. 195–233.
21. Stockholm Convention. *The 12 Initial POPs under the Stockholm Convention*; Stockholm Convention: Geneva, Switzerland, 2008.
22. Poh, C.; McPherson, J.D.; Tuscano, J.; Li, Q.; Parikh-Patel, A.; Vogel, C.F.; Keegan, T. Environmental pesticide exposure and non-Hodgkin lymphoma survival: A population-based study. *BMC Med.* **2022**, *20*, 165. [CrossRef]
23. Georgakopoulou, R.; Fiste, O.; Sergentanis, T.N.; Andrikopoulou, A.; Zagouri, F.; Gav-riatopoulou, M.; Psaltopoulou, T.; Kastritis, E.; Terpos, E.; Dimopoulos, M.A. Occupational exposure and multiple myeloma risk: An updated review of meta-analyses. *J. Clin. Med.* **2021**, *10*, 4179. [CrossRef] [PubMed]
24. Renier, M.; Busson, A.; Boulanger, M.; Piel, C.; Pons, R.; Tual, S.; Amadéo, B.; Meryet-Figuiere, M.; Marcotullio, E.; Clin, B.; et al. Agricultural exposure and risk of soft tissue sarcomas and gastrointestinal stromal sarcoma in the AGRIculture and CANcer (AGRICAN) cohort. *Int. J. Cancer* **2022**, *150*, 1792–1803. [CrossRef] [PubMed]
25. Shankar, A.; Dubey, A.; Saini, D.; Singh, M.; Prasad, C.P.; Roy, S.; Bharati, S.J.; Rinki, M.; Singh, N.; Seth, T.; et al. Environmental and occupational determinants of lung cancer. *Transl. Lung Cancer Res.* **2019**, *8*, S31. [CrossRef]
26. Porta, M.; Gasull, M.; Pumarega, J.; Kiviranta, H.; Rantakokko, P.; Raaschou-Nielsen, O.; Bergdahl, I.A.; Sandanger, T.M.; Agudo, A.; Rylander, C.; et al. Plasma concentrations of persistent organic pollutants and pancreatic cancer risk. *Int. J. Epidemiol.* **2022**, *51*, 479–490. [CrossRef] [PubMed]
27. Perrot-Applanat, M.; Pimpie, C.; Cano-Sancho, G.; Antignac, J.P.; Pocard, M. Detection of Persistent Organic Pollutants in Omental Adipose Tissue from Patients with Diffuse-Gastric Cancer: A Pilot Study. *Cancers* **2021**, *13*, 4874. [CrossRef] [PubMed]
28. VoPham, T. Environmental risk factors for liver cancer and nonalcoholic fatty liver disease. *Curr. Epidemiol. Rep.* **2019**, *6*, 50–66. [CrossRef] [PubMed]
29. Deb, A.A.; Okechukwu, C.E.; Emara, S.; Sami, A.A. Occupational exposure as risk factor for kidney and bladder cancer: A systematic review and meta-analysis. *Urol. Nephrol. Open Access J.* **2019**, *7*, 143–151. [CrossRef]
30. Pérez-Moreno, P.; Riquelme, I.; García, P.; Brebi, P.; Roa, J.C. Environmental and Lifestyle Risk Factors in the Carcinogenesis of Gallbladder Cancer. *J. Pers. Med.* **2022**, *12*, 234. [CrossRef]
31. WHO. *International Agency for Research on Cancer IARC Monographs on the Evaluation of the Carcinogenic Risk of Chemicals to Humans*; Distributed for IARC by WHO; WHO: Geneva, Switzerland, 1980; Volume 22.
32. Madia, F.; Worth, A.; Whelan, M.; Corvi, R. Carcinogenicity assessment: Addressing the challenges of cancer and chemicals in the environment. *Environ. Int.* **2019**, *128*, 417–429. [CrossRef]
33. Hansen, L.G.; Shane, B.S. Xenobiotic metabolism. In *Basic Environmental Toxicology*; CRC Press: Boca Raton, FL, USA, 2019; pp. 49–105.
34. Anzenbacher, P.; Anzenbacherova, E. Cytochromes P450 and metabolism of xenobiotics. *Cell. Mol. Life Sci. CMLS* **2001**, *58*, 737–747. [CrossRef]
35. Esteves, F.; Rueff, J.; Kranendonk, M. The Central Role of Cytochrome P450 in Xenobiotic Metabolism—A Brief Review on a Fascinating Enzyme Family. *J. Xenobiot.* **2021**, *11*, 94–114. [CrossRef] [PubMed]
36. Keswani, C.; Dilnashin, H.; Birla, H.; Roy, P.; Tyagi, R.K.; Singh, D.; Rajput, V.D.; Minkina, T.; Singh, S.P. Global footprints of organochlorine pesticides: A pan-global survey. *Environ. Geochem. Health* **2022**, *44*, 149–177. [CrossRef] [PubMed]
37. Donley, N. The USA lags behind other agricultural nations in banning harmful pesticides. *Environ. Health* **2019**, *18*, 44. [CrossRef] [PubMed]
38. Kumar, B.; Verma, V.K.; Naskar, A.K.; Chakraborty, P.; Kumar, S.; Mukherjee, D. Human health risk from hexachlorocyclohexane and dichlorodiphenyltrichloroethane pesticides, through consumption of vegetables: Estimation of daily intake and hazard quotients. *J. Xenobiot.* **2013**, *3*, e6. [CrossRef]

39. ToxFAQsTM for Aldrin and Dieldrin. Agency for Toxic Substances and Disease Registry (ATSDR)-Toxic Substance portal. Available online: https://wwwn.cdc.gov/tsp/ToxFAQs/ToxFAQsDetails.aspx?faqid=316toxid=56 (accessed on 1 October 2022).
40. Owago, O.J.; Qi, S.; Xin, X.; Yuan, Z.; Sylvie, M.A. Residues of organochlorine pesticides in vegetables from Deyang and Yanting areas of the Chengdu economic region, Sichuan Province, China. *J. Am. Sci.* **2009**, *5*, 91–100.
41. Randhawa, M.A.; Anjum, F.M.; Asi, M.R.; Ahmed, A.; Nawaz, H. Field incurred endosulfan residues in fresh and processed vegetables and dietary intake assessment. *Int. J. Food Prop.* **2014**, *17*, 1109–1115. [CrossRef]
42. Krishnan, A.V.; Stathis, P.; Permuth, S.F.; Tokes, L.; Feldman, D. Bisphenol-A: An estrogenic substance is released from polycarbonate flasks during autoclaving. *Endocrinology* **1993**, *132*, 2279–2286. [CrossRef]
43. Mima, M.; Greenwald, D.; Ohlander, S. Environmental toxins and male fertility. *Curr. Urol. Rep.* **2018**, *19*, 50. [CrossRef]
44. Pan, X.; Xu, T.; Xu, H.; Fang, H.; Yu, Y. Characterization and genome functional analysis of the DDT-degrading bacterium *Ochrobactrum* sp. DDT-2. *Sci. Total Environ.* **2017**, *592*, 593–599. [CrossRef]
45. Gutgesell, R.M.; Tsakiridis, E.E.; Jamshed, S.; Steinberg, G.R.; Holloway, A.C. Impact of pesticide exposure on adipose tissue development and function. *Biochem. J.* **2020**, *477*, 2639–2653. [CrossRef]
46. New-Aaron, M.; Naveed, Z.; Rogan, E.G. Estrogen Disrupting Pesticides in Nebraska Groundwater: Trends between Pesticide-contaminated Water and Estrogen-related Cancers in An Ecological Observational Study. *Water* **2021**, *13*, 790. [CrossRef]
47. Navarrete, I.A.; Tee, K.A.M.; Unson, J.R.S.; Hallare, A.V. Organochlorine pesticide residues in surface water and groundwater along Pampanga River, Philippines. *Environ. Monit. Assess.* **2018**, *190*, 289. [CrossRef] [PubMed]
48. Ślósarczyk, K.; Witkowski, A.J. Preliminary Evaluation of the Possible Occurrence of Pesticides in Groundwater Contaminated with Nitrates—A Case Study from Southern Poland. *Water* **2021**, *13*, 3091. [CrossRef]
49. Sarkar, U.K.; Pathak, A.K.; Srivastava, S.M. Assessing organochlorine pesticide (OCP) residues in water and fish samples from a small perennial river and associated wetlands of Ganga Basin; India for sustainable management. *Sustain. Water Resour. Manag.* **2022**, *8*, 17. [CrossRef]
50. Panis, C.; Candiotto, L.Z.P.; Gaboardi, S.C.; Gurzenda, S.; Cruz, J.; Castro, M.; Lemos, B. Widespread pesticide contamination of drinking water and impact on cancer risk in Brazil. *Environ. Int.* **2022**, *165*, 107321. [CrossRef]
51. Gupta, S.; Gupta, K. Bioaccumulation of pesticides and its impact on biological systems. In *Pesticides in Crop Production Physiological and Biochemical Action*; John Wiley & Sons Ltd.: Hoboken, NJ, USA, 2020; pp. 55–67.
52. Gerber, R.; Smit, N.J.; Van Vuren, J.H.; Nakayama, S.M.; Yohannes, Y.B.; Ikenaka, Y.; Wepener, V. Bioaccumulation and human health risk assessment of DDT and other organochlorine pesticides in an apex aquatic predator from a premier conservation area. *Sci. Total Environ.* **2016**, *550*, 522–533. [CrossRef]
53. Arisekar, U.; Shakila, R.J.; Shalini, R.; Jeyasekaran, G.; Arumugam, N.; Almansour, A.I.; Perumal, K. Bioaccumulation of organochlorine pesticide residues (OCPs) at different growth stages of pacific white leg shrimp (*Penaeus vannamei*): First report on ecotoxicological and human health risk assessment. *Chemosphere* **2022**, *308*, 136459. [CrossRef]
54. Zhang, G.; Pan, Z.; Bai, A.; Li, J.; Li, X. Distribution and bioaccumulation of organochlorine pesticides (OCPs) in food web of Nansi Lake, China. *Environ. Monit. Assess.* **2014**, *186*, 2039–2051. [CrossRef]
55. Uslu, U.; Sandal, S.; Cumbul, A.; Yildiz, S.; Aydin, M.; Yilmaz, B. Evaluation of estrogenic effects of polychlorinated biphenyls and organo-chlorinated pesticides using immature rat uterotrophic assay. *Hum. Exp. Toxicol.* **2013**, *32*, 476–482. [CrossRef]
56. Presibella, K.M.; Kita, D.H.; Carneiro, C.B.; Andrade, A.J.; Dalsenter, P.R. Re-productive evaluation of two pesticides combined (deltamethrin and endosulfan) in female rats. *Reprod. Toxicol.* **2005**, *20*, 95–101. [CrossRef]
57. Soto, A.M.; Sonnenschein, C. Estrogens, xenoestrogens, and the development of neoplasms. In *Endocrine Disruptors*; CRC Press: Boca Raton, FL, USA, 2004; pp. 175–215.
58. Varayoud, J.; Monje, L.; Bernhardt, T.; Muñoz-de-Toro, M.; Luque, E.H.; Ramos, J.G. Endosulfan modulates estrogen-dependent genes like a non-uterotrophic dose of 17β-estradiol. *Reprod. Toxicol.* **2008**, *26*, 138–145. [CrossRef] [PubMed]
59. Vandenberg, L.N.; Najmi, A.; Mogus, J.P. Agrochemicals with estrogenic endocrine disrupting properties: Lessons Learned? *Mol. Cell. Endocrinol.* **2020**, *518*, 110860. [CrossRef] [PubMed]
60. Kalinina, T.S.; Kononchuk, V.V.; Gulyaeva, L.F. Expression of estrogen-, progesterone-, and androgen-responsive genes in MCF-7 and MDA-MB-231 cells treated with o, p′-DDT, p, p′-DDT, or endosulfan. *J. Biochem. Mol. Toxicol.* **2021**, *35*, 1–8. [CrossRef] [PubMed]
61. USEPA Office of Pesticide Programs, Health Effects Division, Science Information Management Branch. *Chemicals Evaluated for Carcinogenic Potential*; USEPA Office of Pesticide Programs, Health Effects Division, Science Information Management Branch: Washington, DC, USA, 2006.
62. Briz, V.; Molina-Molina, J.M.; Sánchez-Redondo, S.; Fernández, M.F.; Grimalt, J.O.; Olea, N.; Rodríguez-Farré, E.; Sunol, C. Differential estrogenic effects of the persistent organochlorine pesticides dieldrin, endosulfan, and lindane in primary neuronal cultures. *Toxicol. Sci.* **2011**, *120*, 413–427. [CrossRef] [PubMed]
63. Eze, U.A.; Huntriss, J.; Routledge, M.N.; Gong, Y.Y.; Connolly, L. The effect of individual and mixtures of mycotoxins and persistent organochloride pesticides on oestrogen receptor transcriptional activation using in vitro reporter gene assays. *Food Chem. Toxicol.* **2019**, *130*, 68–78. [CrossRef]
64. Soto, A.M.; Sonnenschein, C. Environmental causes of cancer: Endocrine disruptors as carcinogens. *Nat. Rev. Endocrinol.* **2010**, *6*, 363–370. [CrossRef]

65. Alarcón, R.; Varayoud, J.; Luque, E.H.; Milesi, M.M. Effect of neonatal exposure to endosulfan on myometrial adaptation during early pregnancy and labor in rats. *Mol. Cell. Endocrinol.* **2019**, *491*, 110435. [CrossRef]
66. Sinha, N.; Narayan, R.; Shanker, R.; Saxena, D.K. Endosulfan-induced bio-chemical changes in the testis of rats. *Vet. Hum. Toxicol.* **1995**, *37*, 547–549.
67. Sebastian, R.; Raghavan, S.C. Exposure to Endosulfan can result in male in-fertility due to testicular atrophy and reduced sperm count. *Cell Death Discov.* **2015**, *1*, 15020. [CrossRef]
68. Katzenellenbogen, J.A. The structural pervasiveness of estrogenic activity. *Environ. Health Perspect.* **1995**, *103*, 99–101. [CrossRef]
69. Bonefeld-Jørgensen, E.C.; Ghisari, M.; Wielsøe, M.; Bjerregaard-Olesen, C.; Kjeldsen, L.S.; Long, M. Biomonitoring and hormone-disrupting effect biomarkers of persistent organic pollutants in vitro and ex vivo. *Basic Clin. Pharmacol. Toxicol.* **2014**, *115*, 118–128. [CrossRef] [PubMed]
70. Xu, X.; Dailey, A.B.; Talbott, E.O.; Ilacqua, V.A.; Kearney, G.; Asal, N.R. Associations of serum concentrations of organochlorine pesticides with breast cancer and prostate cancer in US adults. *Environ. Health Perspect.* **2010**, *118*, 60–66. [CrossRef] [PubMed]
71. Nelles, J.L.; Hu, W.Y.; Prins, G.S. Estrogen action and prostate cancer. *Expert. Rev. Endocrinol. Metab.* **2011**, *6*, 437–451. [CrossRef] [PubMed]
72. Ibarluzea, J.M.; Fernández, M.F.; Santa-Marina, L.; Olea-Serrano, M.F.; Rivas, A.M.; Aurrekoetxea, J.J.; Expósito, J.; Lorenzo, M.; Torné, P.; Villalobos, M.; et al. Breast cancer risk and the combined effect of environmental estrogens. *Cancer Causes Control* **2004**, *15*, 591–600. [CrossRef]
73. Siegel, R.L.; Miller, K.D.; Fuchs, H.E.; Jemal, A. Cancer Statistics. *CA Cancer J. Clin.* **2021**, *71*, 7–33. [CrossRef] [PubMed]
74. Mathur, V.; Bhatnagar, P.; Sharma, R.G.; Acharya, V.; Sexana, R. Breast cancer incidence and exposure to pesticides among women originating from Jaipur. *Environ. Int.* **2002**, *28*, 331–336. [CrossRef]
75. Calaf, G.M.; Ponce-Cusi, R.; Aguayo, F.; Muñoz, J.P.; Bleak, T.C. Endocrine disruptors from the environment affect breast cancer. *Oncol. Lett.* **2020**, *20*, 19–32.
76. Davis, D.L.; Bradlow, H.L.; Wolff, M.; Woodruff, T.; Hoel, D.G.; Anton-Culver, H. Medical hypothesis: Xenoestrogens as preventable causes of breast cancer. *Environ. Health Perspect.* **1993**, *101*, 372–377. [CrossRef]
77. Cohn, B.A.; Wolff, M.S.; Cirillo, P.M.; Sholtz, R.I. DDT and breast cancer in young women: New data on the significance of age at exposure. *Environ. Health Perspect.* **2007**, *115*, 1406–1414. [CrossRef]
78. Cohn, B.A.; Cirillo, P.M.; Terry, M.B. DDT and breast cancer: Prospective study of induction time and susceptibility windows. *J. Natl. Cancer Inst.* **2019**, *111*, 803–810. [CrossRef]
79. Wan, M.L.Y.; Co, V.A.; El-Nezami, H. Endocrine disrupting chemicals and breast cancer: A systematic review of epidemiological studies. *Crit. Rev. Food Sci. Nutr.* **2021**, *62*, 6549–6576. [CrossRef]
80. Chang, S.; El-Zaemey, S.; Heyworth, J.; Tang, M.C. DDT exposure in early childhood and female breast cancer: Evidence from an ecological study in Taiwan. *Environ. Int.* **2018**, *121*, 1106–1112. [CrossRef]
81. Høyer, A.P.; Grandjean, P.; Jørgensen, T.; Brock, J.W.; Hartvig, H.B. Organochlorine exposure and risk of breast cancer. *Lancet* **1998**, *352*, 1816–1820. [CrossRef]
82. Hanley, G.D. *Toxicological Profile for Aldrin/Dieldrin*; Agency for Toxic Substances and Disease Registry: Atlanta, GA, USA, 2002.
83. Pizzatti, L.; Kawassaki, A.C.B.; Fadel, B.; Nogueira, F.C.; Evaristo, J.A.; Woldmar, N.; Teixeira, G.T.; Da Silva, J.C.; Scandolara, T.B.; Rech, D.; et al. Toxicoproteomics disclose pesticides as downregulators of TNF-α, IL-1β and estrogen receptor pathways in breast cancer women chronically exposed. *Front. Oncol.* **2020**, *10*, 1698. [CrossRef]
84. Bradlow, H.L.; Davis, D.L.; Lin, G.; Sepkovic, D.; Tiwari, R. Effects of pesticides on the ratio of 16 alpha/2-hydroxyestrone: A biologic marker of breast cancer risk. *Environ. Health Perspect.* **1995**, *103*, 147–150. [CrossRef]
85. Alleva, R.; Manzella, N.; Gaetani, S.; Bacchetti, T.; Bracci, M.; Ciarapica, V.; Tomasetti, M. Mechanism underlying the effect of long-term exposure to low dose of pesticides on DNA integrity. *Environ. Toxicol.* **2018**, *33*, 476–487. [CrossRef]
86. Miao, Y.; Rong, M.; Li, M.; He, H.; Zhang, L.; Zhang, S.; Zeng, Q. Serum concentrations of organochlorine pesticides; biomarkers of oxidative stress, and risk of breast cancer. *Environ. Pollut.* **2021**, *286*, 117386. [CrossRef]
87. Weiderpass, E.; Adami, H.O.; Baron, J.A.; Wicklund-Glynn, A.; Aune, M.; Atuma, S.; Persson, I. Organochlorines and endometrial cancer risk. *Cancer Epidemiol. Biomark. Prev.* **2000**, *9*, 487–493.
88. Mallozzi, M.; Leone, C.; Manurita, F.; Bellati, F.; Caserta, D. Endocrine disrupting chemicals and endometrial cancer: An overview of recent laboratory evidence and epidemiological studies. *Int. J. Environ. Res. Public Health* **2017**, *14*, 334. [CrossRef] [PubMed]
89. Fénichel, P.; Chevalier, N. Is testicular germ cell cancer estrogen dependent? The role of endocrine disrupting chemicals. *Endocrinology* **2019**, *160*, 2981–2989. [CrossRef] [PubMed]
90. Patisaul, H.B.; Adewale, H.B. Long-term effects of environmental endocrine disruptors on reproductive physiology and behavior. *Front. Behav. Neurosci.* **2009**, *3*, 10. [CrossRef]
91. Rodprasert, W.; Main, K.M.; Toppari, J.; Virtanen, H.E. Associations between male reproductive health and exposure to endocrine-disrupting chemicals. *Curr. Opin. Endocr. Metab. Res.* **2019**, *7*, 49–61. [CrossRef]
92. Nori, F.; Carbone, P.; Giordano, F.; Osborn, J.; Figà-Talamanca, I. Endocrine-disrupting chemicals and testicular cancer: A case-control study. *Arch. Environ. Occup. Health* **2006**, *61*, 87–95. [CrossRef]
93. Bräuner, E.V.; Lim, Y.H.; Koch, T.; Uldbjerg, C.S.; Gregersen, L.S.; Pedersen, M.K.; Frederiksen, H.; Petersen, J.H.; Coull, B.A.; Andersson, A.M.; et al. Endocrine disrupting chemicals and risk of testicular cancer: A systematic review and meta-analysis. *J. Clin. Endocr. Metab.* **2021**, *106*, e4834–e4860. [CrossRef]

94. Kristensen, P.; Andersen, A.; Irgens, L.M.; Bye, A.S.; Sundheim, L. Cancer in offspring of parents engaged in agricultural activities in Norway: Incidence and risk factors in the farm environment. *Int. J. Cancer* **1996**, *65*, 39–50. [CrossRef]
95. Cohn, B.A.; Cirillo, P.M.; Christianson, R.E. Prenatal DDT exposure and testicular cancer: A nested case-control study. *Arch. Environ. Occup. Health* **2010**, *65*, 127–134. [CrossRef]
96. Requena-Mullor, M.; Navarro-Mena, A.; Wei, R.; López-Guarnido, O.; Lozano-Paniagua, D.; Alarcon-Rodriguez, R. Evaluation of gonadal alterations in a population environmentally exposed to a mixture of endocrine active pesticides. *Int. J. Environ. Res. Public Health* **2021**, *18*, 2355. [CrossRef]
97. Diamanti-Kandarakis, E.; Bourguignon, J.P.; Giudice, L.C.; Hauser, R.; Prins, G.S.; Soto, A.M.; Zoeller, R.T.; Gore, A.C. Endocrine-disrupting chemicals: An Endocrine Society scientific statement. *Endocr. Rev.* **2009**, *30*, 293–342. [CrossRef]
98. Rawla, P. Epidemiology of prostate cancer. *World J. Oncol.* **2019**, *10*, 63. [CrossRef]
99. Dobbs, R.W.; Malhotra, N.R.; Greenwald, D.T.; Wang, A.Y.; Prins, G.S.; Abern, M.R. Estrogens and prostate cancer. *Prostate Cancer Prostatic Dis.* **2019**, *22*, 185–194. [CrossRef] [PubMed]
100. Hu, W.Y.; Shi, G.B.; Hu, D.P.; Nelles, J.L.; Prins, G.S. Actions of estrogens and endocrine disrupting chemicals on human prostate stem/progenitor cells and prostate cancer risk. *Mol. Cell. Endocrinol.* **2012**, *354*, 63–73. [CrossRef]
101. Welshons, W.V.; Nagel, S.C.; Thayer, K.A.; Judy, B.M.; Vom Saal, F.S. Low-dose bioactivity of xenoestrogens in animals: Fetal exposure to low doses of methoxychlor and other xenoestrogens increases adult prostate size in mice. *Toxicol. Ind. Health* **1999**, *15*, 12–25. [CrossRef] [PubMed]
102. Kumar, V.; Yadav, C.S.; Singh, S.; Goel, S.; Ahmed, R.S.; Gupta, S.; Grover, R.K.; Banerjee, B.D. CYP 1A1 polymorphism and organochlorine pesticides levels in the etiology of prostate cancer. *Chemosphere* **2010**, *81*, 464–468. [CrossRef]
103. Kumar, V.; Banerjee, B.D.; Datta, S.K.; Yadav, C.S.; Singh, S.; Ahmed, R.S.; Gupta, S. Association of CYP1A1, CYP1B1 and CYP17 gene polymorphisms and organochlorine pesticides with benign prostatic hyperplasia. *Chemosphere* **2014**, *108*, 40–45. [CrossRef]
104. Dhananjayan, V.; Ravichandran, B. Occupational health risk of farmers exposed to pesticides in agricultural activities. *Curr. Opin. Environ. Sci. Health* **2018**, *4*, 31–37. [CrossRef]
105. Barul, C.; Parent, M.E. Occupational exposure to polycyclic aromatic hydrocarbons and risk of prostate cancer. *Environ. Health* **2021**, *20*, 71. [CrossRef]
106. Lewis-Mikhael, A.M.; Bueno-Cavanillas, A.; Guiron, T.O.; Olmedo-Requena, R.; Delgado-Rodríguez, M.; Jiménez-Moleón, J.J. Occupational exposure to pesticides and prostate cancer: A systematic review and meta-analysis. *Occup. Environ. Med.* **2016**, *73*, 134–144. [CrossRef]
107. Ritchie, J.M.; Vial, S.L.; Fuortes, L.J.; Guo, H.; Reedy, V.E.; Smith, E.M. Organochlorines and risk of prostate cancer. *J. Occup. Environ. Med.* **2003**, *45*, 692–702. [CrossRef]
108. Settimi, L.; Masina, A.; Andrion, A.; Axelson, O. Prostate cancer and exposure to pesticides in agricultural settings. *Int. J. Cancer* **2003**, *104*, 458–461. [CrossRef]
109. Pi, N.; Chia, S.E.; Ong, C.N.; Kelly, B.C. Associations of serum organohalogen levels and prostate cancer risk: Results from a case–control study in Singapore. *Chemosphere* **2016**, *144*, 1505–1512. [CrossRef]
110. Gutierrez, É.A.V.; Sánchez, L.E.T.; Salas, R.A.V.; Carrillo, L.L.; Cebrián, M.E. Endosulfan Exposure Is Associated with Prostate Cancer In Mexico. In Proceedings of the 2015 ISEE Conference, Sao Paulo, Brazil, 30 August–3 September 2015.

Journal of *Xenobiotics*

MDPI

Article

Application of Fabric Phase Sorptive Extraction (FPSE) Engaged to Tandem LC-MS/MS for Quantification of Brorphine in Oral Fluid

Dimitra Florou [1], Thalia Vlachou [1,2], Vasilios Sakkas [2] and Vassiliki Boumba [1,*]

[1] Department of Forensic Medicine and Toxicology, Faculty of Medicine, School of Health Sciences, University of Ioannina, University Campus, 5110 Ioannina, Greece
[2] Department of Analytical Chemistry, Faculty of Chemistry, University of Ioannina, 5110 Ioannina, Greece
* Correspondence: vboumba@uoi.gr; Tel.: +30-2651007724 or +30-6974663683

Abstract: Brorphine (1-[1-[1-(4-bromophenyl) ethyl]-piperidin-4-yl]-1,3-dihydro-2H-benzo [d]imidazol-2-one) is one of the most recent novel synthetic opioids (NSOs) on the novel psychoactive substances (NPSs) market, involved in over 100 deaths in 2020. Brorphine is a substituted piperidine-benzimidazolone analogue that retains structural similarities to fentanyl, acting as a full agonist at the μ-opioid receptor. Oral Fluid (OF) is an alternative matrix, frequently analyzed for the detection of NPS. Fabric phase sorptive extraction (FPSE) is a superior, green-sample -preparation technology recently applied for drug analysis. This contribution presents the development and validation of a method, based on the application of FPSE and liquid chromatography–tandem mass spectrometry (LC-MS/MS), to determine/quantitate brorphine in OF. The method's linearity ranged between 0.05 and 50 ng/mL (R^2 = 0.9993), the bias ranged between 12.0 and 16.8%, and inter- and intra-day precisions ranged between 6.4 and 9.9%. Accuracy and extraction efficiency lied between 65 and 75%. LOD/LOQ were 0.015 ng/mL/0.05 ng/mL. Analyte's post-preparative stability was higher than 95%, while no matrix interferences and carryover between runs were observed. This is the first report introducing the application of FPSE for NPS determination, specifically, the quantification of brorphine in OF, thereby presenting a simple, rapid, sensitive, specific, effective, and reliable procedure engaged to LC-MS/MS that is suitable for routine application and the analysis of more NPSs.

Keywords: brorphine; FPSE; LC-MS/MS; NPS; NSO; oral fluid; extraction

Citation: Florou, D.; Vlachou, T.; Sakkas, V.; Boumba, V. Application of Fabric Phase Sorptive Extraction (FPSE) Engaged to Tandem LC-MS/MS for Quantification of Brorphine in Oral Fluid. *J. Xenobiot.* **2022**, *10*, 356–364. https://doi.org/10.3390/jox12040025

Academic Editors: Roel Vermeulen, Lauren Petrick, Maaike van Gerwen and Marcela Segundo

Received: 30 August 2022
Accepted: 22 November 2022
Published: 25 November 2022

1. Introduction

Novel synthetic opioids (NSOs) are a subclass of novel psychoactive substances (NPSs) that poses serious risk to the public health in many countries worldwide [1]. NSO class is subdivided into fentanyl- and non-fentanyl- (e.g., U-44700, MT-45) analogs subclasses [2]. The European Monitoring Centre for Drugs and Drug Addiction (EMCDDA) recorded 63 new NSOs and an increased number of NSO-related intoxications (both nonfatal and fatal) in the United States, Europe, Australia, Japan, and Canada between 2009 and 2020, especially in the past few years [2,3].

One of the most recent NSO entries on the NPS market is brorphine, which has been involved in more than 100 deaths in a few months [4]. Brorphine (1-[1-[1-(4-bromophenyl)ethyl]-piperidin-4-yl]-1,3-dihydro-2H-benzo[d]imidazol-2-one) is a substituted, piperidine- benzimidazolone analogue that retains structural similarities (Figure 1) to fentanyl [5]. Brorphine binds to the μ-opioid receptor as an agonist at MOR, being more potent than morphine and less potent than fentanyl [6]. It was firstly synthesized in 1967 by Janssen [7], and its physicochemical properties have not been fully defined yet.

Brorphine was identified on the United States' NPS market in mid-2019 for the first time; later, in June 2020, its identification in Sweden was reported to the EMCDDA [8].

Occasionally, it was found under the street name, "purple heroin" [9]. Moreover, oxycodone tablets adulterated with brorphine, probably purchased via the internet, were implicated in fatal and non-fatal intoxications [10]. The first reported quantification of brorphine in the serum of an intoxicated individual from Belgium was performed by liquid chromatography–high resolution mass spectrometry (LC-HRMS) in February 2020 [11]. Furthermore, a considerable number of intoxications involving brorphine (approximately 20) were reported in mid-2020, and various metabolism studies (in vivo and in vitro) were performed on biological specimens obtained from these cases using LC-MS/MS [4,9]. Currently, brorphine falls under national legislation targeting fentanyl analogues [8,12], having been placed temporarily in Schedule I of the Controlled Substances Act [13,14].

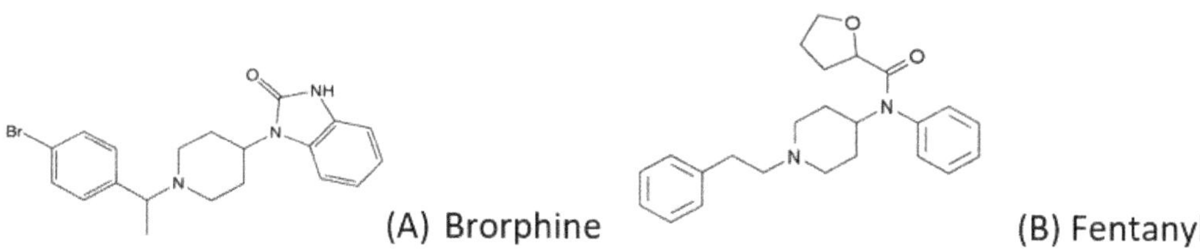

Figure 1. Chemical structure of (**A**) brorphine (MW 436.8, pKa 6.61), and (**B**) fentanyl (for comparison).

A wide array of techniques (including GC-MS, liquid chromatography engaged either to DAD or low- and high-resolution mass spectrometers (LC-DAD, LC-MS/MS, LC-HRMS), spectroscopic methods (FT-IR and ^{1}H- and ^{13}C-NM)) has been applied for the determination and identification of brorphine in powder-exhibits [4,8,11]. Also, analytical procedures involving LC-MS/MS have been reported for the quantification of brorphine in conventional biological specimens (blood/serum and urine) [11]. The application of liquid chromatography quadrupole time-of-flight mass spectrometry (LC-QToF-MS) has been used to determine brorphine's metabolites in blood and urine [5]. However, to the best of our knowledge, an up-to-date quantitation of brorphine in oral fluid (OF) has not been reported.

OF is a conventional biological specimen, secreted by salivary glands and composed mainly of water (99.5%), mucus, cells, (such as white blood cells and epithelial cells), molecules, (such as electrolytes, enzymes, and antimicrobial agents), and lysozymes [15]. OF is produced in the salivary glands after "filtering" of blood; therefore, drug levels in it are considered to correspond to free-drug-plasma concentrations [15]. Additionally, OF is slightly more acidic than blood (pH 5.8–6.8), and this results in the ionization of weak basic drugs and higher drug concentrations in OF than in plasma [15]. The advantages of this matrix over conventional biological fluids have established OF as the most suitable alternative matrix to assess recent exposure to psychoactive drugs [15]. Moreover, OF is one of the main biological fluids analyzed for the assessment of different classes of NPS, taking advantage of the progress in extraction and analysis procedures [16–18].

Fabric phase sorptive extraction (FPSE) is a relatively new sample preparation technique that has the potential of wide applications in bioanalysis [19]. The fabric in this technique consists of a fabric substrate, natural or synthetic, that is chemically modified so that an ultra-thin coating, with a hybrid sol–gel and organic–inorganic sorbent, could be formed [20]. The main determinants of the extraction selectivity are the polarity of the polymer, and the hydrophilicity or hydrophobicity of the media [21]. Depending on the physicochemical features of the sol–gel, this microextraction approach can extract analytes with a wide polarity range [22]. FPSE membrane allows direct analytes extraction from the biological substrate, minimizing sample pre-treatment steps. Also, the analytes are desorpted rapidly into organic solvents, from the FPSE media. The aforementioned characteristics of FPSE have demonstrated its leading position as a green-sample-preparation technology of the 21st century. Previous application of FPSE in the analysis of OF includes the extraction of non-steroidal, anti-inflammatory drugs with the aid of liquid chromatography [23].

The reported implication of brorphine in emergent intoxication cases and the possibility of its widespread use stress the importance of being able to detect this potent NPS in clinical intoxication cases and alternative specimens. The purpose of this contribution is to develop and validate a method based on FPSE and tandem LC-MS/MS, for the identification and quantification of brorphine in OF; the reported procedure has the potential for application of FPSE in the analysis of more NPS in OF.

2. Materials and Methods

2.1. Chemicals and Reagents

All solvents and reagents used for LC-MS analyses were at least of high-performance liquid chromatography (HPLC) grade. Brorphine HCl (≥98% purity) was obtained from Cayman Chemical Company (Ann Arbor, MI, USA). Ammonium acetate (99%) was purchased from Fluka™ Analytical Standards (Steinheim, Germany). Dichloromethane was purchased from Fisher Scientific (part of ThermoFisher Scientific) (Waltham, MA, USA). Acetonitrile, methanol, water (all UHPLC-MS grade), and formic acid (99%) were obtained from CARLO ERBA Reagents GmbH (Cornaredo, Italy). Ultra-pure water was supplied by an Aquatron Water Still A4000D purification system from Bibby Sterilin Limited (Staffordshire, UK). Whatman microfiber glass filters (110 mm) and Whatman cellulose filter papers (125 mm), which were used for FPSE, were purchased from General Electric (Boston, MA, USA). The organic polymer polyethylene glycol (PEG 300) was from Sigma-Aldrich (Athens, Greece). Trimethoxymethylsilane (MTMS), trifluoroacetic acid (TFA), acetone, sodium hydroxide, and hydrochloric acid were supplied by Merck (Darmstadt, Germany).

2.2. Working/Standard Solutions and Calibrators

Stock solution of 1 mg of brorphine dissolved in 1 mL of methanol was stored at −20 °C in the dark until analysis. Standard working solutions were prepared daily in methanol. Calibrators at seven concentration levels, 0.05, 0.25, 1.0, 5.0, 10.0, 25.0, 50.0 ng/mL, were prepared by spiking a pool of blank OF. Quality control samples were prepared at 1, 5, and 50 ng/mL in methanol by fortifying blank OF.

2.3. Pre-Treatment of Fabric for Sol-Gel Coatings

A Whatman Cellulose circular filter of 125-mm diameter and Whatman Microfiber Glass filter of 110 mm were put in a vial with deionized water and were soaked under constant sonication for 15 min. Then, the procedure was repeated with NaOH (1.0 M, 1 h) and HCl (0.1 M, 1 h). Washing with deionized water was followed in each step. Afterwards, the filter was left to dry at 25 °C for 24 h, in ambient air, and then followed by sol-gel sorbent coating. In our study, 5 gr of PEG 300 solution was developed, as the sol–gel precursor, using 5 mL of MTMS, 2 mL of TFA catalyst with 5% water, and a mixture of 10 mL of acetone and dichloromethane (50/50 v/v). The substrate was immersed into the sol solution for 4 h in room temperature. After sol-gel coating, the fabric was left to dry in the residual sorbents for 24 h; then, it was rinsed with a mixture of acetone and dichloromethane (50/50 v/v) under sonication for 30 min, air dried for 30 min, cut into circular pieces with a 1 cm diameter, and stored until use.

2.4. Preliminary Experiments

Two types of Whatman filters were examined in the preliminary experiments as an FPSE substrate: microfiber glass filter (FG) and cellulose filter (WC). The extraction times of 10, 20, and 30 min were tested. PH values were investigated at three different levels: acidic (4.5), neutral (7), and alkaline (9). For the most suitable back extraction, solvent methanol, acetonitrile, and their mixture of 50:50 v/v were tested. Finally, three different time intervals (5, 10, and 15 min) were examined in order to select the optimal back-extraction time.

2.5. Sample Collection and Storage

OF was collected from drug-free donors, by a simple expectoration technique, which allowed it to accumulate in the lower part of the mouth while the subject spat into a pre-weighed test tube every 60 s. Then, the collected OF was vortexed for 30 s and put in an ultrasound bath for 60 s. Finally, aliquots of 10 mL of OF were stored in plastic tubes, frozen until analysis.

2.6. FPSE Procedure

One mL of OF was treated with 100 μL of acetonitrile, then repeating vortex for 30 s and ultrasound bath for 60 s, and, finally, centrifugation at 15,000 rpm for 15 min, were applied. The supernatant was put in a tube with a glass microfiber filter (FG) coated with PEG300, and left for 30 min to interact by stirring at 300 rpm, pH = 7. Finally, 150 μL of methanol were added to achieve desorption under gentle stirring at 300 rpm for 10 min.

2.7. LC Conditions

Analysis was performed on a Dionex UHPLC system (Thermo Scientific, Waltham, MA, USA) comprised of a degasser, a binary pump, an autosampler, and a column oven. The system was coupled to a Q-Trap 5500™ mass spectrometer (Sciex, Darmstadt, Germany), operated in multiple reaction monitoring (MRM) mode, and equipped with an electrospray ionization (ESI) Turbo V Source operated in positive mode. Separation was performed on an Accupore C18 column (50 mm × 3 mm, 2.6 μm particle size) equipped with a precolumn cartridge (2.1 mm × 0.2 μm) (Thermo Scientific, Waltham, MA, USA), both operated at 30 °C. Mobile phases were: 10 mM of aqueous ammonium acetate adjusted to pH 3.5 with 0.1% formic acid (eluent A) and acetonitrile UHPLC-MS grade with 0.1% formic acid (eluent B), degassed by Elmasonic S ultrasonic, Germany. The autosampler temperature was 5 °C; the injector's needle was rinsed with 200 μL of methanol before and after each injection, and the injected sample volume was 5 μL. Gradient program was as follows: 0.00–1.00 min 12% eluent B; 1.00–2.00 min linear gradient to 50% eluent B, 2.00–3.00 from 50% to 100% eluent B; 3.00–5.00 min remained at 100%; 5.50 min returned to initial conditions; total run time of 6.50 min with a flow rate of 0.500 mL/min.

2.8. MS/MS Conditions

The mass spectrometer was equipped with an electrospray ionization source (ESI) operating in positive mode. The applied ESI inlet conditions were as follows: gas 1, nitrogen (55 psi); gas 2, nitrogen (55 psi); ion-spray voltage of 5500 V, positive mode; ion-source temperature, of 550 °C; nitrogen as the curtain gas at 55 psi (Table 1). Optimization of the dwell times and all other settings were performed using the scheduled MRM algorithm incorporated into Sciex Analyst® software version 1.7.1 (Sciex, Darmstadt, Germany), in the automatic quantitative optimization mode. SciexOS 1.6 software 1 (Sciex, Darmstadt, Germany) was applied for data processing.

Table 1. MS parameters for the identification and quantitation of brorphine. The ion in bold (218.2) was used for quantification.

	Q1 Mass (Da)	Q3 Mass (Da)	Dwell Time (ms)	ID	DP (Volts)	EP (Volts)	CE (Volts)	CXP (Volts)
1	399.9	218.2	400	Brorphine 1	106	10	29	20
2	399.9	182.9	400	Brorphine 2	106	10	33	16
3	399.9	104.06	400	Brorphine 3	106	10	59	12

2.9. Method Validation

The guidelines expressed by the Scientific Working Group for Forensic Toxicology (SWGTOX) were applied for validation of the analytical procedure [24]. Evaluated parameters were linearity, selectivity, bias, carryover, precision, sensitivity, recovery, matrix effect, and stability.

Selectivity experiments were carried out with blank OF collected from at least ten different drug-free volunteers. The samples were tested to exclude any interference from the matrices and were then pooled. Three blank OF samples were prepared by fortifying with a mixture of 87 drugs (50 ng/mL of each drug) to study selectivity of the method. The evaluation of carryover was performed by analyzing blank matrix samples immediately after running triplicates of high-concentration-control samples (400 ng/mL). The method's linearity was evaluated using triplicates of one of seven concentration levels, 0.05, 0.25, 1.0, 5.0, 10.0, 25.0, and 50.0 ng/mL of brorphine spiked in OF. Sensitivity was expressed by the limit of detection (LOD) and limit of quantification (LOQ), defined as the lowest concentration giving a signal-to-noise ratio of three (S/N = 3) and ten (S/N = 10), respectively. Criteria for identification were a symmetrical peak that eluted at the expected retention time ±2%, and an ion ratio of quantifier-to-qualifier MRM within ±25% of the established standard compound. Both LOD and LOQ were determined by analyzing three fortified OF samples at decreasing concentrations (1–0.001 ng/mL). Intra-day repeatability and inter-day reproducibility (precision), expressed as RSD %, were studied at three concentration levels (1, 5, and 50 ng/mL). Bias was measured in fortified OF samples at three different concentration pools (1, 5, and 50 ng/mL) using a minimum of three separate samples per concentration over five different runs; the maximum value of acceptable bias was ±20% at each concentration. Matrix effect (%) was evaluated in low and high concentrations of analyte, of 1 and 50 ng/mL, respectively, and positive results expressed ionization enhancement and negative results ionization suppression. Stability experiments were performed by using fortified OF samples (1 ng/mL and 50 ng/mL) to establish time-zero responses. Then, half vials were refrigerated at −20 °C, and the rest were stored on the autosampler at 5 °C. All the vials were then analyzed on LC-MS/MS in triplicate, after two and four weeks, and the average responses were compared to the time-zero ones.

2.10. Statistical Analysis

Statistical analysis of the results was carried out using Microsoft EXCEL for Windows to calculate means and standard deviations, and to conduct Student's *t*-test to compare differences between groups (significant at $p < 0.05$).

3. Results

3.1. Method Development

The applied chromatographic conditions provided a chromatogram of brorphine with a good resolution. A single peak with a retention time of 2.88 min (Figure 2a) and an exact m/z value of 399.9 were obtained; the acquired ion spectrum of brorphine is shown in Figure 2b.

3.2. Optimization of the FPSE Procedure for Sample Preparation

The greatest desorption yields were obtained when glass microfiber filter (FG), coated with PEG300, was used; therefore, this combination was chosen for further optimization (Figure 3a). PEG 300 is widely used and was chosen due to its water solubility and hydrophilic properties. The optimum extraction time of 30 min was found to be the most effective (Figure 3b). The extraction time was linked directly to distribution parameters that determined the interaction between the FPSE material and brorphine, indicating a rapid distribution in our case. Neutral pH was shown to have a milder effect on the sample dilution and, consequently, the concentration variation of the analyte (Figure 3c). Methanol was far more efficient than acetonitrile and their mixture (Figure 3d), and the back-extraction-time of 10 min was selected (Figure 3e). Generally, the differences obtained in parameters when applying the different conditions were not statistically significant ($p > 0.05$), and the best performing parameter was chosen to be applied.

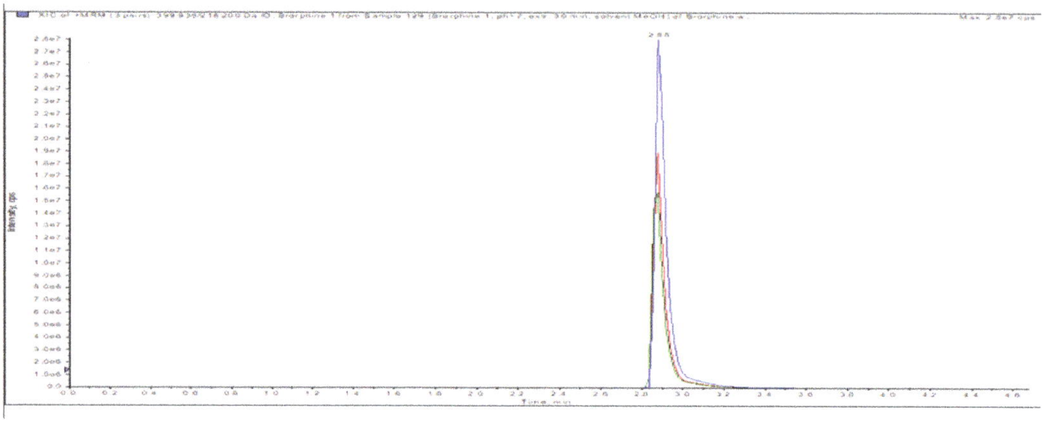

(a)

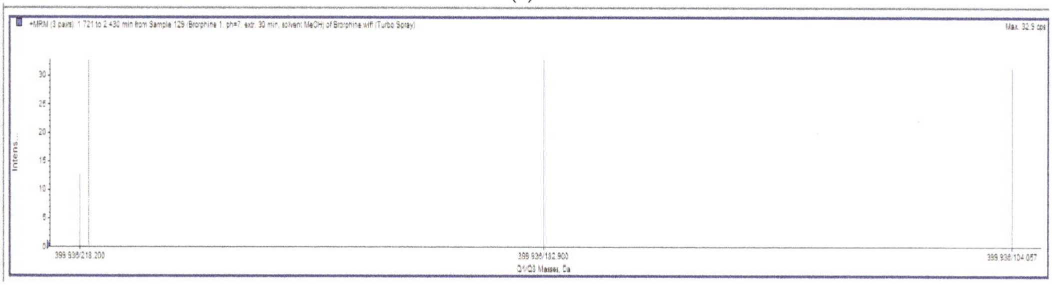

(b)

Figure 2. (**a**) Total ion chromatogram (Full scan) of brorphine. (**b**) Ion spectra of brorphine.

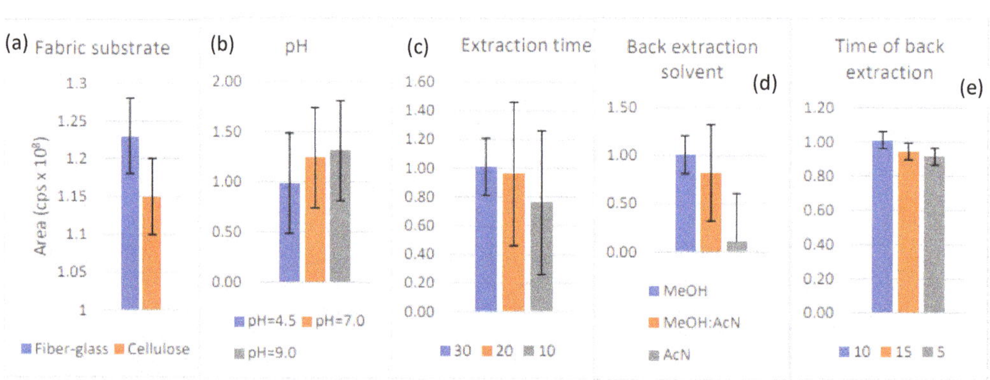

Figure 3. (**a**) Influence of coating material on brorphine's yield; (**b**) Influence of pH on brorphine's yield; (**c**) Influence of extraction on brorphine's yield; (**d**) Influence of solvent on brorphine's yield; (**e**) Influence of back-extraction time on brorphine's yield.

3.3. *Method Validation*

The developed method displayed linearity at the specified range between LOQ = 0.05 and 50 ng/mL (R^2 = 0.9993). The bias of analyte ranged between 12.0 and 16.8%. Relative standard deviations (RSDs %) for intra-day and inter-day precision were found between 6.4 and 9.9% (Table 2). Accuracy (expressed as recovery %) was in the range of 65–75%. Brorphine's limit of detection (LOD) and limit of quantification (LOQ) were 0.015 ng/mL and 0.05 ng/mL, respectively. Post-preparative stability of the analyte, refrigerated at

−20 °C after two weeks, was 98% and, after four weeks, was 97%. When the analyte was stored at 5 °C after two weeks, its post-preparative stability was 97% and, after four weeks, was 95%. In the selectivity experiments, no matrix interferences and no carryover between analyses during the LC-MS/MS runs were observed.

Table 2. Validation parameters of the method for the identification and quantitation of brorphine in OF.

Analyte	LOD (ng/mL)	LOQ (ng/mL)	Matrix Effect		Extraction Efficiency		Intra-Day Precision		Inter-Day Precision	
			Low (%)	High (%)	Low (%)	High (%)	Low (%)	High (%)	Low (%)	High (%)
Brorphine	0.015	0.05	−20	−23	65	75	6.4	7.6	9.9	8.1

4. Discussion

Although we are not aware of any intoxication case or police case involving brorphine in our country yet, the fact that this substance was identified in some European countries [8–11] indicates that it is possible for it to be spread widely to other countries. Therefore, the development of an analytical methodology to identify and quantify this NSO in clinical cases is essential. Moreover, the application of the FPSE sample preparation technique is employed for the first time to the analysis of an NPS, offering a new tool that could advance forensic research.

The current contribution presents the development and validation of a simple, rapid, and sensitive method for the determination of brorphine in OF, by employing FPSE for the extraction of an analyte, engaged to LC-MS/MS for identification and quantification.

In a series of preliminary experiments, the extraction procedure was optimized regarding coating material, extraction time, pH, and extraction solvent. Methanol was proved to be an efficient back-extraction solvent at the extraction time of 10 min. Adsorption time affects the distribution of analytes between the solvent and the FPSE medium. The required time to reach extraction equilibrium is synergistically affected and reduced by factors such as the porous network of the sol-gel coating, the highly active surface of the FPSE apparatus, and the permeable fabric substrate.

According to literature, brorphine acts as a weak acid [24], so it is expected that the ionized form pKa + 2 (8.61) would provide the ideal outcome. However, it was observed that results with pH = 7 and pH = 9 were comparable to each other. The low viscosity of the solvents allows them to easily penetrate the thin layer of the adsorbent, and decompose quickly and quantitatively brorphine. Our results revealed that a shorter elution time did not demonstrate decent desorption capacity, while increasing the elution time slightly decreased the efficiency, which may be due to the reabsorption of the substances by the FPSE medium.

The developed method shows acceptable analytical characteristics, according to the SWGTOX guidelines [25]. The linearity range in the OF is expected to be related to the concentrations of plasma analyte, a consideration that is confirmed by previous reported results [11]. The external-standard-absolute calibration with certified-reference material is used for the validation experiments of this relatively simple extraction technique [26]. The obtained recoveries, although not excellent (>95%), are considered relatively good and acceptable; they are comparable to relevant results from other studies on brorphine [5,8,11], or, on NPS in general [27]. Matrix effect is acceptable, and is recorded as ion suppression for both low and high concentrations. It is worth mentioning that the limits of detection (LOD) and quantification (LOQ) for brorphine for OF reported in this study are lower than those reported previously for serum and urine (0.05/0.03 ng/mL as the LODs and 0.1 ng/mL as the LOQs for serum and urine, respectively) [5]. The short- and long-term stability of brorphine in OF is acceptable at both −20 °C and at 5 °C.

This is the first report of an analytical procedure for the determination of brorphine in OF; the reported analytical procedure introduces the application of an innovative microextraction technique of FPSE engaged to LC-MS/MS for quantification of an NPS. The characteristics of the FPSE technique show potential for application in the analysis of a wider scale of NPS classes. This relatively simple methodology is effective and reliable for identification and quantitation of brorphine in OF in routine clinical casework.

5. Conclusions

The presented method is validated according to international guidelines to identify and quantify brorphine in OF. The application of FPSE for sample preparation engaged to LC-MS/MS resulted in rapid and easy sample preparation, and high sensitivity and selectivity of analysis.

Author Contributions: Conceptualization, V.S. and V.B.; methodology, D.F., T.V. and V.B.; validation, D.F., T.V., V.S. and V.B.; investigation, D.F.; data curation, D.F.; writing—original draft preparation, D.F.; writing—review and editing, V.B.; supervision, V.S. and V.B.; project administration, V.B. All authors have read and agreed to the published version of the manuscript.

Funding: This research received no external funding.

Institutional Review Board Statement: The study was conducted in accordance with the Declaration of Helsinki, and approved by the Institutional Review Board of University of Ioannina/University Hospital of Ioannina (834/18-10-2022).

Informed Consent Statement: Informed consent was obtained from all subjects involved in the study.

Conflicts of Interest: The authors declare no conflict of interest.

References

1. UNODC. Booklet 2: Global Overview of Drug Demand and Drug Supply. In *World Drug Report 2021*; United Nations: Vienna, Austria, 2021; Available online: https://www.unodc.org/res/wdr2021/field/WDR21_Booklet_2.pdf (accessed on 29 June 2021).
2. EMCDDA. *European Drug Report 2020: Trends and Developments*; Publications Office of the European Union: Luxembourg, 2020; Available online: https://www.emcdda.europa.eu/system/files/publications/13236/TDAT20001ENN_web.pdf (accessed on 29 June 2021).
3. EMCDDA. *European Drug Report 2021: Trends and Developments*; Publications Office of the European Union: Luxembourg, 2021; Available online: https://www.emcdda.europa.eu/system/files/publications/13838/TDAT21001ENN.pdf (accessed on 29 June 2021).
4. Krotulski, A.J.; Papsun, D.M.; Noble, C.; Kacinko, S.L.; Logan, B.K. Brorphine—Investigation and quantitation of a new potent synthetic opioid in forensic toxicology casework using liquid chromatography-mass spectrometry. *J. Forensic Sci.* **2021**, *66*, 664–676. [CrossRef]
5. Grafinger, K.E.; Wilde, M.; Otte, L.; Auwärter, V. Pharmacological and metabolic characterization of the novel synthetic opioid brorphine and its detection in routine casework. *Forensic Sci. Int.* **2021**, *327*, 110989. [CrossRef] [PubMed]
6. Kennedy, N.M.; Schmid, C.L.; Ross, N.C.; Lovell, K.M.; Yue, Z.; Chen, Y.T.; Cameron, M.D.; Bohn, L.M.; Bannister, T.D. Optimization of a series of Mu opioid receptor (MOR) agonists with high G protein signaling bias. *J. Med. Chem.* **2018**, *6*, 8895–8907. [CrossRef] [PubMed]
7. Janssen, P.A.J. Derivatives of Benzimidazolinyl Piperidine. U.S. Patent 3,318,900, 9 May 1967.
8. Vandeputte, M.M.; Krotulski, A.J.; Papsun, D.M.; Logan, B.K.; Stove, C.P. The Rise and Fall of Isotonitazene and Brorphine: Two Recent Stars in the Synthetic Opioid Firmament. *J. Anal. Toxicol.* **2021**, *46*, 115–121. [CrossRef] [PubMed]
9. Vohra, V.; King, A.M.; Jacobs, E.; Aaron, C. Death associated with brorphine, an emerging novel synthetic opioid. *Clin. Toxicol.* **2021**, *59*, 851–852. [CrossRef] [PubMed]
10. Razinger, G.; Grenc, D.; Pezdir, T.; Kranvogl, R.; Brvar, M. Severe rhabdomyolysis and acute kidney failure due to synthetic opioid brorphine exposure in combination with chronic sertraline therapy. *Eur. J. Clin. Pharmacol.* **2021**, *77*, 1759–1761. [CrossRef] [PubMed]
11. Verougstraete, N.; Vandeputte, M.M.; Lyphout, C.; Cannaert, A.; Hulpia, F.; Van Calenbergh, S.; Verstraete, A.G.; Stove, C. First Report on Brorphine: The Next Opioid on the Deadly New Psychoactive Substance Horizon? *J. Anal. Toxicol.* **2021**, *44*, 937–946. [CrossRef] [PubMed]
12. Bao, Y.; Meng, S.; Shi, J.; Lu, L. Control of fentanyl-related substances in China. *Lancet Psychiatry* **2019**, *6*, e15. [CrossRef] [PubMed]
13. DEA. CFR Part 1308. In *Schedules of Controlled Substances: Temporary Placement of Brorphine in Schedule I.*; DEA: Springfield, VA, USA, 2021; pp. 78047–78050. Volume 85. Available online: https://www.deadiversion.usdoj.gov/fed_regs/rules/2020/fr1203_2.htm (accessed on 29 June 2021).

14. DEA. CFR Part 1308. In *Schedules of Controlled Substances: Temporary Placement of Brorphine in Schedule I.*; DEA: Springfield, VA, USA, 2021; pp. 11862–11867. Volume 86. Available online: https://deadiversion.usdoj.gov/fed_regs/rules/2021/fr0301_4.htm (accessed on 29 June 2021).
15. de Campos, E.G.; da Costa, B.R.B.; dos Santos, F.S.; Monedeiro, F.; Alves, M.N.R.; Santos Junior, W.J.R.; De Martinis, B.S. Alternative matrices in forensic toxicology: A critical review. *Forensic Toxicol.* **2022**, *40*, 1–18. [CrossRef]
16. Sorribes-Soriano, A.; Valencia, A.; Esteve-Turrillas, F.A.; Armenta, S.; Herrero-Martínez, J.M. Development of pipette tip based poly(methacrylic acid-co-ethylene glycol dimethacrylate) monolith for the extraction of drugs of abuse from oral fluid samples. *Talanta* **2019**, *205*, 120158. [CrossRef] [PubMed]
17. Rocchi, R.; Simeoni, M.C.; Montesano, C.; Vannutelli, G.; Curini, R.; Sergi, M.; Compagnone, D. Analysis of new psychoactive substances in oral fluids by means of microextraction by packed sorbent followed by ultra-high-performance liquid chromatography-tandem mass spectrometry. *Drug Test Anal.* **2018**, *10*, 865–873. [CrossRef] [PubMed]
18. Bianchi, F.; Agazzi, S.; Riboni, N.; Erdal, N.; Hakkarainen, M.; Ilag, L.L.; Anzillotti, L.; Andreoli, R.; Marezza, F.; Moroni, F.; et al. Novel sample-substrates for the determination of new psychoactive substances in oral fluid by desorption electrospray ionization-high resolution mass spectrometry. *Talanta* **2019**, *202*, 136–144. [CrossRef] [PubMed]
19. Mohamed, H.M. Solventless Microextraction Techniques for Pharmaceutical Analysis: The Greener Solution. *Front. Chem.* **2022**, *9*, 785830. [CrossRef] [PubMed]
20. Kabir, A.; Furton, K.G.; Malik, A. Innovations in sol-gel microextraction phases for solvent-free sample preparation in analytical chemistry. *Trends Anal. Chem.* **2013**, *45*, 197–218. [CrossRef]
21. Jiménez-Holgado, C.; Chrimatopoulos, C.; Stathopoulos, V.; Sakkas, V. Investigating the Utility of Fabric Phase Sorptive Extraction and HPLC-UV-Vis/DAD to Determine Antidepressant Drugs in Environmental Aqueous Samples. *Separations* **2020**, *7*, 39. [CrossRef]
22. Aznar, M.; Alfaro, P.; Nerin, C.; Kabir, A.; Furton, K.G. Fabric phase sorptive extraction: An innovative sample preparation approach applied to the analysis of specific migration from food packaging. *Anal. Chim. Acta* **2016**, *936*, 97–107. [CrossRef] [PubMed]
23. Tartaglia, A.; Kabir, A.; D'Ambrosio, F.; Ramundo, P.; Ulusoy, S.; Ulusoy, H.I.; Merone, G.M.; Savini, F.; D'Ovidio, C.; De Grazia, U.; et al. Fast off-line FPSE-HPLC-PDA determination of six NSAIDs in saliva samples. *J. Chromatogr. B* **2020**, *1144*, 122082. [CrossRef] [PubMed]
24. Scientific Working Group for Forensic Toxicology (SWGTOX). Standard Practices for Method Validation in Forensic Toxicology. *J. Anal. Toxicol.* **2013**, *37*, 452–474. [CrossRef] [PubMed]
25. Center for Drug Evaluation and Research (CDER). Bioanalytical Method Validation-Guidance for Industry, Food and Drug Administration. 2018. Available online: https://www.fda.gov/downloads/drugs/guidances/ucm070107.Pdf (accessed on 29 June 2021).
26. Imre, S.; Tero-Vescan, A.; Dogaru, M.T.; Kelemen, L.; Muntean, D.-L.; Curticapean, A.; Szegedi, N.; Vari, C.-E. With or Without Internal Standard in HPLC Bioanalysis. A Case Study. *J. Chromatogr. Sci.* **2019**, *57*, 243–248. [CrossRef] [PubMed]
27. Boumba, V.A.; Di Rago, M.; Peka, M.; Drummer, O.H.; Gerostamoulos, D. The analysis of 132 novel psychoactive substances in human hair using a single step extraction by tandem LC/MS. *Forensic Sci. Int.* **2017**, *279*, 192–202. [CrossRef] [PubMed]

International Journal of
Environmental Research and Public Health

MDPI

Article

The Impact Evaluation of Acid Mine Drainage on Zebrafish (*Danio rerio*) and Water Fleas (*Daphnia magna*) in the Vicinity of the Geum River Basin in Korea

Hyojik Yoon [1,*] and Jonghyun Yoon [2]

1 Institute of Natural and Science, College of Science and Technology, Korea University, Sejong 30019, Republic of Korea
2 National Institute of Environmental Research, Incheon 22689, Republic of Korea
* Correspondence: hyojik88@korea.ac.kr

Abstract: Heavy metals, such as copper, lead, and cadmium, carried by acid mine drainage are pollutants of the aquatic ecosystem, posing a significant health risk to the water resource for humans. Environmental technologies to reduce metal contamination are applied for post-mining prevention and improvement. Despite detailed pollution management, water contaminated by heavy metals still flows into the natural water system. This study investigated the impact of drainage discharged from abandoned mines near the major river in South Korea on aquatic organisms. The toxicity of the field water showed a more significant effect than observed through the experiment for each heavy-metal concentration. Various toxic substances coexisted in the field water around the mine, such that the overall toxic intensity was high even when the concentration of each heavy metal was low. As a result, the inhibition of activity of aquatic organisms was observed at low individual concentrations, and further investigation on the effect of long-term exposure to trace amounts of heavy metals is required.

Keywords: acid mine drainage; bioassay; cognitive function; locomotion; zebrafish

Citation: Yoon, H.; Yoon, J. The Impact Evaluation of Acid Mine Drainage on Zebrafish (*Danio rerio*) and Water Fleas (*Daphnia magna*) in the Vicinity of the Geum River Basin in Korea. *Int. J. Environ. Res. Public Health* **2022**, *19*, 16470. https://doi.org/10.3390/ijerph192416470

Academic Editors: Roel Vermeulen, Lauren Petrick and Maaike van Gerwen

Received: 10 November 2022
Accepted: 6 December 2022
Published: 8 December 2022

1. Introduction

Metal mining produces heavy metals in aquatic ecosystems [1]. After mining, metal leak prevention management should be performed as part of environmental protection [2] to protect surface water and groundwater systems. Despite this effort, acid mine drainage (AMD) can flow into rivers, adversely affecting the entire aquatic ecosystem and, ultimately, humans through biological accumulation [3,4]. Pollution from abandoned mines is determined by various physical, chemical, and geographic factors specific to each site, as well as local meteorological (rainfall) factors. The heavy metals in mine drainage are deposited as sediments or absorbed in the receiving river sediments and are concentrated at extremely high levels [5]. Several studies have observed that as rainfall intensity increases, heavy-metal concentrations naturally increase in the receiving bodies of water [6–8]. Heavy metals carried by AMD are potentially lethal to humans and other organisms, as they are dispersed in the environment through transport media, such as water, soil, and the atmosphere [9–11]. Therefore, the removal and management of AMD sources are necessary to develop countermeasures against the contamination of leachate-containing heavy metals [12]. However, effective post-management programs for these abandoned mines require a clear understanding of the behavior of pollutants based on various factors and approaches in terms of pollutant loads.

Moreover, the current ecological assessment methods are not sufficiently sensitive, particularly for trace concentrations of pollutants. Previous research provided a procedure to estimate the overall health of the local water system by analyzing the conditions and diversity of the collected species [12]. Recently, metabolomics has become an essential tool for understanding the mechanisms of toxicity because it can provide valuable insights into changes in the health of living organisms upon exposure to stimuli with greater sensitivity [13,14]. This study aimed to quantify the heavy-metal pollution load in the bodies of water near three abandoned mines and to assess the toxicity of these heavy metals through exposure tests using zebrafish (*Danio rerio*) and planktonic crustacean (*Daphnia magna*). Zebrafish (*Danio rerio*) are easy to handle embryologically and genetically due to the rapid growth characteristics of the embryo. In addition, they offer economic advantages over mammalian species that are primarily used as genetic models in medicine, biotechnology, and pharmacology. Therefore, zebrafish are used as a vertebrate animal model in various research fields; in particular, they are used for anatomical, physiological, genetic, developmental, and behavioral studies in visual neuroscience [15]. Because the zebrafish egg is transparent and some zebrafish have been genetically manipulated to have a fluorescent organ, they are easily accessible during all stages of development, from embryo to adult [16].

Toxicity was quantified based on the effect of exposure to AMD on the behavioral and cognitive abilities of zebrafish larvae.

2. Materials and Methods

2.1. Site Description and Water Sampling

The Geum River, one of the major rivers in the Republic of Korea, was chosen as the study area. Mines in the Geum River Basin that could affect the water source conservation area (Figure 1 and Table 1) were classified. The G1 site was a gold and silver mining area. The S-site mine contained galena, sphalerite, quartz, and calcite. The G2 mine primarily mined copper, zinc, and lead. Eight water samplings were carried out, five times during the dry period and three times during rainfall. Based on location data, the flow rate and pollutants flowing into the nearby river from the three surveyed mines were investigated. The runoff distances from each mining area to the nearest river were 495 m in G1, 1460 m in S, and 380 m in G2, respectively. The water was sampled from the middle of the stream. In each sampling, the grab-sampling method was used. The water samples were collected in the period from March 2016 to September 2016. The water quality indicators (DO, pH, EC, TDS, and ORP) were measured and recorded onsite. The water (1000 mL) was filled in a sterilized sample bottle and spiked with 2 mL nitric acid and sealed after removing the air. The concentrations of the metals were determined using inductively coupled plasma mass spectrometry (ICP-MS, II Perkin ELAN-DRC, Waltham, MA, USA).

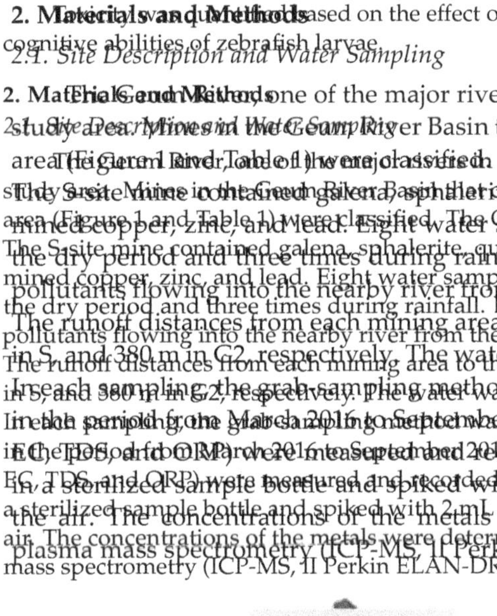

Figure 1. Location of the research area.

Table 1. The latitude and longitude of research areas (Sampling point).

Mining Site	Latitude	Longitude
G1	35°54′22.7″ N	127°44′45.4″ E
G2	36°20′08.6″ N	127°43′52.9″ E
S	35°57′00.3″ N	127°40′15.1″ E

2.2. Zebrafish

Zebrafish (*Danio rerio*) are used in toxicity evaluation because they have a preferred wavelength. Zebrafish hatch within 2 to 3 days after spawning, and 5 days after birth, the visual organ is developed such that the wavelength of color can be distinguished.

Zebrafish that have begun to distinguish colors have a characteristic of preferring a short-wavelength region (blue) rather than a long-wavelength region (yellow), which has been proven experimentally by previous work [17].

2.3. Cognitive Function and Locomotion Change Test

The experimental concentrations were selected based on a river water quality survey. The highest concentration of each metal detected was set as the exposure concentration (Copper: 1000 μg/L; Pb: 100 μg/L; Cd: 1000 μg/L; As: 50 μg/L; Zn: 100,000 μg/L).

The tests were performed with copper, lead, cadmium, arsenic, and zinc, respectively.

Zebrafish can detect visible light domains similar to those of humans. Zebrafish have specific preferences and wavelengths. Using these zebrafish attributes, researchers measure and quantify behavioral changes by analyzing the number of zebrafish larvae that remain at the location of the preferred wavelength for half an hour.

A color maze system was installed to monitor the zebrafish to determine the manner in which behavioral changes are affected by exposure to small amounts of heavy metals.

Zebrafish preferred when the visible light zone was set to blue wavelengths and the secret visible light zone was set to yellow wavelengths (Figure 2). Ten zebrafish larvae were placed in each color maze kit (Genomic Design™, Daejeon, Republic of Korea); the temperature and light intensity were adjusted. Since the larvae for each heavy-metal experiment were obtained from a new embryo tank, a control experiment was required for each tank. As a control group, a clean culture solution was applied. The movement of larvae in a clean culture solution without pollutants was set as a control. A digital camcorder was used for 30 min to record the larvae after they were exposed to heavy metals. The Lolitrack 4.1 (Loligo® systems, Viborg, Denmark) was used to analyze locomotive changes. Additionally, the average velocity, acceleration, and active time were analyzed. The experiment was repeated three times with new larvae.

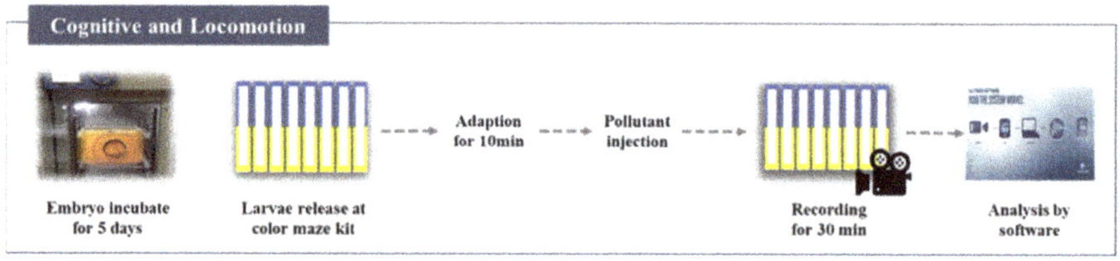

Figure 2. Zebrafish cognitive and locomotive change test model.

2.4. Daphnia magna Bioassay Test

To evaluate the toxicity of *Daphnia magna*, breeding conditions were established according to the standard toxicity test method [18]. The culture medium was aerated overnight after injecting KCl, $MgSO_4$, and $NaHCO_3$ into 18 L DI water. Subsequently, $CaSO_4 \cdot 2H_2O$ was added to 2 L of distilled water and dissolved. The two solutions were mixed and used as the culture medium. The culture conditions were a water temperature of 20 ± 2 °C, a light/dark cycle of 16 h:8 h, and a light intensity of 500–1000 lux. The culture medium was changed three times per week.

The population of *D. magna* increased after continuous subculture. Therefore, in this study, we performed a bioassay by setting the monitoring items for *D. magna*. The bioassay used as a risk assessment in the environment analyzes the effects of biological function reduction, swimming inhibition, and lethality of the target species when exposed to a specific environmental substance.

2.5. Statistical Analysis

All statistical analyses were conducted using SPSS (IBM SPSS 21) software (Chicago, IL, USA). A probability of $p < 0.05$ was accepted as statistically significant using a t-test. ANOVA analysis was applied to analyze the impact of differences between monitoring sites.

3. Results and Discussion

3.1. Analysis of Stream Water

The rivers surrounding the G1 site are contaminated with cadmium and zinc (Table 2). The pollution concentration primarily increases after rainfall (August and September), probably owing to mine runoff. Severe cadmium and zinc contamination was also discovered in river water near the G1 area. However, copper, arsenic, and lead did not exceed the water quality standards. Cadmium and zinc exceeded the standard for river water quality by 7–14% and 7%, respectively. Rivers near G2 were contaminated with cadmium, copper, and zinc, even during the non-rainy season. Copper and zinc exceeded the standard for water quality. In rivers near the S site, cadmium exceeded the water quality standard by 64–78%. The standards were consistently exceeded for lead and zinc, similar to cadmium. Copper, arsenic, mercury, and nickel concentrations either did not exceed water quality standards or were not detected. In the S region, the pollution range was widely distributed compared with the G1 and G2 sites. This is believed to be due to the topographical characteristics of the river flowing along a steep continuation.

Table 2. Analysis of contaminated water samples in each monitoring site.

		Concentration (N.D: Not Detected)										
		Cd (μg/L)	Cu (μg/L)	As (μg/L)	Hg (μg/L)	Pb (μg/L)	Ni (μg/L)	Zn (μg/L)	pH	EC (μS/cm)	DO (mg/L)	TDS (mg/L)
River water quality standards in Korea		5		50		50			6.5–8.5		7.5	
Dry Period	G1	6.42 ± 1.17	40.31 ± 28.56	12.82 ± 3.62	N.D	6.85 ± 2.05	N.D	711.00 ± 58.00	6.07 ± 0.79	302.14 ± 90.33	8.81 ± 0.93	207.99 ± 59.31
	G2	1253.6 ± 67.84	38,942 ± 5229.82	N.D	N.D	83 ± 25.25	N.D	127,994 ± 3533.29	7.31 ± 0.39	181.28 ± 3.94	6.29 ± 1.26	117.91 ± 2.52
	S	31.40 ± 6.50	10.8 ± 3.63	N.D	N.D	61.40 ± 40.78	N.D	4732 ± 1173.82	7.84 ± 0.08	227.2 ± 52.49	7.36 ± 1.14	115.27 ± 20.54
Rainfall period	G1	4.38 ± 0.31	74.20 ± 19.93	7.80 ± 3.97	N.D	39.40 ± 23.31	N.D	488.80 ± 52.38	7.26 ± 0.08	365.28 ± 21.98	9.39 ± 0.23	279.23 ± 7.69
	G2	6.94 ± 1.49	16.96 ± 8.51	14.82 ± 4.21	N.D	3.04 ± 0.91	N.D	614.70 ± 59.45	7.32 ± 0.41	180.18 ± 3.49	5.03 ± 0.90	149.60 ± 39.15
	S	33.20 ± 7.73	7.80 ± 4.62	N.D	N.D	136.00 ± 64.33	N.D	5402.20 ± 785.56	7.52 ± 0.41	242 ± 59.17	7.44 ± 0.89	174.48 ± 42.66

3.2. Behavioral Changes of Zebrafish

3.2.1. Inhibition of Zebrafish Preferred-Wavelength Discrimination by Exposure to Heavy Metals

The zebrafish larvae exposed to copper tended to remain in the non-preferred wavelength region (yellow circle) compared to the control at low concentrations (Figure 3a,b). However, when the wavelength position was switched after 30 min, the cognitive ability normalization rate was higher than that of the relatively high-concentration exposure group; therefore, the position ratio in the preferred wavelength region (blue circle) increased. Exposure to 1000 μg/L of copper caused cognitive impairment in the inability to distinguish preferred wavelengths (blue) over time. Lead exposure changed (Figure 3c,d) swimming behavior.

A relatively unstable swimming time, at the preferred wavelength, was observed at 10 μg/L and 100 μg/L compared to the control group. As the concentration increased to 100 μg/L, the swimming activity reduced, and fish trying to escape from the non-preferred wavelength region were not detected. After 30 min of observation, cognitive impairment was consistently confirmed at concentrations of 100 μg/L or higher, and the recovery of cognitive ability did not occur. Figure 3e,f present the experimental results for cadmium.

As the cadmium concentration increased, the positional trend standard deviation of zebrafish larvae changed significantly. It was observed that increased cadmium exposure over time resulted in an increased fluctuation range for the preferred wavelength of zebrafish, which affects the cognitive ability of larvae. Arsenic is a toxic heavy metal; it severely harms various organs, as well as the central nervous system. Significant side effects that inhibit neurotransmission in the brain occur; therefore, arsenic concentrations should be monitored continuously. Compared to other heavy metals, the standard deviation of the preferred wavelength (blue circle) was relatively large, even below 50 μg/L (Figure 3g,h). No changes in behavior were observed. It is believed that the tendency to move from the initial position did not occur as, based on arsenic exposure, swimming sharply decreased.

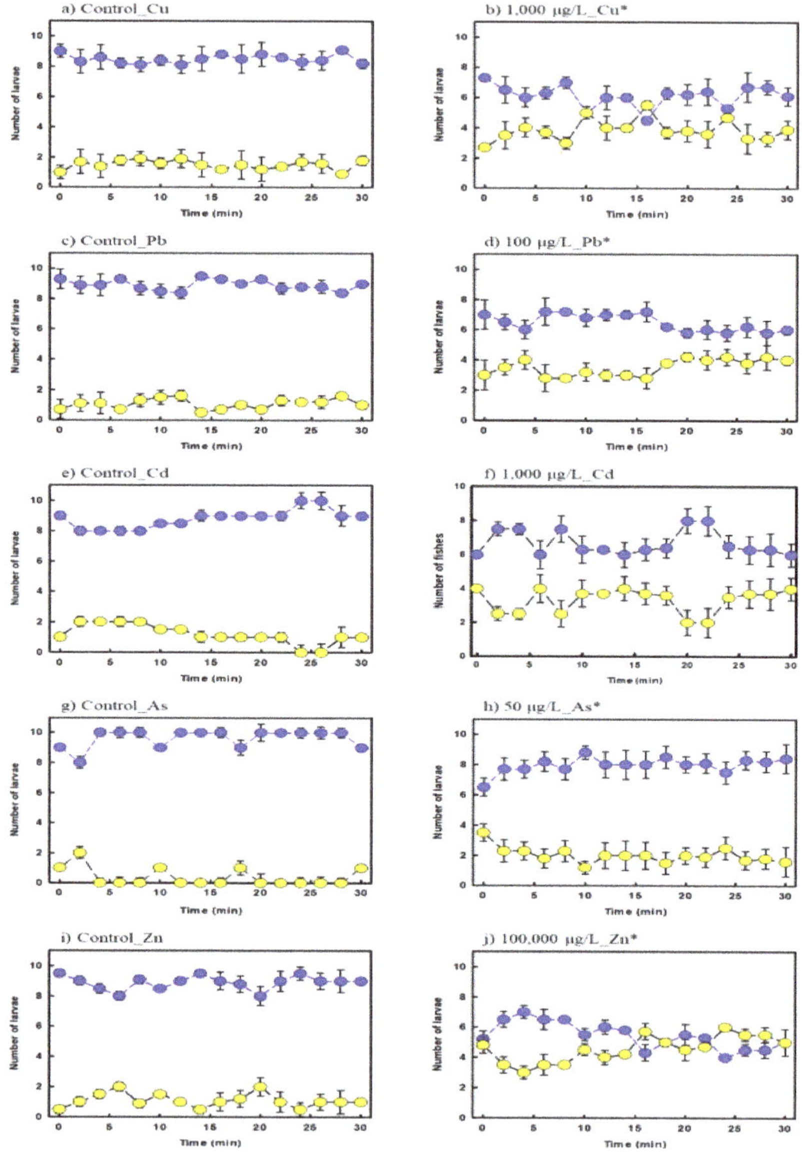

Figure 3. Cognitive function changes of zebrafish larvae after heavy-metal exposure (blue; preferred region, yellow; non-preferred region, * $p < 0.05$).

Zinc is a relatively weak metal. However, excessive zinc influx into nerve cells leads to nerve damage. The zinc concentration in the water sample was more than 100 times higher than that of other metals, which may have an effect on the aquatic ecosystem. Below 1000 μg/L of zinc, no significant cognitive impairment was observed (Figure 3i,j). When exposed to 5000 μg/L or more, cognitive impairment increased by approximately 20%, and at 100,000 μg/L, the incidence of cognitive impairment increased by 50% or more compared to the control group.

3.2.2. Locomotion Changes by Exposure to Heavy Metal

Organisms escape from predators in aquatic ecosystems by changing the velocity, acceleration, and activity time [19]. To analyze the locomotion of the larvae, the average velocity, average acceleration, and active time were measured and compared. The analysis results for each heavy metal substance are as follows. First, copper changed locomotion at concentrations of 10 and 100 μg/L. Compared to the control group, the average velocity increased by 10–56%, acceleration by 16.7–50%, and activity time by 8.3–16.7%. The recovery time after exposure was prolonged, and the overall locomotion was not significantly affected. However, when exposed to 1000 μg/L, it decreased by 19.3, 35.4, and 66.7% for all observation factors, respectively (Figure 4a).

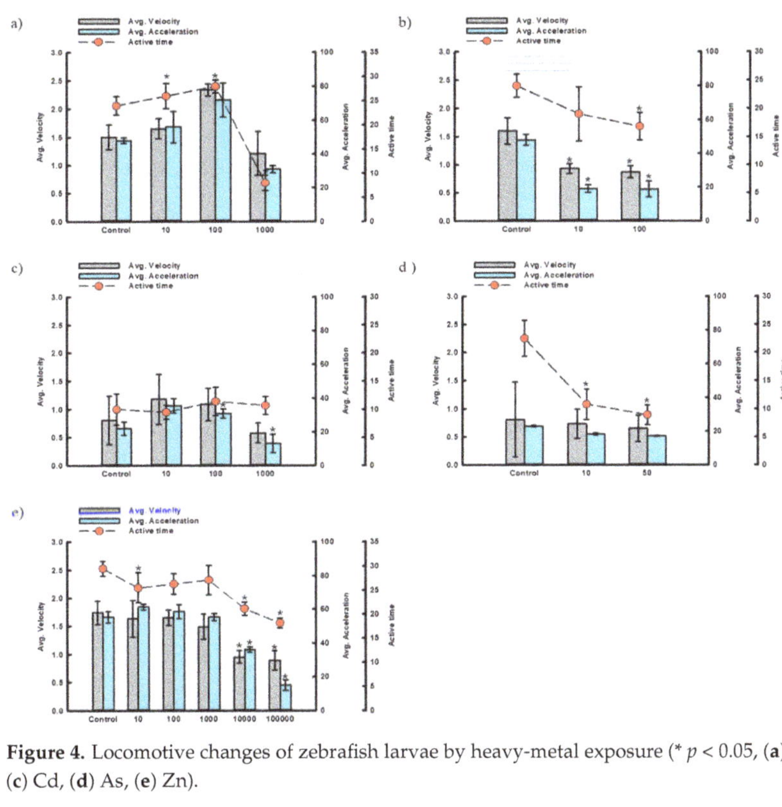

Figure 4. Locomotive changes of zebrafish larvae by heavy-metal exposure (* $p < 0.05$, (**a**) Cu, (**b**) Pb, (**c**) Cd, (**d**) As, (**e**) Zn).

Lead showed a decreasing trend for all items, despite the low exposure concentrations. Compared to the control group, the speed decreased by 41.9–45.6%, acceleration by 60.4–61.0%, and activity time by 20.8–30%, respectively. Lead toxicity reduces brain activity [20]. It was confirmed that it played a significant role in reducing the brain function that controls swimming and changes in cognitive ability (Figure 4b).

In general, cadmium toxicity causes cancer and respiratory, liver, and kidney diseases or interferes with calcium intake. No appreciable change was observed after exposure to 10–100 μg/L concentrations. However, when exposed to concentrations greater than 1000 μg/L, the velocity and acceleration decreased by 28.4% and 40.5%, respectively. It was indirectly confirmed that the effect on the body gradually increased as the concentration or time of exposure to cadmium increased. As mentioned in Section 3.2.1, arsenic showed little change in all areas, with the active time reduced by up to 60.4% (Figure 4c).

The arsenic exposure test showed a relatively low value compared to that of other heavy metals. Thus, the effect of arsenic on the aquatic ecosystem can be stronger than that of other heavy metals (Figure 4d). Compared with other substances, zinc showed toxic effects at high concentrations. When exposed to concentrations greater than 10,000 μg/L, the velocity and acceleration were reduced by more than 30%. Although high zinc concentrations affected cognitive ability, they had a relatively low effect on motility changes (Figure 4e).

3.3. Field Water Exposure Test

3.3.1. Inhibition of Zebrafish Preferred-Wavelength Discrimination by Exposure to Real Water

Lab-scale exposure experiments (Section 3.2) may show different results from actual field experiments. Therefore, the same experimental method was used to examine the effect of river water around the mine, wherein various types of heavy metals are mixed, on zebrafish larvae. For field water, the sample with the highest degree of contamination was selected and used for the experiment. From lab-scale experiments, it was expected that the toxicity evaluation of the field water itself would be complex, as it affected the behavior of zebrafish even when the concentration of heavy metals was 10 μg/L or lower. At the G1 site, there was no significant difference from the control group when exposed to contaminated field water for the first 30 min. However, the discrimination ability for the preferred wavelength decreased as the exposure time increased, and a change in the ability, up to 20%, was observed. In the G2 region, the discrimination ability decreased by 15% with increasing exposure time. However, owing to the strong neurotoxicity of arsenic, the time taken to develop cognitive impairment was reduced by 20%. In the S region, the zinc concentration was approximately 10 times higher than that in the other regions. Nevertheless, cognitive impairment was not observed, as zinc has relatively weak toxicity compared to other heavy metals (Figure 5a).

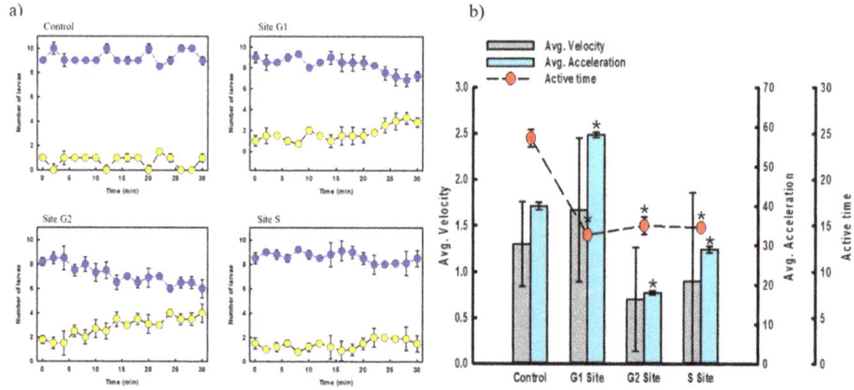

Figure 5. (**a**) Cognitive function, and (**b**) locomotion changes of zebrafish larvae (Real-site sample) (blue; preferred region, yellow; non-preferred region, * $p < 0.05$).

3.3.2. Locomotive Changes by Exposure to Water

At the G1 site, with water exposure, activity time decreased by 42.9%, although the average speed and acceleration increased by 28.5% and 45%, respectively, compared with the control group (Figure 5b). The inhibition of activity showed the most significant effect at the G2 and S sites. A decrease occurred in all items compared with the control group, and a reduction in the acceleration of up to 55% was observed. In the G2 region, more arsenic was detected compared to other mine samples. In the S region, the zinc concentration was 10 times higher than that of the other samples. Among the motion analysis parameters, the activity time item showed the same trend as the results of the cognitive observation experiment. It is believed that a correlation exists between the change in cognitive intelligence and the activity-time relationship of fish; therefore, precise verification is required through further research. As a result of ANOVA analysis, the significance probability for the other items, except average velocity, was lower than 0.05. The contaminated field river water was confirmed to have an impact on zebrafish locomotion.

3.4. *Daphnia magna* Bioassay Test

To conduct a general toxicity test on field water, toxicity sensitivity was assessed using *D. magna* (Table 3). Using the EC50 as the criterion, the same TU results were obtained, and the highest ecotoxicity was confirmed in the G2 and S sites as a result of the TU calculation. In the case of the S site, all *D. magna* swimming inhibition occurred at 100% raw water concentration. As a result, the toxicity unit was 3.00. At the G2 site, similar to the S region, a 3.10 value was obtained. The study results for zebrafish and the swimming inhibition rate of *D. magna* showed similar trends.

Table 3. The bioassay test using *D. magna* by field water exposure.

Mining Site	Exposure Concentration	Immobilization Reduction	EC50 (%)	TU
G1	6.25%	1.33 ± 0.58	55.12	1.81
	12.5%	2.33 ± 0.58		
	25%	4.00 ± 1.00		
	50%	7.33 ± 0.58		
	100%	11.00 ± 1.00		
G2	6.25%	4.00 ± 1.00	32.3	3.10
	12.5%	7.00 ± 0.00		
	25%	10.67 ± 1.15		
	50%	16.67 ± 0.58		
	100%	19.33 ± 0.58		
S	6.25%	3.33 ± 0.58	33.32	3.00
	12.5%	7.67 ± 0.58		
	25%	11.00 ± 1.00		
	50%	17.00 ± 1.00		
	100%	19.67 ± 0.58		

4. Conclusions

Abandoned mines, which are widely distributed around the world, cause environmental pollution due to various factors. Mine drainage pollution could be caused by rainfall, water flow, or mountain slope. Drainage would be introduced into the nearby natural water system. Pollutants in mine drainage pose serious threats to aquatic ecosystems. Contaminants are mostly composed of heavy metals, so the harmful effects on the water environment must be clearly analyzed.

The leakage of heavy metals shows immediate toxicity to the ecosystem; however, it is believed that the environment, over time, can adapt to this toxicity level [21]. Despite adapting, its impact on the ecosystem is an area that requires continuous observation.

In this study, it was confirmed that AMD toxicity that can occur in mines was similarly expressed in zebrafish (*Danio rerio*) and water fleas (*D. magna*), the species to be evaluated for toxicity.

However, although it was detected in Danio rerio, it was confirmed that the TU value for *D. magna* was not detected. This means that the existing ecotoxicity assessment may overlook the toxicity of trace heavy metals in the water system.

Through these results, it is possible to suggest that the environmental impact can be analyzed more accurately if a comprehensive toxicity evaluation of the two species is performed on field water samples containing complex contaminants.

Nevertheless, since this research method also has uncertainty due to statistical values, it is necessary to continuously develop advanced toxicity evaluation methods in order to quantify harmful toxicity.

Author Contributions: Conceptualization, software, validation, formal analysis, data curation, writing—original draft preparation, writing—review and editing, visualization, supervision, funding acquisition, H.Y.; methodology, investigation, visualization, project administration, J.Y. All authors have read and agreed to the published version of the manuscript.

Funding: This research was supported by Basic Science Research Program through the National Research Foundation of Korea (NRF) funded by the Ministry of Education (NRF-2022R1I1A1A01052986).

Institutional Review Board Statement: Not applicable.

Informed Consent Statement: Not applicable.

Data Availability Statement: Not applicable.

Conflicts of Interest: The authors declare no conflict of interest.

References

1. Chan, W.S.; Routh, J.; Luo, C.; Dario, M.; Miao, Y.; Luo, D.; Wei, L. Metal accumulations in aquatic organisms and health risks in an acid mine-affected site in South China. *Environ. Geochem. Health* **2021**, *43*, 4415–4440. [CrossRef] [PubMed]
2. Dong, S.; Xu, B.; Yin, S.; Han, Y.; Zhang, X.; Dai, Z. Water resources utilization and protection in the coal mining area of Northern China. *Sci. Rep.* **2019**, *9*, 1214. [CrossRef] [PubMed]
3. Thisani, S.; Kallon, D.; Byren, P. Geochemical classification of global mine water drainage. *Sustainability* **2020**, *12*, 10244. [CrossRef]
4. Ushakova, E.; Menshikova, E.; Blinov, S.; Osovetsky, B.; Belkin, P. Environmental assessment impact of acid mine drainage from Kizel coal basin on the Kosva bay of the Kama Reservoir (Perm Krai, Russia). *Water* **2022**, *14*, 727. [CrossRef]
5. Choudhury, T.R.; Acter, T.; Uddin, N.; Kamal, M.; Chowdhury, A.M.S.; Rahman, M.S. Heavy metals contamination of river water and sediments in the mangrove forest ecosystems in Bangladesh: A consequence of oil spill incident. *Environ. Nanotechnol. Monit. Manag.* **2021**, *16*, 100484. [CrossRef]
6. Wijngaard, R.R.; van der Perk, M.; van der Grift, B.; de Nijs, T.C.M.; Bierkens, M.F.P. The impact of climate change on metal transport in a lowland catchment. *Water Air Soil Pollut.* **2017**, *228*, 107. [CrossRef]
7. Zhang, L.; Zhao, B.; Xu, G.; Guan, Y. Characterizing fluvial heavy metal pollutions under different rainfall conditions: Implication for aquatic environment protection. *Sci. Total Environ.* **2018**, *635*, 1495–1506. [CrossRef]
8. Iordache, A.M.; Nechita, C.; Voica, C.; Pluháček, T.; Schug, K.A. Climate change extreme and seasonal toxic metal occurrence in Romanian freshwaters in the last two decades-case study and critical review. *NPJ Clean Water* **2022**, *5*, 2. [CrossRef]
9. Seo, J.; Kang, S.W.; Ji, W.; Jo, H.J.; Jung, J. Potential risks of effluent from acid mine drainage treatment plants abandoned coal mines. *Bull. Environ. Contam. Toxicol.* **2012**, *88*, 990–996. [CrossRef]
10. Simate, G.S.; Ndlovu, S. Acid mine drainage: Challenges and opportunities. *J. Environ. Chem. Eng.* **2014**, *2*, 1785–1803. [CrossRef]
11. Chamorro, S.; Barata, C.; Piña, B.; Casado, M.; Schwarz, A.; Sáez, K.; Vidal, G. Toxicological analysis of acid mine drainage by water quality and land use bioassays. *Mine Water Environ.* **2018**, *37*, 88–97. [CrossRef]
12. Macías, F.; Caraballo, M.A.; Nieto, J.M. Environmental assessment and management of metal-rich wastes generated in acid mine drainage passive remediation systems. *J. Hazard. Mater.* **2012**, *229–230*, 107–114. [CrossRef] [PubMed]
13. Yang, Q.; Zhang, A.-H.; Miao, J.-H.; Sun, H.; Han, Y.; Yan, G.-L.; Wu, F.-F.; Wang, X.-J. Metabolomics biotechnology, applications, and future trends: A systematic review. *R. Soc. Chem. Adv.* **2019**, *9*, 37245–37257. [CrossRef] [PubMed]
14. Bedia, C. Metabolomics in environmental toxicology: Applications and challenges. *Trends Environ. Anal. Chem.* **2022**, *34*, e00161. [CrossRef]
15. Stewart, A.M.; Braubach, O.; Spitsbergen, J.; Gerlai, R.; Kalueff, A.V. Zebrafish models for translational neuroscience research: From tank to bedside. *Trends Neurosci.* **2014**, *37*, 264–278. [CrossRef]
16. Veldman, M.B.; Lin, S. Zebrafish as a developmental model organism for pediatric research. *Pediatr. Res.* **2008**, *64*, 470–476. [CrossRef]
17. Yoon, H.; Kim, H.C.; Kim, J.; You, K.; Cho, Y.; Kim, S. Toxicity impact of hydrogen peroxide on the fate of zebrafish and antibiotic resistant bacteria. *J. Environ. Manag.* **2022**, *302*, 114072. [CrossRef]
18. Kim, J.; Kim, S.; Lee, S. Differentiation of the toxicities of silver nanoparticles and silver ions to the Japanese medaka (*Oryzias latipes*) and the cladoceran *Daphnia magna*. *Nanotoxicology* **2011**, *5*, 208–214. [CrossRef]
19. Posthuma, L.; van Straalen, N.M. Heavy metal adapatation in terrestrial invertebrates: A review of occurrence, genetics, physiology and ecological consequences. *Comp. Biochem. Physiol. Part C: Pharmacol. Toxicol. Endocrinol.* **1993**, *106*, 11–38.
20. Gerlai, R.; Fernandes, Y.; Pereira, T. Zebrafish (*Danio rerio*) responds to the animated image of a predator: Towards the development of an automated aversive task. *Behav. Brain Res.* **2009**, *201*, 318–324. [CrossRef]
21. Tiwari, S.S.; Tambo, F.; Agarwal, R. Assessment of lead toxicity on locomotion and growth in a nematode Caenorhabditis elegans. *J. Appl. Nat. Sci.* **2020**, *12*, 36–41. [CrossRef]

toxics

MDPI

Article

Accumulation of Metals in the Environment and Grazing Livestock near A Mongolian Mining Area

Bayartogtokh Bataa [1,†], Kodai Motohira [2,†], Delgermurun Dugar [1], Tsend-Ayush Sainnokhoi [1], Lkhamjav Gendenpil [1], Tserenchimed Sainnokhoi [3], Bolormaa Pelden [1], Yared Beyene Yohannes [2,4], Sumiya Ganzorig [5], Shouta M. M. Nakayama [1,6], Mayumi Ishizuka [1] and Yoshinori Ikenaka [2,7,8,9,*]

1 Department of Pharmacology and Internal Medicine, School of Veterinary Medicine, Mongolian University of Life Sciences, Zaisan, Khan-Uul District, Ulaanbaatar 17024, Mongolia
2 Laboratory of Toxicology, Department of Environmental Veterinary Sciences, Hokkaido University, Kita 18, Nishi 9, Kita-ku, Sapporo 060-0818, Hokkaido, Japan
3 State Central Veterinary Laboratory, Zaisan, Khan-Uul District, Ulaanbaatar 17024, Mongolia
4 Department of Chemistry, College of Natural and Computational Science, University of Gondar, Gondar P.O. Box 196, Ethiopia
5 Spatial Analysis Laboratory, Department of Biology, National University of Mongolia, Baga Toiruu 2, Sukhbaatar District, Ulaanbaatar 14200, Mongolia
6 Biomedical Sciences Department, School of Veterinary Medicine, The University of Zambia, P.O. Box 32379, Lusaka 10101, Zambia
7 Water Research Group, Unit for Environmental Sciences and Management, Potchefstroom Campus, North-West University, Private Bag X6001, Potchefstroom 2520, South Africa
8 Translational Research Unit, Veterinary Teaching Hospital, Faculty of Veterinary Medicine, Hokkaido University, Sapporo 060-0818, Hokkaido, Japan
9 One Health Research Center, Hokkaido University, Sapporo 060-0818, Hokkaido, Japan
* Correspondence: y_ikenaka@vetmed.hokudai.ac.jp; Tel.: +81-11-706-6949; Fax: +81-11-706-5105
† These authors contributed equally to this work.

Citation: Bataa, B.; Motohira, K.; Dugar, D.; Sainnokhoi, T.-A.; Gendenpil, L.; Sainnokhoi, T.; Pelden, B.; Yohannes, Y.B.; Ganzorig, S.; Nakayama, S.M.M.; et al. Accumulation of Metals in the Environment and Grazing Livestock near A Mongolian Mining Area. *Toxics* **2022**, *10*, 773. https://doi.org/10.3390/toxics10120773

Academic Editor: Roel Vermeulen

Received: 26 October 2022
Accepted: 6 December 2022
Published: 10 December 2022

Abstract: The Mongolian economy is supported by rich deposits of natural resources, such as copper, coal, and gold. However, the risk of heavy metal pollution to livestock and human have been recently discussed. This research collected various samples from soil and animal (sheep, goat, horse, cow, and camel), blood and organs (kidney and liver) in the Mongolian countryside. These samples were processed, and the concentration of metals was quantified using inductively coupled plasma-mass spectrometry (ICP/MS). As previously reported, arsenic was found at high levels of accumulation in soil. Selenium is another concern, as median concentration in one area exceeded the maximum allowable level. Cadmium and selenium were found to be highly accumulated in animal kidney. This research revealed the current pollution level in Mongolia based on evaluation of soil and animals. The concentration in animals could not indicate that animals had severe effects because of heavy metal exposure. However, kidney is eaten in Mongolia, and so there is a direct connection to human health, and this research suggested the possible risks posed by each edible animal. In particular, evaluation of metals in livestock is rare in Mongolia. This result can contribute to animal and human health in Mongolian communities.

Keywords: metals; animal health; monitoring; food safety

1. Introduction

Mongolia is a landlocked country located between Russian and China, known for its rich diversity of natural resources, including gold, silver, copper, coal, fluoride, iron, petroleum oil, and uranium [1]. Traditionally, the main economic activity in Mongolia is based on agriculture, livestock, and the mineral industry. Mongolia's economy has been expanding through exploitation of vast mineral resources during the past 20 years. The mining sector has supported economic growth and development in Mongolia, as mining recently contributed approximately 20% of gross domestic product (GDP) and about 80% of total exports in 2017 [2,3].

However, metal contamination derived from mining and industrial activity has been observed by many groups and has been reported worldwide [4–7]. For example, the area near mining site in Zambia recorded high concentration of metals including lead and cadmium, and geographic analysis revealed that the mining activity was possible source of pollution [4]. Heavy metals represent one of the main environmental pollutants that cause adverse effects to human and animal health. These effects, such as neurotoxicity, hepatotoxicity, carcinogenicity, teratogenicity, and nephrotoxicity, have also been reported [8,9]. The International Agency for Research on Cancer (IARC) has classified heavy metals such as arsenic, cadmium and nickel as Group1 (carcinogenic), and lead and cobalt as Group2B (possibly carcinogenic). While metals affect human health, many reports have also indicated heavy metal toxicity in animals. For instance, cadmium is highly toxic, affecting nearly all animal systems. Arsenic shows various toxicities with gastrointestinal and nervous symptoms in cattle and induced weight loss, reduction of milk yield, anemia, and liver and kidney damage by chronic exposure [10]. Another consideration is that heavy metals cannot be degraded through physical processes. Therefore, they persist in the animal body and environment for a long period [11].

Environmental contamination and exposure of humans to metals are practical concerns in Mongolia. Contamination with metals usually results from many different sources, such as industrial waste, chemicals used in agriculture, construction, vehicular exhaust, coal and fossil fuel combustion, and mining [8,12,13]. In Ulaanbaatar, Mongolia, survey sampling in 2007 showed high arsenic contamination in soil. The possible source of this arsenic could be coal combustion and vehicles in the city [12]. Another study of samples collected in 2019 also showed high arsenic contamination in soil in Ulaanbaatar, and arsenic and chromium have been evaluated as metals with high ecological risk [14].

In recent years, environmental problems have gained attention in Mongolia. Economic losses because of metals through toxic effects on livestock and humans have also been found in other countries [15,16]. Concerning human health, various pathways through food, air, and water were considered as intake routes for metals, and animal products are one possible source of metals to humans [8,17]. Traditionally, Mongolians eat a lot of meats, and sheep meat is one of the main protein sources. The major livestock are sheep and cows for meat and milk in Mongolia. Mongolians eat not only muscle, but also liver, kidney, heart, etc. This unique food culture may increase the risk of metal contamination of animal products. As toxicity of metals in humans has been a concern in several countries, monitoring metals in grazing livestock in Mongolia is necessary in order to protect both animal health and human food safety. However, this kind of scientific research has not been pursued in detail in Mongolia at present. Therefore, monitoring environmental pollution that can cause toxicity to animals is desirable in Mongolia. Thus, this research targeted various livestock and soils in Mongolia, analyzing metal levels in collected samples.

2. Materials and Methods

2.1. Study Area

This study was conducted on animal farms in two areas: Dornogovi (Ulaanbadrakh: UB, Zuun-bayan: ZB and Airag: AE soum) and Tuv-Aimag (Zaamar soum). Dornogovi (44°53′N 110°09′E) is one of the 21 aimags (provinces) of Mongolia, located about 456 km southeast of the capital city, Ulaanbaatar. Dornogovi is known for its harsh weather conditions; it is located in the Gobi Desert, where frequent sand- and snowstorms occur. Tuv-aimag is another province of Mongolia, located about 45 km from the capital city. Many animal products, including milk and meat, are produced in both areas and distributed in Ulaanbaatar. Another notable product in Tuv-Aimag is gold from the mining located on the Tuul River. This area is still used for mining, and new mines are continuously being developed. Dornogovi also has mining areas, for fluorspar, uranium, etc. The soil and tissue samples were collected near mining areas. Details are shown in Figure 1.

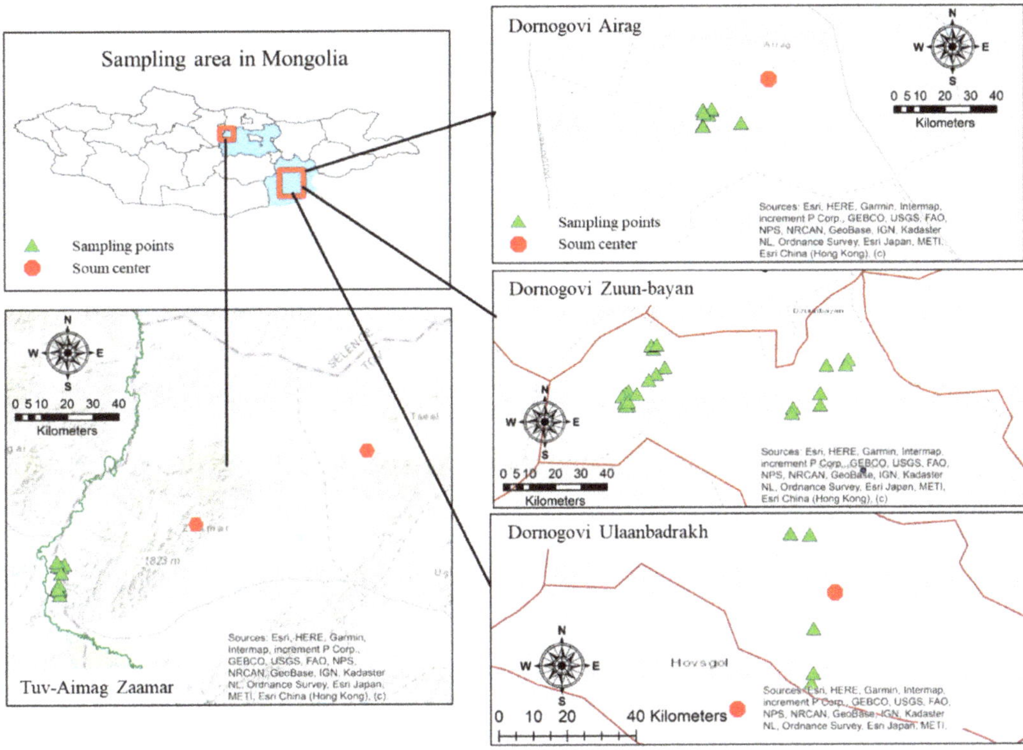

Figure 1. Sampling location. Sampling areas in Tuv-Aimag Zaamar, Dornogovi Airag, Zuun-bayan and Ulaanbadrakh were described by map using ArcGIS 10.7.1 (ESRI Co., Redlands, CA, USA).

2.2. Sample Collection

All samplings were done in June, 2016, in Mongolia. In total, 53 tissue samples (liver and kidney) were collected from various deceased animals (horse, goat, sheep, and cow) in Dornogovi province for metal analysis. In addition, 55 soil samples were collected from the same area. Approximately 50 g of soil from a depth of 0–5 cm was collected for each sample and kept in a plastic tube. At least three composite soil samples were collected from each sampling point. Approximately 150 blood samples were randomly collected from various animals (horse, goat, sheep, cow, and camel) in Dornogovi and Tuv-Aimag. Collected tissue and blood samples were stored in cooler boxes with dry ice during the sampling on-site, and all samples were transported and kept at 4 °C for soil and at −20 °C for blood and organs at the School of Veterinary Medicine, Mongolian University of Life Sciences, before being transported and analyzed for metal concentration at the School of Veterinary Medicine, Hokkaido University, Japan around 20 days after sample collection.

2.3. Sample Preparation and Metal Extraction

All laboratory materials and instruments used in metal extraction were washed in 2% nitric acid (HNO_3) and oven-dried (50 °C). Metals were extracted from the samples (tissue, soil, and blood) by acid digestion with the method used in previous reports for tissue samples [18] and environment samples [19], with minor modifications.

Tissue samples (liver and kidney) were weighed (1 g) and dried for 48 h in an oven at 60 °C. Then, the dried tissue samples were digested using 5 mL 30% nitric acid (HNO_3) and 1 mL 30% hydrogen peroxide (H_2O_2) (analytical grade, Kanto Chemical Corp. Japan) in a speed wave MWS-2 microwave oven system. Digestion of blood samples (1 mL) followed the same process as the tissue samples. The microwave system was programmed to ramp in five min to 160 °C (held for five min), followed by a ramp time of three min increased to 200 °C (held for 20 min), and then hold for 10 min at 75 °C. After digestion, samples were cooled and transferred to a 15 mL polypropylene tube followed by dilution to 10 mL with Milli Q water.

In the case of soil samples, extraction was performed using about 0.5 g of sample and 5 mL 60% HNO_3, and 1 mL 30% H_2O_2. The microwave was programmed at an initial temperature of 150 °C (held for 5 min), then with a ramp time of 2 min increased to 175 °C (held for 10 min) and with a ramp time of three min increased to 190 °C (held for 30 min). After digestion, samples were filtered through ADVANTEC 5C filter paper into 15 mL polypropylene tubes and made up to 10 mL using Milli Q water. Analytical blanks were run for all samples.

Tissues, blood, and soil were assessed for concentrations of ten metals (Cr, Mn, Co, Ni, Cu, Zn, As, Se, Cd, and Pb) using an inductively coupled plasma –mass spectrometer (ICP-MS; 7700 series, Agilent Technologies, Tokyo, Japan) using ^{9}Be, ^{115}In, and ^{205}Tl as internal standards.

2.4. Quality Control and Quality Assurance

All chemicals and standard stock solutions were analytical reagent grade (Wako Pure Chemicals, Osaka, Japan). For quality control, blanks were analyzed after every 10 samples, and a calibration curve for each element was constructed with an R2 value greater than 0.995. Analytical quality control was performed using certified reference materials: DOLT-4 (dogfish liver, National Research Council of Canada), DORM (fish protein, National Research Council of Canada, Ottawa, Canada), SRM-1944 (New/York/New Jersey Waterway Sediment, National Institute of Standards and Technology, New York, USA), and BCR-320 (channel sediment, Community Bureau of Reference of European Commission, Brussels, Belgium). Replicate analysis of these reference materials showed good recoveries, ranging from 90–110%.

2.5. Statistical Analysis

JMP 16 (SAS Institute, NC, USA) software was used for all statistical analyses, with a significance threshold of $p < 0.05$. The regional difference in soils was tested using Steel-Dwass tests. Outliers were detected assessing quantile range outliers (tail quantile 0.05, Q 10). Four metal concentrations in animal blood (from two individual outlier animals) were extracted. These two animals were also defined as the first and second biggest neighborhood distance calculated from K nearest neighbor outliers (K: 2–8, threshold was not decided). Thus, these two animals were excluded from the sample dataset. There was no outlier among soil samples, which were handled in the same way as animal samples.

3. Results

The metal concentration in the soil is shown in Table 1 and Supplementary Table S1. The Tuv-Aimag Zaamar region had generally higher concentrations of most metals; copper and chromium concentrations were significantly higher than in other areas ($p < 0.05$) (Supplementary Table S1). Table 1 shows the median concentration of metals in the highest-concentration area, the highest concentration in this study, and the regulations (such as the maximum allowable level in Mongolia, maximum permissible level summarized in [20] and Ecological Soil Screening Levels (Eco-SSLs) [21–28]). Eco-SSLs were provided from the U.S. Environmental Protection Agency (EPA). The Eco-SSLs suggest screening the level of contaminants in soil to indicate ecological risks. Contaminants exceeding their permissible concentration may require further evaluation for protection against ecological risk. The Eco-SSLs mention four different levels of concentration (plants, soil invertebrates, birds, and mammals); this study used its levels for mammals, as shown in Table 1. Chiroma T. M et al. (2014) summarized the maximum allowable limit using World Health Organization (WHO), Food and Agricultural Organization (FAO), and Ewers, U. (1991), Standard Guidelines in Europe. Some of the highest concentrations of target metals exceeded these maximum levels. In particular, arsenic and selenium concentrations were high, and some soil samples have exceeded these levels (Figure 2). For example, more than 50% of samples in Dornogovi AE exceeded the maximum permissible level of selenium (10 mg/kg) in Mongolia.

Table 1. Metal concentration in soil in this study, previous study in Mongolia, and maximum allowable limits.

	Mongolian Government Limit (mg/kg)	Eco-SSLs (mg/kg)	Maximum Permissible Level (mg/kg) *		This Study (2016) with Recorded Area (mg/kg)		Previous Study in Mongolia (2019)(mg/kg)	Previous Study in Mongolia (2007)(mg/kg)
Cr	150	III 34	100	Median	26.5	(Tuv aimag)	29	20.3
		IV 130		Highest	120.6	(Dornogovi ZB)	110.4	23.9
Mn		4000	2000	Median	377.8	(Tuv aimag)	0.5	
				Highest	3504.0	(Dornogovi ZB)	1.1	
Cu	100	49	100	Median	18.1	(Tuv aimag)	28.9	35.9
				Highest	64.7	(Dornogovi ZB)	78.6	47.4
Zn	300	79	300	Median	34.5	(Tuv aimag)	135.6	158.7
				Highest	209.1	(Dornogovi ZB)	384	203.0
As	20	46	20	Median	8.8	(Tuv aimag)	22.9	14.0
				Highest	48.2	(Dornogovi ZB)	526.8	64.1
Se	10	0.63	10	Median	10.7	(Dornogovi AE)	0.1	
				Highest	86.5	(Dornogovi ZB)	1.5	
Cd	3	0.36	3	Median	0.07	(Tuv aimag)	0.2	0.8
				Highest	0.44	(Dornogovi ZB)	1.9	1.1
Pb	100	56	100	Median	7.5	(Tuv aimag)	34.5	63.9
				Highest	76.5	(Dornogovi ZB)	77.3	119.6

The metal concentration in this study is shown with results of a previous study conducted in Mongolia. Brackets indicate the sampling year in each study. * is the reported maximum allowable level summarized in Chiroma T. M et al. (2014). All concentrations are described in mg/kg.

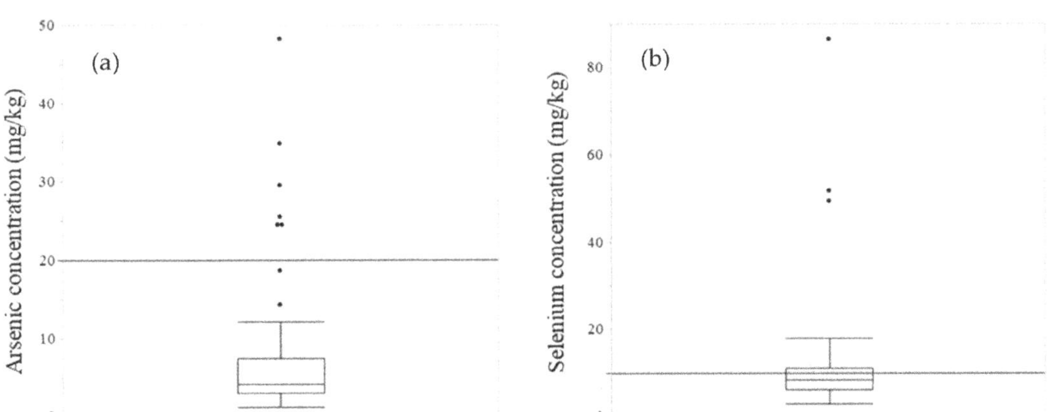

Figure 2. Arsenic and selenium concentration in soil (all samples in this study). Arsenic (**a**) and selenium (**b**) concentration in all soil samples are shown in a box plot. The black line shows the maximum allowable limit, as regulated in Mongolia (arsenic: 20 mg/kg soil, selenium: 10 mg/kg soil).

Blood metal concentration is described in Table 2. Strong species differences were not observed for many metals in the same area. However, the camel accumulated a slightly higher concentration of zinc than the other animals in all areas. The median concentration of zinc in Dornogovi UB was 50.4 mg/kg (dry weight) in camels' blood, while goats and sheep had blood levels of only 29.0 and 20.6 mg/kg, respectively. In the case of copper, the concentration in small ruminants, such as goat and sheep, was a bit higher than in other animals. The order of copper median concentration in Dornogovi ZB was goat > sheep > horse > camel > cow (Table 2). Comparing the regional difference showed that the Tuv-Aimag Zaamar region had a higher concentration of arsenic than other areas, as the soil concentration had already suggested. However, in the case of zinc, selenium, and lead, the median concentration in soil in Dornogovi UB was the lowest, but comparatively high concentrations of metals were detected in blood in that area, as the median zinc and lead concentration (29.0 and 0.1 mg/kg dry weight, respectively) was the highest in goat among all areas.

To give a better understanding of the absolute concentrations of metals in livestock, Supplementary Table S2 summarizes the normal range of metals in animal organs, referring to published values [29]. The concentration in liver and kidney in sheep, goat, and cattle is shown in Table 3. The sample number of organs was not enough to determine regional difference. Thus, Table 3 shows mixed results from all areas (Dornogovi UB, ZB, AE). We also extracted the results of copper, zinc, and selenium in liver and kidney, as shown in Figure 3, with toxic and deficient levels described in a previous report [30]. The relatively high concentration of copper was also seen in sheep liver (median concentrations were 235.1, 92.4, and 148.3 mg/kg dry weight in sheep, goat, and cow, respectively). Arsenic concentration in organs was not as high compared with other high toxic metals, such as cadmium and lead (Table 3).

We additionally collected only two horse kidney samples, but concentrations of cadmium (122.65 and 352.79 mg/kg) in kidney were quite high in both individuals, compared with other animals, as median concentration values in sheep, goat, and cattle were 0.28, 1.02, and 3.78 mg/kg respectively (data not shown).

To summarize our results, each metal concentration was normalized using its sum and standard deviation of all the samples. Then, the mean of each normalized concentration, as shown in Supplementary Figure S1, was used to evaluate the accumulation pattern of each metal.

Table 2. Metal concentration in blood.

Dornogovi AE		Cr	Mn	Cu	Zn	As	Se	Cd	Pb
Goat (15)	median	0.05	0.074	4.0	11.4	0.011	0.95	< 0.001	0.077
(mg/kg)	range	0.023 — 0.14	0.046 — 0.31	[illegible].6 — 4.8	6.8 — 18.3	0.009 — 0.023	0.47 — 1.4	< 0.001 — < 0.001	0.036 — 0.13
Sheep (10)	median	0.0525	0.11	3.5	10.0	0.0125	1.2	< 0.001	0.055
(mg/kg)	range	0.031 — 0.075	0.078 — 0.15	[illegible].4 — 4.5	6.4 — 14.6	0.009 — 0.015	0.59 — 1.5	< 0.001 — < 0.001	0.03 — 0.088
Dornogovi UB		Cr	Mn	Cu	Zn	As	Se	Cd	Pb
Goat (10)	median	0.045	0.10	5.2	29.0	0.013	1.5	0.001	0.10
(mg/kg)	range	0.031 — 0.20	0.069 — 0.37	[illegible].7 — 8.1	9.4 — 148.1	0.008 — 0.049	1.2 — 1.9	< 0.001 — 0.003	0.033 — 0.22
Sheep (10)	median	0.054	0.13	4.9	20.6	0.016	1.6	0.001	0.095
(mg/kg)	range	0.032 — 0.15	0.085 — 0.24	[illegible].7 — 7.9	8.9 — 63.4	0.011 — 0.020	1.3 — 2.2	< 0.001 — 0.004	0.038 — 0.14
Horse (3)	median	0.055	0.09	3.7	9.3	0.013	0.85	0.003	0.036
(mg/kg)	range	0.052 — 0.12	0.067 — 0.12	[illegible].8 — 6.7	7.0 — 10.1	0.008 — 0.013	0.84 — 1.0	0.003 — 0.003	0.031 — 0.039
Camel (13)	median	0.035	0.20	2.9	50.4	0.01	1.4	< 0.001	0.058
(mg/kg)	range	0.015 — 0.18	0.078 — 0.80	[illegible].8 — 19.2	15.1 — 211.2	0.005 — 0.072	0.88 — 10.8	< 0.001 — 0.004	0.038 — 0.37
Dornogovi ZB		Cr	Mn	Cu	Zn	As	Se	Cd	Pb
Goat (21)	median	0.041	0.07	4.8	13.3	0.012	0.96	< 0.001	0.062
(mg/kg)	range	0.014 — 0.14	0.017 — 0.35	[illegible].0 — 8.4	5.2 — 62.7	0.003 — 0.063	0.54 — 2.7	< 0.001 — 0.004	0.018 — 0.19
Sheep (20)	median	0.040	0.12	4.5	11.3	0.019	2.3	0.001	0.055
(mg/kg)	range	0.019 — 0.10	0.073 — 0.29	[illegible].9 — 6.0	7.4 — 56.4	0.008 — 0.050	1.2 — 3.4	< 0.001 — 0.003	0.026 — 0.15
Horse (10)	median	0.044	0.11	3.6	9.4	0.015	2.0	0.002	0.047
(mg/kg)	range	0.015 — 0.12	0.027 — 0.31	1.3 — 6.5	3.7 — 14.0	0.003 — 0.037	0.60 — 3.4	< 0.001 — 0.005	0.019 — 0.33
Camel (15)	median	0.037	0.17	2.8	31.1	0.012	1.2	0.001	0.044
(mg/kg)	range	0.009 — 0.080	0.015 — 0.49	1.0 — 4.2	5.9 — 59.5	0.002 — 0.029	0.47 — 4.0	< 0.001 — 0.003	0.017 — 0.16
Cow (3)	median	0.026	0.04	1.4	2.6	0.003	0.62	< 0.001	0.024
(mg/kg)	range	0.015 — 0.031	0.018 — 0.058	0.028 — 2.5	0.21 — 7.7	0.001 — 0.006	0.004 — 1.1	< 0.001 — 0.002	0.019 — 0.025
Tuv Zaamar		Cr	Mn	Cu	Zn	As	Se	Cd	Pb
Goat (9)	median	0.053	0.13	5.5	21.1	0.047	0.27	0.001	0.065
(mg/kg)	range	0.032 — 0.089	0.059 — 0.32	4.4 — 6.3	8.0 — 44.9	0.007 — 0.094	0.18 — 0.41	< 0.001 — 0.005	0.042 — 0.13
Sheep (10)	median	0.044	0.13	4.4	15.7	0.042	0.26	< 0.001	0.052
(mg/kg)	range	0.021 — 0.17	0.056 — 0.28	1.9 — 5.0	4.3 — 32.4	0.004 — 0.079	0.076 — 0.42	< 0.001 — 0.004	0.015 — 0.18
Cow (5)	median	0.068	0.10	4.2	34.0	0.078	0.14	0.002	0.15
(mg/kg)	range	0.042 — 0.079	0.063 — 0.25	4.0 — 4.7	23.6 — 101.2	0.067 — 0.39	0.11 — 0.18	0.001 — 0.005	0.091 — 0.21

Metal concentration is shown, separated by region and animal species. All concentrations are given in mg/kg (dry weight). The number of samples in each area is indicated in parentheses.

Table 3. Metal concentration in liver (3a) and kidney (3b).

(3a)		Cr	Mn	Cu	Zn	As	Se	Cd	Pb
Sheep (10) (mg/kg)	median	0.21	12.4	235.1	99.5	0.06	1.9	0.28	0.38
	range	0.03 — 0.57	3.7 — 26.6	20.8 — 814.9	65.0 — 239.1	0.03 — 0.16	0.22 — 4.8	0.002 — 0.83	0.22 — 2.2
Goat (11) (mg/kg)	median	0.28	10.2	92.4	176.2	0.06	1.9	0.11	0.24
	range	0.07 — 1.1	4.0 — 33.4	10.5 — 734.5	59.9 — 406.6	0.01 — 0.27	0.40 — 3.6	0.002 — 0.72	0.08 — 2.8
Cattle (4) (mg/kg)	median	0.28	16.2	148.3	305.0	0.19	4.4	1.02	0.31
	range	0.11 — 0.60	2.7 — 32.9	8.8 — 365.5	127.8 — 885.9	0.06 — 0.42	2.4 — 10.5	0.320 — 1.8	0.20 — 0.35
(3b)		Cr	Mn	Cu	Zn	As	Se	Cd	Pb
Sheep (8) (mg/kg)	median	0.27	3.9	16.6	88.8	0.07	4.3	0.28	0.21
	range	0.10 — 0.52	2.5 — 5.9	8.7 — 43.8	45.9 — 231.2	0.03 — 0.13	1.8 — 11.9	0.003 — 4.1	0.10 — 0.67
Goat (11) (mg/kg)	median	0.29	4.8	15.0	87.7	0.07	5.0	1.02	0.24
	range	0.08 — 0.61	1.9 — 22.3	8.4 — 279.1	49.3 — 777.1	0.01 — 0.27	1.1 — 9.6	0.004 — 7.0	0.06 — 0.99
Cattle (4) (mg/kg)	median	0.32	4.4	39.2	306.4	0.24	8.2	3.8	0.60
	range	0.18 — 2.2	4.3 — 6.3	11.6 — 99.8	95.3 — 408.5	0.15 — 0.38	7.4 — 10.1	0.99 — 6.9	0.25 — 1.1

Metal concentration in liver and kidney is shown. All concentrations are given in mg/kg (dry weight). The number of samples in each area is indicated in parentheses.

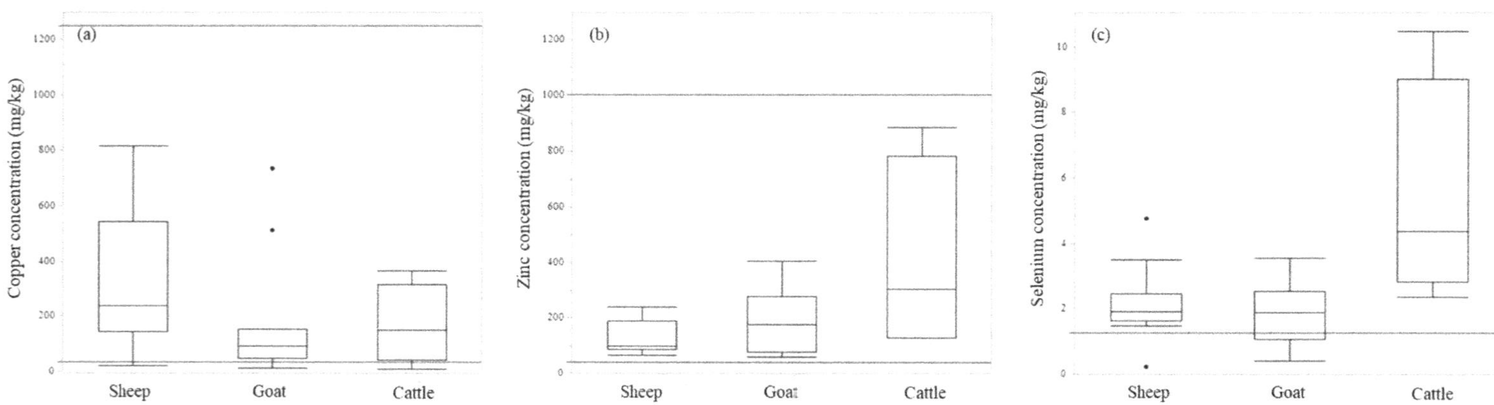

Figure 3. Copper, zinc and selenium concentration in liver. Copper (**a**), zinc (**b**) and selenium (**c**) concentration in liver samples are shown in a box plot. The black line shows toxic and deficient level in (**a**) and (**b**), and marginally deficient level in (**c**). Deficient lines in (**b**) and (**c**) indicated 40 and 1.25 mg/kg although they had range 20—40 and 0.6—1.25 mg/kg respectively.

4. Discussion

4.1. Regulations Related to Heavy Metal and Overview of Results

Table 1 shows two previous studies conducted in Mongolia. The sampling location was different from this study. This study targeted the countryside, while previous reports focused on the city [12,14]. In the current study, notable accumulation was seen in arsenic and selenium, as their concentration in some samples exceeded both Mongolian regulations and Eco-SSLs. The Mongolian regulation level was almost the same as the maximum permissible level summarized in Chiroma T. M et al. (2014), except in the cases of chromium (manganese was not evaluated in the Mongolian regulation). Thus, we used Mongolian regulations to discuss the concentration of metals in soil, as these correspond better to local people in Mongolia. Firstly, the concentration of arsenic and selenium in soil was higher than Mongolian regulation in some samples, and these metals are highly toxic. Thus, Sections 4.2 and 4.3 mentions arsenic and selenium concentrations in Mongolia and compared with previous studies. Section 4.4 focuses on copper and zinc, which are essential metals in animal body. The concentrations of cadmium and lead were not noticeable in the soil sample. However, these metals were known as one of the most toxic metals to animal body. Thus, we mentioned the toxicities of these metals in Section 4.5 with the current results. At the end, we discuss the possible sources of heavy metal contamination in livestock and humans in Section 4.6.

4.2. Arsenic Contamination in Mongolia

This study showed the same trend of high arsenic concentration in soil as the results reported in other studies [12,14]. Thus, arsenic pollution could be a problem all around Mongolia. Arsenic was one of main hazardous inorganic chemical in farm animals, and many toxicities in animals were reported. Daily administration of 20% of the acute lethal dose 50 (ALD50) of arsenic for 12 weeks to goats caused body weight loss, increased respiratory and heart rate, and animal death [31]. A study targeting farm animals in an arsenic-polluted area in India showed significantly higher liver markers, such as aspartate aminotransaminase (AST) and alanine aminotransaminase (ALT), lipid peroxides, and lower superoxide dismutase (SOD), which is one of difference mechanisms against oxidative damage, compared with a control area [32]. Oxidative stress derived from arsenic has been demonstrated using experimental rodents [33]. In this study, in spite of the high concentration of arsenic in soil, its contamination in blood and organs was less than that of other toxic metals. This result followed previous research conducted in Galicia, in which the majority of sheep samples did not accumulate arsenic, even in an arsenic-contaminated area [34]. The monitoring of only one environment (soil) and animals (blood and organs) can mask the risk to animals, as the accumulation patterns of metals and metalloids can be different, depending on the metals (Supplementary Figure S1). The lack of usefulness of blood arsenic concentration as a biomarker for long-term exposure to arsenic has already been reported; this is due to its earlier excretion from the blood compared to other metals and metalloids [35]. Thus, investigating both animals and environments would be beneficial to understand the effects of metals on livestock.

4.3. Selenium Contamination in Mongolia

This study revealed similar concentrations of metals to those described by a previous study conducted in Mongolia examining many metals; however, selenium concentration was high in soil in this study (Table 1). Selenium is distributed variably in the world, depending on geological conditions. Selenium is an essential element for both human and animal bodies, but there is a narrow margin between toxic and deficient levels [36]. For instance, high doses of selenium in lambs were associated with myocardial necrosis, pulmonary edema, and hemorrhage in one histopathological study [37]. An exposure experiment conducted using buffalo calves found that adverse effects and mortality were observed when blood concentrations of selenium became more than 2 and 3.4 μg/mL, respectively. That study investigated relatively acute toxicity during a 30-day observation [38].

Extrapolation to the current study should be done carefully, but the maximum selenium concentration found in this study was around 0.44 μg/mL. A previous review suggested that a selenium concentration between 1.25 and 2.5 mg/kg in the liver is adequate level for cattle, and between 2.3 and 8.0 mg/kg dry weight for newborn calves [30]. The median concentration in sheep and goat was around 2.0 mg/kg, which is an adequate level in liver in this study (Figure 3). Another review mentioned that selenium levels of more than 2 ppm in liver caused acute toxicity, and in many cases reported that sheep died due to acute/subacute intoxication [36]. We did not use the wet weight for calculation because we collected liver samples from the dead body, and conditions such as water content can vary depending on when and where animals have died. However, considering the normal water content in liver (around 3.5—4 times higher concentration in dry weight than that of wet weight [29]), most liver samples would have levels of selenium less than 2 ppm. Thus, even though the selenium concentration in soil exceeded Mongolian regulation levels, we cannot conclude that animals showed an intolerable toxic effect from selenium in Mongolia. However, in cows, the median concentration was higher than in small ruminants. The concentration in three out of four cows exceeded adequate levels. Monitoring selenium in cows is thus required.

4.4. Copper and Zinc Contamination in Mongolia

Copper and zinc are essential elements for the animal body [39]. Thus, the high absolute concentrations of these metals in animal blood, liver, and kidney in this study were reasonable. Susceptibility to copper is higher in sheep > cattle > monogastric animals [40]. The results indicate that susceptible animals were exposed to copper more frequently. The season and hemoglobin type differentiate copper concentrations in sheep. Thus, simple comparison was difficult, but the copper concentration in blood in the current study was relatively low compared with other previous studies [41]. In this area, copper deficiency might be a risk to animals, rather than intoxication. When focusing on liver concentration, previous review report summarized that the cutoff for deficient and adequate copper concentration in liver would be less than 33 and more than 125 mg/kg dry weight, respectively [30]. Most individuals collected in this study demonstrated more than a deficient level (Figure 3). The same report mentioned that copper at a level more than 1250 mg/kg is toxic in cattle liver, but no animal exceeded this value in this study. In the case of zinc, most samples ranged between adequate and high accumulation level. The toxic level (more than 1000 mg/kg in cattle) was not seen in the current dataset (Figure 3) [30] and the risk related to zinc would be relatively low in the current sampling area. Both copper and zinc had broad range of concentration from the lowest to the highest in samples. We did not have their ages data, and this is our limitation. For more proper diagnosis of deficiency, the sampling of liver and kidney from not only dead body but also from healthy slaughter animals with their enough information such as age are needed. Overall, considering the concentration of copper and zinc in blood and animal organs, the risk of intoxication was relatively low compared with arsenic and selenium.

However, zinc concentration in camel blood was high in two areas where we took camel samples. An exposure study in cows showed significant induction of the enzymes which have protective effects against oxidative stress such as SOD at around 2.5—3.0 ug/mL blood zinc concentration [42]. The current study revealed that some camels had more than 5.0 ug/mL of zinc in their blood (data not shown, Table 2 shows only dry weight). Camels are a unique animal living in a severe desert environment. The camel has a unique antioxidant defense system that uses copper; zinc superoxide dismutase was also found [43]. This mechanism allows a high concentration of zinc in camels. However, a previous report also showed that camels maintain a low plasma zinc concentration compared with cows, even when the two species ate similar feed containing multiple metals [44], contrary to the current study. Reports describing camels are fewer in number than those describing other livestock. Further study and measurement of blood parameters, including oxidative stress, in camel are needed to evaluate the effect of metals on camels in this area.

4.5. Cadmium and Lead Contamination in Mongolia

The median concentration of lead and cadmium in soil was less than levels allowed by Mongolian regulation. However, values exceeded Eco-SSLs in some samples. These metals are well known to have high toxicity to humans. The high concentration of cadmium in horses was derived from its high accumulative ability in these animals [45]. The previous report showed much higher concentration of cadmium in horses than in other farm animals, even from the same area [46]. The concentration in horse kidney in this study also incommensurably high compared with other animals in same area. This study has only two kidney samples in horse, but the result might provide hints for communities in Mongolia to carefully consider the use of horse kidneys as food. Itai-itai disease was a notorious incident resulting from cadmium pollution, causing severe symptoms in humans in Japan [47]. The cadmium concentration in soil was quite low in Mongolia, and most of samples had less than 0.1 mg/kg in this study. However, cadmium is highly accumulated in cow kidneys (median concentration: 3.8 mg/kg). Lead also induces severe symptoms in mammals. For instance, previous in vivo Pb exposure study on mice and environmental Pb exposure effect on human showed inhibited activity of d-aminolevulinic acid dehydratase (d-ALAD), a porphobilinogen synthase enzyme which is responsible for heme synthesis [48,49]. The activity of ALAD was estimated to be inhibited by lead with less than around 0.55 mg/kg dry weight in cattle [50]. When focusing on ALAD activity, the cattle in the current sampling area would not get a significant effect of lead as the highest median concentration in cattle was 0.15 mg/kg in blood in Tuv-Aimag Zaamar. However, in particular, cadmium and lead induce severe toxicity to mammals at low concentrations [51]. Thus, even if the detected concentration was less than allowed by Mongolian regulation, we should continue to monitor these metals in order to secure good animal health.

4.6. Source of Heavy Metal Contamination into Livestock and Human

When comparing regional differences in selenium accumulation, the Tuv-Aimag Zaamar area was noticeably lower than other areas, as median concentration in goat and sheep was 0.27 and 0.26 mg/kg in blood, respectively (Table 2). However, the soil accumulated selenium equally between Dornogovi and Tuv-Aimag Zaamar (Supplementary Table S1). As a result, other pathways for heavy metals to reach livestock must also be considered. One of the important sources of heavy metal in livestock is water [10]. For example, a report from India revealed that livestock drinking water containing mercury and manganese at levels higher than their permissible limits. The relationship of metals in livestock to metals in their feed has been also frequently reported, as feed is a major pathway of toxic metal intake for livestock [15,52]. By free grazing, which is the usual form in the Mongolian countryside, sheep and goats eat wild plants. The regional difference in selenium in animal blood could be derived from the local plants. Unusual contamination patterns were also found, such as an animal living in a low-contamination area (Dornogovi UB) accumulating a relatively high concentration of zinc and lead in its blood. As a result of these contaminations, the animals may suffer from effects of the accumulation of heavy metals in their feed (wild plants) and drinking water. Species differences can cause different accumulation patterns of metals in plant [53]. Thus, further investigation of vegetation at sampling sites and metal accumulation in livestock drinking water and plants would be needed to identify the source of contaminants in animals.

Both soil and animal products (liver and kidney) can be sources of metal contamination in human. Previous research monitored the heavy metal concentration in soil in Mongolia and alerted the community to the possible risk on human health [14]. In addition, this research will contribute to evaluation of the impact of metals on human health, particularly involving animal products. The results in this study clearly showed the difference of accumulation pattern of metals among animal and soil samples. For example, arsenic and chromium were much higher in soil than animal samples (Supplementary Figure S1). Even between liver and kidney, for example, selenium and cadmium preferably accumulated in kidney. This might help the identification of possible sources of metals in human. However, for the evaluation of possible risk in human, the daily intake of kidney and liver in Mongolia is necessary, but unfortunately these kinds of information were not found. Thus, further investigation of liver and kidney intake is desirable.

5. Conclusions

The international scientific report targeting metals in Mongolia was limited to the veterinary field. This paper revealed that arsenic and selenium were highly accumulated in Mongolian soil. They should be considered carefully as a risk, but the current concentration results in animal organs were not enough to conclude that animals suffer severe adverse effects because of exposure to these metals. This research also indicates that screening of all of soils, blood, and organs can be effective to monitor metals and to understand the inclusive polluted levels in Mongolia, as arsenic highly accumulated in soil but not in animals. For a better understanding of sources of heavy metal contamination, the monitoring of heavy metals in wild plant and livestock drinking water is desirable. We also focused on animal organs (liver and kidney), which are eaten in the countryside in Mongolia. Metals can be transferred to humans via animal products. Our results contribute to the protection of not only animal health, but also human food safety. This research will be beneficial to protect animal health, inform future studies targeting humans, and help reveal the source of metal contamination.

Supplementary Materials: The following supporting information can be downloaded at: https://www.mdpi.com/article/10.3390/toxics10120773/s1, Table S1: Metal concentration in this study in four areas (mg/kg dry weight); Table S2: The normal range of metals in livestock; Figure S1: Metal concentration in all the samples.

Author Contributions: Conceptualization, B.B.; methodology, B.B., D.D., T.-A.S., L.G., B.P., Y.B.Y. and Y.I.; investigation, B.B., D.D., T.-A.S., L.G., B.P. and S.G.; formal analysis, B.B. and K.M.; funding acquisition, L.G., S.M.M.N., M.I. and Y.I.; supervision, L.G., B.P., S.M.M.N., M.I. and Y.I.; writing—original draft, B.B., K.M. and D.D.; writing—review & editing, K.M., T.S., Y.B.Y. and Y.I. All authors have read and agreed to the published version of the manuscript.

Funding: This work was carried out with the financial support of JICA, with the "The Project for Strengthening the Capacity for Human Resource Development in the Field of Veterinary and Animal Husbandry" in Mongolia. This work was supported by the Grants-in-Aid for Scientific Research from the Ministry of Education, Culture, Sports, Science and Technology of Japan awarded to M. Ishizuka (Nos. 16H0177906, 18K1984708, 18KK028708, 21H02351), Y. Ikenaka (18H0413208, 22K1845), and S.M.M. Nakayama (Nos. 17KK0009 and 20K20633). This work was also supported by the foundation of JSPS Bilateral Open Partnership Joint Research Projects (JPJSBP120209902; SMMN). This work was also supported by JSPS CORE to CORE program (MI), The Japan Prize Foundation; Hokkaido University's SOUSEI Support Program for Young Researchers in FY2020 (SMMN); and Hokkaido University Specific Research Projects (MI). We also acknowledge the financial support from the Sumitomo Foundation and the Nihon Seimei Foundation. This research was also supported by NRF/JST's AJ-CORE program, JST/JICA, SATREPS (Science and Technology Research Partnership for Sustainable Development; No. JPMJSA1501) and aXis (Accelerating Social Implementation for SDGs Achievement; No. JPMJAS2001) funded by JST as well as Program for supporting introduction of the new sharing system (JPMXS0420100619).

Institutional Review Board Statement: This study did not sacrifice any animals. Organs were collected from dead animals. Experienced veterinarians collected blood from grazing livestock managed by herdsmen. Thus, animals did not endure undue stress from blood collection. We always explained and received permission from owners and herdsmen when obtaining blood from animals. We also obtained permission from the Mongolian government for transporting samples from Mongolia to Japan. We did not use any human sample or private individual information in this research.

Data Availability Statement: The datasets used in the study are available in the Supplementary Materials at www.mdpi.com/xxx/s1 and from the corresponding author on reasonable request.

Acknowledgments: We are grateful to JICA for not only their financial support ("The Project for Strengthening the Capacity for Human Resource Development in the Field of Veterinary and Animal Husbandry"), but also support in producing a better environment for our activity in the university, as well as language translation. We are sincerely grateful to all veterinarians and herdsmen in Dornogovi and Tuv provinces who cooperated in this research. We wish to thank our colleagues in the Toxicology Laboratory of Hokkaido University of Japan, as well as those in the Department of Pharmacology and Internal Medicine, School of Veterinary Medicine, MULS. We would also like to thank Editage for English-language editing.

Conflicts of Interest: The authors declare no conflict of interest.

References

1. AZOMINING. Available online: https://www.azomining.com/ (accessed on 17 August 2022).
2. The Observatory of Economic Complexity (OEC). Available online: https://oec.world/ (accessed on 17 August 2022).
3. The Organisation for Economic Co-Operation and Development (OECD) ILibrary. Available online: https://www.oecd-ilibrary.org/ (accessed on 17 August 2022).
4. Nakayama, S.M.M.; Ikenaka, Y.; Hamada, K.; Muzandu, K.; Choongo, K.; Teraoka, H.; Mizuno, N.; Ishizuka, M. Metal and Metalloid Contamination in Roadside Soil and Wild Rats around a Pb-Zn Mine in Kabwe, Zambia. *Environ. Pollut.* **2011**, *159*, 175–181. [CrossRef] [PubMed]
5. Swarup, D.; Naresh, R.; Varshney, V.P.; Balagangatharathilagar, M.; Kumar, P.; Nandi, D.; Patra, R.C. Changes in Plasma Hormones Profile and Liver Function in Cows Naturally Exposed to Lead and Cadmium around Different Industrial Areas. *Res. Vet. Sci.* **2007**, *82*, 16–21. [CrossRef] [PubMed]
6. Angelovičová, L.; Fazekašová, D. Contamination of the Soil and Water Environment by Heavy Metals in the Former Mining Area of Rudňany (Slovakia). *Soil Water Res.* **2014**, *9*, 18–24. [CrossRef]
7. Orisakwe, O.E.; Oladipo, O.O.; Ajaezi, G.C.; Udowelle, N.A. Horizontal and Vertical Distribution of Heavy Metals in Farm Produce and Livestock around Lead-Contaminated Goldmine in Dareta and Abare, Zamfara State, Northern Nigeria. *J. Environ. Public Health* **2017**, *2017*, 3506949. [CrossRef] [PubMed]
8. Duruibe, J.O.; Ogwuegbu, M.O.C.; Egwurugwu, J.N. Heavy Metal Pollution and Human Biotoxic Effects. *Int. J. Phys. Sci.* **2007**, *2*, 112–118.
9. Briffa, J.; Sinagra, E.; Blundell, R. Heavy Metal Pollution in the Environment and Their Toxicological Effects on Humans. *Heliyon* **2020**, *6*, e04691. [CrossRef]
10. Mukesh, K.R.; Puneet, K.; Manoj, S.; Anand, S. Toxic Effect of Heavy Metals in Livestock Health. *Vet. World* **2008**, *1*, 28–30.
11. Ali, H.; Khan, E.; Ilahi, I. Environmental Chemistry and Ecotoxicology of Hazardous Heavy Metals: Environmental Persistence, Toxicity, and Bioaccumulation. *J. Chem.* **2019**, *2019*, 6730305. [CrossRef]
12. Batjargal, T.; Otgonjargal, E.; Baek, K.; Yang, J.S. Assessment of Metals Contamination of Soils in Ulaanbaatar, Mongolia. *J. Hazard Mater.* **2010**, *184*, 872–876. [CrossRef]
13. Vardhan, K.H.; Kumar, P.S.; Panda, R.C. A Review on Heavy Metal Pollution, Toxicity and Remedial Measures: Current Trends and Future Perspectives. *J. Mol. Liq.* **2019**, *290*, 111197. [CrossRef]
14. Battsengel, E.; Murayama, T.; Fukushi, K.; Nishikizawa, S.; Chonokhuu, S.; Ochir, A.; Tsetsgee, S.; Davaasuren, D. Ecological and Human Health Risk Assessment of Heavy Metal Pollution in the Soil of the Ger District in Ulaanbaatar, Mongolia. *Int. J. Environ. Res. Public Health* **2020**, *17*, 4668. [CrossRef] [PubMed]
15. Rajaganapathy, V.; Xavier, F.; Sreekumar, D.; Mandal, P.K. Heavy Metal Contamination in Soil, Water and Fodder and Their Presence in Livestock and Products: A Review. *J. Environ. Sci. Technol.* **2011**, *4*, 234–249. [CrossRef]
16. Anees, M.; Jamal, Q.; Durani, P.; Khan, K.; Munir, S.; Hussain, S.; Munir, K. Heavy Metals Accumulation and Their Toxic Effects: Review. *J. Bio-Mol. Sci. (JBMS)* **2013**, *1*, 27–36.
17. Pareja-Carrera, J.; Mateo, R.; Rodríguez-Estival, J. Lead (Pb) in Sheep Exposed to Mining Pollution: Implications for Animal and Human Health. *Ecotoxicol. Environ. Saf.* **2014**, *108*, 210–216. [CrossRef] [PubMed]
18. Yohannes, Y.B.; Ikenaka, Y.; Nakayama, S.M.; Mizukawa, H.; Ishizuka, M. Trace Element Contamination in Tissues of Four Bird Species from the Rift Valley Region. *Bull. Environ. Contam. Toxicol.* **2017**, *98*, 172–177. [CrossRef]
19. Yohannes, Y.B.; Ikenaka, Y.; Saengtienchai, A.; Watanabe, K.P.; Nakayama, S.M.M.; Ishizuka, M. Occurrence, Distribution, and Ecological Risk Assessment of DDTs and Heavy Metals in Surface Sediments from Lake Awassa-Ethiopian Rift Valley Lake. *Environ. Sci. Pollut. Res.* **2013**, *20*, 8663–8671. [CrossRef]
20. Chiroma, T.M.; Ebewele, R.O.; Hymore, F.K. Comparative Assessment of Heavy Metal Levels in Soil, Vegetables and Urban Grey Waste Water Used for Irrigation in Yola and Kano. *Int. Refereed J. Eng. Sci.* **2014**, *3*, 1–9.
21. U.S. Environmental Protection Agency. *Ecological Soil Screening Levels for Zinc*; U.S. Environmental Protection Agency: Washington, DC, USA, 2007.
22. U.S. Environmental Protection Agency. *Ecological Soil Screening Levels for Cadmium*; U.S. Environmental Protection Agency: Washington, DC, USA, 2005.

23. U.S. Environmental Protection Agency. *Ecological Soil Screening Levels for Arsenic*; U.S. Environmental Protection Agency: Washington, DC, USA, 2005.
24. U.S. Environmental Protection Agency. *Ecological Soil Screening Levels for Manganese*; U.S. Environmental Protection Agency: Washington, DC, USA, 2007.
25. U.S. Environmental Protection Agency. *Ecological Soil Screening Levels for Selenium*; U.S. Environmental Protection Agency: Washington, DC, USA, 2007.
26. U.S. Environmental Protection Agency. *Ecological Soil Screening Levels for Copper*; U.S. Environmental Protection Agency: Washington, DC, USA, 2006.
27. U.S. Environmental Protection Agency. *Ecological Soil Screening Levels for Lead*; U.S. Environmental Protection Agency: Washington, DC, USA, 2005.
28. U.S. Environmental Protection Agency. *Ecological Soil Screening Levels for Chromium*; U.S. Environmental Protection Agency: Washington, DC, USA, 2005.
29. Puls, R. *Mineral Levels in Animal Health: Diagnostic Data*; Sherpa International: Aldergrove, BC, Canada, 1994.
30. Kincaid, R.L. Assessment of Trace Mineral Status of Ruminants: A Review. *J. Anim. Sci.* **1999**, *77*, 1–10. [CrossRef]
31. Biswas, U.; Sarkar, S.; Bhowmik, M.K.; Samanta, A.K.; Biswas, S. Chronic Toxicity of Arsenic in Goats: Clinicobiochemical Changes, Pathomorphology and Tissue Residues. *Small Rumin. Res.* **2000**, *38*, 229–235. [CrossRef]
32. Rana, T.; Bera, A.K.; Das, S.; Bhattacharya, D.; Bandyopadhyay, S.; Pan, D.; Kumar Das, S. Effect of Chronic Intake of Arsenic-Contaminated Water on Blood Oxidative Stress Indices in Cattle in an Arsenic-Affected Zone. *Ecotoxicol. Environ. Saf.* **2010**, *73*, 1327–1332. [CrossRef]
33. Nandi, D.; Patra, R.C.; Swarup, D. Effect of Cysteine, Methionine, Ascorbic Acid and Thiamine on Arsenic-Induced Oxidative Stress and Biochemical Alterations in Rats. *Toxicology* **2005**, *211*, 26–35. [CrossRef] [PubMed]
34. Alonso, M.L.; Benedito, J.L.; Miranda, M.; Castillo, C.; Hernández, J.; Shore, R.F. Cattle as Biomonitors of Soil Arsenic, Copper, and Zinc Concentrations in Galicia (NW Spain). *Arch. Environ. Contam. Toxicol.* **2002**, *43*, 103–108. [CrossRef] [PubMed]
35. Jomova, K.; Jenisova, Z.; Feszterova, M.; Baros, S.; Liska, J.; Hudecova, D.; Rhodes, C.J.; Valko, M. Arsenic: Toxicity, Oxidative Stress and Human Disease. *J. Appl. Toxicol.* **2011**, *31*, 95–107. [CrossRef] [PubMed]
36. Raj Khanal, D.; Knight, A.P. Selenium: Its Role in Livestock Health and Productivity. *J. Agric. Environ.* **2010**, *11*, 101–106. [CrossRef]
37. Tiwary, A.K.; Stegelmeier, B.L.; Panter, K.E.; James, L.F.; Hall, J.O. Comparative Toxicosis of Sodium Selenite and Selenomethionine in Lambs. *J. Vet. Diagn. Investig.* **2006**, *18*, 61–70. [CrossRef] [PubMed]
38. Deore, M.D.; Srivastava, A.K.; Sharma, S.K. Effect of Reduced Glutathione Treatment on Selenosis, Blood Selenium Concentration and Glutathione Peroxidase Activity after Repeated Short-Term Selenium Exposure in Buffalo Calves. *Toxicology* **2005**, *213*, 169–174. [CrossRef]
39. Osredkar, J.; Sustar, N. Copper and Zinc, Biological Role and Significance of Copper/Zinc Imbalance. *Clin. Toxicol.* **2011**, *s3*, 1. [CrossRef]
40. Todd, J.R. Chronic Copper Toxicity of Ruminants. *Proc. Nutr. Soc.* **1969**, *28*, 189–198. [CrossRef] [PubMed]
41. Wiener, G.; Hall, J.G.; Hayter, S. An Association between the Concentration of Copper in Whole Blood and Haemoglobin Type in Sheep. *Anim. Prod.* **1973**, *17*, 1–7. [CrossRef]
42. Sobhanirad, S.; Naserian, A.A. Effects of High Dietary Zinc Concentration and Zinc Sources on Hematology and Biochemistry of Blood Serum in Holstein Dairy Cows. *Anim. Feed Sci. Technol.* **2012**, *177*, 242–246. [CrossRef]
43. Chafik, A.; Essamadi, A.; Çelik, S.Y.; Mavi, A. Purification and Biochemical Characterization of a Novel Copper, Zinc Superoxide Dismutase from Liver of Camel (Camelus Dromedarius): An Antioxidant Enzyme with Unique Properties. *Bioorg. Chem.* **2019**, *86*, 428–436. [CrossRef] [PubMed]
44. Bengoumi, M.; Essamadi, A.K.; Tressol, J.C.; Faye, A.B. Comparative Study of Copper and Zinc Metabolism in Cattle and Camel. *Biol. Trace Elem. Res.* **1998**, *63*, 81–94. [CrossRef] [PubMed]
45. Farmer, A.A.; Farmer, A.M.; Stan, T. Concentrations of Cadmium, Lead and Zinc in Livestock Feed and Organs around a Metal Production Centre in Eastern Kazakhstan. *Sci. Total Environ.* **2000**, *257*, 53–60. [CrossRef]
46. Jorhem, L. Lead and Cadmium in Tissues from Horse, Sheep, Lamb and Reindeer in Sweden. *Z Lebensm. Unters Forsch.* **1999**, *208*, 106–109. [CrossRef]
47. Janette, O. Itai-Itai Disease Caused by Cadmium Poisoning. *J. Heavy Met. Toxic. Dis.* **2021**, *6*, 455–463.
48. Nakata, H.; Nakayama, S.M.M.; Kataba, A.; Yohannes, Y.B.; Ikenaka, Y.; Ishizuka, M. Evaluation of the Ameliorative Effect of Spirulina (Arthrospira Platensis) Supplementation on Parameters Relating to Lead Poisoning and Obesity in C57BL/6J Mice. *J. Funct. Foods* **2021**, *77*, 104344. [CrossRef]
49. Nakata, H.; Nakayama, S.M.M.; Yabe, J.; Muzandu, K.; Toyomaki, H.; Yohannes, Y.B.; Kataba, A.; Zyambo, G.; Ikenaka, Y.; Choongo, K.; et al. Clinical Biochemical Parameters Associated with the Exposure to Multiple Environmental Metals in Residents from Kabwe, Zambia. *Chemosphere* **2021**, *262*, 127788. [CrossRef]
50. Nakata, H.; Nakayama, S.M.M.; Ikenaka, Y.; Mizukawa, H.; Ishii, C.; Yohannes, Y.B.; Konnai, S.; Darwish, W.S.; Ishizuka, M. Metal Extent in Blood of Livestock from Dandora Dumping Site, Kenya: Source Identification of Pb Exposure by Stable Isotope Analysis. *Environ. Pollut.* **2015**, *205*, 8–15. [CrossRef]
51. Pandey, G.; Madhuri, S. Heavy Metals Causing Toxicity in Animals and Fishes. *Res. J. Anim. Vet. Fish. Sci.* **2014**, *2*, 17–23.

52. Reglero, M.M.; Monsalve-González, L.; Taggart, M.A.; Mateo, R. Transfer of Metals to Plants and Red Deer in an Old Lead Mining Area in Spain. *Sci. Total Environ.* **2008**, *406*, 287–297. [CrossRef]
53. Filipović-Trajković, R.; Ilić, Z.S.; Šunić, L.; Andjelković, S. The Potential of Different Plant Species for Heavy Metals Accumulation and Distribution. *J. Food Agric. Environ.* **2005**, *10*, 959–964.

toxics

MDPI

Article

Spleen Toxicity of Organophosphorus Flame Retardant TDCPP in Mice and the Related Mechanisms

Lanqin Cao [1], Lai Wei [2], Qiaoyun Du [2], Ying Su [2], Shuzi Ye [2,*] and Kaihua Liu [3,*]

1 Xiangya Hospital, Central South University, Changsha 410078, China
2 Xiangya School of Public Health, Central South University, Changsha 410078, China
3 Department of Anatomy and Cell Biology, Carver College of Medicine, University of Iowa, Iowa City, IA 52242, USA
* Correspondence: shuziyemickey@163.com (S.Y.); kaihua-liu@uiowa.edu.cn (K.L.)

Abstract: Tris(1,3-dichloro-2-propyl) phosphate (TDCPP) is an organophosphorus flame retardant that has been utilized in recent years as a primary replacement for polybrominated diphenyl ethers (PBDEs) in a wide variety of fire-sensitive applications. However, the impact of TDCPP on the immune system has not been fully determined. As the largest secondary immune organ in the body, the spleen is considered to be an important study endpoint for determining immune defects in the body. The aim of this study is to investigate the effect of TDCPP toxicity on the spleen and its possible molecular mechanisms. In this study, for 28 consecutive days, TDCPP was administered intragastrically (i.g), and we assessed the general condition of mice by evaluating their 24 h water and food intake. Pathological changes in spleen tissues were also evaluated at the end of the 28-day exposure. To measure the TDCPP-induced inflammatory response in the spleen and its consequences, the expression of the critical players in the NF-κB pathway and mitochondrial apoptosis were detected. Lastly, RNA-seq was performed to identify the crucial signaling pathways of TDCPP-induced splenic injury. The results showed that TDCPP intragastric exposure triggered an inflammatory response in the spleen, likely through activating the NF-κB/IFN-γ/TNF-α/IL-1β pathway. TDCPP also led to mitochondrial-related apoptosis in the spleen. Further RNA-seq analysis suggested that the TDCPP-mediated immunosuppressive effect is associated with the inhibition of chemokines and the expression of their receptor genes in the cytokine–cytokine receptor interaction pathway, including four genes of the CC subfamily, four genes of the CXC subfamily, and one gene of the C subfamily. Taken together, the present study identifies the sub-chronic splenic toxicity of TDCPP and provides insights on the potential mechanisms of TDCPP-induced splenic injury and immune suppression.

Keywords: TDCPP; splenic injury; NF-κB pathway; mitochondrial apoptotic pathway; chemokines and receptors; immunosuppression

Citation: Cao, L.; Wei, L.; Du, Q.; Su, Y.; Ye, S.; Liu, K. Spleen Toxicity of Organophosphorus Flame Retardant TDCPP in Mice and the Related Mechanisms. *Toxics* **2023**, *11*, 231. https://doi.org/10.3390/toxics11030231

Academic Editors: Roel Vermeulen, Lauren Petrick and Maaike van Gerwen

Received: 2 January 2023
Revised: 19 February 2023
Accepted: 22 February 2023
Published: 27 February 2023

1. Introduction

Tris(1,3-dichloro-2-propyl) phosphate (TDCPP) is one of the commonly used organophosphorus flame retardants (OPFRs) in a broad range of products, including furniture, buildings, textiles, cars, clothing, and other items [1]. Some polybrominated diphenyl ethers (PBDEs), such as Penta- and Octa-BDE, have been restricted or banned in the EU due to overwhelming evidence that they are environmentally persistent organic pollutants (POPs) [2–4]. With its widespread use, TDCPP has been universally distributed to the environment—in the atmosphere, soil, surface water, groundwater and even glaciers that migrate long distances to the poles [1,5]. Besides its prevalence in the environment, TDCPP and its metabolites have also been found in human biological samples such as placenta, breast milk, plasma, and urine in recent years [6,7]. Furthermore, emerging evidence suggests that TDCPP exposure can cause neurological, reproductive, endocrine, hepatic and renal damage [1,8,9]. Notably, it was listed as a known carcinogen on the Proposition 65 list of substances [10,11].

A number of studies have shown that TDCPP induces cytotoxicity through the activation of the mitochondrial apoptosis pathway [12]. Meanwhile, there is also evidence suggesting that TDCPP can induce neuronal damage by triggering the microglia-mediated inflammatory response [13] and secretion of inflammation-like adipokines [14]. In addition, it was demonstrated that TDCPP can disrupt the phagocytosis of human THP-1-derived macrophages through immunosuppression and promote cytokine/chemokine secretion and inflammatory responses [15].

The spleen, a secondary lymphoid organ with hematological and immunological activities, is the body's largest immune organ. It has the immunological function of identifying pathogens and aberrant cells and removing them; hence, it plays a significant role in the innate and adaptive immune response [16]. Patients treated clinically with splenectomy have a reduced clearance of malaria-hosted red blood cells and a higher risk of other infectious diseases [17]. Therefore, the spleen, as a key immune organ, stores many immune cells and is often regarded as an endpoint for assessing immunotoxin injury [18].

Despite the advances in TDCPP research over the past few years, there is still a lack of adequate understanding of the splenic immunotoxicological properties of this emerging contaminant. Transcriptome analysis techniques are currently an important research tool for conducting toxicity explorations of emerging contaminants. In this study, we identified differential genes based on a transcriptome sequencing platform to provide a basis for understanding the mechanisms of TDCPP-induced inflammatory responses and immune damage in the mice spleen.

2. Materials and Methods

2.1. Animals and Exposure

Male, 4-week-old C57BL/6J mice (initial weight: 18 ± 2 g) were purchased from the Hunan SJA Laboratory Animal Co., Ltd. (SCXK(XIANG)2019-0004). All animal experiments were approved by the Laboratory Animal Welfare Ethical Committee of Central South University (Ethics Approval Code: XYGW-2021-113). The mice were housed in a climate control and pathogen-free room (6 animals/cage, 12 h dark–light cycle, 24 ± 2 °C, 65 ± 25% humidity).

After at least 1 week of acclimatation, animals were randomly allocated into two groups using random number tables: the control and the TDCPP groups. The TDCPP group was intragastric administrated (i.g) with TDCPP (purity: 95%; Bide Pharmaceutical Technology Co., Ltd., Shanghai, China) of 300 mg/kg a day for 28 consecutive days. Corn oil was used in this study as the TDCPP solvent. TDCPP had a reported oral LD50 of 2670 mg/kg in male mice [19]. A rather high dose of 300 mg/kg (approximately 1/10 LD50) was used as the oral dose in this investigation to examine the subacute toxic effects. The 24 h water and food intakes on the 7th, 14th, 21st and 28th days of exposure were measured. Every three days during the same period, the weight of the mice was measured.

2.2. Tissue Collection

Anesthesia was induced in mice using sodium pentobarbital (i.p. 40 mg/kg). Peripheral blood sampling was performed after the mice were anesthetized. Upon tissue harvest, the spleen was divided into two pieces. One piece was instilled with cold 0.9% saline, fixed in 4% paraformaldehyde for 4 h at 4 °C and stored for further processing. The other piece of spleen was flash frozen in liquid nitrogen and stored at −80 °C until further processing.

2.3. Hematoxylin-Eosin (H&E) Staining

All specimens underwent 4% paraformaldehyde fixation at 4 °C for 4 h before being moved to an ethanol gradient for dehydration. Dehydrated samples were then placed in a mixed solution of xylene and paraffin for paraffin permeation, followed with dewaxing with xylene and ethanol gradients in turn. After soaking and staining with hematoxylin and eosin dye, respectively, ethanol, xylene and neutral resin were used to dehydrate and seal the sections and make them transparent. An OLYMPUS BX53 digital camera and the DP73 controller software were used to take all of the digital pictures (×200). According to the standard classification, normal (0), mild (1), moderate (2), severe (3), and very severe (4) categories were used to determine the degree of splenic tissue damage. Six visual fields/mouse pathology scores were used to compute the severity of the damage for each group of at least three mice. The specific operation is based on the methods used in a prior investigation [20].

2.4. Immunohistochemistry (IHC) Staining

During the IHC staining of spleen tissue, the sample was paraffin-embedded, sectioned, and dewaxed in the same manner as the previous H&E staining. Then, the sections were heated in a microwave oven for antigen retrieval. Prior to the incubation of the primary antibody, sections were immersed in 3% hydrogen peroxide solution for 15 min at room temperature and incubated in 1% BSA for 15 min at room temperature to block. Primary antibody solutions at different dilutions were prepared and incubated overnight at 4 °C in humid boxes; then, the appropriate secondary antibody was added, and the mixture was left to sit at room temperature for 60 min in the dark. This was followed by DAB coloring, counterstaining with hematoxylin, dehydration with ethanol, and making it transparent with xylene, and resin that is neutral was used for sealing. Finally, under a 400-fold microscope (OLYMPUS, Tokyo, Japan, BX53 & DP73), at least three regions of the spleens of each group of mice were observed and statistically analyzed. The intensity of the stained signals was measured and analyzed using Image-Pro Plus 6.0 image analysis software (Media Cybernetics, Inc. Silver Spring, MD, USA) according to previous studies and the average density of the digital images was calculated [21]. The mean density of the digital images (×400) was designated as representative NF-κB/IFN-γ/TNF-α/IL-1β staining intensity (indicative of relative NF-κB/IFN-γ/TNF-α/IL-1β expression levels). All picture capture sessions used the same microscope settings, and the experimenters were unaware of the treatment groups in order to perform image quantification.

Antibodies used in this study: anti-NF-κB p65 (Wanleibio, China, WL01980), anti-IFN-γ (Wanleibio, China, WL02440), anti-TNF-α (Wanleibio, China, WL01896), anti-IL-1β (Wanleibio, China, WLH3903), and HRP Goat Anti-Rabbit IgG (H + L) (thermoFisher, Waltham, MA, USA, #31460).

2.5. Terminal dUTP Nick-End Labeling (TUNEL) Assay

The principle of the TUNEL assay is that, when apoptosis occurs, some DNA endonucleases will be activated. These endonucleases will cut off genomic DNA between nucleosomes. When genomic DNA breaks, exposed 3′-OH can be added with fluorescein dUTP under the catalysis of terminal deoxynucleotidyl transferase (TdT), Td, which can be detected with a fluorescence microscope and through flow cytometry. The specific operation was carried out according to previous research methods [22]. To quantify the image (×400) of the final spleen tissue, unified microscope settings (OLYMPUS, Tokyo, Japan, BX53 & DP73) were maintained in all image acquisition processes, with an excitation wavelength of 450 nm and emission wavelength of 520 nm. The spleen apoptosis of each group of three animals was examined for quantitative analysis by TUNEL assay as follows: The number of apoptotic cells and the total number of cells in at least three fields of view were counted for each animal, and the apoptotic rate in the three fields of view was calculated and the mean value was taken. The apoptosis rate was calculated according to the following formula: Apoptosis rate = number of apoptotic cells/total number of cells.

2.6. Western Blot (WB) Analysis

The radio immunoprecipitation assay (RIPA) strong lysate (Beyotime Biotechnology, Shanghai, China, P0013B) was used to lyse the spleen tissues for 30 min, after which they were centrifuged at 12,000× g for 15 min at 4 °C. The quantification of protein concentrations was performed using a BCA protein concentration detection kit (Beyotime Biotechnology, Shanghai, China, P0010). The tissue proteins were first incubated to primary antibodies at 4 °C overnight, and then to matching secondary antibodies for 1 h, and finally they were incubated using the LumiBest ECL substrate solution kit for visualization (ShareBio, Shanghai, China, SB-WB011). The protein band intensity was measured using Image-J software.

Antibodies used in this study: anti-TNF-α (Immunoway, Suzhou, China, YT4689, 1:1000), anti-IL-1β (Beyotime Biotechnology, Shanghai, China, AF7209), anti-NF-κB p65 (ABclone, Wuhan, China, A2547, 1:1000), anti-IFN-γ (Immunoway, Suzhou, China, YT2279, 1:1000), anti-Caspase-3 (Proteintech, Wuhan, China, 19677-1-AP, 1:1000) anti-Bcl-2 (ABclone, China, A19693, 1:1000), anti-Bax (Proteintech, China, 50599-2-lg, 1:1000), anti-Cytochrome C (Proteintech, China, 10993-1-AP, 1:1000), anti-β-actin (MultiSciences, Hangzhou, China, #HA-R1207-1-200, 1:1000), HRP Goat Anti-Rabbit IgG (H + L) (Abclone, China, #AS014, 1:4000), and HRP Goat Anti-Mouse IgG (H + L) (Abclone, China, #AS003, 1:4000).

2.7. RNA Extraction and cDNA Library Generation

Total RNA was isolated from each group of two mice spleen tissues, 100 mg/spleen tissue, using the TRIzol® Reagent (Magen) in accordance with the manufacturer's instructions. The Nanodrop ND-2000 system from Thermo Scientific (Waltham, MA, USA) was used to measure RNA concentrations based on the A260/A280 absorbance ratio, and the Agilent Bioanalyzer 4150 system was used to determine the RNA integrity number (RIN) (Agilent Technologies Inc., Santa Clara, CA, USA). Only approved samples were utilized to build the library. Following the manufacturer's instructions, paired-end libraries were created using an ABclonal mRNA-seq Lib Prep Kit (ABclonal, Wuhan, China). Using oligo (dT) magnetic beads and divalent cations at high temperatures in ABclonal First Strand Synthesis Reaction Buffer, the mRNA was isolated from 1 μg of total RNA. Then, using mRNA fragments as templates, first-strand cDNAs were created using random hexamer primers and Reverse Transcriptase (RNase H), and second-strand cDNAs were created using DNA polymerase I, RNAseH, buffer, and dNTPs. The prepared paired-end library was created by adapter-ligating the generated double stranded cDNA fragments. For PCR amplification, adaptor-ligated cDNA was utilized. On the Agilent Bioanalyzer 4150 system, library quality was evaluated after PCR products were purified to use the AMPure XP system. Finally, 150 bp paired-end reads were produced using the Illumina Novaseq 6000 (Illumina, San Diego, CA, USA) to sequence the library preparations.

2.8. RNA-Seq Data Analysis

On the Illumina platform, RNA-Seq data analysis was carried out. Shanghai Applied Protein Technologies was used for all analyses. These are the main software elements and parameters.

2.8.1. Quality Control

Fastq-format raw data (or raw reads) were initially processed using custom Perl scripts. In this step, the adapter sequence must be removed, followed by the filtering out of low-quality read (low-quality reads are those where the proportion of lines with a string quality value of less than or equal to 25 accounts for more than 60% of the entire reading) and N read (N reads are those where the base information cannot be determined) ratios greater than 5% in order to produce clean reads suitable for further analysis.

2.8.2. Mapping

HISAT2 software (version: 2.1.0, http://daehwankimlab.github.io/hisat2/ format: 29 November 2022.) was then used to align clean reads sequentially to the reference genome in orientation mode to produce mapped reads [23].

2.8.3. Quantification of Gene Expression Level

The amount of reads that were mapped to each gene was counted using Feature Counts (version: 2.0.0, http://subread.sourceforge.net/ format: 29 November 2022.). The length of each gene and the number of reads mapped to it were used to compute the FPKM of each gene.

2.8.4. Differential Expression Analysis

Differentially expressed genes (DEGs) with |log2FC| > 1 and *p*-value 0.05 were regarded to be significantly different expressed genes. Differential expression analysis was carried out using DESeq2 (version: 1.34.0, format: 29 November 2022.) [24].

2.8.5. Enrichment Analysis

The functional enrichment of differential genes can be explained through the Gene Ontology (GO) and KEGG enrichment analysis of differential genes, which can also shed light on the variations between samples at the level of gene function. For KEGG pathway enrichment analysis and GO function enrichment, we used the cluster Profiler R software package. The GO or KEGG function is deemed considerably enriched when $p < 0.05$ [25,26].

2.9. Quantitative Reverse-Transcription PCR(RT-qPCR) Analysis

Using RNA isolater Total RNA Extraction Reagent (Vazyme, Nanjing, China, R401-01) and the FastKing RT kit (with gDNA) (Tiangen, Beijing, China, KR116), spleen tissue from six mice in each group at 100 mg/spleen tissue was extracted from total RNA samples and reverse-transcribed into cDNA. The reverse-transcription reaction conditions were 42 °C for 3 min, 42 °C for 15 s, 95 °C for 30 min, and 0 °C throughout. SuperReal PreMix Plus (SYBR Green) (Tiangen, China, FP205) was used to perform qPCR on the CFX96 System (Bio-Rad, Hercules, CA, USA). The main mixtures were prepared and the cycle conditions were set (predenaturation stage at 95 for 15 min × 1 cycles; PCR reaction stage denaturation at 95 for 15 s; annealing/elongation at 63 for 30 s × 45 cycles; and melting curve stage at 95 °C for 15 s, 60 °C for 60 s, and 95 °C for 15 s × 1 cycles). The sequences of the primers used for the qPCR validation of RNA-Seq Data are listed in Table 1. The $2^{-\Delta\Delta CT}$ method was used to compute the relative expressions of genes, and they were then normalized to the Glyceraldehyde-3-phosphate dehydrogenase reference gene (Gapdh).

Table 1. Primer sequences for the qPCR validation of RNA-Seq data.

Gene Name	Gene ID	Primer Sequence (5′–3′)	Tm (°C)	Length (bp)
		Chemokine		
		CC subfamily		
Ccrl1l	12770	GGCATCATCAGTAGCATCA GAGGAAGAAGAGAAGCGTAA	55	303
Ccr3	12771	GCTTTGAGACCACACCCTATG ACCATTATGTTGCCCAGGAG	58	136
Ccr10	12777	ATGCTTCACTCGGTCTCT CACTACTGGATAGCGATAGG	55	127
Ccl24	56221	GGTTCAGAGGCACATACAA AGAGATGGACAGACAGACA	55	492
		CXC subfamily		
Ppbp	57349	TGTGCTGATGTGGAAGTGATAG CTGAGCAGGAAAATGGTTTGG	58	150
Pf4	56744	GAGGTGATCAAGGCAGGAC AGCTGATACCTAACTCTCCAGG	58	141
Cxcl10	15945	GATGGATGGACAGCAGAG GGAAGATGGTGGTTAAGTTC	54	417
Ackr3	12778	TTCATCAACCGCAACTACA TCTCCTCTTCATACCACTCA	55	260

Table 1. Cont.

Gene Name	Gene ID	Primer Sequence (5′–3′)	Tm (°C)	Length (bp)
		Chemokine		
		C subfamily		
Xcr1	23832	ATCTTCTTCCTCCTCTCCAT ATCCACTTCTCCTTGTCTTC	55	473
		Glyceraldehyde-3-phosphate dehydrogenase protein family		
Gapdh	14433	TCTCCTGCGACTTCAACA TGTAGCCGTATTCATTGTCA	56	117

2.10. Statistical Analysis

In this work, all data were collected from at least three different experiments and are expressed as mean ± standard deviation (SD). A two-sample *t*-test was used to statistically analyze differences between the two groups. At a level of p-values < 0.05, differences were deemed significant. The experimental data were analyzed and plotted using SPSS 26 and GraphPad 9.0 software.

3. Results

3.1. Oral Exposure to TDCPP Caused Systematic Response and Induced Splenic Damage

In Figure 1A, the chemical formula for TDCPP is displayed. One of the most crucial measures of an animal's health status is its body weight. Therefore, at the same time of day, we measured the weight of mice in each group on Days 0, 3, 6, 9, 12, 15, 18, 21, 24 and 27 of TDCPP exposure. The body weights of all animals were similar at the beginning before TDCPP exposure. Starting from day 3, compared to the control group, the average body weight of mice in the TDCPP group was significantly lower (Figure 1B). Similarly, animals in the TDCPP group started to have significantly reduced 24 h water and food intake since day 14 and day 7, respectively (Figure 1C,D). In addition, as shown by the H&E staining of spleen tissue (Figure 1E), the white pulp of mice spleen in the TDCPP group was expanded and fused, with a blurry boundary, and the red pulp color was significantly deepened due to vascular expansion. Additionally, there was a significant difference in the splenic pathological damage scores between the two groups of animals, which showed a significantly higher TDCPP than the control group.

3.2. Exposure to TDCPP Activated the NF-kB Pathway and Induced Inflammation in the Spleen

The NF-κB signaling pathway is involved in physiological and pathological processes, such as infection, immune regulation, inflammatory reaction and tumor formation [27]. Therefore, to investigate the role of potential mechanisms of inflammatory response in splenic tissue injury by TDCPP, we analyzed the protein expression of NF-κB/IFN-γ/TNF-α/IL-1β in mice spleens using IHC and WB, respectively. As indicated by the IHC staining and quantification in Figure 2A, the expressions of NF-κB, IFN-γ, TNF-α and IL-1β were all more highly expressed in the TDCPP group compared to the control group. Similar trends were found by immunoblotting NF-κB, IFN-γ, TNF-α and IL-1β in total protein lysates of spleen tissues from control or TDCPP-treated mice (Figure 2B). IL-1β is a pro-inflammatory cytokine produced mainly by activated monocytes and epithelial cells. Pro-IL-1β is cleaved by caspase-1 into cleaved-IL-1β, which is the mature form of IL-1β and a good indicator of caspase-1 activity. Overall, these results indicate that the subchronicintragastric exposure of TDCPP may trigger an inflammatory response in the spleen through the activation of the NF-κB/IFN-γ/TNF-α/IL-1β pathway.

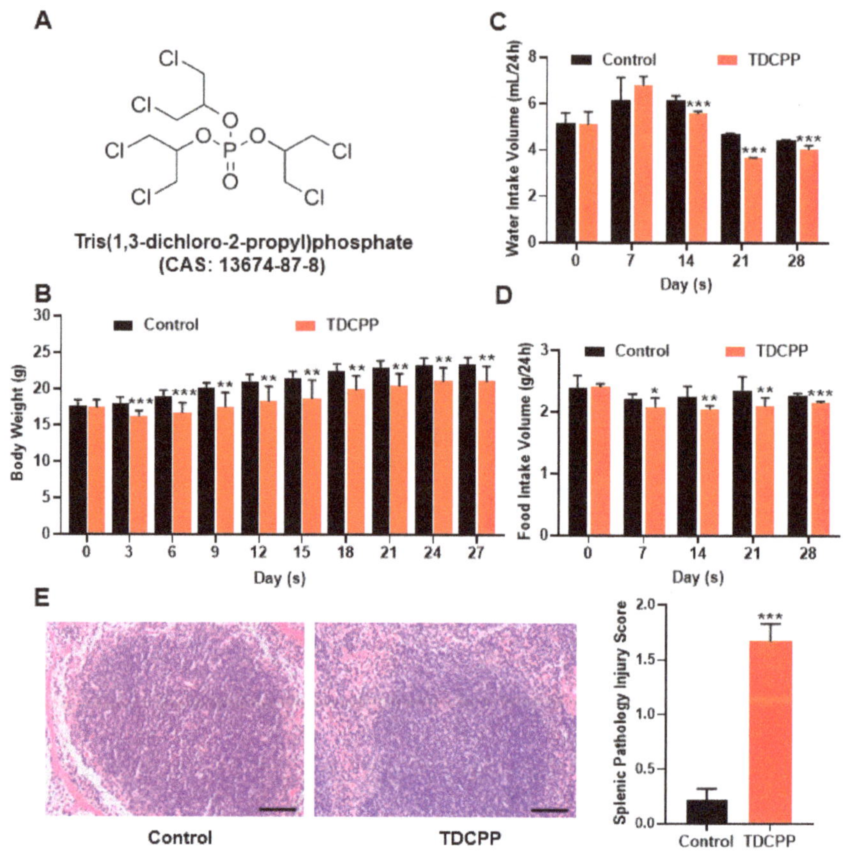

Figure 1. Oral exposure to TDCPP caused systematic response and induced splenic damage. (**A**) Chemical structural formula of TDCPP. (**B**) Body weights of mice in each group were measured on days 0, 3, 6, 9, 12, 15, 18, 21, 24 and 27 under oral administration of 0 or 300 mg/kg TDCPP per day for 28 consecutive days. (**C**) Measurement of 24 h water intake on days 7, 14, 21 and 28 of TDCPP exposure. (**D**) Twenty-four-hour food intake was measured at days 7, 14, 21 and 28 of TDCPP exposure. (**E**) Examination of the effect of TDCPP on spleen histomorphology by fluorescence microscopy of H&E staining of spleen tissue. * $p < 0.05$, ** $p < 0.01$, *** $p < 0.001$ comparing with control group.

3.3. *TDCPP-Induced Apoptotic Cell Death in the Spleen*

TDCPP was shown to increase the activities of caspase-3 and caspase-9, promote the expression of Bax, inhibit the expression of Bcl-2, and finally lead to apoptosis through the mitochondrial apoptosis pathway [1]. Apoptosis can also be induced through the activation of death receptors, such as the interaction of TNF-α with TNF-αR [28]. As shown in Figure 3A, the proportion of TUNEL-positive cells in the spleen tissues from the TDCPP group was significantly higher than that of the control group. Consistent with the TUNEL staining, pro-caspase 3 and cleaved-caspase 3 expressions were increased in the TDCPP group in comparison to the control group. The expression of BAX and cytochrome C was also increased in TDCPP group. Further, there was a decrease in Bcl-2 expression in the TDCPP group. The above results show that TDCPP exposure leads to the apoptosis of mice spleen tissue cells via the mitochondrial apoptotic pathway.

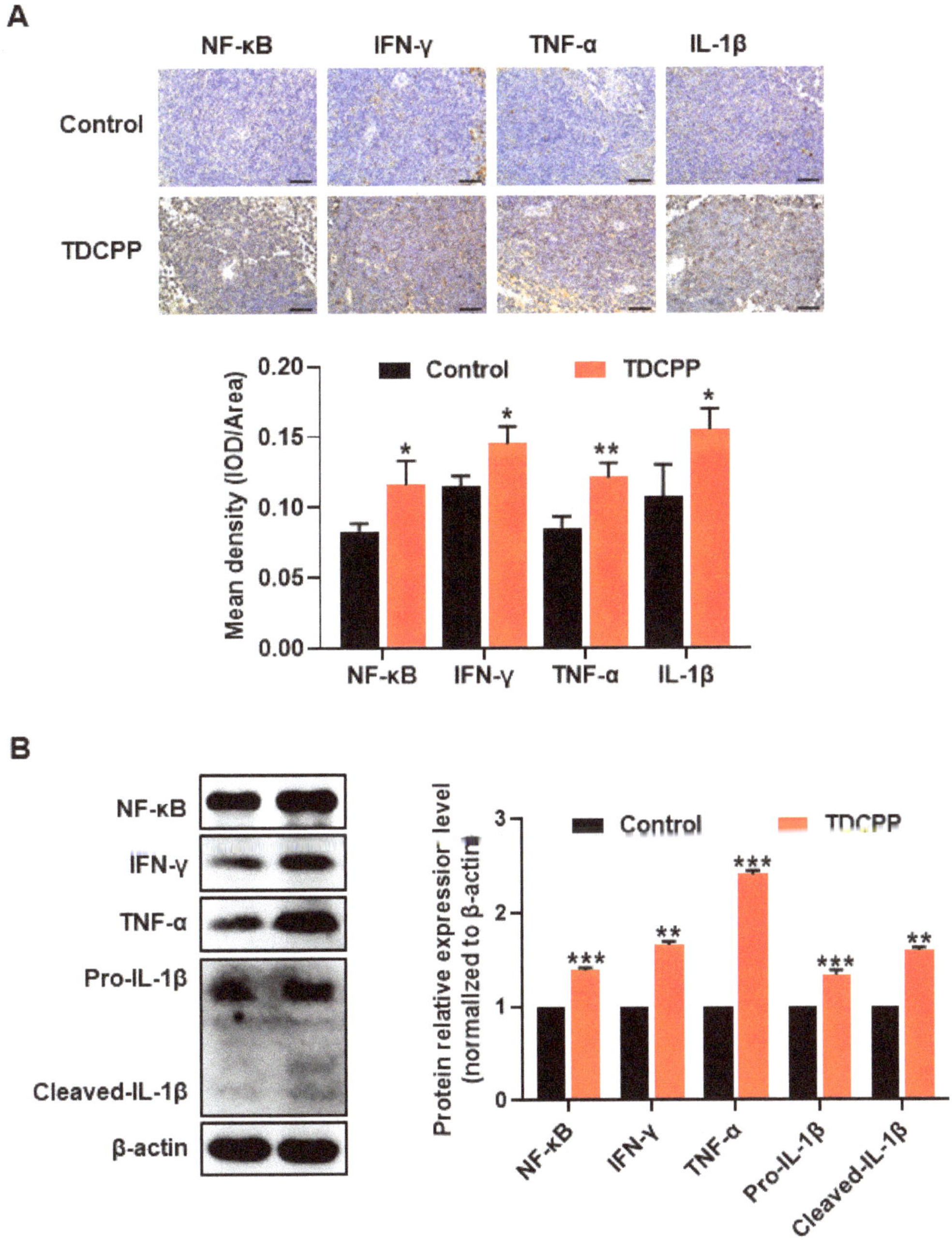

Figure 2. Exposure to TDCPP activated the NF-kB pathway and induced inflammation in the spleen. (**A**) To explore the changes in NF-κB signaling in the inflammatory response to TDCPP-induced splenic tissue injury, NF-κB/IFN-γ/TNF-α/IL-1β expression and tissue localization were detected in mice spleen using IHC. (**B**) The protein expressions of NF-κB/IFN-γ/TNF-α/IL-1β were detected using Western blotting. * $p < 0.05$, ** $p < 0.01$, *** $p < 0.001$, compared with control group.

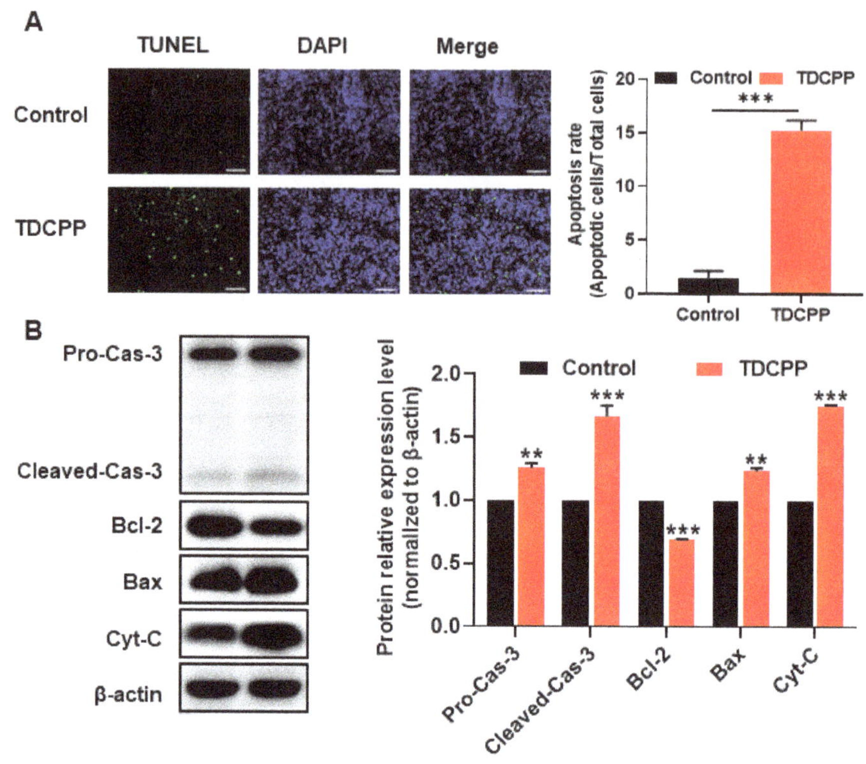

Figure 3. TDCPP-induced apoptotic cell death in the spleen. (**A**) The spleens of both groups of mice were examined by TUNEL assay, and the apoptotic rates of TUNEL-positive cells were calculated. (**B**) Detection of caspase-3/Bcl-2/Bax/Cyt-C protein expression levels using Western blotting to determine that TDCPP exposure leads to apoptosis in mice spleen tissue via the mitochondrial apoptosis pathway. ** $p < 0.01$, *** $p < 0.001$, compared with control group.

3.4. RNA-Seq Analysis of Mice Spleen Tissue

To get a broader picture of the changes post TDCPP exposure in the spleen at the transcriptome level, we performed RNA-seq analysis on RNAs isolated from the spleens. A total of 47,708,220 genes were identified, 97.29% of the sequence bases could be aligned to the genome, and the junction reads added up to more than 27.71% (Table S1). Gene expression levels were calculated for each sample in both groups, and the overall distribution of sample gene expression and inter-sample correlation were analyzed. As a result, 1025 DEGs in mice spleen were identified after TDCPP stimulation (Table S1). Overall, 23 upregulated and 1002 downregulated DEGs were identified in the TDCPP vs. control groups, as is shown in the DEG volcano plot (Figure 4A), statistical table of differential gene expression (Table S2), and DEG cluster heatmap (Figure 4B).

3.5. GO Analysis

To identify the function of DEGs in the spleen of mice after TDCPP exposure, a GO enrichment analysis of DEGs was carried out based on the GO database. The results show the top 10 GO terms according to secondary classification terms, biological process (BP), molecular function (MF) and cellular component (CC). In the BP annotation category, the upregulated genes were mainly associated with cell adhesion, while the downregulated genes were enriched for the immunologic process in the TDCPP vs. control groups (Figure 5A–C). In the MF and CC categories, the upregulated genes were in the intracellular junctional components, while the downregulated genes were enriched for cell membrane components in the TDCPP vs. control groups (Figure 5D–F).

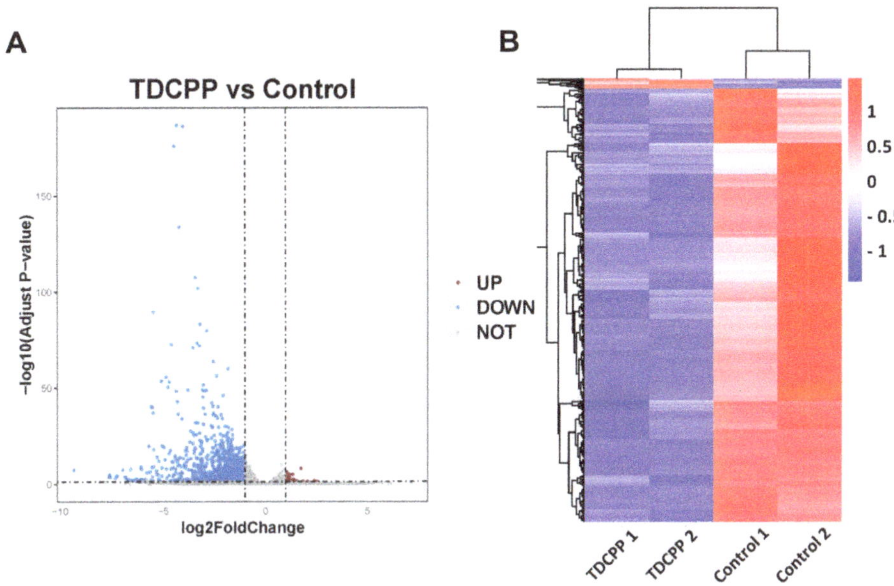

Figure 4. RNA-Seq analysis of mice spleen tissue. (**A**) DEGs were analyzed and volcano plots were drawn for mice treated with different doses of TDCPP. (**B**) Hierarchical clustering heat maps were plotted for DEGs between the control and TDCPP groups, and increases in expression are indicated in red, while decreases are indicated in blue.

3.6. KEGG Enrichment Analysis

We found 264 markedly impacted metabolic and signaling pathways using the KEGG pathway as a unit and the reference genome as the background. The KEGG enrichment pathway includes the top 40 up- and downregulated genes, as shown in Figure 6A, including: environmental information processing (signal transduction, signaling molecules and interaction), cellular processes (cellular community–eukaryotes, transport and catabolism, cellular community–eukaryotes), human diseases (infectious disease: parasitic, cardiovascular disease, environmental adaptation, cancer: overview, immune disease, infectious disease: bacterial), organismal systems (development and regeneration, immune system, digestive system, circulatory system, endocrine system), and metabolism (lipid metabolism, xenobiotics biodegradation and metabolism).

To further explore the effects of TDCPP on splenic immune and inflammatory responses, we noted the following ten KEGG pathways: complement and coagulation cascades, hematopoietic cell lineage, platelet activation, chemokine signaling pathway, neutrophil extracellular trap formation, ECM-receptor interaction, viral protein interaction with cytokine and cytokine receptor, cytokine-cytokine receptor interaction. Chemokines play a key role in inflammation and immunity by interacting with receptors to regulate the targeted migration of immune cells, remove sources of infection, and promote wound healing, among other functions [29]. Therefore, to investigate the effect of TDCPP on the spleen, we chose the cytokine–cytokine receptor interaction pathway for validation (Figure 6B,C). We validated nine candidate genes for the cytokine–cytokine receptor interaction pathway, including: Ccr1l1, Ccr3, Ccr10, Ccl24, Ppbp, Pf4, Cxcl10, Ackr3, and Xcr1.

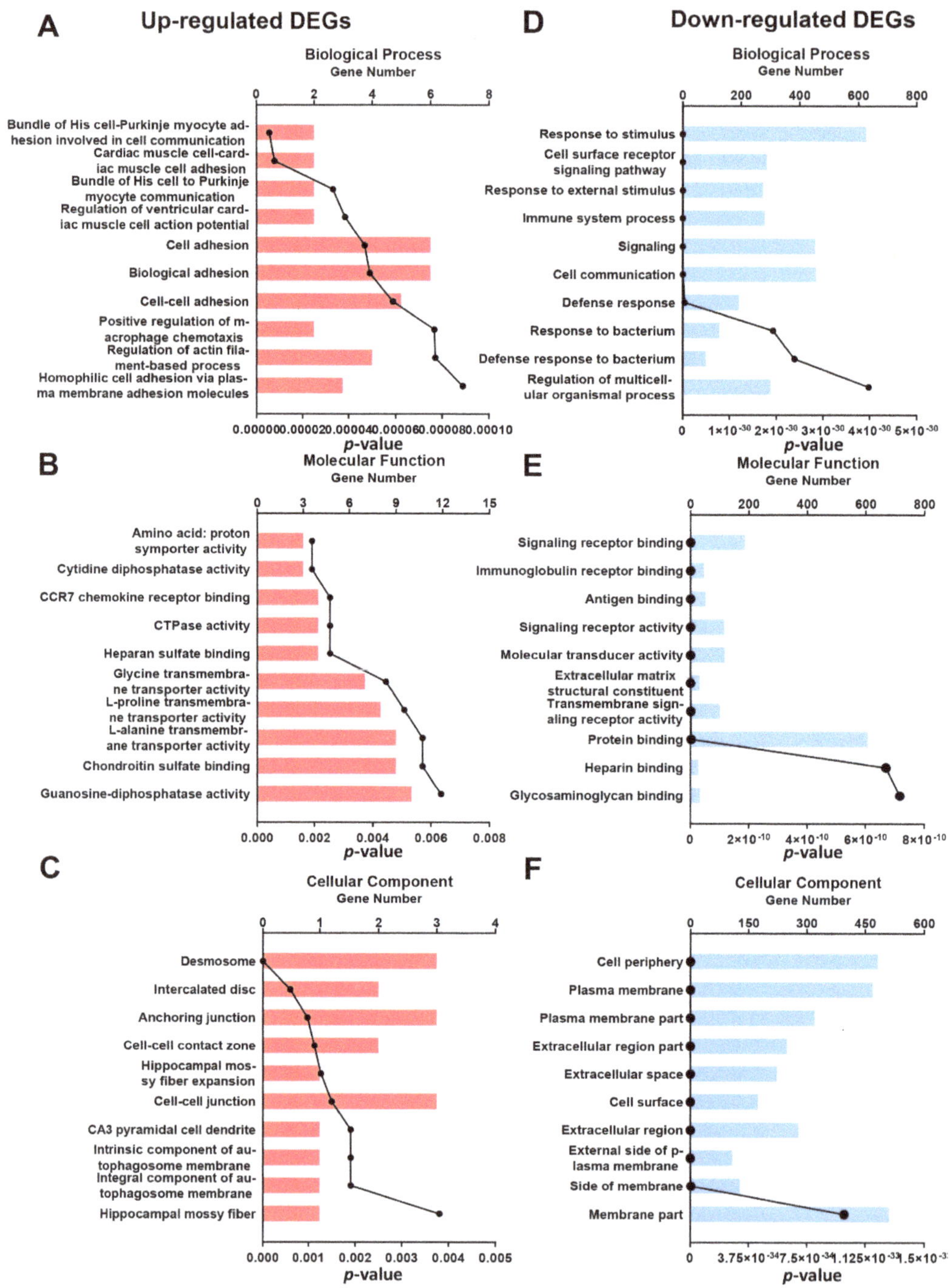

Figure 5. GO analysis of DEGs in TDCPP vs. control. The top 10 most significant p-values for the GO terms were upregulated by TDCPP compared to controls in terms of biological processes (BP) (**A**), molecular functions (MF) (**B**), and cellular components (CC) (**C**). The top 10 most significant p-values for the GO terms were downregulated by TDCPP compared to controls in terms of biological processes (BP) (**D**), molecular functions (MF) (**E**) and cellular components (CC) (**F**).

3.7. Validation of Gene Expression

As shown in Figure 6D, TDCPP significantly inhibited the expressions of Ccr1l1, Ccr3, Ccr10, Ccl24, Ppbp, Pf4, Cxcl10, Ackr3, and Xcr1 mRNAs in splenic tissue transcriptome sequencing, demonstrating the ability of TDCPP to inhibit the cytokine–cytokine receptor interaction pathway. As noted, we validated nine candidate genes. Consistent with RNA-Seq data, according to qRT-PCR analysis, compared to the control group, the expressions of chemokine-related genes, Ccr1l1, Ccr3, Ccr10, Ccl24, Ppbp, Pf4, Cxcl10, Ackr3, and Xcr1 in spleen tissue were significantly reduced ($p < 0.05$) (Figure 6D).

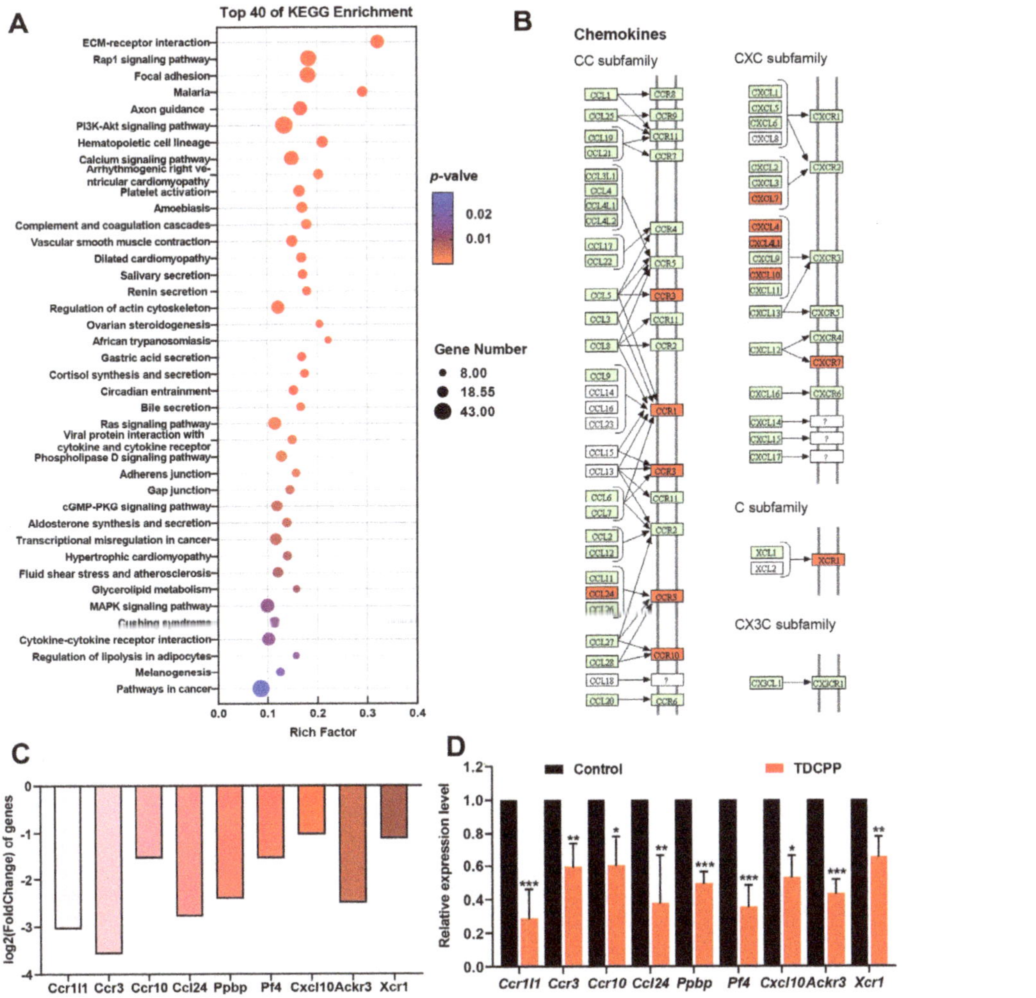

Figure 6. KEGG enrichment analysis and validation of gene expression of TDCPP vs. control DEGs. (**A**) The most significant p-values for the KEGG enrichment pathways were up- and downregulated by TDCPP vs. control. (**B**) Gene expression of chemokines in the cytokine–cytokine receptor interaction pathway. Red markers indicate downregulated genes. (**C**) The RNA sequencing results showed that the DEGs of chemokines in the cytokine–cytokine receptor interaction pathway in spleen tissue of TDCPP vs. control groups. (**D**) Validation of DEGs of chemokines in the cytokine–cytokine receptor interaction pathway in splenic tissues of TDCPP vs. control groups through qRT-PCR. * $p < 0.05$, ** $p < 0.01$, *** $p < 0.001$, compared with control group.

4. Discussion

TDCPP has been widely used in fire-susceptible items as one of the main PBDE alternatives over the past 20 years [1]. TDCPP, however, could leak out of items into different environmental media over the course of their lifetimes, in the same way that PBDEs do. For instance, through the industrial processing of food products, TDCPP entered food and was eaten by humans as a result [30]. Over the past few decades, studies on the toxicological effects of this product have mainly focused on reproduction [31], development [32], nerve toxicity [33], and endocrine disruption [34]. Research on immune system damage is still unclear, especially regarding the mechanisms of injury. In the present study, we employed environmental toxicology and transcriptome analysis techniques to determine the immunotoxicity of TDCPP exposure to mice spleen. As the largest secondary immune organ in our body, the spleen is an important line of defense for organisms against disease and invasion by harmful foreign bacteria and viruses, and it is susceptible to environmental perturbations. The spleen histomorphology of TDCPP-treated mice showed pathological changes. These pathomorphological alterations in the spleen are similar to previous findings, which showed that BDE-209 suppresses splenic cell immune and physiological functions by inducing inflammation and apoptosis, ultimately leading to splenic atrophy [35] and cadmium-induced spleen toxicity in mice [36], suggesting that these environmental pollutants cause splenic toxicity in a similar way. Body weight is one of the important indicators of the health of organisms and is influenced by several factors, such as excessive or insufficient energy intake and abnormal energy consumption, such as excessive exercise and disease. Therefore, when monitoring the weight of each group of mice, we also monitored their 24 h water intake and food intake every 7 days, respectively. After excluding the effects of drinking and feeding from the i.g operation, we found that TDCPP treatment might inhibit drinking and feeding in mice. This might be one of the main reasons for the weight change of TDCPP mice in a short time after treatment. Similarly, Sun et al. also noted in their study that rapid weight loss is primarily caused by a decrease in energy intake and not by an increase in energy expenditure or cachexia [37].

The inflammatory cytokine TNF-α can promote the inflammatory response by activating the NF-κB signaling pathway and further regulating the expression of downstream interleukin-related genes, adhesion factor-related genes, and other genes [38–40]. Our results are supported by previous reports of increased expression of NF-κB, TNF-α, and IL-1β in splenic tissue after TDCPP treatment, which further leads to splenic inflammation [39]. Moreover, some studies have also carried out in vivo and ex vivo tests that showed that TDCPP can increase the expressions of the pro-inflammatory factors Il-1beta and Tnf-mRNA in hippocampus neuronal cells, resulting in the induction of neurological inflammation [13]. IFN-γ is a cytokine with immune-regulatory, antivirus, antitumor, and antiparasitic effects, and it plays an essential role in maintaining the cellular immunity of the body, secreted by activated NK cells and T cells [41]. In a mouse model of aplastic anemia induced by irradiation and spleen–thymus lymphocyte infusion, the levels of the cytokines IL-6, IL-8, IL-17, TNF-α, and IFN-γ were increased in peripheral blood and bone marrow, and an immunoinflammatory response occurred [42]. Interestingly, this is consistent with our finding of increased IFN-γ and TNF-α expression in the spleen when the spleen was subjected to TDCPP. The above evidence suggests that TDCPP-induced inflammatory responses in the spleen are accompanied by the activation of the NF-κB signaling pathway.

In addition, TDCPP caused apoptosis by altering the transcriptional levels of bcl-2, bax, and caspase 3 genes, which ultimately led to apoptosis in SH-SY5Y cells [13]. In normal human-skin keratinocytes, TDCPP caused the protein expression of D1, CDK2, CDK6, and Bcl-2, in addition to boosting Bax and Caspase-3 expression to cause apoptosis and cell cycle arrest [43]. Corresponding to this, in our study, we discovered that in the spleen tissues of mice given TDCPP, the relative expressions of caspase 3, BAX, and cytochrome C proteins were elevated, whereas the relative expressions of Bcl-2 proteins were lowered. Thus, the activation of mitochondrial apoptosis pathways is important for TDCPP-induced cytotoxicity. It is worth noting that some studies suggest that the impairment of splenic immune function may be via immunosuppression and the activation of the NF-B signaling pathway [44,45].

According to the GO analysis, the DEGs were mainly related to the immunologic process, metabolic process, and cellular components. Therefore, these results are almost consistent with our conjecture that TDCCP induces immunoinflammatory toxicity in mouse spleen. It also provides a foundation for the selection of candidate genes to further study the molecular mechanism of immunoinflammatory-related toxicity of TDCPP in the spleen. In addition, the KEGG pathway analysis identified eight pathways related to immunoinflammatory responses. Considering the purpose of this study and the number of differentially enriched genes, we selected the cytokine–cytokine receptor interaction pathway as the subsequent point of interest. We only focused on nine candidate genes for the cytokine–cytokine receptor interaction pathway, including four genes of the CC subfamily, four genes of the CXC subfamily, and one gene of the C subfamily. The development and homeostasis of the immune system depend on chemokines, which are involved in all immunological and inflammatory responses, whether they are protective or damaging [46]. CD4+ and CD8+ lymphocytes, dendritic cells, eosinophils, macrophages, monocytes, and NK cells are immune system cells that depend on CC family chemokines for survival [47], as does tumor development. In individuals with immunological thrombocytopenia, the expression of the Th1- and Th2-associated chemokine CCR3 gene was reduced [48], and animals lacking the chemokine receptor CCR1 had elevated Th1 responses and glomerular damage in nephritis [49]. In addition, CCL24 can combine with CCR3 to recruit eosinophils and tumor-associated macrophages, and immune function is limited when CCR3/CCL24 is suppressed [50]. Moreover, research has demonstrated that CXC chemokines either promote or inhibit immunity, which in turn affects the development of cancer [51]. The XCL1-XCR1 axis plays an important role in ensuring effective CD8 T cell-mediated cytotoxic immune responses, such as the ability of XCR1 to promote the CD8 DC activation of early CD8 T cell-mediated defense against intracellular pathogenic bacteria [52]. In this study, TDCPP significantly inhibited Ccr1l1, Ccr3, Ccr10, Ccl24, Ppbp, Pf4, Cxcl10, Ackr3, and Xcr1 genes expression in spleen tissue homogenates according to RNA-Seq and qRT-PCR, which indicates that the cytokine–cytokine receptor interaction pathway is a critical toxic target of TDCPP.

5. Conclusions

In summary, by combining cell biological assessment and transcriptome analysis techniques, we revealed the gene expression changes that were affected by TDCPP. First, mice treated with TDCPP i.g for 28 consecutive days were found to have altered general conditions and induced pathological changes in the spleen. Second, analysis of relevant inflammatory factors in splenic tissues showed that TDCPP can activate mitochondrial apoptotic pathways and alter apoptosis-related proteins. Third, RNA-seq analysis showed that TDCPP induced changes in eight KEGG pathways associated with immune and inflammatory responses in the spleen. Subsequently, qRT-PCR was used to validate 17 genes of the cytokine–cytokine receptor interaction pathway and revealed that TDCPP caused immunosuppression and induced an inflammatory response to toxins in the spleen. Accordingly, our study confirmed the immunotoxicity of TDCPP exposure to mice characterized by an inflammatory reaction, the activation of mitochondrial apoptosis pathways, and the inhibition of the expression of chemokines and their related receptors, causing an immunosuppressive effect.

Supplementary Materials: The following supporting information can be downloaded at: https://www.mdpi.com/article/10.3390/toxics11030231/s1, Table S1: Statistics of Reads and Reference Genome Comparison; Table S2: Statistical Table of Differential Gene Expression.

Author Contributions: L.C.: Investigation, Writing—original draft, Writing—review and editing, and Data curation. L.W.: Investigation, Software, and Formal analysis. Q.D. and Y.S.: Software, Resources and Formal analysis. S.Y. and K.L.: Writing—review and editing, Conceptualization, and Funding acquisition. All authors have read and agreed to the published version of the manuscript.

Funding: The current work was supported by Postgraduate Independent Exploration and Innovation Project of Hunan Province, China (CX20200117) and Postgraduate Independent Exploration and Innovation Project of Central South University, China (2022ZZTS0999).

Institutional Review Board Statement: All animal experiments were approved by the Laboratory Animal Welfare Ethical Committee of Central South University (Ethics Approval Code: XYGW-2021-113, Approval Date: 28 December 2021).

Informed Consent Statement: Not applicable.

Data Availability Statement: The data provided in this study are available upon request from the corresponding authors. Due to privacy issues, these data cannot be made public.

Conflicts of Interest: The authors declare no conflict of interest.

References

1. Wang, C.; Chen, H.; Li, H.; Yu, J.; Wang, X.; Liu, Y. Review of emerging contaminant tris(1,3-dichloro-2-propyl)phosphate: Environmental occurrence, exposure, and risks to organisms and human health. *Environ. Int.* **2020**, *143*, 105946. [CrossRef]
2. Sharkey, M.; Harrad, S.; Abdallah, M.A.-E.; Drage, D.S.; Berresheim, H. Phasing-out of legacy brominated flame retardants: The UNEP Stockholm Convention and other legislative action worldwide. *Environ. Int.* **2020**, *144*, 106041. [CrossRef]
3. Akortia, E.; Okonkwo, J.O.; Lupankwa, M.; Osae, S.D.; Daso, A.P.; Olukunle, O.I.; Chaudhary, A. A review of sources, levels, and toxicity of polybrominated diphenyl ethers (PBDEs) and their transformation and transport in various environmental compartments. *Environ. Rev.* **2016**, *24*, 253–273. [CrossRef]
4. Alcock, R.E.; Sweetman, A.J.; Prevedouros, K.; Jones, K.C. Understanding levels and trends of BDE-47 in the UK and North America: An assessment of principal reservoirs and source inputs. *Environ. Int.* **2003**, *29*, 691–698. [CrossRef]
5. Rodgers, T.F.M.; Truong, J.; Jantunen, L.M.; Helm, P.A.; Diamond, M.L. Organophosphate Ester Transport, Fate, and Emissions in Toronto, Canada, Estimated Using an Updated Multimedia Urban Model. *Environ. Sci. Technol.* **2018**, *52*, 12465–12474. [CrossRef]
6. Ding, J.; Xu, Z.; Huang, W.; Feng, L.; Yang, F. Organophosphate ester flame retardants and plasticizers in human placenta in Eastern China. *Sci. Total. Environ.* **2016**, *554–555*, 211–217. [CrossRef]
7. Wang, X.L.; Liu, Q.; Zhong, W.J.; Yang, L.P.; Yang, J.; Covaci, A.; Zhu, L. Estimating renal and hepatic clearance rates of organophosphate esters in humans: Impacts of intrinsic metabolism and binding affinity with plasma proteins. *Environ. Int.* **2020**, *134*, 105321. [CrossRef]
8. Carignan, C.C.; Mínguez-Alarcón, L.; Butt, C.M.; Williams, P.L.; Meeker, J.D.; Stapleton, H.M.; Toth, T.L.; Ford, J.B.; Hauser, R. Erratum: "Urinary Concentrations of Organophosphate Flame Retardant Metabolites and Pregnancy Outcomes among Women Undergoing in Vitro Fertilization". *Environ. Health Perspect.* **2017**, *125*, 119001. [CrossRef]
9. Fu, J.; Han, J.; Zhou, B.; Gong, Z.; Santos, E.M.; Huo, X.; Zheng, W.; Liu, H.; Yu, H.; Liu, C. Toxicogenomic Responses of Zebrafish Embryos/Larvae to Tris(1,3-dichloro-2-propyl) Phosphate (TDCPP) Reveal Possible Molecular Mechanisms of Developmental Toxicity. *Environ. Sci. Technol.* **2013**, *47*, 10574–10582. [CrossRef]
10. Carignan, C.C.; McClean, M.D.; Cooper, E.M.; Watkins, D.J.; Fraser, A.J.; Heiger-Bernays, W.; Stapleton, H.M.; Webster, T.F. Predictors of tris(1,3-dichloro-2-propyl) phosphate metabolite in the urine of office workers. *Environ. Int.* **2013**, *55*, 56–61. [CrossRef]
11. Saquib, Q.; Al-Salem, A.M.; Siddiqui, M.A.; Ansari, S.M.; Zhang, X.; Al-Khedhairy, A.A. Organophosphorus Flame Retardant TDCPP Displays Genotoxic and Carcinogenic Risks in Human Liver Cells. *Cells* **2022**, *11*, 195. [CrossRef]
12. Li, R.; Zhou, P.; Guo, Y.; Lee, J.-S.; Zhou, B. Tris (1,3-dichloro-2-propyl) phosphate-induced apoptotic signaling pathways in SH-SY5Y neuroblastoma cells. *Neurotoxicology* **2017**, *58*, 1–10. [CrossRef]
13. Zhong, X.; Wu, J.; Ke, W.; Yu, Y.; Ji, D.; Kang, J.; Qiu, J.; Wang, C.; Yu, P.; Wei, Y. Neonatal exposure to organophosphorus flame retardant TDCPP elicits neurotoxicity in mouse hippocampus via microglia-mediated inflammation in vivo and in vitro. *Arch. Toxicol.* **2020**, *94*, 541–552. [CrossRef]
14. Liu, Y.; Le, Y.; Xu, M.; Wang, W.; Chen, H.; Zhang, Q.; Wang, C. Remodeling on adipocytic physiology of organophosphorus esters in mature adipocytes. *Environ. Pollut.* **2022**, *305*, 119287. [CrossRef]
15. Li, X.; Li, N.; Rao, K.; Huang, Q.; Ma, M. In Vitro Immunotoxicity of Organophosphate Flame Retardants in Human THP-1-Derived Macrophages. *Environ. Sci. Technol.* **2020**, *54*, 8900–8908. [CrossRef]

16. Lewis, S.M.; Williams, A.; Eisenbarth, S.C. Structure and function of the immune system in the spleen. *Sci. Immunol.* **2019**, *4*, eaau6085. [CrossRef]
17. Bronte, V.; Pittet, M.J. The Spleen in Local and Systemic Regulation of Immunity. *Immunity* **2013**, *39*, 806–818. [CrossRef]
18. Chen, J.; Xu, Y.; Han, Q.; Yao, Y.; Xing, H.; Teng, X. Immunosuppression, oxidative stress, and glycometabolism disorder caused by cadmium in common carp (*Cyprinus carpio* L.): Application of transcriptome analysis in risk assessment of environmental contaminant cadmium. *J. Hazard. Mater.* **2019**, *366*, 386–394. [CrossRef]
19. Kamata, E.; Naito, K.; Nakaji, Y.; Ogawa, Y.; Suzuki, S.; Kaneko, T.; Takada, K.; Kurokawa, Y.; Tobe, M. Acute and subacute toxicity studies of tris (1,3-dichloro-2-propyl) phosphate on mice. *Eisei Shikenjo Hokoku. Bull. Natl. Inst. Hyg. Sci.* **1989**, *107*, 36–43.
20. Momtazi-Borojeni, A.A.; Banach, M.; Tabatabaei, S.A.; Sahebkar, A. Preclinical toxicity assessment of a peptide-based antiPCSK9 vaccine in healthy mice. *Biomed. Pharmacother.* **2023**, *158*, 114170. [CrossRef]
21. Fernández-Carrobles, M.M.; Bueno, G.; García-Rojo, M.; González-López, L.; López, C.; Déniz, O. Automatic quantification of IHC stain in breast TMA using colour analysis. *Comput. Med. Imaging Graph.* **2017**, *61*, 14–27. [CrossRef]
22. Wang, H.; Yang, L.-L.; Ji, Y.-L.; Chen, Y.-H.; Hu, J.; Zhang, C.; Zhang, J.; Xu, D.-X. Different fixative methods influence histological morphology and TUNEL staining in mouse testes. *Reprod. Toxicol.* **2016**, *60*, 53–61. [CrossRef]
23. Kim, D.; Langmead, B.; Salzberg, S.L. HISAT: A fast spliced aligner with low memory requirements. *Nat. Methods* **2015**, *12*, 357–360. [CrossRef]
24. Love, M.I.; Huber, W.; Anders, S. Moderated estimation of fold change and dispersion for RNA-seq data with DESeq2. *Genome Biol.* **2014**, *15*, 550. [CrossRef]
25. Gu, Z.; Eils, R.; Schlesner, M. Complex heatmaps reveal patterns and correlations in multidimensional genomic data. *Bioinformatics* **2016**, *32*, 2847–2849. [CrossRef]
26. Kanehisa, M.; Araki, M.; Goto, S.; Hattori, M.; Hirakawa, M.; Itoh, M.; Katayama, T.; Kawashima, S.; Okuda, S.; Tokimatsu, T.; et al. KEGG for linking genomes to life and the environment. *Nucleic Acids Res.* **2008**, *36*, D480–D484. [CrossRef]
27. Luo, H.; Wang, J.; Qiao, C.; Ma, N.; Liu, D.; Zhang, W. Pycnogenol attenuates atherosclerosis by regulating lipid metabolism through the TLR4–NF-κB pathway. *Exp. Mol. Med.* **2015**, *47*, e191. [CrossRef]
28. Van Herreweghe, F.; Festjens, N.; Declercq, W.; Vandenabeele, P. Tumor necrosis factor-mediated cell death: To break or to burst, that's the question. *Cell. Mol. Life Sci.* **2010**, *67*, 1567–1579. [CrossRef]
29. Vilgelm, A.E.; Richmond, A. Chemokines Modulate Immune Surveillance in Tumorigenesis, Metastasis, and Response to Immunotherapy. *Front. Immunol.* **2019**, *10*, 333. [CrossRef]
30. Poma, G.; Sales, C.; Bruyland, B.; Christia, C.; Goscinny, S.; Van Loco, J.; Covaci, A. Occurrence of Organophosphorus Flame Retardants and Plasticizers (PFRs) in Belgian Foodstuffs and Estimation of the Dietary Exposure of the Adult Population. *Environ. Sci. Technol.* **2018**, *52*, 2331–2338. [CrossRef]
31. Liu, X.; Ji, K.; Jo, A.; Moon, H.-B.; Choi, K. Effects of TDCPP or TPP on gene transcriptions and hormones of HPG axis, and their consequences on reproduction in adult zebrafish (*Danio rerio*). *Aquat. Toxicol.* **2013**, *134–135*, 104–111. [CrossRef]
32. McGee, S.P.; Cooper, E.M.; Stapleton, H.M.; Volz, D.C. Early Zebrafish Embryogenesis Is Susceptible to Developmental TDCPP Exposure. *Environ. Health Perspect.* **2012**, *120*, 1585–1591. [CrossRef]
33. Moser, G.; McDaniel, K.; Phillips, P.; Hedge, J. Neurobehavioral and thyroid evaluations of rats developmentally exposed to tris(1,3-dichloro-2-propyl)phosphate (TDCPP). *Neurotoxicol. Teratol.* **2014**, *43*, 97. [CrossRef]
34. Liu, X.; Cai, Y.; Wang, Y.; Xu, S.; Ji, K.; Choi, K. Corrigendum to "Effects of tris(1,3-dichloro-2-propyl) phosphate (TDCPP) and triphenyl phosphate (TPP) on sex dependent alterations of thyroid hormones in adult zebrafish" [Ecotoxicol. Environ. Saf. 170 (2019) 25–32]. *Ecotoxicol. Environ. Saf.* **2020**, *201*, 110867. [CrossRef]
35. Cheng, L.; Rao, Q.; Zhang, Q.; Song, W.; Guan, S.; Jiang, Z.; Wu, T.; Zhao, Z.; Song, W. The immunotoxicity of decabromodiphenyl ether (BDE-209) on broiler chicks by transcriptome profiling analysis. *Ecotoxicol. Environ. Saf.* **2022**, *232*, 113284. [CrossRef]
36. Hao, R.; Jiang, Y.; Li, F.; Sun-Waterhouse, D.; Li, D. MiR-182-5p/TLR4/NF-κB axis contributes to the protective effect of caffeic acid phenethyl ester against cadmium-induced spleen toxicity and associated damage in mice. *Food Chem. Toxicol.* **2021**, *158*, 112654. [CrossRef]
37. Sun, H.D.; Malabunga, M.; Tonra, J.R.; DiRenzo, R.; Carrick, F.E.; Zheng, H.; Berthoud, H.-R.; McGuinness, O.P.; Shen, J.; Bohlen, P.; et al. Monoclonal antibody antagonists of hypothalamic FGFR1 cause potent but reversible hypophagia and weight loss in rodents and monkeys. *Am. J. Physiol. Metab.* **2007**, *292*, E964–E976. [CrossRef]
38. Cao, Y.; Li, Y.; He, C.; Yan, F.; Li, J.-R.; Xu, H.-Z.; Zhuang, J.-F.; Zhou, H.; Peng, Y.-C.; Fu, X.-J.; et al. Selective Ferroptosis Inhibitor Liproxstatin-1 Attenuates Neurological Deficits and Neuroinflammation After Subarachnoid Hemorrhage. *Neurosci. Bull.* **2021**, *37*, 535–549. [CrossRef]
39. Hayden, M.S.; Ghosh, S. Regulation of NF-κB by TNF family cytokines. *Semin. Immunol.* **2014**, *26*, 253–266. [CrossRef]
40. Napetschnig, J.; Wu, H. Molecular Basis of NF-κB Signaling. *Annu. Rev. Biophys.* **2013**, *42*, 443–468. [CrossRef]
41. Du, W.; Frankel, T.L.; Green, M.; Zou, W. IFNγ signaling integrity in colorectal cancer immunity and immunotherapy. *Cell. Mol. Immunol.* **2022**, *19*, 23–32. [CrossRef]
42. Ni, R.; Fan, L.; Zhang, L.; Song, Y.; Wang, H.; Wang, A.; Liu, B. A mouse model of irradiation and spleen-thymus lymphocyte infusion induced aplastic anemia. *Hematology* **2022**, *27*, 932–945. [CrossRef]
43. Cui, D.; Bi, J.; Zhang, Z.-N.; Li, M.-Y.; Qin, Y.-S.; Xiang, P.; Ma, L.Q. Organophosphorus flame retardant TDCPP-induced cytotoxicity and associated mechanisms in normal human skin keratinocytes. *Sci. Total. Environ.* **2020**, *726*, 138526. [CrossRef]

44. Wan, F.; Tang, L.; Rao, G.; Zhong, G.; Jiang, X.; Wu, S.; Huang, R.; Tang, Z.; Ruan, Z.; Chen, Z.; et al. Curcumin activates the Nrf2 Pathway to alleviate AFB1-induced immunosuppression in the spleen of ducklings. *Toxicon* **2022**, *209*, 18–27. [CrossRef]
45. Yan, N.; Xu, G.; Zhang, C.; Liu, X.; Li, X.; Sun, L.; Wang, D.; Duan, X.; Li, B. Chronic arsenic exposure induces the time-dependent modulation of inflammation and immunosuppression in spleen. *Cell Biosci.* **2020**, *10*, 91. [CrossRef]
46. Hughes, C.E.; Nibbs, R.J.B. A guide to chemokines and their receptors. *FEBS J.* **2018**, *285*, 2944–2971. [CrossRef]
47. Korbecki, J.; Grochans, S.; Gutowska, I.; Barczak, K.; Baranowska-Bosiacka, I. CC Chemokines in a Tumor: A Review of Pro-Cancer and Anti-Cancer Properties of Receptors CCR5, CCR6, CCR7, CCR8, CCR9, and CCR10 Ligands. *Int. J. Mol. Sci.* **2020**, *21*, 7619. [CrossRef]
48. Zhou, S.-F.; Ma, J.; Qu, H.-T.; Liu, Z.-T.; He, W.-D.; Wang, J.-D.; Dou, A.-X.; Zhang, N.; Liu, J.-L.; Guo, C.-S.; et al. Characterization of Th_1- and Th_2-associated Chemokine Receptor Expression in Spleens of Patients with Immune Thrombocytopenia. *J. Clin. Immunol.* **2013**, *33*, 938–946. [CrossRef]
49. Topham, P.S.; Csizmadia, V.; Soler, D.; Hines, D.; Gerard, C.J.; Salant, D.J.; Hancock, W.W. Lack of chemokine receptor CCR1 enhances Th1 responses and glomerular injury during nephrotoxic nephritis. *J. Clin. Investig.* **1999**, *104*, 1549–1557. [CrossRef]
50. Korbecki, J.; Kojder, K.; Simińska, D.; Bohatyrewicz, R.; Gutowska, I.; Chlubek, D.; Baranowska-Bosiacka, I. CC Chemokines in a Tumor: A Review of Pro-Cancer and Anti-Cancer Properties of the Ligands of Receptors CCR1, CCR2, CCR3, and CCR4. *Int. J. Mol. Sci.* **2020**, *21*, 8412. [CrossRef]
51. Cabrero-de las Heras, S.; Martínez-Balibrea, E. CXC family of chemokines as prognostic or predictive biomarkers and possible drug targets in colorectal cancer. *World J. Gastroenterol.* **2018**, *24*, 4738–4749. [CrossRef]
52. Lei, Y.; Takahama, Y. XCL1 and XCR1 in the immune system. *Microbes Infect.* **2012**, *14*, 262–267. [CrossRef]

toxics

MDPI

Article

Urinary Cypermethrin Metabolites among Conventional and Organic Farmers in Thailand

Atima Tremongkoltip [1], Sumate Pengpumkiat [1], Pornpimol Kongtip [1,*], Noppanun Nankongnab [1], Sukhontha Siri [2] and Susan Woskie [3]

[1] Department of Occupational Health and Safety, Faculty of Public Health, Mahidol University, 420/1 Rajvidhi Road, Bangkok 10400, Thailand; atima.tre@gmail.com (A.T.); sumate.pen@mahidol.ac.th (S.P.); noppanun.nan@mahidol.ac.th (N.N.)

[2] Department of Epidemiology, Mahidol University, 420/1 Rajvidhi Road, Bangkok 10400, Thailand; sukhontha.sir@mahidol.ac.th

[3] Department of Public Health, University of Massachusetts Lowell, 61 Wilder St., Lowell, MA 01854, USA; susan_woskie@uml.edu

* Correspondence: pornpimol.kon@mahidol.ac.th; Tel.: +66-2644-4070; Fax: +66-2354-8561

Abstract: Cypermethrin, a pyrethroid insecticide, is frequently spread on agricultural farmlands and is also used in households in Thailand. Conventional pesticide-using farmers (n = 209) were recruited from the Phitsanulok and Nakornsawan provinces. Certified organic farmers (n = 224) were also recruited in the Yasothorn province. The farmers were interviewed via a questionnaire and the urine from their first morning void was collected. The urine samples were analyzed for 3-phenoxybenzoic acid (3-PBA), cis-3-(2,2-dichlorovinyl)-2,2-dimethylcyclopropane carboxylic acid (cis-DCCA), and trans-3-(2,2-dichlorovinyl)-2,2-dimethylcyclopropane carboxylic acid (trans-DCCA). The results showed no significant difference in the urinary cypermethrin metabolites between the conventional farmers and the organic farmers, for whom the usage of cypermethrin was not accounted for. However, when conventional farmers who used cypermethrin on the farm and in the home were compared with conventional farmers who did not use any cypermethrin or with organic farmers, a significant difference was noted for all metabolites except for trans-DCCA. These findings show that the most significant exposures to cypermethrin are among conventional farmers who use the insecticide on their farms or in their homes. However, measurable levels of all metabolites were found among both conventional and organic farmers who only used cypermethrin in the home or not at all, suggesting that the at-home use of pyrethroids and other possible exposures from pyrethroid residues on market-bought food may contribute to urinary levels of pyrethroids that exceed those of the general population in the US and Canada.

Keywords: urinary metabolite; cypermethrin; conventional farmers; organic farmers

Citation: Tremongkoltip, A.; Pengpumkiat, S.; Kongtip, P.; Nankongnab, N.; Siri, S.; Woskie, S. Urinary Cypermethrin Metabolites among Conventional and Organic Farmers in Thailand. *Toxics* **2023**, *11*, 507. https://doi.org/10.3390/toxics11060507

Academic Editors: Roel Vermeulen, Lauren Petrick and Maaike van Gerwen

Received: 16 May 2023
Revised: 28 May 2023
Accepted: 1 June 2023
Published: 5 June 2023

1. Introduction

Agriculture is a cornerstone of the economy of Thailand [1]. Many pesticides are currently imported into the country for use in farming [2]. Over 125 pesticides are listed as potential endocrine-disrupting chemicals (EDC) [3,4]. Around the world, chronic exposure to organophosphate and pyrethroid pesticides has been associated with adverse neurological outcomes, including neurodevelopmental delays in children, endocrine-related effects including metabolic disruption, obesity, and diabetes, and reproductive effects such as reduced sperm quality, impacts on reproductive hormones, and pregnancy outcomes such as lower gestational age and birth weight [5–7].

Cypermethrin is an insecticide in the pyrethroid group that is often used as an alternative to some of the organophosphate pesticides. It is used in agricultural fields and in households [8,9]. In 2020, 1262 tons of cypermethrin and profenofos were imported into Thailand [2]. Cypermethrin can enter the body mainly through inhalation and skin contact; it is metabolized very quickly by cytochrome P450 into three main metabolites, 3-phenoxybenzoic acid (3-PBA), cis-3-(2,2-dichlorovinyl)-2,2-dimethylcyclopropane carboxylic acid (cis-DCCA), and trans-3-(2,2-dichlorovinyl)-2,2-dimethylcyclopropane carboxylic acid (trans-DCCA), and is then excreted into the urine [10,11]. These metabolites can be used as biomarkers of cypermethrin exposure [12,13]. However, 3-PBA, cis-DCCA, and trans-DCCA are non-specific metabolites of cypermethrin as they are also metabolites of other pyrethroids [12]. For example, 3-PBA was detected in more than 70% of urine samples in the general population of the US and Spain [14,15]. Previously, 3-PBA and cis-trans-DCCA have been detected in the urine samples of conventional pesticide-using farmers in Thailand [16,17]. However, the focus of this study was to evaluate cypermethrin metabolites in the urine of both conventional and organic farmers, while accounting for at-home and occupational usage.

2. Materials and Methods

2.1. Study Population and Data Collection

This was a cross-sectional study collecting information from conventional and organic farmers. All subjects gave their informed consent before participating in the study. The study was conducted in accordance with the Declaration of Helsinki, and the protocol was approved by the Ethics Committee of Human Research, Faculty of Public Health, Mahidol University (MUPH 2015-146). All participants signed informed consent forms before participating in the study. We recruited conventional farmers in two provinces in Thailand: rice and vegetable farmers in Phitsanulok province, and sugarcane and rice farmers in Nakornsawan province. One farmer who sprayed his crops with pesticides was selected in each conventional farming household. The organic farmers were certified as using organic farming practices for all their crops. Organic farmers were recruited from rice, vegetable, and fruit farms in the Yasothorn province. One participant who mainly worked on the farm was selected from each organic farming household. Health-promoting hospital staff, community leaders, and hired site officers helped to recruit farmers. The studied farmers could be male or female and 18 years of age or older but they must not have been diagnosed with or be undergoing treatment for diabetes, high blood pressure, heart disease, or thyroid disease.

A questionnaire was employed to collect data on the demographic characteristics of farmers, risk factors regarding pesticide exposure, home location and home pesticide use, self-reported health problems, agricultural activities, and history of pesticide use. The farmers were interviewed at the health-promoting hospital or in their homes by site officers or trained research team members. The farmers were trained on how to collect a urine sample and were provided with a sterile polyethylene bottle to collect the urine sample. They were asked to collect the first morning void of urine and bring the urine sample to the research team at the health-promoting hospital in their area. The urine samples were kept at −45 °C before analysis.

2.2. Urine Sample Analysis

All urine samples were analyzed following the method of Leng and Gries [18] using gas chromatography-mass spectrometry (GC-MS). The calibration curves of cypermethrin metabolites including cis-DCCA, trans-DCCA, and 3-PBA were set up at concentrations ranging from 0 to 100 ng/mL. The accuracy of the 3-PBA analysis ranged from 97.94 to 105.07% at the concentrations of 10 and 60 ng/mL, with a relative standard deviation (RSD) of less than 2. The accuracy of cis- and trans-DCCA was 97.95–106.95% and 95.50–104.11% at the concentrations of 10 and 60 ng/mL, respectively, with an RSD of less than 5. The detection limits of 3-PBA, cis-DCCA, and trans-DCCA were 0.5, 0.5, and 0.5 ng/mL, respectively. For urinary concentrations below the detection limit, we substituted the detection limit by dividing it by the square root of 2 for metabolites where the geometric standard deviation (GSD) was <3, or we substituted the detection limit divided by two for metabolites where the GSD was ≥ 3 [19]. The creatinine (mg/dL) level in urine was analyzed using an enzymatic method with a linear concentration range of 1–500 mg/dL and a detection limit of 0.16 mg/dL [20]. All urinary metabolites were corrected for creatinine levels by dividing the metabolite concentration by the creatinine concentration, producing units of ng metabolite/g creatinine. This figure was subsequently converted to nmole metabolite/g creatinine for presentation. The results were presented as 3-PBA, cis-DCCA, trans-DCCA, cis,trans-DCCA = cis-DCCA + trans-DCCA, and total cypermethrin = cis,trans-DCCA + 3PBA.

2.3. Statistical Analysis

All statistical analyses were performed in SPSS for Windows, version 23 (IBM Thailand Co., Ltd., Bangkok, Thailand). A descriptive analysis of the demographic characteristics was carried out using the chi-square test, Fisher's exact test, independent *t*-tests, and a one-way ANOVA. Due to the skewed distribution of cypermethrin metabolites, a natural logarithm was used in all analyses.

3. Results

The conventional farmers were more commonly male (74.2%) than female (25.8%), while the organic farmers showed a similar percentage of males (51.3%) and females (48.7%) (Table 1).

Table 1. Demographic characteristics of conventional (n = 209) and organic farmers (n = 224).

Variables		Conventional Farmers n (%)	Organic Farmers n (%)	p-Value
Sex [1]				
	Male	155 (74.2)	115 (51.3)	<0.001 *
	Female	54 (25.8)	109 (48.7)	
Age [3]				
	Mean (SD)	50.10 (11.1)	53.22 (10.3)	0.003 *
	Min–max	18–69	28–79	
Education level [1]				
	Below elementary	14 (6.7)	4 (1.8)	0.036 *
	Elementary	119 (56.9)	125 (55.8)	
	High school	71 (34.0)	84 (37.5)	
	Bachelor's degree or higher	5 (2.4)	11 (4.9)	
Marital status [2]				
	Single	21 (10.0)	13 (5.8)	0.027 *
	Married	180 (86.1)	188 (83.9)	
	Widowed/divorced	8 (3.8)	23 (10.3)	
Body mass index (BMI) (kg/m^2) [1]				
	Normal (<18.49–24.99)	118 (56.7)	167 (74.6)	<0.001 *
	Abnormal (>25.00)	90 (43.3)	57 (25.4)	
	Missing	1 (0.2)		
Expense adequacy [1]				
	Not in debt	135 (64.6)	143 (63.08)	0.870
	In debt	74 (35.4)	81 (36.2)	
Agricultural work time (h/week) [3]				
	Mean (SD)	26.5 (3.4)	28.8 (17.2)	0.149

Table 1. *Cont.*

Variables		Conventional Farmers n (%)	Organic Farmers n (%)	p-Value
Has a second job [1]				
	Yes	160 (76.6)	96 (42.9)	<0.001 *
	No	49 (23.4)	128 (57.1)	
Second job work time (h/week) [3]				
	Mean (SD)	24.9 (13.6)	26.6 (17.8)	0.537
Main types of crop plant				
	Rice [1]	162 (77.5)	213 (96.8)	<0.001 *
	Sugarcane [1]	80 (38.3)	14 (6.6)	<0.001 *
	Banana [1]	14 (6.7)	118 (55.9)	<0.001 *
	Vegetables [1]	78 (37.3)	180 (84.8)	<0.001 *
	Fruits [1]	10 (4.8)	102 (48.6)	<0.001 *
Smoking [1]				
	Current smoker	57 (27.3)	36 (16.1)	0.001 *
	Former smoker	9 (4.3)	25 (11.2)	
	Non-smoker	143 (68.4)	163 (72.8)	
Alcohol intake [1]				
	Current drinker	134 (64.1)	92 (41.1)	<0.001 *
	Former drinker	7 (3.3)	45 (20.1)	
	Non-drinker	68 (32.5)	87 (38.8)	

[1] chi-square test, [2] Fisher's exact test, [3] independent t-test, * significant at a p-value < 0.05.

The average age and education level of the organic farmers was significantly higher than those of the conventional farmers. The organic farmers were significantly more likely to have a normal BMI and second jobs than the conventional farmers. However, the hours of agricultural work and second job working time were not significantly different. The conventional farmers' group included significantly more smokers and alcohol drinkers than the organic farmers' group.

The conventional farmers were significantly more likely than the organic farmers to live near farmland and use insecticides in their homes over the past year (89.5 % vs. 14.8%) (Table 2).

Table 2. Risk factors of conventional (n = 209) and organic farmers (n = 224).

Risk Factors		Conventional Farmers n (%)	Organic Farmers n (%)	p-Value
Living near farm [1]				
	Yes	177 (84.7)	103 (46.4)	<0.001 *
	No	32 (15.3)	117 (52.7)	
Insecticide use at home in the past 1 year [1]				
	Yes	187 (89.5)	33 (14.8)	<0.001 *
	No	22 (10.5)	190 (85.2)	
Type of insecticide used at home				
Aswin or Atsawin [1,2]				
	Yes	2 (1.1)	1 (0.6)	1.000
	No	187 (98.9)	170 (99.4)	
Ars [1,2]				
	Yes	25 (13.2)	3 (1.7)	<0.001 *
	No	164 (86.8)	169 (98.3)	

Table 2. *Cont.*

Risk Factors		Conventional Farmers n (%)	Organic Farmers n (%)	p-Value
Baygon [1,2]				
	Yes	107 (56.6)	11 (6.4)	<0.001 *
	No	82 (43.4)	161 (93.6)	
Shieldtox [1,2]				
	Yes	25 (13.2)	4 (2.3)	<0.001 *
	No	164 (86.8)	169 (97.7)	
Mosquito coil [1,2]				
	Yes	148 (78.3)	19 (10.9)	<0.001 *
	No	41 (21.7)	156 (89.1)	
Cypermethrin used in farm [1]				
	Yes	95 (45.9)	0 (0)	<0.001 *
	No	113 (54.1)	224 (100)	
Water consumption source [1]				
	House and community well	9 (4.3)	6 (2.7)	<0.001 *
	Bottled water	14 (6.7)	142 (64.5)	
	Rainwater	5 (2.4)	64 (29.1)	
	Filtered tap water	181 (86.6)	8 (3.6)	
Daily vegetable intake [1]				
	Yes	121 (57.9)	148 (66.7)	0.060
	No	88 (42.1)	74 (33.3)	
Did not take a shower after spraying [1]				
	Yes	137 (65.5)	0 (0)	<0.001 *
	No	72 (34.4)	224 (100)	
Keeping all used containers at home for disposal [1]				
	Yes	105 (50.2)	0 (0)	<0.001 *
	No	104 (49.8)	224 (100)	

[1] chi-square test, * significant at a p-value < 0.05. [2] Main ingredients of insecticide used at home. Aswin sprays contain tetramethrin 0.2%, permethrin 0.1% [21]. ARS sprays contain imiprothrin 0.2%, cypermethrin 0.1% [21]. Baygon sprays contain cypermethrin (0.1%) and other pyrethroids, imiprotin (0.03%), and prallethrin (0.03%) [22]. Shieldtox sprays contain other pyrethroids, prallethrin 0.076%, and D-phenothrin 0.046% [23]. Mosquito coils have different brands with different active ingredients, but the main active ingredients are also pyrethroids, dimefluthrin, allethrin, metofluthrin, etc. [24,25].

As a result, conventional farmers were also more likely to use the insecticide brands of Ars, Baygon, and Shieldtox, and mosquito coils in their homes over the past year. About 45.9 percent of the conventional farmers used cypermethrin on their farms, while the organic farmers did not use any pesticides on their crops, including cypermethrin. Most of the conventional farmers (57.9%) and organic farmers (66.7%) ate vegetables every day. Sixty-five percent of the conventional farmers did not take a shower after spraying pesticides and 50.2% of conventional farmers kept all their used pesticide containers at home for disposal.

A comparison of the urinary cypermethrin metabolites found that there was no significant difference between the metabolite levels of conventional and organic farmers (Table 3).

Table 3. Urinary cypermethrin metabolites (geometric mean (GM), geometric standard deviation (GSD), median, nmole/g creatinine), compared between conventional (n = 209) and organic farmers (n = 224).

Cypermethrin Metabolites	Conventional Farmers (nmole/g Creatinine)		Organic Farmers (nmole/g Creatinine)		p-Value [1]
	GM (GSD)	Median	GM (GSD)	Median	
3-PBA	6.01 (3.31)	6.42	5.12 (3.13)	4.74	0.159
Cis-DCCA	17.78 (2.46)	16.08	15.45 (2.54)	12.48	0.112
Trans-DCCA	11.15 (2.40)	10.7	10.03 (2.48)	8.14	0.219
Cis,Trans-DCCA	30.03 (2.32)	26.65	26.33 (2.45)	21.49	0.113
Total Cypermethrin	37.30 (2.36)	33.78	32.30 (2.46)	26.72	0.089

[1] t-test on the natural logs of metabolite concentration.

However, among conventional farmers, those who used cypermethrin both on the farm and at home had significantly higher levels of urinary 3PBA, cis-DCCA, cis,trans-DCCA, and total cypermethrin compared to those who only used it in the home (Table 4).

Table 4. Urinary cypermethrin metabolites (GM (GSD), nmole/g creatinine) of conventional farmers, classified by insecticide use.

	1. Farm and Home Use (n = 93)	2. Farm Use Only (n = 3)	3. Home Use Only (n = 95)	4. No Farm or Home Use (n = 18)	p-Value of Overall ANOVA	p-Value for Multiple Comparisons [1]
3PBA	9.58 (2.62)	17.93 (4.38)	3.69 (3.41)	5.88 (3.06)	<0.001	(1–3) * $p < 0.001$
Cis-DCCA	24.63 (2.27)	17.58 (2.97)	13.51 (2.33)	14.10 (2.92)	<0.001	(1–3) * $p < 0.001$, (1–4) $p = 0.059$
Trans-DCCA	12.86 (2.36)	9.97 (4.21)	9.86 (2.34)	10.34 (2.68)	0.212	
Cis,Trans-DCCA	38.19 (2.23)	27.85 (3.38)	24.54 (2.21)	25.85 (2.32)	0.003	(1–3) * $p = 0.002$
Total Cypermethrin	48.60 (2.23)	46.06 (3.76)	29.24 (2.26)	33.13 (2.59)	0.001	(1–3) * $p < 0.001$

[1] Multiple comparisons with the Tukey on natural logs of metabolite concentration, * significant at a p-value < 0.05.

When comparing the samples from conventional farmers who used cypermethrin on their farms and in their homes with the samples from those who did not use cypermethrin in either location, we found that the cis-DCCA level was borderline significantly lower among the non-users. When conventional farmers who used insecticides only at home or not at all were compared with organic farmers who used insecticides only at home or not at all, we found non-significant differences for all cypermethrin metabolites (Table 5).

Table 5. Urinary cypermethrin metabolites (GM (GSD), nmole/g creatinine) of conventional and organic farmers, classified by insecticide use.

	Conventional Farmers		Organic Farmers		p-Value of Overall ANOVA *
	Home Use Only (n = 95)	No Farm or Home Use (n = 18)	Home Use Only (n = 33)	No Farm or Home Use (n = 191)	
3PBA	3.69 (3.41)	5.88 (3.06)	6.72 (1.25)	4.89 (1.08)	0.045
Cis-DCCA	13.51 (2.33)	14.10 (2.92)	20.66 (1.21)	14.69 (1.07)	0.146
Trans-DCCA	9.86 (2.34)	10.34 (2.68)	10.89 (1.20)	9.89 (1.07)	0.945
Cis,Trans-DCCA	24.54 (2.21)	25.85 (2.32)	32.69 (1.20)	25.37 (1.06)	0.420
Total Cypermethrin	29.24 (2.26)	33.13 (2.59)	40.29 (1.20)	31.09 (1.06)	0.337

* No subgroup multiple comparisons were significant.

With regard to organic farmers, the urinary concentrations of cypermethrin metabolites for organic farmers of different ages were not significantly different (Table 6).

Table 6. Urinary cypermethrin metabolites (GM (GSD), nmole/g creatinine) of organic farmers, classified by age, sex, insecticide use at home in the past 1 year, home use of insecticides (Aswin, Ars, Baygon, Shieldtox, and mosquito coils), and daily vegetable intake.

	3-PBA		Cis-DCCA		Trans-DCCA		Cis,trans-DCCA		Total Cypermethrin	
	GM (GSD)	*p*-Value	GM (GSD)	*p*-Value	GM (GSD)	*p*-Value	GM (GSD)	*p*-Value	GM (GSD)	*p*-Value
Age [1]										
1. <35	4.52 (1.45)	0.129	9.42 (1.58)	0.073	7.94 (1.66)	0.311	18.54 (1.33)	0.185	25.15 (1.37)	0.170
2. 35–60	5.57 (3.28)		16.63 (2.68)		10.54 (2.60)		27.89 (2.58)		35.44 (2.56)	
3. >60	3.85 (2.75)		12.53 (2.09)		8.59 (2.15)		22.25 (2.04)		27.94 (2.07)	
Sex [2]										
Male	3.91 (2.97)	<0.001 *	13.13 (2.38)	0.007 *	8.99 (2.41)	0.064	23.02 (2.30)	0.021 *	27.66 (2.29)	0.008 *
Female	6.81 (3.11)		18.35 (2.65)		11.26 (2.54)		30.34 (2.56)		38.04 (2.56)	
Insecticide use at home in the past 1 year [2]										
Yes	6.72 (3.59)	0.145	20.66 (3.00)	0.055	10.89 (2.79)	0.584	32.69 (2.84)	0.138	40.29 (2.84)	0.131
No	4.91 (3.05)		14.72 (2.45)		9.91 (2.44)		25.43 (2.38)		31.18 (2.38)	
Type of insecticide used at home										
Baygon [2,3]										
Yes	6.67 (3.58)	0.607	18.83 (2.86)	0.544	10.45 (2.92)	0.924	29.69 (2.82)	0.723	36.93 (2.87)	0.716
No	5.56 (3.07)		15.68 (2.61)		0.16 (2.53)		26.79 (2.51)		33.25 (2.50)	
Shieldtox [2,3]										
Yes	6.27 (2.28)	0.839	18.04 (1.54)	0.785	10.78 (2.25)	0.899	29.52 (1.70)	0.841	36.32 (1.73)	0.854
No	5.58 (3.11)		15.79 (2.64)		10.15 (2.55)		26.86 (2.54)		33.32 (2.53)	
Mosquito coil [2,3]										
Yes	7.12 (3.75)	0.320	21.59 (3.30)	0.137	11.72 (2.85)	0.503	34.72 (3.07)	0.208	43.06 (3.05)	0.205
No	5.43 (2.99)		15.27 (2.51)		10.07 (2.50)		26.20 (2.44)		32.47 (2.43)	
Daily vegetable intake [2]										
Yes	9.18 (2.30)	0.011 *	15.56 (2.68)	0.196	9.19 (2.95)	0.316	25.47 (2.69)	0.226	35.37 (2.54)	0.114
No	6.66 (2.03)		13.14 (2.31)		8.31 (2.50)		22.24 (2.27)		29.34 (2.17)	

[1] Multiple comparisons with the Tukey on the natural logs of metabolite concentration, [2] *t*-test on the natural logs of metabolite concentration. [3] Main ingredients of insecticide used at home. Baygon sprays contain cypermethrin (0.1%) and other pyrethroids, imiprotin (0.03%), and prallethrin (0.03%) [22]. Shieldtox sprays contain other pyrethroids, prallethrin 0.076%, and D-phenothrin 0.046% [23]. Mosquito coils are of different brands with different active ingredients, but the main active ingredients are also pyrethroids, dimefluthrin, allethrin, metofluthrin, etc. [24,25]. * Significant at a *p*-value < 0.05.

Female organic farmers showed significantly higher urinary metabolites of 3-PBA, cis-DCCA, cis,trans-DCCA, and total cypermethrin than male organic farmers. Organic farmers who used insecticides at home in the past year had borderline significantly higher urinary metabolites of cis-DCCA than those who did not use insecticides at home. However, organic farmers who reported the use of Baygon, Shieldtox, or mosquito coils in their homes did not have significantly different urinary metabolite levels compared to those who did not use these home insecticides. Organic farmers who ate vegetables daily had significantly higher urinary 3-PBA levels than those who did not.

For conventional farmers, the urinary concentrations of trans-DCCA, cis,trans-DCCA, and total cypermethrin were significantly higher for those aged 35 to 60 compared to those below the age of 35. The urinary trans-DCCA levels for those over 60 were significantly higher than for those below the age of 35 (Table 7).

Female conventional farmers had significantly higher urinary metabolites of cis-DCCA than male farmers. Farmers who used cypermethrin on their farms had significantly higher 3-PBA, cis-DCCA, trans-DCCA, cis,trans-DCCA, and total cypermethrin levels than those who did not. Farmers who used Baygon in their homes in the past year had significantly higher urinary metabolites of 3-PBA, cis-DCCA, trans-DCCA, cis,trans-DCCA, and total cypermethrin levels than those who did not. Farmers who used ARS in their homes had significantly higher urinary metabolite levels of cis-DCCA than those who did not, while the metabolite levels of those who used Aswin, Shieldtox, and mosquito coils in the home were not significantly different from those who did not. When they sprayed pesticides, those who reported that they did not take a shower or wash immediately after spraying had significantly higher 3-PBA, cis-DCCA, trans-DCCA, cis,trans-DCCA, and total cypermethrin levels than those who did. Conventional farmers who kept all their used pesticide containers at home for disposal had significantly higher 3-PBA, cis-DCCA, cis,trans-DCCA, and total cypermethrin levels than those who did not. Those conventional farmers who ate vegetables daily had significantly higher 3-PBA, cis-DCCA, trans-DCCA, cis,trans-DCCA, and total cypermethrin levels than those who did not report daily vegetable consumption.

Table 7. Urinary cypermethrin metabolites (GM(GSD), nmole/g creatinine) of conventional farmers, classified by age, sex, insecticide use at home in the past 1 year, cypermethrin use on their farm, and the use of insecticides at home (Aswin, Ars, Baygon, Shieldtox, and the mosquito coil) did not report taking a bath immediately after spraying, kept all used chemical containers at home for disposal, and had a daily intake of vegetables.

	3-PBA GM (GSD)	*p*-Value	Cis-DCCA GM (GSD)	*p*-Value	Trans-DCCA GM (GSD)	*p*-Value	Cis,trans-DCCA GM (GSD)	*p*-Value	Total Cypermethrin GM (GSD)	*p*-Value
Age [1]										
1. <35	4.14 (3.30)	0.287	12.07 (2.50)	0.090	6.82 (2.52)	0.016,	20.16 (2.26)	0.053,	24.94 (2.33)	0.056,
2. 35–60	6.33 (3.25)		18.76 (2.41)		11.84 (2.41)	(1–2) *p* = 0.013,	31.74 (2.31)	(1–2)	39.45 (2.33)	(1–2)
3. >60	6.09 (3.58)		18.22 (2.57)		11.88 (2.13)	(1–3) *p* = 0.046	30.96 (2.29)	*p* = 0.042	38.15 (2.38)	*p* = 0.044
Sex [2]										
Male	5.67 (3.41)	0.241	16.35 (2.50)	0.022 *	10.87 (2.47)	0.477	28.46 (2.36)	0.112	35.32 (2.40)	0.119
Female	7.08 (3.02)		22.61 (2.25)		12.00 (2.22)		35.18 (2.19)		43.63 (2.20)	
Insecticide use at home in the past 1 year [2]										
Yes	5.89 (3.33)	0.480	18.11 (2.26)	0.389	11.21 (2.37)	0.791	30.43 (2.29)	0.544	37.46 (2.33)	0.831
No	7.13 (3.26)		15.21 (2.84)		10.63 (2.73)		27.11 (2.66)		35.94 (2.64)	
Cypermethrin use on the farm [2]										
Yes	9.77 (2.66)	<0.001 *	24.37 (2.28)	<0.001 *	12.76 (2.39)	0.040 *	37.82 (2.25)	<0.001 *	48.52 (2.33)	<0.001 *
No	3.97 (3.38)		13.60 (2.41)		9.94 (2.38)		24.74 (2.27)		29.83 (2.30)	
Type of insecticide used at home										
ARS [2,3]										
Yes	9.24 (2.54)	0.052	25.31 (2.18)	0.043 *	12.46 (2.27)	0.515	38.02 (2.18)	0.151	48.04 (2.17)	0.123
No	5.58 (3.45)		17.22 (2.44)		11.02 (2.41)		29.41 (2.31)		36.20 (2.36)	
Baygon [2,3]										
Yes	8.03 (2.93)	<0.001 *	21.02 (2.46)	0.008 *	12.54 (2.36)	0.043 *	34.74 (2.29)	0.012 *	43.83 (2.31)	0.004 *
No	4.05 (3.57)		14.92 (2.30)		9.67 (2.40)		25.59 (2.25)		30.75 (2.30)	
Shieldtox [2,3]										
Yes	5.89 (3.61)	0.958	17.12 (2.46)	0.734	11.63 (2.49)	0.816	29.98 (2.34)	0.924	36.94 (2.42)	0.914
No	5.98 (3.33)		18.27 (2.43)		11.14 (2.38)		30.49 (2.30)		37.68 (2.34)	
Mosquito coil [2,3]										
Yes	6.09 (3.29)	0.662	18.69 (2.43)	0.358	11.61 (2.37)	0.289	31.57 (2.29)	0.245	38.86 (2.32)	0.307
No	5.54 (3.64)		16.18 (2.41)		9.89 (2.46)		26.61 (2.33)		33.31 (2.43)	
Did not take a bath immediately after spraying [2]										
Yes	11.12 (2.20)	<0.001 *	25.32 (2.25)	<0.001 *	13.69 (2.32)	0.014 *	40.07 (2.18)	<0.001 *	52.01 (2.15)	<0.001 *
No	4.34 (3.49)		14.77 (2.44)		10.01 (2.41)		25.85 (2.31)		31.32 (2.35)	
Kept all used chemical containers at home for disposal [2]										
Yes	9.25 (2.48)	<0.001 *	22.44 (2.27)	<0.001 *	12.34 (2.30)	0.092	35.41 (2.22)	0.005 *	45.42 (2.22)	<0.001 *
No	3.88 (3.66)		14.06 (2.52)		10.06 (2.49)		25.49 (2.36)		30.57 (2.40)	
Daily vegetable intake [2]										
Yes	7.25 (3.04)	0.007 *	20.28 (2.47)	0.013 *	12.40 (2.39)	0.039 *	34.16 (2.30)	0.010 *	42.65 (2.32)	0.008 *
No	4.63 (3.56)		14.84 (2.39)		9.63 (2.38)		25.22 (2.28)		31.02 (2.33)	

[1] Multiple comparisons with the Tukey scores on the natural logs of metabolite concentration; *p*-value refers to the overall ANOVA and *p*-value of each subgroup; [2] *t*-test on the natural logs of metabolite concentration. [3] Main ingredients of insecticide used at home. ARS sprays contain imiprothrin 0.2%, cypermethrin 0.1% [21]. Baygon sprays contain cypermethrin (0.1%) and other pyrethroids, imiprotin (0.03%), and prallethrin (0.03%) [22]. Shieldtox sprays contain other pyrethroids, prallethrin 0.076%, and D-phenothrin 0.046% [23]. Mosquito coils have different brands with different active ingredients, but the main active ingredients are also pyrethroids, dimefluthrin, allethrin, metofluthrin, etc. [24,25]. * Significant at a *p*-value < 0.05.

4. Discussion

Among the conventional farmers, there were more males than females because our inclusion criteria specified that the conventional farmers had to spray pesticides on their own farms [26]. Other studies have reported that most pesticide applicators are male, while female farmers are more likely to choose sowing, hand-picking pests, watering, and harvesting [27]. The organic farmers were older and had a higher level of education than the conventional farmers. This study is similar to that of Chouichom, which showed that longer farm experience (a higher age) and higher education levels were supporting factors in changing over to an organic farming system [28,29]. The organic farmers were more likely to have a normal BMI value than the conventional farmers. This may be because organic farming is more labor-intensive, while conventional farmers use more agricultural machinery in their operations [28,30]. The conventional farmers were significantly more likely to be smokers and alcohol drinkers than the organic farmers. This is probably because there were more male than female conventional farmers and these behaviors are more typical among Thai males [31,32].

Most conventional farmers lived near their farmlands and used insecticides at home; therefore, they may have experienced multiple routes of exposure to pesticides (direct spraying, spray drift, and home use) compared to the organic farmers. The majority of organic farmers (65%) drank bottled water, while the majority (86.6%) of the conventional farmers drank filtered tap water. Pesticide contamination levels in commercial bottled water or in government-provided tap water have not been reported.

There was no significant difference in cypermethrin metabolite levels between the conventional farmers and the organic farmers when cypermethrin usage was not considered. This may be because the conventional farmers generally apply several types of pesticides, including insecticides, herbicides, and fungicides at different times of the year; within these categories, they use a variety of brands [33]. In the current cohort, only 45.9 percent of the conventional farmers reported the use of cypermethrin on their farms. The urinary cypermethrin metabolite levels (3PBA, cis-DCCA, cis,trans-DCCA, and total cypermethrin) from those who used cypermethrin on their farms and in their homes were significantly higher than in samples from those who only used insecticide at home. In most cases, cypermethrin use on both the farm and in the home produced the highest urinary metabolite levels, followed by those who used cypermethrin on the farm only. However, this was generally a non-significant difference, due to the low number of farmers who only used cypermethrin on their crops ($n = 3$). Those who used cypermethrin on their farms could be exposed to cypermethrin during the mixing and spraying of cypermethrin, cleaning and fixing the spraying equipment, and spilling and splashing the cypermethrin solution during spraying [34]. The cypermethrin metabolites found in conventional farmers who used cypermethrin only at home or not at all were not significantly different; this was most likely due to the low number of farmers who did not use cypermethrin at all ($n = 18$) and the high GSD values of the metabolite levels. When we compared the cypermethrin metabolites of conventional farmers and organic farmers who used insecticides only at home or not at all, they were also not significantly different. The similar and non-significant differences between those who only used cypermethrin at home or not at all suggest that the cypermethrin metabolite levels measured may be due to the use of other pyrethroids in the home, from residues on market-purchased foods [17,35–37], or from pets treated with insecticides or cleaning products that contain pyrethroids [38,39].

Although we have found no other work that has compared metabolite levels between farmers who used cypermethrin on their farmland and those who used it in their homes, one study looked at pyrethroid exposures among pest-control operators and residents in rural areas. They found that the urinary concentrations of 3-PBA for professional sprayers were significantly higher than for rural residents [40]. In this study, 3-PBA, cis-DCCA, cis,trans-DCCA, and total cypermethrin urinary metabolites among farmers who used cypermethrin on their farms and in their homes were significantly higher than those who only used cypermethrin at home. It is well known that 3-PBA is a non-specific metabolite that is found in many pyrethroids, including cyhalohathrin, cypermethrin, permethrin, deltramethrin, ethonfenprox, esfenvalerate, fenpropathrin, and phenothrin [41]; therefore, if farmers used products containing these agents on their farms or in their homes, the differences that were recorded in 3-PBA levels might be confounded.

Regarding organic farmers, female farmers had significantly higher 3-PBA, cis-DCCA, cis-trans-DCCA, and total cypermethrin levels than male organic farmers. This is probably because female farmers usually take care of the house, cleaning, cooking, and maintaining the home for the whole family, which includes using insecticides at home to get rid of insects, cockroaches, termites, etc. They may, thus, have more contact with the insecticides that are used at home, which mostly contain pyrethroids including cypermethrin, permethrin, cyfluthrin, etc. Organic farmers who reported using insecticides at home in the past year (14.8%), had borderline significantly higher cis-DCCA values than those who were not using them. The organic farmers reported using Aswin (0.6%), ARS (1.7%), Baygon (6.4%), Shieldtox (2.3%), and mosquito coils (10.9%). Most Baygon sprays contain cypermethrin (0.1%) and other pyrethroids, imiprotin (0.03%), and prallethrin (0.03%) [22]. Shieldtox sprays contain other pyrethroids, prallethrin 0.076%, and D-phenothrin 0.046% [23]. Mosquito coils have different brand names with different active ingredients, but the main active ingredients are also pyrethroids, dimefluthrin, allethrin, metofluthrin, etc. [24,25]. When different types of pyrethroids enter the body, they may produce the same or different urinary metabolites. For example, cypermethrin, permethrin, and cyfluthrin have the same urinary metabolites of 3-PBA, cis, and trans-DCCA in humans [41]. Transfluthrin is also metabolized into cis-DCCA, trans-DCCA, and other metabolites, such as 2,3,5,6-tetrafluorobenzyl alcohol and 2,3,5,6-tetrafluorobenzoic acid [42]. The urinary 3-PBA levels of organic farmers who ate vegetables daily were significantly higher than in the samples from those who did not. Vegetables could be sources of exposure for organic farmers if residues persisted after insecticides were used by other farmers who sold their vegetables in the market [43].

With regard to conventional farmers aged 35–60, the concentration levels of trans-DCCA, cis,trans-DCCA, and total cypermethrin were significantly higher than those of farmers below the age of 35. Urinary trans-DCCA levels for those over the age of 60 were significantly higher than for those below the age of 35. Pyrethroids are removed from the body by liver esterase enzymes. As age increases, the liver often loses some functionality, which may be one explanation for the higher metabolite levels recorded among older farmers [44]. It is also possible that older farmers are less careful regarding safe work practices or personal protective equipment when spraying pesticides. This study found that female farmers had higher urinary cis-DCCA levels than male farmers, although none of the other metabolites differed according to gender. A study by Chetna et al. found that female sprayers did not take into account the wind direction while applying pesticides and did not read and understand the pesticide labels, which may explain their higher metabolite levels [45]. Those who used Baygon at home had significantly higher urinary 3-PBA, Cis-DCCA, Trans-DCCA, cis,trans-DCCA, and total cypermethrin levels than those who did not use Baygon at home. This is because Baygon has cypermethrin as its main active ingredient (0.1%) [22]. Those who used ARS had significantly higher urinary Cis-DCCA than those who did not. The main active ingredients of ARS are tetramethrin, permethrin, allethrin, and cypermethrin [21]. Cypermethrin, permethrin, and cyfluthrin produce the same urinary metabolites of 3-PBA, cis-DCCA, and trans-DCCA in humans [41]. We asked the farmers using pesticides whether they took a shower or washed immediately after spraying pesticides; those who said yes demonstrated significantly lower levels of all cypermethrin metabolites than those who did not. When pesticides are mixed, the liquid form is prone to splashing and spillage, resulting in direct or indirect skin contact through clothing contamination [43]. When farmers sprayed the pesticide alachlor, the overspray could be measured on their foreheads, arms, legs, and backs [27]. Showering after spraying can reduce dermal exposure and reduce the pesticide metabolites in the urine. When farmers used up the pesticides, we asked where they kept the empty pesticide bottles. Those who kept the empty bottles at home for disposal had significantly higher 3-PBA, cis-DCCA, cis,trans-DCCA, and total cypermethrin levels than those who did not. The empty pesticide bottles could be a source of pesticide exposure in or around the homes of conventional farmers. The empty pesticide bottles still have residues of pesticides in them. Exposure could occur through the inhalation of residual air concentrations or exposure to the residues found on surfaces, clothing, food, dust, discarded pesticide containers, or application equipment [43]. The cypermethrin metabolites of conventional farmers who ate vegetables daily had significantly higher 3-PBA, cis-DCCA, trans-DCCA, cis,trans-DCCA, and total cypermethrin levels than those who did not. It is important to note that, as described above, the cypermethrin metabolites of organic farmers who ate vegetables daily were significantly higher for only 3-PBA, compared to organic farmers who did not eat vegetables daily. These differences may reflect the finding that the most common source of pesticides is from the vegetables that are being eaten. In the case of conventional farmers, their own or the market vegetables are treated with insecticides. In the case of organic farmers, it is more likely that they mostly ate organic vegetables, with only some supplemental vegetables that could contain pesticide residues from the market. Currently, 3-PBA is a non-specific metabolite of many pyrethroids, including cyhalohathrin, cypermethrin, permethrin, deltramethrin, ethonfenprox, esfenvalerate, fenpropathrin, and phenothrin [41]; finding 3-PBA in the urine of organic farmers suggests that general pyrethroid residues may be found on supplemental market vegetables.

When we compared the concentrations of 3-PBA, cis-DCCA, and trans-DCCA in conventional farmers' urine to the figures reported in other studies, the metabolite levels measured in this current study were consistent with studies involving farmers, made by Panuwet et al. in Thailand [16] and López-Gálvez et al. in Mexico [44]. These authors studied farmers who applied permethrin and cypermethrin, which can be metabolized into 3-PBA, cis-DCCA, and trans-DCCA. However, the studies by Wang et al. and Kimata et al., regarding pest-control operators in Japan who applied pyrethroid frequently during the summer, reported higher levels of pyrethroid metabolites [40,46]. The pest-control operators in Kimata's study applied pyrethroid indoors, where there was less ventilation.

For organic farmers, the concentration levels of 3-PBA and trans-DCCA were comparable to those reported in the study conducted by Qi et al. in China. Our study reported that organic farmers used insecticides at home, which was similar to the findings in Qi's study of pregnant women, where, although they were not exposed occupationally, about half of them reported applying indoor insecticides [47]. However, the levels in our study were higher compared to the general adult populations of the US [48] and Canada [49], where insecticides are used less frequently in the home. These results and those from the conventional farmers who did not use cypermethrin in the home or on their farms suggest other sources of exposure, such as food purchased at the market.

The limitations of this study include the fact that there was only one spot-checked urine sample for each subject. We did not collect data regarding the last time that they applied any specific insecticides on their crops or in the home. The use of cypermethrin on the farm depends on the season and the types of crops, while the use of insecticides in the home also varies with the season. The cypermethrin metabolites have very short half-lives in human urine (3BPA (5.7 h) and DCCA (4.5–5.4 h)) [50]; thus, reports of usage may not have been linked closely enough in time regarding the collection of urine samples to show strong associations. We did not collect information on whether the farmers had pets or animals at home, or whether the farmers used showering or cleaning products such as Chaingard, which contains permethrin (0.5–1%) or cypermethrin (0.1–0.15%), on their pets or animals [51]. Nor did we ask about how much of their food was organic versus how much was not organic. All of these activities could be additional sources of exposure.

5. Conclusions

Conventional farmers who use cypermethrin on their farms showed the highest levels of associated metabolites, while both organic and conventional farmers who only used cypermethrin at home or who did not use any cypermethrin demonstrated the lowest urinary levels. Nevertheless, the finding that the levels of these metabolites among organic farmers and conventional farmers who do not use cypermethrin exceeded those in the general population of Western countries suggests that other sources, such as market produce, pets, cleaning products, or the use of other pyrethroids may be contributing to this higher background level.

Author Contributions: Conceptualization, S.P., P.K., N.N. and S.W.; methodology, A.T., N.N., S.S. and S.P.; data curation, A.T., S.P., P.K. and S.S.; writing—original draft preparation, A.T. and P.K.; writing—review and editing, P.K., N.N., S.P., A.T. and S.W.; supervision, P.K., S.P. and S.W.; project administration, P.K. and S.P.; funding acquisition, P.K. and S.W. All authors have read and agreed to the published version of the manuscript.

Funding: This research was financially supported by NIH, the Fogarty International Center, the National Institute of Environmental Health Science, and the Center for Disease Control, under the award numbers U01 TW010091 and U2R TW010088.

Institutional Review Board Statement: The study was conducted according to the guidelines of the Declaration of Helsinki and was approved by the Institutional Review Board (or Ethics Committee) of the Faculty of Public Health, Mahidol University (protocol code MUPH 2015-146 and date of approval, 28 August 2015).

Informed Consent Statement: Informed consent was obtained from all subjects involved in the study.

Data Availability Statement: The data presented in this study are available on request from the corresponding author.

Acknowledgments: We would also like to thank all the participants and health-promoting hospital staff helping in recruitment of the subjects in the Nakornsawan, Phitsanulok, and Yasothon Provinces.

Conflicts of Interest: The authors declare no conflict of interest. The funders had no role in the design of the study; in the collection, analyses, or interpretation of data; in the writing of the manuscript, or in the decision to publish the results.

References

1. Pongsrihadulchai, A. Thailand Agricultural Policies and Development Strategies. Available online: https://ap.fftc.org.tw/article/1393 (accessed on 12 May 2023).
2. Office of Agricultural Economics, Ministry of Agriculture and Cooperatives. Top Ten Pesticides Imported in 2021. Available online: https://www.doa.go.th/ard/?page_id=386 (accessed on 12 May 2023). (In Thai)
3. McKinlay, R.; Plant, J.A.; Bell, J.N.B.; Voulvoulis, N. Endocrine disrupting pesticides: Implications for risk assessment. *Environ. Int.* **2008**, *34*, 168–183. [CrossRef] [PubMed]
4. Mnif, W.; Hassine, A.I.H.; Bouaziz, A.; Bartegi, A.; Thomas, O.; Roig, B. Effect of endocrine disruptor pesticides: A review. *Int. J. Environ. Res. Public Health* **2011**, *8*, 2265–2303. [CrossRef] [PubMed]
5. Hertz-Picciotto, I.; Sass, J.B.; Engel, S.; Bennett, D.H.; Bradman, A.; Eskenazi, B.; Lanphear, B.; Whyatt, R. Organophosphate exposures during pregnancy and child neurodevelopment: Recommendations for essential policy reforms. *PLoS Med.* **2018**, *15*, e1002671. [CrossRef]
6. Costa, L.G. Organophosphorus Compounds at 80: Some Old and New Issues. *Toxicol. Sci.* **2018**, *162*, 24–35. [CrossRef] [PubMed]
7. Saillenfait, A.M.; Ndiaye, D.; Sabaté, J.P. Pyrethroids: Exposure and health effects—An update. *Int. J. Hyg. Environ. Health* **2015**, *218*, 281–292. [CrossRef] [PubMed]
8. Ratelle, M.; Coté, J.; Bouchard, M. Time profiles and toxicokinetic parameters of key biomarkers of exposure to cypermethrin in orally exposed volunteers compared with previously available kinetic data following permethrin exposure. *J. Appl. Toxicol.* **2015**, *35*, 1586–1593. [CrossRef]
9. Siriwat, S.; Ong-artborirak, P.; Nganchamung, T.; Robson, M.G.; Siriwong, W. Health risk assessment of residential exposure to cypermethrin among young children in agricultural communities in northeastern Thailand. *Hum. Ecol. Risk Assess.* **2019**, *25*, 614–623. [CrossRef]
10. Cote, J.; Bonvalot, Y.; Carrier, G.; Lapointe, C.; Fuhr, U.; Tomalik-Scharte, D.; Wachall, B.; Bouchard, M. A Novel Toxicokinetic Modeling of Cypermethrin and Permethrin and Their Metabolites in Humans for Dose Reconstruction from Biomarker Data. *PLoS ONE* **2014**, *9*, e88517. [CrossRef]
11. Woollen, B.H.; Marsh, J.R.; Laird, W.J.D.; Lesser, J.E. The metabolism of cypermethrin in man: Differences in urinary metabolite profiles following oral and dermal administration. *Xenobiotica* **1991**, *22*, 983–991. [CrossRef]

12. Klimowska, A.; Amenda, K.; Rodzaj, W.; Wilenska, M.; Jurewicz, J.; Wielgomas, B. Evaluation of 1-year urinary excretion of eight metabolites of synthetic pyrethroids, chlorpyrifos, and neonicotinoids. *Environ. Int.* **2020**, *145*, 106119. [CrossRef]
13. Singleton, S.T.; Lein, P.J.; Farahat, F.M.; Farahat, T.; Bonner, M.R.; Knaak, J.B.; Olson, J.R. Characterization of cypermethrin Exposure in Egyptian Agricultural Workers. *Int. J. Hyg. Environ. Health* **2014**, *217*, 538–545. [CrossRef]
14. Barr, D.B.; Olsson, A.O.; Wong, L.Y.; Udunka, S.; Baker, S.E.; Whitehead, R.D.; Magsumbol, J.M.S.; Williams, B.L.; Needham, L.L. Urinary Concentrations of Metabolites of Pyrethroid Insecticides in the General U.S. Population: National Health and Nutrition Examination Survey 1999–2002. *Environ. Health Perspect* **2010**, *118*, 742–748. [CrossRef]
15. Garí, M.; González-Quinteiro, Y.; Bravo, N.; Grimalt, J.O. Analysis of metabolites of organophosphate and pyrethroid pesticides in human urine from urban and agricultural populations (Catalonia and Galicia). *Sci. Total Environ.* **2018**, *622–623*, 526–533. [CrossRef]
16. Panuwet, P.; Prapamontol, T.; Chantara, S.; Olsson, A.O.; Barr, D.B. A Pilot Survey of Pesticide-Specific Urinary Metabolites among Farmers in Chiang Mai Highland Agricultural Area. *CMU J.* **2004**, *3*, 25–34.
17. Thiphom, S.; Prapamontol, T.; Chantara, S.; Mangklabruks, A.; Suphavilai, C.; Ahn, K.C.; Gee, S.J.; Hammock, B.D. Determination of the pyrethroid insecticide metabolite 3-PBA in plasma and urine samples from farmer and consumer groups in northern Thailand. *J. Environ. Sci. Health* **2014**, *49*, 15–22. [CrossRef]
18. Leng, G.; Gries, W. Determination of Pyrethroids in Blood Plasma and Pyrethroid/Pyrethrin Metabolites in Urine by Gas Chromatography-Mass Spectrometry and High-Resolution GC-MS. In *Pesticide Protocols: Methods in Biotechnology*; Martínez Vidal, J.L., Frenich, A.G., Eds.; Humana Press: Cham, Switzerland, 2006; Volume 19, pp. 17–33.
19. Hornung, R.W.; Reed, L.D. Estimation of average concentration in the presence of nondetectable values. *Appl. Occup. Environ. Hyg.* **1990**, *5*, 46–51. [CrossRef]
20. Beckmann Coulter, Instruction for Use. Creatinine. Available online: https://www.beckmancoulter.com/wsrportal/techdocs?docname=/cis/A69463/%%/EN (accessed on 13 May 2023).
21. Chareonviriyaphap, T.; Bangs, M.J.; Suwonkerd, W.; Kongmee, M.; Corbel, V.; Ngoen-Klan, R. Review of insecticide resistance and behavioral avoidance of vectors of human diseases in Thailand. *Parasites Vectors* **2013**, *6*, 280. [CrossRef]
22. S. C. Johnson & Son, Inc. Information for Baygon Products. Available online: https://www.baygon.co.th/th-th/products/baygon-multi-insect-killer-dlinonene/tabs/ingredients (accessed on 12 May 2023). (In Thai).
23. Pantry Express. Shieldtox Mosquito Spray. Available online: https://www.pantryexpress.my/pest-control-repellents/1138-shieldtox-mosquito-spray-600ml-x2-9556111175270.html (accessed on 12 May 2023).
24. Mosquito Coil. Available online: https://en.wikipedia.org/wiki/Mosquito_coil (accessed on 12 May 2023).
25. Norkaew, S.; Taneepanichskul, N.; Siriwong, W.; Siripattanakul, S.; Robson, M. Household pesticide use in agricultural community, Northeastern Thailand. *J. Med. Med. Sci.* **2012**, *3*, 631–637.
26. Kongtip, P.; Nankongnab, N.; Tipayamongkholgul, N.; Bunngamchairat, A.; Yimsabai, J.; Pataitiemthong, A.; Woskie, S. Across-sectional investigation of cardiovascular and metabolic biomarkers among conventional and organic farmers in Thailand. *Int. J. Environ. Res. Public Health* **2018**, *15*, 2590. [CrossRef] [PubMed]
27. Mahaboonpeeti, R.; Kongtip, P.; Nankongnab, N.; Tipayamongkholgul, M.; Bunngamchairat, A.; Yoosook, W.; Woskie, S. Evaluation of Dermal Exposure to the Herbicide Alachlor Among Vegetable Farmers in Thailand. *Ann. Work. Expo. Health* **2018**, *62*, 1147–1158. [CrossRef] [PubMed]
28. Chouichom, S.; Yamao, M. Comparing opinions and attitudes of organic and non-organic farmers towards organic rice farming system in north-eastern Thailand. *J. Org. Syst.* **2010**, *5*, 25–35.
29. Kings, D.; Ilbery, B. Organic and Conventional Farmers' Attitudes towards Agricultural Sustainability. Available online: https://www.intechopen.com/books/organic-farming-and-food-production/organic-and-conventional-farmers-attitudes-towards-agricultural-sustainability (accessed on 13 May 2023).
30. Nankongnab, N.; Kongtip, P.; Tipayamongkholgul, M.; Bunngamchairat, A.; Sitthisak, S.; SusanWoskie, S. Difference in accidents, health symptoms and ergonomic problems between chemical use and organic farmers. *J. Agromed.* **2020**, *25*, 158–165. [CrossRef] [PubMed]
31. Aungkulanon, S.; Pitayarangsarit, S.; Bundhamcharoen, K.; Akaleephan, C.; Chongsuvivatwong, V.; Phoncharoen, R.; Tangcharoensathien, V. Smoking prevalence and attributable deaths in Thailand: Predicting outcomes of different tobacco control interventions. *BMC Public Health* **2019**, *19*, 984. [CrossRef] [PubMed]
32. Wakabayashi, M.; McKetin, R.; Banwell, C.; Yiengprugsawan, V.; Kelly, M.; Seubsman, S.; Iso, H.; Sleigh, A.; Thai Cohort Study Team. Alcohol consumption patterns in Thailand and their relationship with non-communicable disease. *BMC Public Health* **2015**, *15*, 1297. [CrossRef] [PubMed]
33. Kongtip, P.; Nankongnab, N.; Kallayanatham, N.; Pundee, R.; Yimsabai, J.; Woskie, S. Longitudinal Study of Metabolic Biomarkers among Conventional and Organic Farmers in Thailand. *Int. J. Environ. Res. Public Health* **2020**, *17*, 4178. [CrossRef]
34. Damalas, C.A.; Koutroubas, S.D. Farmers' Exposure to Pesticides: Toxicity Types and Ways of Prevention. *Toxics* **2016**, *4*, 1. [CrossRef]
35. Morgan, M.K.; Jones, P.A. Dietary predictors of young children's exposure to current-use pesticides using urinary biomonitoring. *Food Chem. Toxicol.* **2013**, *62*, 131–141. [CrossRef]

36. Kimata, A.; Kondo, T.; Ueyama, J.; Yamamoto, K.; Kamijima, M.; Suzuki, K.; Inoue, T.; Ito, Y.; Hamajima, N. Relationship between dietary habits and urinary concentrations of 3-phenoxybonzoic acid in a middle-aged and elderly general population in Japan. *Environ. Health Prev. Med.* **2009**, *14*, 173–179. [CrossRef]
37. Ye, M.; Beach, J.; Martin, J.W.; Senthilselvan, A. Associations between dietary factors and urinary concentrations of organophosphate and pyrethroid metabolites in a Canadian general population. *Int. J. Hyg. Environ. Health* **2015**, *218*, 616–626. [CrossRef]
38. Li, Z.M.; Robinson, M.; Kannan, K. An assessment of exposure to several classes of pesticides in pet dogs and cats from New York, United States. *Environ. Int.* **2022**, *169*, 107526. [CrossRef]
39. Wise, C.F.; Hammel, C.S.; Herkert, N.J.; Ospina, M.; Calafat, A.M.; Breen, M.; Stapleton, H.M. Comparative Assessment of Pesticide Exposures in Domestic Dogs and Their Owners Using Silicone Passive Samplers and Biomonitoring. *Environ. Sci. Technol.* **2022**, *56*, 1149–1161. [CrossRef] [PubMed]
40. Kimata, A.; Kondo, T.; Ueyama, J.; Yamamoto, K.; Yoshitake, J.; Takagi, K.; Suzuki, K.; Inoue, T.; Ito, Y.; Hamajima, N.; et al. Comparison of urinary concentrations of 3-phenoxybenzoic acid among general residents in rural and suburban areas and employees of pest control firms. *Int. Arch. Occup. Environ. Health* **2009**, *82*, 1173–1178. [CrossRef] [PubMed]
41. Wielgomas, B.; Klimowska, A.; Rodzaj, W. Analytical Methods for Determination Urinary Metabolites of Synthetic Pyrethroids. In *Pyrethroid Insecticides*; Eljarrat, E., Ed.; Springer: Cham, Switzerland, 2020; Volume 92, pp. 47–80.
42. Yoshida, T. Biomarkers for monitoring transfluthrin exposure: Urinary excretion kinetics of transfluthrin metabolites in rats. Environ. *Toxicol. Pharmacol.* **2014**, *37*, 103–109. [CrossRef] [PubMed]
43. Damalas, C.A.; Eleftherohorinos, I.G. Pesticide exposure, safety issues, and risk assessment indicators. *Int. J. Environ. Res. Public Health* **2011**, *8*, 1402–1419. [CrossRef] [PubMed]
44. López-Gálvez, N.; Wagoner, R.; Beamer, P.; de Zapien, J.; Rosales, C. Migrant Farmworkers' Exposure to Pesticides in Sonora, Mexico. *Int. J. Environ. Res. Public Health* **2018**, *15*, 2651. [CrossRef] [PubMed]
45. Chetna, D.; Vaibhav, K.; Pallavi, N.; Jitendra, R. Gender differences in knowledge, attitude and practices regarding the pesticide use among farm workers: A questionnaire based study. *RJPBCS Res. J. Pharm. Biol. Chem. Sci.* **2012**, *3*, 632–639.
46. Wang, D.; Kamijima, M.; Imai, R.; Suzuki, T.; Kameda, Y.; Asai, K.; Okamura, A.; Naito, H.; Ueyama, J.; Saito, I.; et al. Biological Monitoring of Pyrethroid Exposure of Pest Control Workers in Japan. *J. Occup. Health* **2007**, *49*, 509–514. [CrossRef]
47. Qi, X.; Zheng, M.; Wu, C.; Wang, G.; Feng, C.; Zhou, Z. Urinary pyrethroid metabolites among pregnant women in an agricultural area of the Province of Jiangsu, China. *Int. J. Hyg. Environ. Health* **2012**, *215*, 487–495. [CrossRef]
48. Trunnelle, K.J.; Bennett, D.H.; Tulve, N.S.; Clifton, M.S.; Davis, M.D.; Calafat, A.M.; Moran, R.; Tancredi, D.J.; Hertz-Picciotto, I. Urinary Pyrethroid and Chlorpyrifos Metabolite Concentrations in Northern California Families and Their Relationship to Indoor Residential Insecticide Levels, Part of the Study of Use of Products and Exposure Related Behavior (SUPERB). *Environ. Sci. Technol.* **2014**, *48*, 1931–1939. [CrossRef]
49. Health Canada, Second Report on Human Biomonitoring of Environmental Chemicals in Canada: Results of the Canadian Health Measures Survey Cycle 2 (2009–2011). Available online: https://www.canada.ca/content/dam/hc-sc/migration/hc-sc/ewh-semt/alt_formats/pdf/pubs/contaminants/chms-ecms-cycle2/chms-ecms-cycle2-eng.pdf (accessed on 12 May 2023).
50. Ratelle, M.; Côté, J.; Bouchard, M. Toxicokinetics of permethrin biomarkers of exposure in orally exposed volunteers. *Toxicol. Lett.* **2015**, *232*, 369–375. [CrossRef]
51. Chaingard, Thailand. Information for Chaingard Products. Available online: https://chaingardthailand.com/product-category/shampoo-th/shampoo-for-dog (accessed on 12 May 2023). (In Thai)

pollutants

MDPI

Article

Mercury Intake Estimation in Adult Individuals from Trieste, Italy: Hair Mercury Assessment and Validation of a Newly Developed Food Frequency Questionnaire

Andrea De Giovanni [1,2], Vincenzo Iannuzzi [3], Gianni Gallello [4], Cristina Giuliani [3], Mauro Marini [2,5,*], M. Luisa Cervera [6] and Donata Luiselli [1,2]

1 Department of Cultural Heritage, University of Bologna, Via degli Ariani 1, 48121 Ravenna, Italy; andrea.degiovanni5@unibo.it (A.D.G.); donata.luiselli@unibo.it (D.L.)
2 Fano Marine Center, The Inter-Institute Center for Research on Marine Biodiversity, Resources and Biotechnologies (FMC), Viale Adriatico 1/N, 61032 Fano, Italy
3 Laboratory of Molecular Anthropology, Centre for Genome Biology, Department of Biological, Geological and Environmental Sciences, University of Bologna, Via Selmi 3, 40126 Bologna, Italy; vincenzo.iannuzzi2@unibo.it (V.I.); cristina.giuliani2@unibo.it (C.G.)
4 Department of Prehistory, Archaeology and Ancient History, University of Valencia, Avenida de Blasco Ibañez 28, 46010 Valencia, Spain; gianni.gallello@uv.es
5 Institute for Biological Resources and Marine Biotechnologies, National Research Council (IRBIM, CNR), Largo Fiera della Pesca 2, 60125 Ancona, Italy
6 Department of Analytical Chemistry, University of Valencia, Dr. Moliner 50, 46100 Burjassot, Spain; m.luisa.cervera@uv.es
* Correspondence: mauro.marini@cnr.it

Citation: De Giovanni, A.; Iannuzzi, V.; Gallello, G.; Giuliani, C.; Marini, M.; Cervera, M.L.; Luiselli, D. Mercury Intake Estimation in Adult Individuals from Trieste, Italy: Hair Mercury Assessment and Validation of a Newly Developed Food Frequency Questionnaire. *Pollutants* **2023**, *3*, 320–336. https://doi.org/10.3390/pollutants3030022

Academic Editors: Roel Vermeulen, Lauren Petrick and Maaike van Gerwen

Received: 28 April 2023
Revised: 24 June 2023
Accepted: 29 June 2023
Published: 5 July 2023

Abstract: Seafood constitutes the primary source of exposure to the organic form of mercury in the general population, and the Trieste Gulf is considered a hotspot of mercury contamination. We used a newly developed quantitative food frequency questionnaire to obtain an estimation of the intake of mercury through seafood consumption in a sample of 32 individuals from Trieste. Then, we validated the results obtained from the questionnaire against those of the analysis of total mercury measured in the hair of the same individuals through Spearman rank correlation coefficients, Cohen's weighted Kappa statistic, and a Bland–Altman plot. The Spearman rank correlation coefficient and Cohen's weighted Kappa statistic were 0.76 and 0.69, respectively. In the Bland–Altman plot, 93.75% of the data points lay within the acceptability range. The plot revealed an ever-increasing overestimation of mercury intake by the questionnaires as the hair mercury increased. By applying a standardized filtering procedure to the results of the questionnaires, we obtained a Spearman rank correlation coefficient and Cohen's weighted Kappa statistic of 0.69 and 0.57, respectively. In this Bland–Altman plot, 93.75% of the data points lay within the acceptability range. In this latter plot, the proportionality between the mean difference and the magnitude of the measurement was more subtle compared to that observed in the plot built upon the non-filtered questionnaires. This preliminary study shows the high accuracy of the reported questionnaire in the estimation of habitual mercury intake, similar to the one measured through the analysis of hair.

Keywords: mercury; food frequency questionnaire; validation; Bland–Altman plot; hair mercury; Trieste

1. Introduction

Among the general population, seafood is the main source of exposure to the organic form of mercury, i.e., methylmercury (MeHg) [1,2], with large, long-lived, predatory fishes (e.g., Atlantic bluefin tuna, swordfish, and sharks) showing higher concentrations of this contaminant as an effect of biomagnification along the food chain [3] and MeHg uptake being a lifelong process of bioaccumulation [4].

MeHg is a well-established neurotoxicant, and exposure to MeHg is associated with nervous system damage in adults and impaired neurological development in infants and children [5], as well as with an increased risk of cardiovascular disease in adults [6].

Based on epidemiological evidence accumulated over the years, national and international agencies have set limits and given recommendations in order to protect citizens' health. In 2001, the U.S. Environmental Protection Agency (US EPA) derived an oral reference dose (RfD) for MeHg—that is, the maximum oral dose that is likely to be without appreciable risk of deleterious effects during a lifetime—of 0.1 $\mu g\ kg^{-1}\ day^{-1}$ [7,8]. Then, in 2012, the European Food Safety Authority (EFSA) Panel on Contaminants in the Food Chain (CONTAM) established a tolerable intake of 1.3 $\mu g\ kg^{-1}$ body weight (b.w.) per week [9,10], corresponding to an apparent no-observed-effect level (NOEL) of ~11.5 mg kg^{-1} and ~46 $\mu g\ L^{-1}$ in the hair and blood, respectively. This threshold value has been adopted for all classes of consumers, even though adults may be less sensitive to the adverse effects of MeHg [11].

Hair mercury level is frequently used as a biomarker of endogenous exposure to MeHg. Indeed, MeHg is permanently incorporated into the growing hair through diffusion from the blood, and the ratio between the total mercury (THg) in the hair and that in the blood is around 250:1 [12,13]. While blood can be used to document MeHg short-term exposure, hair provides an estimation of the long-term average exposure [12].

As several investigations have shown, hair THg level often depends on habitual seafood consumption [14–26]. In the study by Elhamri and colleagues on a Moroccan coastal community [15], the authors found a linear relationship between the log-transformed data on hair THg levels and the fish consumption frequency (FCF) in times per week. Moreover, they found that subjects with an FCF of three to five times per week showed greater hair THg levels (geometric mean = 5.30 $\mu g\ g-1$) compared to those of subjects with an FCF of one to two times per week (geometric mean = 1.04 $\mu g\ g-1$). Accordingly, when analyzing 237 adults from Naples (Italy), Díez and colleagues [16] found a strong positive correlation (rs = 0.536; $p < 0.05$) between the THg in hair and the fish consumption rate.

The Mediterranean Sea is generally considered a geological hot spot for Hg [11], as it is characterized by large deposits of mercury sulfur (HgS) that account for about 65% of the global mercury reserves [27], and the highest concentrations of Hg in Europe tend to be found in fish caught in the Mediterranean Sea [28]. Moreover, data have shown a more marked Hg bioavailability in the Tyrrhenian and the Adriatic coastal waters compared to the rest of the Mediterranean [29], and Hg levels higher than the legal limit have been discovered in seafood caught in both areas [4,30–35].

Considering all the above, and also the great amount of local seafood consumed by Mediterranean communities [11,20,36], it is not surprising that several studies revealed high hair THg levels in people from Mediterranean regions [23,37,38].

Among the methods aimed at estimating habitual seafood consumption [39,40] or the intake of specific nutrients [41,42] and/or contaminants [43], the food frequency questionnaire (FFQ) is one of the most used. The FFQ is a retrospective direct method for dietary assessment in that it collects information on the foods and beverages already consumed, assessing the frequency with which foods and/or food groups are eaten over a certain time period [44]. Whenever a new FFQ is developed, whether de novo or from pre-existing questionnaires, it should be validated in the investigated population on which the study is going to be conducted, which means that the results should be compared with those obtained applying at least another method [45], such as a dietary record [39,46], a 24 h recall [47,48], or the use of biomarkers [49,50].

Here, we used a newly developed quantitative FFQ—i.e., an FFQ that asks respondents' usual portion size based on a specified measure—coupled with a database on Hg in Mediterranean seafood [51], to obtain an estimation of the intake of Hg through seafood consumption in a sample of 32 individuals from a coastal city in northern Italy, i.e., Trieste, which is considered a Hg contamination hotspot [52]. Then, we compared the results obtained from the questionnaire with those of the analysis of THg measured in the hair of the same individuals with the aim of (1) validating the questionnaire and (2) collecting preliminary data on Hg exposure among the population of Trieste.

2. Materials and Methods

2.1. Sampling Location

The Italian city of Trieste is located in the region of Friuli-Venezia Giulia, facing the northwestern part of the Adriatic Sea (Figure 1a). Trieste Gulf is considered a hotspot of Hg contamination, mostly due to the Hg input from the polluted Isonzo River (Figure 1b), which crosses the cinnabar-rich deposits of the Idrija mine [52].

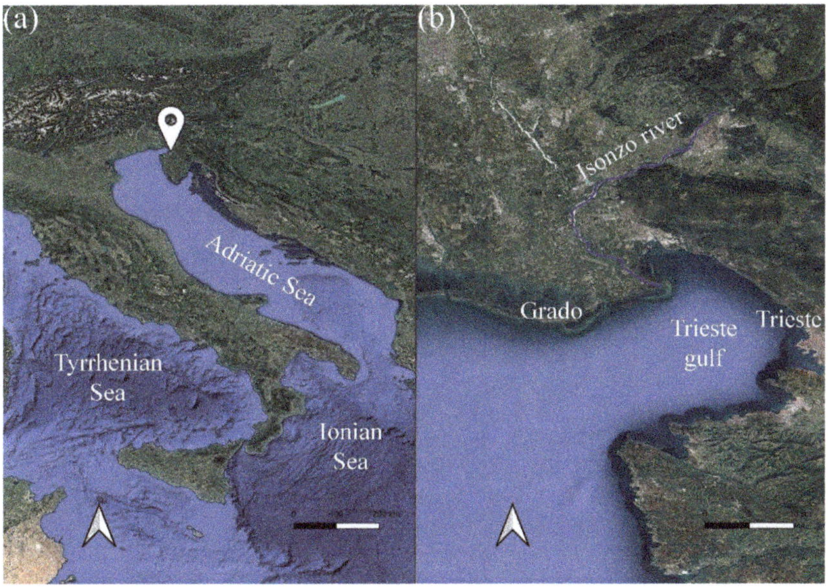

Figure 1. (**a**) Italian peninsula. The white waypoint indicates the location of Trieste. (**b**) The Trieste Gulf with the Isonzo River highlighted in blue.

2.2. FFQ Development and Administration

Following recommendations made in [44,45], we developed a quantitative FFQ consisting of close-ended questions on the habitual consumption of 52 aquatic species. These species were selected through the following procedure: we examined the Hg concentration data retrieved from the literature [4,30,31,33–35,52–75], coupled with a large database on Hg in Mediterranean biota [51], and calculated the average concentrations of Hg in the edible tissues of more than 350 Mediterranean aquatic species. Among these, we selected those species that are included in the ministerial decree of 22 September 2017 (n. 19105), listing species that are commercially relevant in Italy that show an average Hg concentration exceeding the maximum allowable level of Hg in seafood of 0.5 mg kg^{-1} set by EU [76] in their edible tissue, thus obtaining a list of 41 commercially relevant aquatic species contaminated by Hg in the Mediterranean Sea. Finally, we further expanded the above list by adding several species that are most consumed by the Italian population (ISMEA, 2011).

For each of the 52 species, the questionnaire assessed respondents' habitual consumption frequency and portion size over a period spanning the previous six months before the interview. To ease species recognition, photos were shown to the participants, and species names in local dialects were provided. Apart from the questions on seafood consumption, the questionnaire included questions on other routes of exposition to Hg—i.e., breakage of Hg thermometers, occupation, dental amalgam fillings—smoking, and general state of health. The questionnaire was interviewer-administered through a voluntary response sampling carried out among the general population. The only two exclusion criteria pertain to the state of health of the participant—i.e., participants were required not to suffer from any specific illnesses and were asked to report their general state of health—and to the age of the participant—i.e., participants had to be 18 years of age and over. The sampling was approved by the bioethics committee of the University of Bologna.

2.3. Hg Intake Estimation from the FFQ

Once collected, the questionnaires were used to calculate the weekly Hg intake of the participants via the following formula:

$$\frac{\Sigma\left(F\left(times\ week^{-1}\right)\times \mathrm{C}\left(\mathrm{mg\ kg^{-1}}\right)\right)}{BW\left(\mathrm{kg}\right)}, \quad (1)$$

where F is the consumption frequency of each species in times week^{-1}, C is the Hg concentration in the edible tissues of each species in mg kg^{-1}, and BW is the body weight of the participant in kg.

After discovering, through comparison with hair THg levels, that the questionnaire tended to overestimate Hg intake in individuals consuming higher amounts of seafood, we applied a standardized filtering process to all 32 questionnaires; namely, for each participant, the estimation of Hg weekly intake was achieved using only those species that had been consumed at the two highest frequencies. For example, if a participant had consumed a given species twice a week, another species twice a month, and another one once a month, we excluded the last species from the calculation of the Hg weekly intake of that participant.

2.4. Hair Sampling and THg Measurement

Hairs of the particpants were collected following the procedure recommended by the World Health Organization [12]. After a cleaning step involving rinsing the hair with 70% isopropyl alcohol, a bundle of hair approximately 0.75–1.0 cm in diameter was cut from the occipital region of each participant, using blunt-tipped, clean stainless steel. To make sure that an adequate amount (50 mg) of hair had been collected, samples were also weighed with a precision scale. Once cut, the hair bundle was wrapped close to the scalp end with a small Post-it note and held together with a plastic clip. Finally, the hair sample was placed in a marked paper envelope, which was in turn stored in a sealed plastic bag.

The measurement of THg concentration in hair samples was carried out using a Mercury Analyzer DMA-80 Milestone, which is a spectrophotometer based on the atomic absorption determination after the thermal decomposition of the samples.

The analysis process consists of introducing the sample into a small oven that can reach 650 °C. After the drying and incineration of the sample, the volatile compounds are carried by a flow of O_2 to a catalyst, where the released mercury is reduced to Hg0, which is later retained in a small gold amalgam. The amalgam is heated to 900 °C, and the volatilized mercury passes through two measurement cells where it is quantified by detection using atomic absorption spectroscopy (AAS), with a detection limit of 0.5 μg kg^{-1} of Hg.

Once the amount of mercury expressed in ng is deduced, it is divided by the mass in grams of the sample, thus obtaining the result in concentration (μg kg^{-1}).

1633c Trace Elements in Coal Fly Ash from NIST were used as certified reference material.

A solvent-free, easy, fast, and waste-free method was employed [77] using the EPA 7473 (SW-846) method. Each hair bundle was cut into pieces of ~1.5 cm in order to assess the monthly Hg intake, as each 1–1.5 cm segment of hair incorporates the Hg assimilated during the preceding month [78]. For each segment obtained, a quantity of ~10 mg of sample was weighted into a quartz cuvette. After that, 50 μL of ultrapure water was added to the sample in the cuvette, which was then introduced into the DMA-80. Samples were analyzed using the same temperature and time parameters as in [77], and for each 1.5 cm of a sample, we made two replicates, using their mean for statistical analyses.

2.5. Hg Intake Estimation from Hair THg

For each participant, to estimate their Hg intake from their hair THg level, we used the average of the THg concentrations measured in each ~1.5 cm segment of the hair sample. First, we derived the blood THg level using the above-mentioned ratio of 250:1 between the THg levels in the two matrices; hence, we obtained the estimate of the Hg dietary intake with the formula reported in [9]:

$$\frac{C\ (\mu g\ L^{-1}) \times b \times V\ (L)}{A \times f \times BW\ (kg)} \tag{2}$$

where C is Hg concentration in blood in μg L^{-1}, b is the elimination constant, which is equal to the ratio between ln (2) and Hg half-life in blood, which was assumed to be equal to 50 days as in [9], V is the blood volume, which was assumed to be equal to 5 L as has been assumed by WHO and US EPA [9], A is the gastrointestinal absorption factor (0.95), f is the fraction of absorbed dose distributed to blood, which was assumed to be equal to 0.05 as in [9], and BW is the body weight of the participant in kg.

2.6. Statistical Analysis

Statistical analyses were carried out using RStudio version 3.6.1.

After assessing the normality of data via Shapiro–Wilk normality test (R function: shapiro.test) and QQ-plots (R function: qqPlot), we used the Kruskal–Wallis rank sum test (R function: kruskal.test) to assess the statistical significance of the differences between the Hg intakes or hair THg concentrations in the two sexes, and we calculated Spearman rank correlation coefficients (R function: cor.test; method = "spearman") to evaluate the association between the Hg intakes or hair THg concentrations and the age of participants.

To assess the validity of the FFQ, we compared the Hg intakes obtained with the two methods—i.e., the FFQ and the hair THg measurement—by calculating the Spearman rank correlation coefficients between the two and using Cohen's weighted Kappa statistic (R function: cohen.kappa) which measures the extent to which the two methods assign the same participant to the same quintile. Finally, for the FFQ validation, we also used the Bland–Altman plot (R function: blandr.draw) [79,80], which is a graphical method that is recommended in conjunction with correlation coefficients when agreement between two quantitative methods has to be evaluated [45]. However, the correlation between results obtained with two different methods measuring the same quantity does not necessarily imply an agreement between the two methods [81]. Briefly, the Bland–Altman plot consists of a scatter plot, in which, the Y-axis shows the difference between each paired measurement (A and B)—i.e., the difference between the Hg intake estimated from the FFQ and that derived from the hair THg measurement—while the X-axis indicates the average of these measurements ((A + B)/2) [81]. When the difference between the two paired measurements is plotted against their mean, the two methods can be deemed in agreement if at least 95% of the data points lie within ±1.96s of the mean difference or within the non-parametric lower and upper limits of agreement (LOA) [82].

3. Results

3.1. Sample Characteristics

Information on participants is reported in Table 1 and Tables S1 and S2 of Supplementary File S1. The 32 participants were all adults and residents of Trieste and include 13 males and 19 females. The mean age was 35 among males (median = 32, SD = 10.90, and range: 21–55) and 48 among females (median = 51, SD = 11.63, and range: 20–65). Nine individuals, seven males and two females, were current smokers, and six participants, one male and five females, had dental amalgam fillings, while two female individuals were not able to provide this latter information.

Table 1. Information on participants from Trieste.

Variables	Male Participants (n = 13)	Female Participants (n = 19)	Overall (n = 32)
Mean age ± SD (median, range)	35 ± 10.90 (32, 21–55)	48 ± 11.63 (51, 20–65)	42.81 ± 12.82 (45.50, 20–65)
Current smokers	7	2	9
Dental amalgam fillings	1	5	6
Mean number of seafood servings (times a week)	1.32	1.25	1.28
Mean Hg w.i. [1] from n.f. [2] FFQ (μg kg^{-1} b.w.) ± SD (median, range)	4.03 ± 3.13 (3.46, 0.52–9.55)	4.11 ± 5.28 (1.99, 0.00–17.64)	4.08 ± 4.47 (2.62, 0.00–17.21)
Mean Hg w.i. [1] from f. [3] FFQ (μg kg^{-1} b.w.) ± SD (median, range)	1.98 ± 1.08 (2.12, 0.36–3.53)	2.57 ± 2.63 (1.23, 0.00–8.01)	2.33 ± 2.13 (1.70, 0.00–8.00)
Mean hair THg (mg kg^{-1}) ± SD (median, range)	2.91 ± 2.16 (2.07, 0.72–7.43)	1.42 ± 0.48 (0.48, 0.02–9.63)	2.02 ± 2.42 (1.08, 0.01–9.63)
Mean Hg w.i. [1] from hair THg (μg kg^{-1} b.w.) ± SD (median, range)	2.16 ± 1.61 (1.54, 0.53–5.52)	1.05 ± 1.82 (0.36, 0.01–7.15)	1.50 ± 1.80 (0.80, 0.01–7.15)

[1] w.i., weekly intake, [2] n.f., non-filtered, and [3] f., filtered.

3.2. Seafood Consumption

The average consumption of seafood products in our sample was 1.28 times a week, with minimal difference between males (1.32 times a week) and females (1.25 times a week) (Table 1). The species consumed in the largest quantity was the Gilthead seabream (*Sparus aurata*), with an average consumption rate of 161.82 g $week^{-1}$, followed by common sole (*Solea solea*), with an average consumption rate of 79.62 g $week^{-1}$, and squid (*Loligo vulgaris*), with an average consumption rate of 67.19 g $week^{-1}$. The most frequently consumed species was Gilthead seabream (0.42 times $week^{-1}$), followed by tuna (*Thunnus* spp.) (0.39 times $week^{-1}$), and common sole (0.27 times $week^{-1}$).

3.3. Exposure Assessment

3.3.1. FFQ

Including all the species consumed by participants, without any filtering procedure, we obtained a median Hg weekly intake of 2.62 μg kg^{-1} b.w. (mean = 4.08 μg kg^{-1} b.w., SD = 4.47, and range: 0.00–17.21 μg kg^{-1} b.w.), which is more than twice the EFSA maximum recommended intake. Among males, the median weekly intake was equal to 3.46 μg kg^{-1} b.w. (mean = 4.03 μg kg^{-1} b.w., SD = 3.13, and range: 0.52–9.55 μg kg^{-1} b.w.), while among females it was equal to 1.99 μg kg^{-1} b.w. (mean = 4.11 μg kg^{-1} b.w., SD = 5.28, and range: 0.00–17.64 μg kg^{-1} b.w.). Hg weekly intakes calculated from non-filtered FFQs were not normally distributed (Table S3 of Supplementary File S1). The difference between the Hg weekly intake of females and males was not statistically significant (Table S4 of Supplementary File S1), moreover, we found no statistically significant correlation between the inferred weekly intake and age, nor in the overall sample, nor in the two sexes separately (Table S5).

The first three species which contribute most to the habitual Hg intake in our sample were the European seabass (*Dicentrarchus labrax*), with an average intake of 31.92 μg $week^{-1}$, Gilthead seabream, with an average intake of 30.85 μg $week^{-1}$, and the Norway lobster (*Nephrops norvegicus*), with an average intake of 21.64 μg $week^{-1}$.

After applying the above-described filtering procedure, i.e., including only those species that had been consumed at the two highest frequencies in the estimation of Hg intake of each participant, and assuming a negligible intake for the others, we obtained a median Hg weekly intake of 1.70 μg kg^{-1} b.w. (mean = 2.33 μg kg^{-1} b.w., SD = 2.13, and range: 0.00–8.00 μg kg^{-1} b.w.), which is still slightly higher than the EFSA maximum recommended intake. Among males, the median weekly intake after filtering of the questionnaires was equal to 2.12 μg kg^{-1} b.w. (mean = 1.98 μg kg^{-1} b.w., SD = 1.08, and range: 0.36–3.53 μg kg−1 b.w.), while among females it was equal to 1.23 μg kg^{-1} b.w. (mean = 2.57 μg kg−1 b.w., SD = 2.63, and range: 0.00–8.01 μg kg^{-1} b.w.). The Hg weekly intakes calculated from the filtered FFQs were not normally distributed (Table S3 of Supplementary File S1). Additionally, in this case, we found no statistically significant difference between the Hg weekly intake of males and females (Table S4 of Supplementary File S1). Moreover, we found no statistically significant correlation between the inferred weekly intake and age, nor in the overall sample, nor among female participants, while the inferred weekly intake and age showed a statistically significant negative correlation ($\tau = -0.5$, $p < 0.05$) among male subjects (Table S5). After the filtering procedure, we found that Gilthead seabream was the species that contributed the most to the habitual intake of Hg in our sample, with an average intake of 26.03 μg week−1, followed by the European seabass, with an average intake of 23.13 μg $week^{-1}$, and tuna, with an average intake of 18.73 μg $week^{-1}$.

3.3.2. Hair Analysis

As can be seen from Figures S1 and S2 and Table S2 of Supplementary File S1, not all participants had long enough hairs to assess the Hg intake during the preceding six months, and for six male participants, only the first 1.5 cm of hair was available.

After averaging the THg concentrations measured in every ~1.5 cm segment of each participant's hair, we found a median hair THg level in the analyzed sample of 1.08 mg kg^{-1} (mean = 2.02 mg kg^{-1}, SD = 2.42, and range: 0.01–9.63 mg kg^{-1}), which was well below the NOEL of ~11.5 mg kg^{-1} set by the EFSA. Concerning the difference between the two sexes, the results of the analysis of the hair are consistent with those of the FFQ, with male individuals being more exposed to Hg compared to female participants, and this difference was statistically significant ($p < 0.05$) (Table S4 of Supplementary File S1). Indeed, among males, the median hair THg level was equal to 2.07 mg kg^{-1} (mean = 2.91 mg kg^{-1}, SD = 2.16, and range: 0.72–7.43 mg kg^{-1}), while among females it was equal to 0.48 mg kg^{-1} (mean = 1.42 mg kg^{-1}, SD = 0.48, and range: 0.02–9.63 mg kg^{-1}). Consistently, with the results obtained from the FFQ analysis, we found no statistically significant correlation between the hair THg level and age, nor in the overall sample, nor the two sexes separately (Table S5). Moreover, we found no statistically significant correlation between the hair THg concentration and the number of amalgam fillings, even after controlling for the number of cigarettes smoked a day and for seafood consumption frequency. Conversely, we found a weak but statistically significant positive correlation between the hair THg concentration and the number of cigarettes smoked a day, but only after controlling for the effect of the number of amalgam fillings and for seafood consumption frequency (Table S6 of Supplementary File S1). However, as shown by the Kruska–Wallis rank sum test, there was no statistically significant difference between smokers' and non-smokers' hair THg concentration, nor between participants with and without amalgam fillings (Table S4 of Supplementary File S1). Moreover, most of the participants—i.e., 23 out of 32—were non-smokers, and only 6 participants had dental amalgam fillings, therefore, these results must be interpreted with caution.

The hair THg level was used to derive the Hg dietary intake of each participant using (2). The median Hg weekly intake derived from hair THg level was equal to 0.80 μg kg^{-1} b.w. (mean = 1.50 μg kg^{-1} b.w., SD = 1.80, and range: 0.01–7.15 μg kg^{-1} b.w.), which is lower than the EFSA maximum recommended intake and the values obtained from the FFQ, both before and after filtering. Among males, the median weekly intake was equal to 1.54 μg kg^{-1} b.w. (mean = 2.16 μg kg^{-1} b.w., SD= 1.61, and range: 0.53–5.52 μg kg^{-1} b.w.), while among females it was equal to 0.36 μg kg^{-1} b.w. (mean = 1.05 μg kg^{-1} b.w., SD = 1.82, and range: 0.01–7.15 μg kg^{-1} b.w.).

3.4. FFQ Validation

The Spearman rank correlation coefficient showed that the Hg weekly intakes derived from the non-filtered FFQs were positively correlated with those derived from hair THg concentrations (Table 2; Figure S3 of Supplementary File S1) and that this positive correlation is strong and statistically significant (R = 0.76, $p < 0.05$).

Table 2. Spearman rank correlation coefficient (ρ) and Cohen's weighted Kappa values.

	ρ (*p*-Value)	**Cohen's Weighted Kappa**
Hg w.i. [1] from n.f. [2] FFQ—Hg w.i. [1] from hair THg	0.76 (5.54×10^{-7})	0.69
Hg w.i. [1] from f. [3] FFQ—Hg w.i. [1] from hair THg	0.61 (1.90×10^{-4})	0.57

[1] w.i., weekly intake, [2] n.f., non-filtered, and [3] f., filtered.

The weighted Kappa statistic was equal to 0.69, pointing to a moderate agreement between the two methods [83]. Then, we built the Bland–Altman plot upon the two series of data. In the case of a few participants, the FFQ and hair THg measurement returned intake values that were markedly more discordant than the other paired values. Accordingly, the Shapiro–Wilk normality test showed that the differences between each paired measurement were not normally distributed. As stated by Bland and Altman, when there are one or more extreme discrepancies between the method—i.e., the difference between one or more pair of measurements differs considerably from the others—a nonparametric approach may be preferable. Therefore, as suggested by Frey and colleagues [84], we used quantile estimation based on one and two-order statistics—i.e., Harrell–Davis quantiles, sample quantiles, and the Sfakianakis–Verginis quantiles estimator—to derive the non-parametric limits of agreement. The mean difference between the two methods was equal to 2.57. The resulting plot (Figure 2a) showed that 93.75% of the data points—i.e., 30 out of 32—lay within the acceptability range determined by the sample quantiles (lower LOA = −1.13, upper LOA = 12.51), by Harrell–Davis quantiles (lower LOA = −0.99, upper LOA = 11.55), and by Sfakianakis–Verginis quantiles estimator (lower LOA = −1.02, upper LOA = 11.76).

The Bland–Altman plot in Figure 2a also reveals an ever-increasing overestimation of Hg intakes by the FFQ as the hair THg level increases. In other words, the mean difference between the two methods was approximately proportional to the magnitude of the measurement, and this is made apparent by the substantial slope of the red line shown in the plot. Therefore, we tried to build the plot using the logarithms of the two data series, as suggested by Bland and Altman [79], after excluding two female participants for whom their weekly Hg intake from the FFQ was equal to zero. In this new plot (Figure 2b), 93.75% of the data points—i.e., 30 out of 32—lay within the acceptability range determined by the sample quantiles (lower LOA = −0.59, upper LOA = 2.77), while 90% of the data points—i.e., 27 out of 30—lay within the acceptability range determined by the Harrell–Davis quantiles (lower LOA = −0.56, upper LOA = 2.67), and 96.67% of the data points—i.e., 29 out of 30—lay within the acceptability range determined by the Sfakianakis–Verginis quantiles estimator (lower LOA = −0.62, upper LOA = 2.65). Additionally, in this case, the proportionality between the mean difference and the magnitude of the measurement persisted, even if at a lower level.

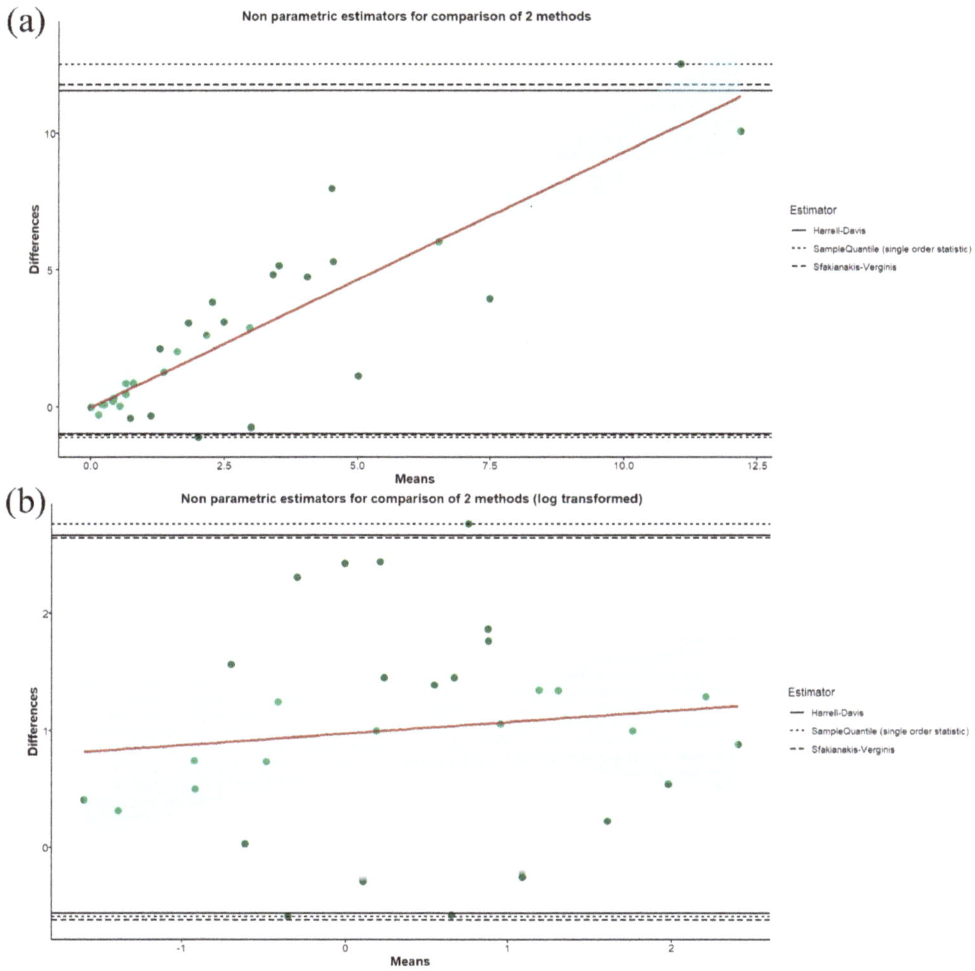

Figure 2. (**a**) Bland–Altman plot built on the weekly Hg intakes from the non-filtered FFQ and hair THg concentration. The Y−axis indicates the difference between each paired measurement—i.e., the difference between the Hg intake estimated from the non-filtered FFQ and that derived from the hair THg measurement—while the X−axis indicates the average of these measurements. (**b**) The Bland–Altman plot built on the log-transformed weekly Hg intakes from the non−filtered FFQ and hair THg concentration.

Given the bias that the non-filtered FFQ showed in the assessment of the Hg weekly intake, we filtered participants' answers through several procedures, each time assessing the resulting improvement through the Bland–Altman plot. In this way, we found that the best results were obtained when only those species that had been consumed at the two highest frequencies were included in the calculation of the Hg weekly intake of each participant. Furthermore, in this case, the Spearman rank correlation coefficient showed that the Hg weekly intakes derived from the filtered FFQ were positively correlated with those derived from the hair THg concentrations (Table 2; Figure S4 of Supplementary File S1) and that this positive correlation is strong and statistically significant ($R = 0.61$, $p < 0.05$). The weighted Kappa statistic was equal to 0.57, pointing to a weak agreement between the two methods [83]. The Bland–Altman plot built upon the same filtered data (Figure 3a) showed that 93.75% of the data points—i.e., 30 out of 32—lay within the acceptability range determined by the sample quantiles (lower LOA = −4.06, upper LOA = 6.79), by the Harrell–Davis quantiles (lower LOA = −3.35, upper LOA = 6.18), and by the Sfakianakis–Verginis quantiles estimator (lower LOA = −3.16, upper LOA = 6.31).

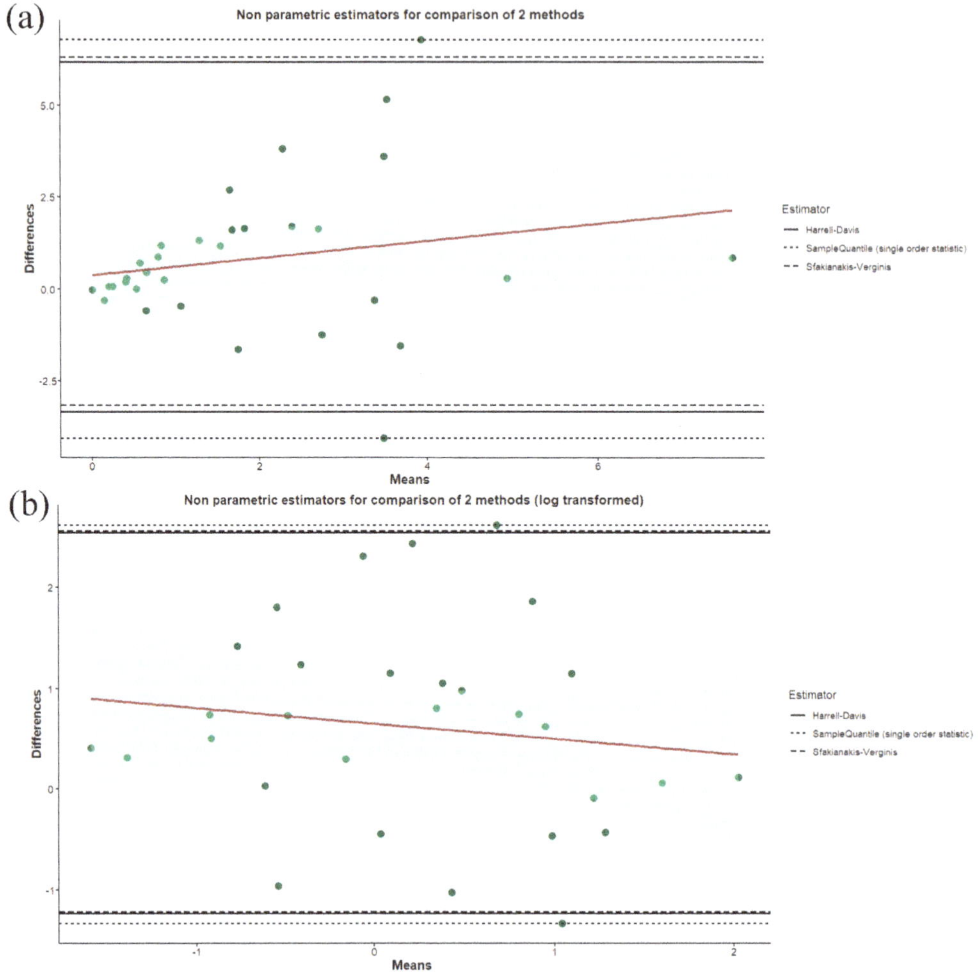

Figure 3. (**a**) Bland–Altman plot built on the weekly Hg intakes from the filtered FFQ and hair THg concentration. The Y-axis indicates the difference between each paired measurement—i.e., the difference between the Hg intake estimated from the filtered FFQ and that derived from the hair THg measurement—while the X-axis indicates the average of these measurements. (**b**) The Bland–Altman plot built on log-transformed weekly Hg intakes from the filtered FFQ and hair THg concentration.

The proportionality between the mean difference and the magnitude of the measurement was more subtle compared to that observed in the plot built upon the non-filtered FFQs. Then, we built the Bland–Altman plot using the log-transformed hair THg concentrations and Hg intakes from filtered FFQs. In this case (Figure 3b), 93.75% of the data points—i.e., 30 out of 32—lay within the acceptability range determined by the sample quantiles (lower LOA = −1.33, upper LOA = 2.62), by the Harrell–Davis quantiles (lower LOA = −1.23, upper LOA = 2.55), and by the Sfakianakis–Verginis quantiles estimator (lower LOA = −1.22, upper LOA = 2.56). In this latter plot, the mean difference between the two methods slightly decreased as the magnitude of the measurement increased.

4. Discussion

4.1. FFQ Reliability

It has been suggested that the correlation coefficients in validation studies should not be below 0.3–0.4 [39,45], and results derived from both the non-filtered and the filtered questionnaires developed in the present study meet this requirement.

Regarding the weighted Kappa statistic, in his original paper, Cohen suggested that values of Kappa ≤ 0 indicate no agreement, and interpreted values between 0.01 and 0.20 as none to slight, values between 0.21 and 0.40 as fair, values between 0.41 and 0.60 as moderate, values between 0.61 and 0.80 as substantial, and values between 0.81 and 1.00 as almost perfect agreement [85]. However, other authors have suggested that any Kappa below 0.60 should be interpreted as an indication of the inadequate agreement between the two methods, especially in a clinical context [83]. On the basis of such premises, we conclude that the FFQ developed in the present study, both before and after filtering, shows a moderate to a substantial agreement with the hair THg measurement.

A strong correlation does not imply an agreement between the two methods [80], and that is why correlation should be used alongside the Bland–Altman plot [45]. Results of the Bland–Altman plots with non-parametric LOAs showed that the filtered FFQ is more reliable in assessing the Hg weekly intake compared to the non-filtered version, as the latter tends to overestimate the Hg intake in individuals consuming higher amounts of seafood. However, it is important to note that the requirement for the validation—i.e., at least 95% of the data points lying within $\pm 1.96s$ of the mean difference or the non-parametric lower and upper LOA—was met only in the case of the log-transformed data series from the non-filtered FFQs and hair THg concentration, using the Sfakianakis–Verginis quantiles as LOA. In most other cases, only 93.75% of the data points lay within the non-parametric range of acceptability, with the above percentage dropping to 90% in the case of the log-transformed data series from the non-filtered FFQs and hair THg concentration, and using the Harrell–Davis quantiles as LOA.

Overall, the results of the present study are promising and point to the fair efficacy of the newly developed FFQ in assessing habitual Hg intake.

As stated by Cade and colleagues in their review [45], for the Bland–Altman method, a sample size of at least 50 is desirable, while, for validation studies using correlation coefficients, this number rises to 100. Therefore, administering the FFQ to a larger sample may strengthen the validity assessment.

Another limitation of the present study is that we did not check the accuracy of the answers because each participant answered the questionnaire only once. Indeed, it is advisable to administer the questionnaire twice, in order to compare the answers given the first time with those given the second.

Finally, a potential source of error in the present investigation stems from the fact that we do not know the exact geographic origin of the seafood consumed by the participants. This may be a source for errors in the estimation of Hg intake via the FFQ, as seafood contamination also depends on local oceanographic factors and the vicinity of emission sources.

4.2. Hg Exposure

The FFQ, both before and after filtering, returned a median Hg weekly intake that was higher than the value derived from the hair THg measurement. In particular, the non-filtered FFQ returned a value (2.62 μg kg^{-1} b.w.) that is more than twice the EFSA maximum recommended intake of 1.3 μg kg^{-1} b.w. per week [9,10], and the filtered FFQ returned a value (1.70 μg kg^{-1} b.w.) that is only slightly higher than the same limit. On the other hand, the median Hg weekly intake derived from the hair THg measurement was equal to 0.80 μg kg^{-1} b.w., which is lower than EFSA's maximum recommended intake.

The median hair THg concentration in the analyzed sample from Trieste is equal to 1.08 mg kg^{-1}, which is well below the NOEL of ~11.5 mg kg^{-1} set by the EFSA. However, it is important to note that there is no consensus on the actual hair THg level above which health risk may occur, as several investigations have led to the establishment of threshold values ranging from 1, set by the US EPA [86], to 14 mg kg^{-} set by the Joint FAO/WHO Expert Committee on Food Additives (JECFA) [87]. In total, 16 out of the 32 participants in this study exhibited hair THg concentrations higher than 1 mg kg^{-1}, and 6 participants exhibited concentrations higher than 3.75 mg kg^{-1}, which is the concentration at which adverse health effects are possible according to the German Human Biomonitoring Commission [88], and only 1 female participant exhibited a THg concentration slightly greater than the EFSA NOEL of 11.5 mg kg^{-1} in one single ~1.5 cm segment of her hair, roughly corresponding to the period from February to March 2021.

The median hair THg concentration obtained in the present study is greater than that found by Basu and colleagues [23] for the populations of Africa (0.69 mg kg^{-1}) and Europe (0.30 mg kg^{-1}), for the populations living on the coast of the Atlantic (0.62 mg kg^{-1}) and Arctic (0.74 mg kg^{-1}) oceans, and those on the coast of the Mediterranean Sea (0.88 mg kg^{-1}), while it is lower than that found in the same study for populations in the Americas (2.02 mg kg^{-1}), eastern Mediterranean (1.68 mg kg^{-1}), southeast Asia (3.10 mg kg^{-1}) and western Pacific (1.40 mg kg^{-1}), and for the coastal populations of the Pacific Ocean (1.75 mg kg^{-1}) and fish consumers—i.e., non-Indigenous or non-Arctic groups who consume relatively high amounts of seafood—(3.04 mg kg^{-1}).

The coastal town of Grado, in Friuli-Venezia Giulia, is less than 30 km from Trieste (Figure 1b) and is strongly impacted by Hg pollution due to maritime traffic, a local chlor-alkali plant, and the former Idrija mercury mine in Slovenia. An investigation carried out on 19 inhabitants of this town found a hair THg concentration ranging from 1.13 to 20.16 mg kg^{-1}, the highest values being measured among fishermen, with a median value (3.90 mg kg^{-1}) well above that found in the present study. On the other hand, in another study on the same geographic area [89], the authors found a mean hair THg concentration of 0.83 mg kg^{-1} among mothers of children with advanced fine motor skills, and of 1.24 mg kg^{-1} among mothers of children who showed normal or delayed skills. Both of these values are lower than the mean hair THg concentration exhibited by the sample analyzed in our investigation (2.02 mg kg^{-1}), and, accordingly, the two groups of mothers in that study showed a fresh fish intake during pregnancy of 0.47 and 0.66 servings per week, respectively, while the 32 participants from Trieste showed a mean seafood intake of 1.28 servings per week.

Compared to our results, lower hair THg concentrations are also shown by the general population of several towns on the coast of Sicily (mean concentration = 0.23 mg kg^{-1}, SD = 0.4) [21], by people living in Priolo, near the chlor-alkali plant of Augusta (Sicily), one of the largest in Europe (mean concentration = 1.37 mg kg^{-1}, median concentration = 1.00 mg kg^{-1}), and by the general population of Naples (mean concentration = 0.638 mg kg^{-1}). On the contrary, higher THg concentrations were found in fishermen of several coastal towns of Sicily (mean concentration = 6.45 mg kg^{-1}, SD = 7.03), in people from Augusta (mean concentration = 2.61 mg kg^{-1}, median concentration = 1.90 mg kg^{-1}), and in tuna consumers from Carloforte (Sardinia) (median concentration = 9.6 mg kg^{-1}, range: 1.4–34.5 mg kg^{-1}).

It is important to note that, due to its small size, the sample analyzed in the present study may not be representative of the entire population of Trieste.

Finally, it comes as no surprise the fact that, as revealed by the questionnaires, Gilthead seabream, European seabass, squid, and tuna are the species that contribute the most to the habitual Hg intake of the participants, since the first three species are among those that are the most consumed in Italy, and tuna is one of the most contaminated species.

5. Conclusions

To gain insights into the Hg exposure level among a Mediterranean community living in an area that is considered a contamination hotspot—i.e., the Trieste Gulf—we developed and validated a new FFQ and compared its results against those of the analysis of the total mercury measured in the hair of the same individuals. The reported newly developed questionnaire shows high accuracy in the estimation of habitual mercury intake, similar to the one measured through the analysis of the hair. Moreover, the questionnaire allowed us to obtain a first glimpse of the exposure to Hg from seafood among the population of Trieste.

Supplementary Materials: The following supporting information can be downloaded at: https://www.mdpi.com/article/10.3390/pollutants3030022/s1, Supplementary-File S1: Supplementary materials. Figure S1: Hair THg concentrations (Y-axis) in the first four ~1.5-cm segments of hair (X-axis) in female participants, starting from the nearest (1) to the farthest (4) from the scalp. Each line color corresponds to a participant, with bigger colored dots representing measurements. On top of the graph are shown the periods around which the detected Hg was ingested. The red transparent bar represents the evolution of the mean hair THg level. Figure S2: Hair THg concentrations (Y-axis) in the first four ~1.5-cm segments of hair (X-axis) in male participants, starting from the nearest (1) to the farthest (4) from the scalp. Each line color corresponds to a participant, with bigger colored dots representing measurements. On top of the graph are shown the periods around which the detected Hg was ingested. The red transparent bar represents the evolution of the mean hair THg level. Figure S3: Scatterplot for the weekly Hg intake from the hair THg concentration and non-filtered FFQ, with a regression line based on the Spearman rank correlation coefficient. Figure S4: Scatterplot for the weekly Hg intake from the hair THg concentration and filtered FFQ, with a regression line based on the Spearman rank correlation coefficient. Table S1: Information and results on each study participant. Reported hair THg was calculated by averaging the THg concentrations measured in every ~1.5 cm segment of hair and is expressed in mg kg^{-1}, while intakes are in μg kg−1 body weight. Table S2: Hair THg concentrations (mg kg^{-1}) in the first four ~1.5-cm segments of hair of each study participant. Table S3: Results of the Shapiro–Wilk normality test on several data series (first column). In bold are the *p*-values indicating a normal distribution (*p*-value > 0.05). Table S4: Results of the Kruskal–Wallis rank sum test aiming to assess the statistical significance of differences between the Hg intakes or hair THg concentrations of the two sexes. Table S5: Spearman rank correlation coefficients between several data series and the age of participants. Supplementary-File S2: original FFQ in Italian. Supplementary-File S3: FFQ translated in English.

Author Contributions: Conceptualization, A.D.G., G.G., C.G., M.L.C. and D.L.; methodology, A.D.G., V.I., G.G., C.G. and M.L.C.; software, A.D.G. and V.I.; validation, G.G., M.M., C.G., M.L.C. and D.L.; formal analysis, A.D.G., V.I., G.G. and M.L.C.; investigation, A.D.G., V.I., G.G. and M.L.C.; resources, G.G., C.G., M.L.C. and D.L.; data curation, A.D.G., V.I., G.G. and M.L.C.; writing—original draft preparation, A.D.G.; writing—review and editing, G.G., M.M., C.G., M.L.C. and D.L.; visualization, A.D.G. and V.I.; supervision, G.G., M.M., C.G., M.L.C. and D.L.; project administration, G.G., M.L.C. and D.L.; funding acquisition, G.G., M.L.C. and D.L. All authors have read and agreed to the published version of the manuscript.

Funding: This study was supported by the Beatriz Galindo Fellowship (2018) funded by the Spanish Ministry of Science and Innovation and Ministry of Universities [Project BEA-GAL18/00110 "Development of analytical methods applied to archaeology"] to G.G., by the project "Smartphone and Green Analytical Chemistry" funded by the Valencian Regional Government—Conselleria d'Innovacio, Universitats, Ciència i Societat Digital [PROMETEO 2019-056] to M.L.C. and by the Project funded under the National Recovery and Resilience Plan (NRRP), Mission 4 Component 2 Investment 1.4—Call for tender No. 3138 of 16 December 2021, rectified by Decree n.3175 of 18 December 2021 of Italian Ministry of University and Research funded by the European Union—NextGenerationEU; Project code CN_00000033, Concession Decree No. 1034 of 17 June 2022 adopted by the Italian Ministry of University and Research, Project title "National Biodiversity Future Center—NBFC" to D.L.

Data Availability Statement: All new data are available in the paper in the Supplementary Materials.

Acknowledgments: The research leading to these results was conceived under the collaboration between the University of Bologna and the National Research Council for the implementation of the International Ph.D. Program "Innovative Technologies and Sustainable Use of Mediterranean Sea Fishery and Biological Resources" (https://site.unibo.it/laboratorio-biologia-marina-pesca-fano/en/ accessed on 29 April 2022). G.G. acknowledges the financial support of the Beatriz Galindo Fellowship (2018) funded by the Spanish Ministry of Science and Innovation and the Ministry of Universities (Project BEA-GAL18/00110 "Development of analytical methods applied to archaeology"). M.L.C. acknowledges the Valencian Regional Government—Conselleria d'Innovacio, Universitats, Ciència i

Societat Digital for funding the project "Smartphone and Green Analytical Chemistry" (PROMETEO 2019-056). D.L. acknowledges the Italian Ministry of Education, Universities and Research for funding the project "1000 Ancient Italian Genomes: Evidence from ancient biomolecules for unravelling past human population Dynamics (AGED)" (PRIN 2017, prot. 20177PJ9XF). The authors would also like to thank the participants for their generous participation and sharing to make this project possible.

Conflicts of Interest: The authors declare no conflict of interest. The funders had no role in the design of the study; in the collection, analyses, or interpretation of data; in the writing of the manuscript; or in the decision to publish the results.

References

1. Szefer, P. Safety assessment of seafood with respect to chemical pollutants in European Seas. *Oceanol. Hydrobiol. Stud.* **2013**, *42*, 110–118. [CrossRef]
2. Cossa, D.; Knoery, J.; Bănaru, D.; Harmelin-Vivien, M.; Sonke, J.E.; Hedgecock, I.M.; Bravo, A.G.; Rosati, G.; Canu, D.; Horvat, M.; et al. Mediterranean Mercury Assessment 2022: An Updated Budget, Health Consequences, and Research Perspectives. *Environ. Sci. Technol.* **2022**, *56*, 3840–3862. [CrossRef] [PubMed]
3. Storelli, A.; Barone, G.; Garofalo, R.; Busco, A.; Storelli, M.M. Determination of Mercury, Methylmercury and Selenium Concentrations in Elasmobranch Meat: Fish Consumption Safety. *Int. J. Environ. Res. Public Health* **2022**, *19*, 788. [CrossRef] [PubMed]
4. Annibaldi, A.; Truzzi, C.; Carnevali, O.; Pignalosa, P.; Api, M.; Scarponi, G.; Illuminati, S. Determination of Hg in Farmed and Wild Atlantic Bluefin Tuna (*Thunnus thynnus* L.) Muscle. *Molecules* **2019**, *24*, 1273. [CrossRef]
5. Rice, K.M.; Walker, E.M., Jr.; Wu, M.; Gillette, C.; Blough, E.R. Environmental Mercury and Its Toxic Effects. *J. Prev. Med. Public Health* **2014**, *47*, 74–83. [CrossRef]
6. Landrigan, P.J.; Stegeman, J.J.; Fleming, L.E.; Allemand, D.; Anderson, D.M.; Backer, L.C.; Brucker-Davis, F.; Chevalier, N.; Corra, L.; Czerucka, D.; et al. Human Health and Ocean Pollution. *Ann. Glob. Health* **2020**, *86*, 151. [CrossRef]
7. Rice, D.C.; Schoeny, R.; Mahaffey, K. Methods and Rationale for Derivation of a Reference Dose for Methylmercury by the U.S. EPA. *Risk Anal.* **2003**, *23*, 107–115. [CrossRef]
8. Rice, D.C. The US EPA reference dose for methylmercury: Sources of uncertainty. *Environ. Res.* **2004**, *95*, 406–413. [CrossRef]
9. EFSA Panel on Contaminants in the Food Chain (CONTAM). Scientific Opinion on the risk for public health related to the presence of mercury and methylmercury in food. *EFSA J.* **2012**, *10*, 2985. [CrossRef]
10. EFSA Scientific Committee. Statement on the Benefits of Fish/Seafood Consumption Compared to the Risks of Methylmercury in Fish/Seafood. *EFSA J.* **2015**, *13*, 3982. [CrossRef]
11. Brambilla, G.; Abete, M.C.; Binato, G.; Chiaravalle, E.; Cossu, M.; Dellatte, E.; Miniero, R.; Orletti, R.; Piras, P.; Roncarati, A.; et al. Mercury occurrence in Italian seafood from the Mediterranean Sea and possible intake scenarios of the Italian coastal population. *Regul. Toxicol. Pharmacol.* **2013**, *65*, 269–277. [CrossRef]
12. WHO Guidance for Identifying Populations at Risk from Mercury Exposure 2008. Available online: https://Wedocs.Unep.Org/Bitstream/Handle/20.500.11822/11786/IdentifyingPopnatRiskExposuretoMercury_2008Web.Pdf?Sequence=1&isAllowed=y (accessed on 18 March 2022).
13. Berglund, M.; Lind, B.; Björnberg, K.A.; Palm, B.; Einarsson, Ö.; Vahter, M. Inter-individual variations of human mercury exposure biomarkers: A cross-sectional assessment. *Environ. Health* **2005**, *4*, 20. [CrossRef]
14. Weihe, P.; Grandjean, P.; Debes, F.; White, R. Health implications for Faroe Islanders of heavy metals and PCBs from pilot whales. *Sci. Total Environ.* **1996**, *186*, 141–148. [CrossRef]
15. Elhamri, H.; Idrissi, L.; Coquery, M.; Azemard, S.; El Abidi, A.; Benlemlih, M.; Saghi, M.; Cubadda, F. Hair mercury levels in relation to fish consumption in a community of the Moroccan Mediterranean coast. *Food Addit. Contam.* **2007**, *24*, 1236–1246. [CrossRef]
16. Díez, S.; Montuori, P.; Pagano, A.; Sarnacchiaro, P.; Bayona, J.M.; Triassi, M. Hair mercury levels in an urban population from southern Italy: Fish consumption as a determinant of exposure. *Environ. Int.* **2008**, *34*, 162–167. [CrossRef]
17. Endo, T.; Haraguchi, K. High mercury levels in hair samples from residents of Taiji, a Japanese whaling town. *Mar. Pollut. Bull.* **2010**, *60*, 743–747. [CrossRef]
18. Tian, W.; Egeland, G.M.; Sobol, I.; Chan, H.M. Mercury hair concentrations and dietary exposure among Inuit preschool children in Nunavut, Canada. *Environ. Int.* **2011**, *37*, 42–48. [CrossRef]
19. Fang, T.; Aronson, K.J.; Campbell, L.M. Freshwater Fish—Consumption Relations with Total Hair Mercury and Selenium Among Women in Eastern China. *Arch. Environ. Contam. Toxicol.* **2012**, *62*, 323–332. [CrossRef]
20. Višnjevec, A.M.; Kocman, D.; Horvat, M. Human mercury exposure and effects in Europe. *Environ. Toxicol. Chem.* **2014**, *33*, 1259–1270. [CrossRef]
21. Giangrosso, G.; Cammilleri, G.; Macaluso, A.; Vella, A.; D'orazio, N.; Graci, S.; Dico, G.M.L.; Galvano, F.; Giangrosso, M.; Ferrantelli, V. Hair Mercury Levels Detection in Fishermen from Sicily (Italy) by ICP-MS Method after Microwave-Assisted Digestion. *Bioinorg. Chem. Appl.* **2016**, *2016*, 5408014. [CrossRef]

22. Okati, N.; Esmaili-Sari, A. Determination of Mercury Daily Intake and Hair-to-Blood Mercury Concentration Ratio in People Resident of the Coast of the Persian Gulf, Iran. *Arch. Environ. Contam. Toxicol.* **2017**, *74*, 140–153. [CrossRef] [PubMed]
23. Basu, N.; Horvat, M.; Evers, D.C.; Zastenskaya, I.; Weihe, P.; Tempowski, J. A State-of-the-Science Review of Mercury Biomarkers in Human Populations Worldwide between 2000 and 2018. *Environ. Health Perspect.* **2018**, *126*, 106001. [CrossRef] [PubMed]
24. Barbone, F.; Rosolen, V.; Mariuz, M.; Parpinel, M.; Casetta, A.; Sammartano, F.; Ronfani, L.; Brumatti, L.V.; Bin, M.; Castriotta, L.; et al. Prenatal mercury exposure and child neurodevelopment outcomes at 18 months: Results from the Mediterranean PHIME cohort. *Int. J. Hyg. Environ. Health* **2019**, *222*, 9–21. [CrossRef] [PubMed]
25. Seo, J.-W.; Kim, B.-G.; Hong, Y.-S. The Relationship between Mercury Exposure Indices and Dietary Intake of Fish and Shellfish in Women of Childbearing Age. *Int. J. Environ. Res. Public Health* **2020**, *17*, 4907. [CrossRef] [PubMed]
26. Sekovanić, A.; Piasek, M.; Orct, T.; Grgec, A.S.; Sarić, M.M.; Stasenko, S.; Jurasović, J. Mercury Exposure Assessment in Mother—Infant Pairs from Continental and Coastal Croatia. *Biomolecules* **2020**, *10*, 821. [CrossRef]
27. Droghini, E.; Annibaldi, A.; Prezioso, E.; Tramontana, M.; Frapiccini, E.; De Marco, R.; Illuminati, S.; Truzzi, C.; Spagnoli, F. Mercury Content in Central and Southern Adriatic Sea Sediments in Relation to Seafloor Geochemistry and Sedimentology. *Molecules* **2019**, *24*, 4467. [CrossRef]
28. EEA. *Mercury in Europe's Environment, a Priority for European and Global Action*; European Environment Agency: Copenhagen, Denmark, 2018.
29. Copat, C.; Conti, G.O.; Fallico, R.; Sciacca, S.; Ferrante, M. Heavy Metals in Fish from the Mediterranean Sea: Potential Impact on Diet. In *The Mediterranean Diet An Evidence-Based Approach*; Elsevier: Amsterdam, The Netherlands, 2015; pp. 547–562; ISBN 978-0-12-407849-9.
30. Gallo, P.; De Carlo, E.; Marigliano, L.; Maglio, P.; Amato, A.; Improta, A.; Caruso, C.; De Roma, A. Food safety assessment of heavy metals in uncommon and abyssal fish and cephalopod from the Tyrrhenian Sea. *J. Consum. Prot. Food Saf.* **2018**, *13*, 399–402. [CrossRef]
31. Perugini, M.; Visciano, P.; Manera, M.; Abete, M.C.; Gavinelli, S.; Amorena, M. Contamination of different portions of raw and boiled specimens of Norway lobster by mercury and selenium. *Environ. Sci. Pollut. Res.* **2013**, *20*, 8255–8262. [CrossRef]
32. Perugini, M.; Visciano, P.; Manera, M.; Zaccaroni, A.; Olivieri, V.; Amorena, M. Heavy metal (As, Cd, Hg, Pb, Cu, Zn, Se) concentrations in muscle and bone of four commercial fish caught in the central Adriatic Sea, Italy. *Environ. Monit. Assess.* **2013**, *186*, 2205–2213. [CrossRef]
33. Perugini, M.; Zezza, D.; Tulini, S.M.R.; Abete, M.C.; Monaco, G.; Conte, A.; Olivieri, V.; Amorena, M. Effect of cooking on total mercury content in Norway lobster and European hake and public health impact. *Mar. Pollut. Bull.* **2016**, *109*, 521–525. [CrossRef]
34. Storelli, M.M.; Barone, G. Toxic Metals (Hg, Pb, and Cd) in Commercially Important Demersal Fish from Mediterranean Sea: Contamination Levels and Dietary Exposure Assessment: Toxic Metals in Fish. *J. Food Sci.* **2013**, *78*, T362–T366. [CrossRef]
35. Di Lena, G.; Casini, I.; Caproni, R.; Orban, E. Total mercury levels in crustacean species from Italian fishery. *Food Addit. Contam. Part B* **2018**, *11*, 175–182. [CrossRef]
36. De Giovanni, A.; Giuliani, C.; Marini, M.; Luiselli, D. Methylmercury and Polycyclic Aromatic Hydrocarbons in Mediterranean Seafood: A Molecular Anthropological Perspective. *Appl. Sci.* **2021**, *11*, 11179. [CrossRef]
37. Hond, E.D.; Govarts, E.; Willems, H.; Smolders, R.; Casteleyn, L.; Kolossa-Gehring, M.; Schwedler, G.; Seiwert, M.; Fiddicke, U.; Castaño, A.; et al. First Steps toward Harmonized Human Biomonitoring in Europe: Demonstration Project to Perform Human Biomonitoring on a European Scale. *Environ. Health Perspect.* **2015**, *123*, 255–263. [CrossRef]
38. Junqué, E.; Garí, M.; Llull, R.M.; Grimalt, J.O. Drivers of the accumulation of mercury and organochlorine pollutants in Mediterranean lean fish and dietary significance. *Sci. Total. Environ.* **2018**, *634*, 170–180. [CrossRef]
39. Birgisdottir, B.; Kiely, M.; Martinez, J.; Thorsdottir, I. Validity of a food frequency questionnaire to assess intake of seafood in adults in three European countries. *Food Control* **2008**, *19*, 648–653. [CrossRef]
40. Miklavčič, A.; Casetta, A.; Tratnik, J.S.; Mazej, D.; Krsnik, M.; Mariuz, M.; Sofianou, K.; Špirić, Z.; Barbone, F.; Horvat, M. Mercury, arsenic and selenium exposure levels in relation to fish consumption in the Mediterranean area. *Environ. Res.* **2013**, *120*, 7–17. [CrossRef]
41. Woods, R.K.; Stoney, R.M.; Ireland, P.D.; Bailey, M.J.; Raven, J.M.; Thien, F.C.; Walters, E.H.; Abramson, M.J. A valid food frequency questionnaire for measuring dietary fish intake. *Asia Pac. J. Clin. Nutr.* **2002**, *11*, 56–61. [CrossRef]
42. Sullivan, B.L.; Brown, J.; Williams, P.G.; Meyer, B.J. Dietary validation of a new Australian food-frequency questionnaire that estimates long-chain-*n*-3 polyunsaturated fatty acids. *Br. J. Nutr.* **2008**, *99*, 660–666. [CrossRef]
43. Filippini, T.; Malavolti, M.; Cilloni, S.; Wise, L.A.; Violi, F.; Malagoli, C.; Vescovi, L.; Vinceti, M. Intake of arsenic and mercury from fish and seafood in a Northern Italy community. *Food Chem. Toxicol.* **2018**, *116*, 20–26. [CrossRef]
44. FAO. *Dietary Assessment: A Resource Guide to Method Selection and Application in Low Resource Settings*; FAO: Roma, Italy, 2018; ISBN 978-92-5-130635-2.
45. Cade, J.; Thompson, R.; Burley, V.; Warm, D. Development, validation and utilisation of food-frequency questionnaires—A review. *Public Health Nutr.* **2002**, *5*, 567–587. [CrossRef] [PubMed]
46. Buscemi, S.; Rosafio, G.; Vasto, S.; Massenti, F.M.; Grosso, G.; Galvano, F.; Rini, N.; Barile, A.M.; Maniaci, V.; Cosentino, L.; et al. Validation of a food frequency questionnaire for use in Italian adults living in Sicily. *Int. J. Food Sci. Nutr.* **2015**, *66*, 426–438. [CrossRef] [PubMed]

47. Athanasiadou, E.; Kyrkou, C.; Fotiou, M.; Tsakoumaki, F.; Dimitropoulou, A.; Polychroniadou, E.; Menexes, G.; Athanasiadis, A.P.; Biliaderis, C.G.; Michaelidou, A.-M. Development and Validation of a Mediterranean Oriented Culture-Specific Semi-Quantitative Food Frequency Questionnaire. *Nutrients* **2016**, *8*, 522. [CrossRef] [PubMed]
48. Aoun, C.; Daher, R.B.; El Osta, N.; Papazian, T.; Khabbaz, L.R. Reproducibility and relative validity of a food frequency questionnaire to assess dietary intake of adults living in a Mediterranean country. *PLoS ONE* **2019**, *14*, e0218541. [CrossRef] [PubMed]
49. Garneau, V.; Rudkowska, I.; Paradis, A.-M.; Godin, G.; Julien, P.; Pérusse, L.; Vohl, M.-C. Omega-3 fatty acids status in human subjects estimated using a food frequency questionnaire and plasma phospholipids levels. *Nutr. J.* **2012**, *11*, 46. [CrossRef]
50. Shen, W.; Weaver, A.M.; Salazar, C.; Samet, J.M.; Diaz-Sanchez, D.; Tong, H. Validation of a Dietary Questionnaire to Screen Omega-3 Fatty Acids Levels in Healthy Adults. *Nutrients* **2019**, *11*, 1470. [CrossRef]
51. Cinnirella, S.; Bruno, D.E.; Pirrone, N.; Horvat, M.; Živković, I.; Evers, D.C.; Johnson, S.; Sunderland, E.M. Mercury concentrations in biota in the Mediterranean Sea, a compilation of 40 years of surveys. *Sci. Data* **2019**, *6*, 205. [CrossRef]
52. Bajt, O.; Ramšak, A.; Milun, V.; Andral, B.; Romanelli, G.; Scarpato, A.; Mitrić, M.; Kupusović, T.; Kljajić, Z.; Angelidis, M.; et al. Assessing chemical contamination in the coastal waters of the Adriatic Sea using active mussel biomonitoring with *Mytilus galloprovincialis*. *Mar. Pollut. Bull.* **2019**, *141*, 283–298. [CrossRef]
53. Storelli, M.M.; Giacominelli-Stuffler, R.; Storelli, A.; D'Addabbo, R.; Palermo, C.; Marcotrigiano, G.O. Survey of total mercury and methylmercury levels in edible fish from the Adriatic Sea. *Food Addit. Contam.* **2003**, *20*, 1114–1119. [CrossRef]
54. Jureša, D.; Blanuša, M. Mercury, arsenic, lead and cadmium in fish and shellfish from the Adriatic Sea. *Food Addit. Contam.* **2003**, *20*, 241–246. [CrossRef]
55. Licata, P.; Trombetta, D.; Cristani, M.; Naccari, C.; Martino, D.; Caló, M.; Naccari, F. Heavy Metals in Liver and Muscle of Bluefin Tuna (*Thunnus thynnus*) Caught in the Straits of Messina (Sicily, Italy). *Environ. Monit. Assess.* **2005**, *107*, 239–248. [CrossRef]
56. Storelli, M.M.; Giacominelli-Stuffler, R.; Marcotrigiano, G.O. Relationship between Total Mercury Concentration and Fish Size in Two Pelagic Fish Species: Implications for Consumer Health. *J. Food Prot.* **2006**, *69*, 1402–1405. [CrossRef]
57. Storelli, M.M. Potential human health risks from metals (Hg, Cd, and Pb) and polychlorinated biphenyls (PCBs) via seafood consumption: Estimation of target hazard quotients (THQs) and toxic equivalents (TEQs). *Food Chem. Toxicol.* **2008**, *46*, 2782–2788. [CrossRef]
58. Perugini, M.; Visciano, P.; Manera, M.; Zaccaroni, A.; Olivieri, V.; Amorena, M. Levels of Total Mercury in Marine Organisms from Adriatic Sea, Italy. *Bull. Environ. Contam. Toxicol.* **2009**, *83*, 244–248. [CrossRef]
59. Conti, G.O.; Copat, C.; Ledda, C.; Fiore, M.; Fallico, R.; Sciacca, S.; Ferrante, M. Evaluation of Heavy Metals and Polycyclic Aromatic Hydrocarbons (PAHs) in Mullus barbatus from Sicily Channel and Risk-Based Consumption Limits. *Bull. Environ. Contam. Toxicol.* **2012**, *88*, 946–950. [CrossRef]
60. Praticò, V.; Normanno, G.; Dambrosio, A.; Anaclerio, D.; Serpe, L.; Gallo, P.; Federico, G.; Borruto, S. Heavy Metals Contamination in Fish Caught in the Strait of Messina. *Ind. Aliment.* **2013**, *52*, 12–16.
61. Miniero, R.; Beccaloni, E.; Carere, M.; Ubaldi, A.; Mancini, L.; Marchegiani, S.; Cicero, M.; Scenati, R.; Lucchetti, D.; Ziemacki, G.; et al. Mercury (Hg) and methyl mercury (MeHg) concentrations in fish from the coastal lagoon of Orbetello, central Italy. *Mar. Pollut. Bull.* **2013**, *76*, 365–369. [CrossRef]
62. Storelli, M.M.; Barone, G.; Perrone, V.G.; Storelli, A. Risk characterization for polycyclic aromatic hydrocarbons and toxic metals associated with fish consumption. *J. Food Compos. Anal.* **2013**, *31*, 115–119. [CrossRef]
63. Copat, C.; Vinceti, M.; D'Agati, M.G.; Arena, G.; Mauceri, V.; Grasso, A.; Fallico, R.; Sciacca, S.; Ferrante, M. Mercury and selenium intake by seafood from the Ionian Sea: A risk evaluation. *Ecotoxicol. Environ. Saf.* **2014**, *100*, 87–92. [CrossRef]
64. Conte, F.; Copat, C.; Longo, S.; Conti, G.O.; Grasso, A.; Arena, G.; Brundo, M.V.; Ferrante, M. First data on trace elements in *Haliotis tuberculata* (*Linnaeus*, 1758) from southern Italy: Safety issues. *Food Chem. Toxicol.* **2015**, *81*, 143–150. [CrossRef]
65. Barone, G.; Storelli, A.; Garofalo, R.; Busco, V.P.; Quaglia, N.C.; Centrone, G.; Storelli, M.M. Assessment of mercury and cadmium via seafood consumption in Italy: Estimated dietary intake (EWI) and target hazard quotient (THQ). *Food Addit. Contam. Part A* **2015**, *32*, 1277–1286. [CrossRef]
66. Di Lena, G.; Casini, I.; Caproni, R.; Fusari, A.; Orban, E. Total mercury levels in commercial fish species from Italian fishery and aquaculture. *Food Addit. Contam. Part B* **2017**, *10*, 118–127. [CrossRef]
67. Signa, G.; Mazzola, A.; Tramati, C.D.; Vizzini, S. Diet and habitat use influence Hg and Cd transfer to fish and consequent biomagnification in a highly contaminated area: Augusta Bay (Mediterranean Sea). *Environ. Pollut.* **2017**, *230*, 394–404. [CrossRef] [PubMed]
68. Capolupo, M.; Franzellitti, S.; Kiwan, A.; Valbonesi, P.; Dinelli, E.; Pignotti, E.; Birke, M.; Fabbri, E. A comprehensive evaluation of the environmental quality of a coastal lagoon (Ravenna, Italy): Integrating chemical and physiological analyses in mussels as a biomonitoring strategy. *Sci. Total. Environ.* **2017**, *598*, 146–159. [CrossRef] [PubMed]
69. Bajc, Z.; Kirbiš, A. Trace Element Concentrations in Mussels (*Mytilus galloprovincialis*) from the Gulf of Trieste, Slovenia. *J. Food Prot.* **2019**, *82*, 429–434. [CrossRef] [PubMed]
70. Ariano, A.; Marrone, R.; Andreini, R.; Smaldone, G.; Velotto, S.; Montagnaro, S.; Anastasio, A.; Severino, L. Metal Concentration in Muscle and Digestive Gland of Common Octopus (*Octopus vulgaris*) from Two Coastal Site in Southern Tyrrhenian Sea (Italy). *Molecules* **2019**, *24*, 2401. [CrossRef]

71. Arienzo, M.; Toscanesi, M.; Trifuoggi, M.; Ferrara, L.; Stanislao, C.; Donadio, C.; Grazia, V.; Gionata, D.V.; Carella, F. Contaminants bioaccumulation and pathological assessment in *Mytilus galloprovincialis* in coastal waters facing the brownfield site of Bagnoli, Italy. *Mar. Pollut. Bull.* **2019**, *140*, 341–352. [CrossRef]
72. Ancora, S.; Mariotti, G.; Ponchia, R.; Fossi, M.C.; Leonzio, C.; Bianchi, N. Trace elements levels in muscle and liver of a rarely investigated large pelagic fish: The Mediterranean spearfish *Tetrapturus belone* (Rafinesque, 1810). *Mar. Pollut. Bull.* **2020**, *151*, 110878. [CrossRef]
73. La Torre, G.L.; Cicero, N.; Bartolomeo, G.; Rando, R.; Vadalà, R.; Santini, A.; Durazzo, A.; Lucarini, M.; Dugo, G.; Salvo, A. Assessment and Monitoring of Fish Quality from a Coastal Ecosystem under High Anthropic Pressure: A Case Study in Southern Italy. *Int. J. Environ. Res. Public Health* **2020**, *17*, 3285. [CrossRef]
74. Grgec, A.S.; Kljaković-Gašpić, Z.; Orct, T.; Tičina, V.; Sekovanić, A.; Jurasović, J.; Piasek, M. Mercury and selenium in fish from the eastern part of the Adriatic Sea: A risk-benefit assessment in vulnerable population groups. *Chemosphere* **2020**, *261*, 127742. [CrossRef]
75. Ramon, D.; Morick, D.; Croot, P.; Berzak, R.; Scheinin, A.; Tchernov, D.; Davidovich, N.; Britzi, M. A survey of arsenic, mercury, cadmium, and lead residues in seafood (fish, crustaceans, and cephalopods) from the south-eastern Mediterranean Sea. *J. Food Sci.* **2021**, *86*, 1153–1161. [CrossRef]
76. EEA Commission Regulation (EC) No 1881/2006 of 19 December 2006. Setting Maximum Levels for Certain Contaminants in Foodstuffs. 2006. Available online: https://eur-lex.europa.eu/legal-content/EN/ALL/?uri=celex%3A32006R1881 (accessed on 30 June 2023).
77. Vicent, A.; Cervera, M.; Morales-Rubio, A. Fast and accurate methodology for direct mercury determination in hair and nails. *Sample Prep.* **2016**, *2*, 91–98. [CrossRef]
78. Legrand, M.; Passos, C.J.S.; Mergler, D.; Chan, H.M. Biomonitoring of Mercury Exposure with Single Human Hair Strand. *Environ. Sci. Technol.* **2005**, *39*, 4594–4598. [CrossRef]
79. Bland, J.M.; Altman, D.G. Measuring agreement in method comparison studies. *Stat. Methods Med. Res.* **1999**, *8*, 135–160. [CrossRef]
80. Bland, J.M.; Altman, D.G. Statistical methods for assessing agreement between two methods of clinical measurement. *Lancet* **1986**, *327*, 307–310. [CrossRef]
81. Giavarina, D. Understanding Bland Altman analysis. *Biochem. Med.* **2015**, *25*, 141–151. [CrossRef]
82. Kalra, A. Decoding the Bland–Altman plot: Basic review. *J. Pract. Cardiovasc. Sci.* **2017**, *3*, 36. [CrossRef]
83. McHugh, M.L. Interrater reliability: The kappa statistic. *Biochem. Med.* **2012**, *22*, 276–282. [CrossRef]
84. Frey, M.E.; Petersen, H.C.; Gerke, O. Nonparametric Limits of Agreement for Small to Moderate Sample Sizes: A Simulation Study. *Stats* **2020**, *3*, 343–355. [CrossRef]
85. Cohen, J. A Coefficient of Agreement for Nominal Scales. *Educ. Psychol. Meas.* **1960**, *20*, 37–46. [CrossRef]
86. US EPA. Water Quality Criterion for the Protection of Human Health: Methylmercury Chapter 4: Risk Assessment for Methylmercury. Office of Science and Technology, Office of Water. 2001. Available online: https://clu-in.org/Download/Contaminantfocus/Mercury/Water-Quality-Criterion-Methyl-Mercury.Pdf (accessed on 6 April 2022).
87. Joint FAO/WHO Expert Committee on Food Additives. Meeting (67th: 2006: Geneva, S.; International Programme on Chemical Safety. Evaluation of Certain Food Additives and Contaminants. Prepared by the Sixty-Seventh Meeting of the Joint FAO/WHO Expert Committee on Food Additives (JEFCA). 2007. Available online: https://apps.who.int/iris/handle/10665/43592 (accessed on 30 June 2023).
88. Schulz, C.; Angerer, J.; Ewers, U.; Kolossa-Gehring, M. The German Human Biomonitoring Commission. *Int. J. Hyg. Environ. Health* **2007**, *210*, 373–382. [CrossRef] [PubMed]
89. Barbone, F.; Valent, F.; Pisa, F.; Daris, F.; Fajon, V.; Gibicar, D.; Logar, M.; Horvat, M. Prenatal low-level methyl mercury exposure and child development in an Italian coastal area. *Neurotoxicology* **2020**, *81*, 376–381. [CrossRef] [PubMed]

toxics

MDPI

Article

Air, Dermal, and Urinary Metabolite Levels of Backpack and Tractor Sprayers Using the Herbicide Acetochlor in Thailand

Nichcha Kallayanatham [1], Sumate Pengpumkiat [1], Pornpimol Kongtip [1,*], Ritthirong Pundee [2], Noppanun Nankongnab [1], Amarin Kongtawelert [1] and Susan R. Woskie [3]

[1] Department of Occupational Health and Safety, Faculty of Public Health, Mahidol University, 420/1 Rajvidhi Road, Bangkok 10400, Thailand; nichcha.kal@gmail.com (N.K.); sumate.pen@mahidol.ac.th (S.P.); noppanun.nan@mahidol.ac.th (N.N.); amarin.kon@mahidol.ac.th (A.K.)
[2] Nakhonsawan Campus, Mahidol University, Nakhonsawan 60130, Thailand; rtg.pun@gmail.com
[3] Department of Public Health, University of Massachusetts Lowell, Lowell, MA 01854, USA; susan_woskie@uml.edu
* Correspondence: pornpimol.kon@mahidol.ac.th; Tel.: +66-2644-4070; Fax: +66-2354-8561

Abstract: Acetochlor is a chloroacetanilide selective pre-emergent herbicide used for controlling grass and broadleaf weeds in crops. This study compared the acetochlor exposures of backpack and tractor sprayers and assessed whether dermal or air exposures were more important contributors to the overall body burden as measured by urinary metabolites. Sixty sugarcane farmers in Nakhonsawan province, Thailand participated in the study, and breathing zone air and dermal patch samples were collected during spraying. Urine samples were collected before spraying, at the end of the spraying task, and on the day after spraying. For backpack and tractor sprayers, there was no significant difference in their breathing zone air concentrations, total body dermal samples, or urinary 2-methy-6-methyaniline (EMA) concentrations on the day after spraying. In addition, although most backpack and tractor sprayers wore long pants and long sleeve shirts, they were still exposed to acetochlor, as evidenced by a significant increase in the urinary EMA from before spraying (GM = 11.5 μg/g creatinine) to after spraying (GM = 88.5 μg/g creatinine) to the next day (GM = 111.0 μg/g creatinine). Breathing zone air samples were significantly correlated with those of total body dermal patch samples and with urinary EMA concentrations after spraying. This suggests that both air and dermal exposure contribute to urinary EMA levels.

Keywords: acetochlor; herbicide; breathing zone air sample; dermal patch sample; urinary metabolite; backpack sprayer; tractor sprayer; sugarcane farmer

Citation: Kallayanatham, N.; Pengpumkiat, S.; Kongtip, P.; Pundee, R.; Nankongnab, N.; Kongtawelert, A.; Woskie, S.R. Air, Dermal, and Urinary Metabolite Levels of Backpack and Tractor Sprayers Using the Herbicide Acetochlor in Thailand. *Toxics* **2023**, *11*, 622. https://doi.org/10.3390/toxics11070622

Academic Editors: Maaike van Gerwen, Roel Vermeulen and Lauren Petrick

Received: 18 June 2023
Revised: 11 July 2023
Accepted: 14 July 2023
Published: 18 July 2023

1. Introduction

Agriculture is considered the main occupation for the majority of the population in Thailand, with approximately 47% of the country's total land area being used for agricultural purposes [1]. However, most farmers are still considered informal workers, meaning they can rarely access workers' compensation or a government retirement. According to a report in 2022 from the National Statistical Office, 52.5% of informal workers were farmers [2]. It is the largest category of informal work, which makes up 51% of the total Thai workforce. A major hazard for agricultural workers is exposure to hazardous chemicals, including pesticides [2]. According to a 2017 report from the Bureau of Occupational and Environmental Diseases in the Ministry of Public Health, there were 10,312 pesticide poisoning cases, with a rate of 17.12 per 100,000 population [3]. Additionally, the importation of pesticides increased in 2021 compared to 2020, with herbicides comprising the largest volume of pesticides imported into Thailand [4].

Acetochlor (2-chloro-N-(ethoxymethyl)-N-(2-ethyl-6-methylphenyl)-acetamide) is a chloroacetanilide selective pre-emergent herbicide in the chloroacetanilide group used for controlling grass and broadleaf weeds in crops such as corn [5]. Acetochlor was registered in the US in 1994 [6]. Thailand imported about 2661 tons of Acetochlor in 2021 [7] and it is widely used by sugarcane farmers in Thailand, especially in the Nakhonsawan province [8].

The US EPA has classified acetochlor as likely to be carcinogenic to humans because it induces nasal tumors in rats at a maximum tolerated dose of 1000 parts per million [9,10]. Previous human studies from North Carolina in 2010 and Iowa in 2011 showed an increased risk of lung, colorectal, and pancreatic cancers, as well as melanoma, among acetochlor-exposed applicators [11]. Previous research in animals has shown that acetochlor may have genetic toxicity and alter thyroid hormone-dependent gene expression in frog tadpoles [12]. In addition, the long-term exposure of rats to acetochlor caused liver and kidney damage and dysfunction of the antioxidant system [13]. In animals, after absorption, acetochlor was also widely distributed in well-perfused organs and showed a low potential for bioaccumulation. Most (66–72%) acetochlor was eliminated in the urine in about 48 h and 12–21% was eliminated in the feces in 48 h [14].

Conventional farmers who use acetochlor choose from several types of available spraying equipment for their agricultural work. Backpack equipment in this study consisted of a manually operated or motorized backpack, which is carried on the back of the operator. The capacity of these sprayers is mostly around 20 L or less [15]. On the other hand, a tractor is a large vehicle that is used for farming and other agricultural activities. Tractors are equipped with mechanized spraying attachments and a maximum of 20 nozzles that can cover a large area. The capacity of these sprayers is 200–400 L [16]. The main difference between a backpack sprayer and a tractor sprayer is the scale of operation. Backpack sprayers are suitable for small-scale operations, while tractors are usually used for large-scale operations [17].

The goal of this study was to compare the exposures of farmers using backpack versus tractor spraying of the herbicide acetochlor and assess whether dermal exposure or air exposure were more important contributors to the overall body burden of farmers.

2. Materials and Methods

2.1. Study Population and Data Collection

This study was approved by the Ethical Review Committee for Human Research, Faculty of Public Health, Mahidol University (Approval No. MUPH 2015-146, date of approval: (28 August 2015)). From our longitudinal study of conventional farmers (n = 213) and organic farmers (n = 225), we recruited seventy-eight sugarcane conventional farmers from Khao Thong Subdistrict, Phayuha Khiri District in Nakhonsawan province in Thailand [18]. Informed consent was obtained from all farmers. Data collection for this study was carried out on the recruited conventional sugarcane farmers who sprayed acetochlor during March–May 2017 (n = 60). The field staff set an appointment with the farmers for when they planned to spray acetochlor. The type of acetochlor used was 2-chloro-N-ethoxymethyl-6-ethylaceto-o-toluidide 50% w/v emulsifiable concentrate with a tradename of Razer (in Thai). The backpack sprayers usually mixed 200–4000 mL of acetochlor in 40–200 L of water (concentration range of 5–20 mL/L); the tractor sprayers mixed 1500–8000 mL of acetochlor in 300–1000 L of water (concentration range of 5–8 mL/L). In this study, we selected subjects who sprayed only acetochlor (61.7%) as well as some who sprayed acetochlor mixed with other chemicals, such as paraquat, 2,4-dichlorophenoxyacetic acid, cypermethrin, ametryn, chlorpyrifos, and diuron (38.3%).

Before the spraying day, urine samples were collected at waking in a polyethylene bottle as the first morning void and stored in an insulated ice box, which was held until collection by field staff later in the day. On the spray day, the field staff set up an air sample collection and put the dermal patches on the subjects. The field staff also observed the mixing and spraying and interviewed the subject using a questionnaire. The subject then provided a urine sample at the end of the spraying event. The questionnaire covered demographic information, agricultural activities, history of pesticide mixing and spraying, use of personal protective equipment, and health symptoms after spraying pesticides. Following the day of spraying, the subjects collected their first morning void urine and stored it in an insulated icebox awaiting field staff pickup later in the day. The urine samples were frozen at −20 °C until analysis.

2.1.1. Personal Breathing Zone Air Sampling

While the farmer was spraying acetochlor, personal air samples were collected according to NIOSH method 5602 using an XAD-2 sorbent tube connected to a glass fiber filter placed in the breathing zone of the farmer [19]. The personal pump was calibrated at 1 L/min before sampling and checked after sampling with a primary air flow calibrator. The personal air sampler was active for the whole time the farmers were spraying acetochlor. The XAD-2 and glass fiber filters were separated and frozen at −20 °C until analysis.

2.1.2. Dermal Patch Sampling

Dermal contact samples were collected by using a cotton patch (10 cm × 10 cm) sewn on top of an aluminum foil pad that was taped in place on the bare skin of farmers before spraying. Patches were located in 11 positions, including one patch on the face, chest, and back and two patches on the upper arms (left and right), forearms (left and right), thighs (left and right), and lower legs (left and right) (Figure 1). The farmer's typical clothing was worn over the dermal patch samplers. After spraying, the dermal patch samples were removed by forceps and the aluminum foil was removed. The dermal patch samples were stored in sealed bottles and were frozen at −20 °C until analysis.

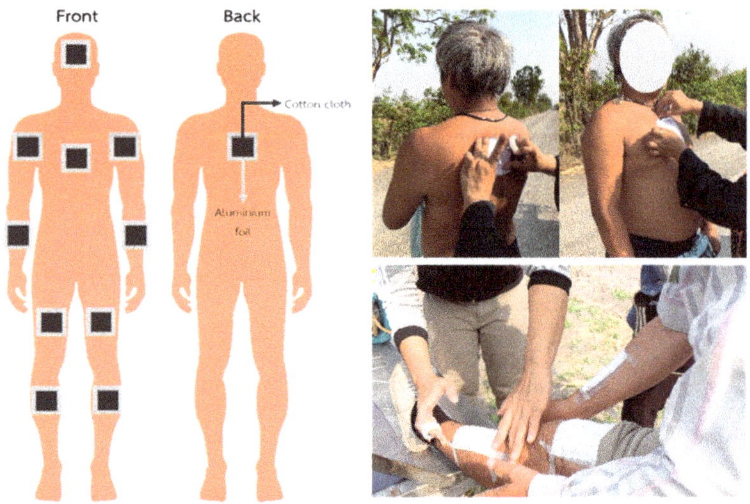

Figure 1. Patches were placed in different locations on bare skin.

2.2. Analysis of Personal Breathing Zone Air Samples and Dermal Patch Samples

2.2.1. Chemical Reagents

Acetochlor and dimethachlor (internal standard for air sample analysis) were obtained from LGC Dr. Ehrenstorfer (Augsburg, Germany). Hexane and acetone were obtained from Merck (Darmstadt, Germany), and Acetochlor-ethane sulfonic acid (Acetochlor ESA) was obtained from LGC Dr. Ehrenstorfer (Augsburg, Germany). Additionally, 2,4,6-Trimethylaniline (internal standard for urine sample analysis) was obtained from Alfa Aesar (Lancashire, UK). Dichloromethane and 50% NaOH and HCl were obtained from Merck (Darmstadt, Germany).

2.2.2. Analysis of Personal Breathing Zone Air Samples

The analysis of air samples was carried out following the method for analysis of dermal patch samples by Mahaboonpeeti et al. [20]. The front and back section of XAD-

2 and the glass fiber filter were separated into different tubes. Following this, 1 mL of hexane and acetone (1:1 *v*/*v*) were pipetted into the tubes, and, subsequently, the cap was closed and sealed with parafilm. Then, the samples were mixed and sonicated for 30 min. Next, 200 μL of the extracted solutions and 30 μL of dimethachlor as the internal standard were injected for gas chromatography-mass spectrometry (GC-MS). The GC-MS conditions were as follows: HP-5MS column (30 m length × 250 μm id × 0.25 μm film thickness), splitless injection mode, and inlet temperature of 250 °C. The temperature of the column was initiated at 50 °C for 1 min, then raised at 10 °C/min to 200 °C, followed by an increase at 3 °C/min to a final temperature of 230 °C for 4 min and post run at 250 °C for 2 min. The flow rate was set at 1 mL/min using helium as the carrier gas. The fixed electron energy was set at 70 eV relative to the standard autotune. The retention times of acetochlor and dimethachlor were 18.2 and 18.0 min, respectively. The quantitation ions were m/z 146 and m/z 223 for acetochlor and m/z 134 and m/z 197 for dimethachlor. Additionally, the calibration curve of acetochlor in the air sample was prepared at concentrations of 25, 50, 100, 300, 500, 1000, 1500, and 2000 ng. The recovery of acetochlor in air samples of 100 ng and 1000 ng ranged from 99.3 to 103.3% with a relative standard deviation (RSD) of less than 4%. The limit of detection (LOD) for acetochlor in the air samples was 1.7 ng.

2.2.3. Analysis of Dermal Patch Samples

For dermal sampling, the cotton patch was put into a tube with 8 mL of hexane and acetone (1:1 *v*/*v*), and the cap was closed. Then, the sample was mixed and sonicated for 20 min. Subsequently, 300 μL of the extracted solution and 20 μL of the dimethachlor internal standard were injected into GC-MS using the same conditions implemented for the air sample analysis. Acetochlor concentrations of cotton patch samples were calculated following the US Environmental Protection Agency guidelines [21]. The length of spraying in hours (h) was used to estimate the dermal contact exposure in μg/h. The concentration of acetochlor on cotton patch samples was expressed as micrograms of acetochlor per square centimeter (100 cm^2) of cotton pad per hour of exposure. After that, the dermal patch concentration ($\mu g/cm^2/h$) was multiplied by the adult body surface areas [21] to obtain dermal contact exposure in μg/h. The estimated total dermal contact exposure was calculated by summing the μg/h levels for the 11 dermal pad samples taken on each individual. The calibration curve of acetochlor in the patch sample was prepared at concentrations of 0.2, 0.4, 0.8, 2.4, 4, 8, 12, and 16 μg. The recovery of acetochlor in patch samples of 0.8 μg and 8 μg ranged from 97.9 to 104.5% with RSD less than 4.5%. Lastly, the limit of detection (LOD) of acetochlor in the patch samples was 0.01 μg.

2.2.4. Analysis of Urine Sample

In monkeys, three main metabolites are formed from acetochlor which are excreted in urine: tert-amide mercapturic acid (19.9–31.6%), O-glucuronic acid conjugate of O-dealkylated acetochlor (12.7–22.8%), and sec-amide mercapturic acid (more than 5%). There were also two minor metabolites identified: sec-amide chloride (3.9%) and tert-thioacetic acid (3.5%) [22]. The analysis of urinary acetochlor metabolites can be conducted by hydrolysis of the metabolites into 2-ethyl-6-methylaniline (EMA), which includes up to 70% of the acetochlor metabolites [22]. The structural formulae of acetochlor, acetochlor metabolites, and EMA are shown in Figure 2. The analysis method for 2-ethyl-6-methylaniline (EMA) was modified from Driskell et al. [23]. Urine samples (3 mL) were hydrolyzed following the Driskell method for 1.5 h. The mixture was neutralized, and 25 μL of 2,4,6-trimethylaniline was added as an internal standard. Then, 1.5 mL of dichloromethane was added to extract the 2-ethyl-6-methylaniline (EMA) by shaking for 5 min and centrifugation at 5000 rpm for 5 min. The extraction was performed twice. The solution was concentrated to 300 μL in a nitrogen evaporator because the 2-ethyl-6-methylaniline (EMA) is volatile and, therefore, concentration to dryness could be avoided using this method. The solution (300 μL) was injected into the GC-MS for analysis. The GC-MS conditions were as follows: DB-200 column (30 m length × 250 μm id × 0.25 μm film thickness), splitless injection mode, and the inlet temperature of 250 °C. The temperature of the column was initiated at 70 °C for 2 min and then raised at 20 °C/min to 100 °C, at 5 °C/min to 130 °C, and at 20 °C/min to a final temperature of 200 °C and post run at 250 °C for 2 min. The flow rate was set at 1 mL/min using helium as the carrier gas. The fixed electron energy was set at 70 eV relative to the standard autotune. The retention times of EMA and 2,4,6-trimethylaniline were 9.0 and 9.2 min, respectively. The quantitation ions were m/z 120 and m/z 135 for EMA and m/z 120, m/z 134, and m/z 135 for 2,4,6-trimethylaniline. In addition, the calibration curve for the EMA in urine used concentrations of 50, 250, 500, 1000, and 1500 ng/mL, and the recovery of EMA concentrations of 100 and 1000 ng/mL ranged from 94.5 to 101.2% with an RSD of less than 5%. The limit of detection (LOD) of EMA was 2 ng/mL. Moreover, all urinary metabolites were corrected for creatinine levels by dividing the metabolite concentration by the creatinine concentration, producing units of ng metabolite/g for the creatinine. Furthermore, the creatinine in urine was analyzed using an enzymatic method with a linear concentration range of 1–500 mg/dL and a detection limit of 0.16 mg/dL [24]. EMA concentrations were presented as μg/g creatinine, and the geometric mean (GM) and geometric standard deviation (GSD) were calculated in μg/g creatinine.

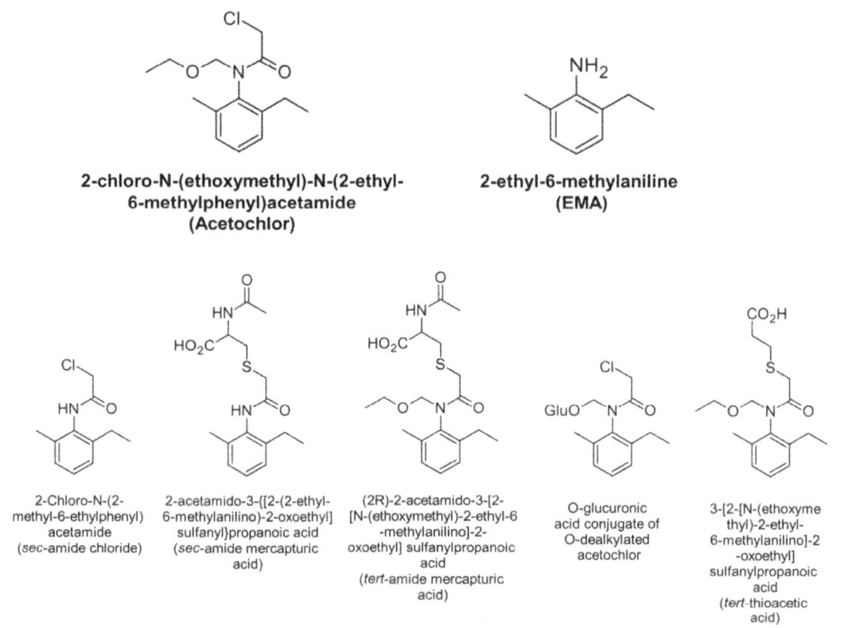

Figure 2. Structural formulae of acetochlor, acetochlor metabolites, and 2-ethyl-6-methylaniline (EMA).

2.3. Statistical Analysis

All statistical analyses were conducted using SPSS for Windows, version 23 (IBM Thailand Co., Ltd., Bangkok, Thailand). A descriptive analysis of the demographic characteristics was carried out using the mean, standard deviation, and independent t-tests. The breathing zone air samples, dermal patch samples, and urinary EMA concentrations were skewed, so the natural logarithm of the concentration was used in all analyses. Furthermore, the geometric mean (GM) was calculated as the exponentiated mean of the natural log values, and the geometric standard deviation (GSD) was calculated as the exponentiated standard deviation of the natural log values. For concentrations below the detection limit, we substituted the detection limit by dividing the detection limit with the square root of 2 when the GSD was <3, and when the GSD was $\geq$3, the detection limit divided by two was used [25]. The natural log values for the urinary EMA concentrations at morning void for the day before spraying, end of spraying task, and the day after spraying were compared using a repeated measures ANOVA with post hoc Fisher's LSD test. The urinary EMA concentrations between backpack and tractor sprayers were compared using the Mann–Whitney U test. The correlation between the natural logarithm of the acetochlor concentrations in the breathing zone air samples with those of total body dermal patch samples and urinary EMA concentration was analyzed using Pearson correlation.

3. Results

Based on the characteristics collected, the average age of the farmers was 49.7 years (Table 1). In addition, there were more male (75%) participants and most had graduated from high school or higher (51.7%). The majority were current alcohol drinkers (66.7%), though only 18.3% were smokers. Many participants also had a BMI that qualified as obese or overweight (45%), and the majority (76.7%) have been spraying pesticides for more than 10 years. Two types of spraying equipment were used, with 83.3% using backpack sprayers and 16.7% using tractor sprayers. The average time of spraying was 40.2 min and ranged from 20 to 120 min.

Table 1. Characteristics of farmers (n = 60).

Variables	Total (n)	Conventional Farmers n (%)
Age		
Min–max	60	18–69
Mean (SD)		49.7 (12.9)
Sex		
Male	45	45 (75.0)
Female	15	15 (25.0)
Educational level		
Elementary	29	29 (48.3)
High school or higher	31	30 (51.7)
Marital status		
Single	7	7 (11.7)
Married	51	51 (85.0)
Widowed/divorced	2	2 (3.3)
Alcohol intake		
Current drinker	40	40 (66.7)
Nondrinker	20	20 (33.3)
Smoking		
Current smoker	11	11 (18.3)
Nonsmoker	49	49 (81.7)
BMI (kg/m^2)		
Underweight (<18.5)	7	7 (11.7)
Healthy Weight (18.5–<25.00)	26	26 (43.3)
Overweight (25.00–<30.00)	19	19 (31.7)
Obesity (30.0 or higher)	8	8 (13.3)
Spraying pesticides in agricultural fields (years)		
≤10	14	14 (23.3)
>10	46	46 (76.7)
Spraying equipment		
Backpack sprayers	50	50 (83.3)
Tractor sprayers	10	10 (16.7)

While spraying, all farmers wore long sleeve shirts (100%), 80% wore shoes, and 60–62% wrapped a cloth around their face when spraying. However, backpack sprayers were significantly more likely to wear long pants (98%) than tractor sprayers (80%). Few farmers also wore gloves, cotton masks, goggles, or boots (Table 2).

Table 2. Personal protective equipment (PPE) used while spraying (n = 60).

PPE	Backpack (n = 50) n (%)	Tractor (n = 10) n (%)
Goggles	2 (4)	0
Cloth gloves	12 (24)	0
Latex gloves	2 (4)	1 (10)
Long sleeve shirts	50 (100)	10 (100)
Shoes	40 (80)	8 (80)
Boots	6 (12)	0
Socks	7 (14)	0
Short pants	1 (2)	2 (20)
Long pants	49 (98)	8 (80)
Large brim straw hat	16 (32)	1 (10)
Cloth wrapped around face	31 (62)	6 (60)
Cotton mask	3 (6)	2 (20)
Balaclava	11 (22)	3 (30)

The detection frequency of urinary EMA concentrations was 66.7, 90.0, and 93.3% on the day before spraying, at the end of the spraying task, and on the day after spraying, respectively (Table 3). The urinary EMA concentrations at the end of spraying and on the day after spraying were significantly higher than on the day before spraying, but the EMA concentrations at the end of the spraying event and the day after spraying were not significantly different. Additionally, the urinary EMA concentrations between backpack and tractor sprayers were not significantly different on the day before spraying, the spraying day, and the day after spraying. The urinary EMA concentrations on the day after spraying were also not significantly different based on smoking or drinking status, gender or BMI category, nor were they significantly different between sprayers who sprayed only acetochlor and those who sprayed acetochlor mixed with other pesticides.

Table 3. Comparison of the natural log of EMA concentrations at different times of urine collection (µg/g creatinine) using repeated measures ANOVA with post hoc Fisher's LSD test and comparison of EMA concentrations at different times between backpack and tractor sprayers using the Mann–Whitney U test on the natural log of EMA.

Urine Collection	EMA (µg/g Creatinine)			Comparison of EMA between Backpack vs. Tractor [1]
	Total Farmers (n = 60)	Backpack (n = 50)	Tractor (n = 10)	
The day before spraying				
Detection frequency (%)	40 (66.7)	31 (62.0)	9 (90.0)	
GM (GSD) [2]	11.5 (8.6)	10.2 (9.2)	20.7 (5.7)	$p = 0.242$
Range (Min–Max)	0.6–1099.0	0.6–1099.0	1.4–308.6	
The end of spraying task				
Detection frequency (%)	54 (90.0)	44 (88.0)	10 (100.0)	
GM (GSD) [2]	88.5 (8.6)	84.0 (10.0)	115.1 (3.2)	$p = 0.937$
Range (Min–Max)	1.2–3849.8	1.2–3849.8	11.4–1155.6	
The day after spraying				
Detection frequency (%)	56 (93.3)	46 (92.0)	10 (100.0)	
GM (GSD) [2]	111.0 (6.4)	117.3 (6.6)	84.3 (5.8)	$p = 0.579$
Range (Min–Max)	1.3–1865.3	1.3–1865.3	4.6–823.4	
Comparison of total farmers' EMA [3]				
The day before spraying and the end of spraying task			$p < 0.001$ *	
The day before and after spraying			$p < 0.001$ *	
The end of spraying task and the day after spraying			$p = 0.350$	

[1] Mann–Whitney U test on natural log of EMA; [2] GM = geometric mean, GSD = geometric standard deviation; [3] Repeated measures ANOVA with post hoc Fisher's LSD test on natural log of EMA; * $p < 0.05$.

The acetochlor concentrations in the breathing zone air samples of backpack sprayers were not significantly different from those of tractor sprayers (Table 4).

Moreover, for dermal patch samples, the acetochlor concentrations of tractor sprayers were not significantly different from those of backpack sprayers. However, the sprayed area, acetochlor solution used, and spraying duration of tractor sprayers were significantly higher in tractor sprayers than those of backpack sprayers.

Table 4. Comparison on acetochlor air concentrations, dermal patch levels, and spraying factors between backpack and tractor spraying.

Acetochlor/EMA Concentrations	Backpack (n = 50)	Tractor (n = 10)	p-Value
Breathing zone air sample (μg/m^3)			0.403
Detection frequency (%)	50 (100.0)	10 (100.0)	
GM (GSD) [1]	13.7 (3.4)	9.3 (6.1)	
Range (Min–Max)	0.3–92.2	0.1–83.9	
Dermal patch sample (μg/h)			
Forehead			0.051
Detection frequency (%)	50 (100.0)	10 (100.0)	
GM (GSD) [1]	5.3 (3.5)	12.9 (3.9)	
Range (Min–Max)	0.8–1398.7	1.2–105.2	
Chest			0.052
Detection frequency (%)	50 (100.0)	10 (100.0)	
GM (GSD) [1]	24.1 (2.8)	52.6 (5.1)	
Range (Min–Max)	3.4–1358.8	6.4–3353.0	
Back			0.814
Detection frequency (%)	50 (100.0)	10 (100.0)	
GM (GSD) [1]	41.0 (6.3)	35.2 (7.6)	
Range (Min–Max)	2.8–10,472.8	5.9–9945.2	
Arms			0.400
Detection frequency (%)	50 (100.0)	10 (100.0)	
GM (GSD) [1]	33.4 (2.9)	45.6 (2.8)	
Range (Min–Max)	4.5–4368.1	19.0–697.1	
Legs			0.798
Detection frequency (%)	50 (100.0)	10 (100.0)	
GM (GSD) [1]	201.8 (4.7)	231.6 (4.4)	
Range (Min–Max)	10.5–32,403.7	36.9–6309.8	
Total body			0.884
Detection frequency (%)	50 (100.0)	10 (100.0)	
GM (GSD) [1]	503.0 (4.5)	466.0 (4.6)	
Range (Min–Max)	24.7–37,645.8	75.6–19,788.2	
Spraying information			
Sprayed area (Ha)			<0.001 *
GM (GSD) [1]	0.5 (1.8)	1.6 (1.7)	
Range (Min–Max)	0.2–3.0	0.5–2.4	
Acetochlor solution used (L)			<0.001 *
GM (GSD) [1]	83.4 (1.5)	823.8 (1.5)	
Range (Min–Max)	40–200	300–1000	
Spraying duration (min)			0.027 *
GM (GSD) [1]	35.5 (1.4)	45.1 (1.8)	
Range (Min–Max)	20–120	25–120	

p-values were calculated by using independent t-Test. * p < 0.05; [1] GM = geometric mean, GSD = geometric standard deviation.

The acetochlor concentrations in the breathing zone air samples were significantly correlated with those of the total body dermal patch samples and the urinary EMA concentrations after spraying (Table 5).

Table 5. The relationship between acetochlor concentrations in the breathing zone air samples, total body dermal patch levels, and EMA concentrations in urine after spraying using Pearson correlation (n = 60). Correlations were performed on the natural log of all metrics.

Acetochlor/EMA Concentrations	Correlation	p-Value
Breathing zone air sample (μg/m^3)		
Total body dermal patch samples (μg/h)	0.269	0.037 *
Urinary EMA concentrations the day after spraying (μg/g creatinine)	0.381	0.003 *
Total body dermal patch samples (μg/h)		
Urinary EMA concentrations the day after spraying (μg/g creatinine)	0.022	0.867

* p < 0.05.

4. Discussion

The average age of farmers in this study was 49.7 years old, which was similar to the average age of 49.5 years in Bootsikeaw et al.'s [26] study in Thailand, and much higher than the 29.8 years average reported age in the study of Jallow et al. [27] in Kuwait, Asia or the 35.2 years reported in the study of Mergia et al. [28] in Ethiopia, Africa. This current study had only 25% female sprayers. It is a tough and difficult job for females to carry a backpack sprayer on their shoulders; however, there was no significant difference in exposures between male and female sprayers. Other studies have also found that men have more responsibility for planting and pesticide application than women [29]. In the Ethiopia study, more than 98% of small-scale farmers were men because farm activities and pesticide spraying were primarily performed by men [28]. In Kuwait, farm activities, especially pesticide use, were performed exclusively by men [27]. Additionally, in this current study, a larger fraction of our farmers had sprayed pesticides for more than 10 years (76.7%) compared to 10.4% in the Ethiopian study and 41.9% in the Kuwait study [26,27]. The sprayers were also less likely to smoke cigarettes (18.3%) than the average Thai population (22.2%) [30], but the percentage of sprayers who reported drinking alcohol (66.7%) was similar to the occasional or regular drinkers (65%) reported in the general Thai population [31].

Moreover, this study found that the percentage of backpack sprayers wearing PPE (primarily long sleeve shirts, long pants, shoes, and face cloths) was higher than those reported by Mergia et al. [28] and Mahaboonpeeti et al. [20]. However, it has been reported that farmers do not have sufficient knowledge of pesticide toxicity and how to use pesticides safely [27].

This study selected 2-ethyl-6-methylaniline (EMA) as the acetochlor metabolite to estimate acetochlor exposure in spraying farmers. EMA is the hydrolysis product of the five identified metabolites of acetochlor: tert-amide mercapturic acid, O-glucuronic acid conjugate of O-dealkylated acetochlor, sec-amide mercapturic acid, sec-amide chloride, and tert-thioacetic acid, which makes up approximately 59.5% to 70.2% of all acetochlor metabolites [22]. Barr et al. [32] studied acetochlor mercapturate (ACM) and EMA, and proposed another new biomarker. In this study, we selected subjects who sprayed only acetochlor (61.7%) and acetochlor mixed with other chemicals, such as paraquat, 2,4-dichlorophenoxyacetic acid, cypermethrin, ametryn, chlorpyrifos, and diuron (38.3%). The comparison of EMA concentrations between subjects who sprayed pure acetochlor and acetochlor mixed with other chemicals was not significantly different. The sprayers in this study did not use metolachlor or chloroacetanilide herbicides, which could be changed to EMA after the hydrolysis process [32]. The detection frequency of urinary EMA was 66.7, 90.0, and 93.3% on the day before spraying, at the end of the spraying task and the day after spraying, respectively. On the day before spraying, urinary EMA concentrations were detected in only 66.7% farmers because the farmers might not have used acetochlor for a while before our visit, but it was still measurable in the urine, perhaps due to field contamination or storage near the home. Gustin et al. [33] reported the detection frequency of urinary EMA (24 h composite urine) of farmers spraying acetochlor as 15, 100, and 100% the day before spraying, the spraying day, and the day after spraying using open- and closed-tractor cab spraying. Furthermore, the detection frequency of Gustin et al.'s [33] study was higher than the current study on the spray days because of the longer duration of spraying, higher spraying volume, and spraying continuously for 5 days. Gustin et al.'s [33] study also reported urinary EMA concentrations of open- and closed-tractor cabins of 0.004 and 0.002 mg/kg bw/day, respectively. Moreover, the Curwin et al. study in the US reported a GM of urinary ACM of 8.0 μg/L in four sprayers [34]. The GM of urinary EMA on the day after spraying in this current study was 111.0 μg/g creatinine (947.4 nmole/L) ranging from 1.3 to 1865.3 μg/g creatinine (10.6–10,202.5 nmole/L). Barr et al. [32] reported the range of urinary EMA of 7.2–1805 nmole/L for seven sprayers, where urine samples were collected within 24–48 h of acetochlor application. The levels in the Barr et al. study were likely lower because the applicators collected a 24 h composite urine sample and had closed cabin tractors with air conditioning and dust and charcoal filters [32,35,36]. In this study, the urinary EMA on the day after application was probably the highest because the body had to metabolize the exposure, and the half-life of acetochlor in urine was reported to be 1.2 days in monkeys [33].

The acetochlor concentrations in the breathing zone air samples of backpack sprayers were slightly higher than tractor sprayers but not significantly different. Since the vapor pressure of acetochlor is low (1.67×10^{-7} mm Hg at 20 °C), it is present more as a particle than vapor. Currently, there is no exposure limit for acetochlor, although it is widely used for agricultural work in many countries, such as the United States [11], China [13], and Thailand [7]. An airborne exposure limit for acetochlor should be established for workers' health and safety in the workplace.

Regarding acetochlor on the dermal patches, the acetochlor concentrations of tractor sprayers were not significantly different from those of backpack sprayers. Since the acetochlor concentrations in the breathing zone between the tractor and backpack sprayers were not significantly different, it is possible that contamination occurred during the period of mixing acetochlor with water and loading it into the tractor tank. We found that tractor sprayers load the acetochlor solution at the height of the farmer's head and chest, which could lead to spilling and splashing of the acetochlor solution onto the head and chest (Figure 3). We also noted that the forehead and chest patch samples for tractor sprayers were higher than those of backpack sprayers.

(a) (b)

Figure 3. The backpack and tractor sprayers while mixing and spraying acetochlor. (**a**) The backpack sprayer; (**b**) the tractor sprayer.

Like the air concentration, the acetochlor concentration on the back, arms, legs, and total body patch samples was not significantly different between the backpack sprayers and the tractor sprayers. The factors affecting exposures between the backpack and tractor sprayers included the area of the field being sprayed onto, the amount of acetochlor solution used, and the spraying duration, as well as other likely factors, such as nozzle size, spraying pressure, application maintenance, application height, and the angle of spraying [17]. Additionally, the backpack sprayers were more easily exposed to acetochlor than tractor sprayers since the sprayer points the nozzle at the soil close to their feet. However, the tractor sprayers used a larger amount of acetochlor solution and sprayed a larger area for a longer period of time than the backpack sprayers. Because both groups mostly wore similar clothing (long pants and long sleeve shirts and a cloth around the face), the closer proximity and smaller volumes of spraying may have resulted in equivalent exposures (μg/h) compared to the larger volume and downwind spray of the tractor drivers.

The back, arms, and leg patches were the most exposed body areas to acetochlor; these results are similar to Mahaboonpeeti et al. [20] and Bootsikeaw et al. [26]. Regarding the backpack sprayers, their back exposure to acetochlor (2.8–10,472.8 μg/h) could occur due to acetochlor leakage of an old sprayer tank during spraying. Arm exposure to acetochlor (4.5–4368.1 μg/h) could also be caused by spilling of the acetochlor solution from the tank during loading and mixing or blowback of the spray while applying. Lastly, high leg concentrations of acetochlor (10.5–32,403.7 μg/h) are most likely caused by the position and angle of the nozzle used to spray on the ground. Even though 98% of the backpack and 80% of tractor sprayers wore long pants and 100% wore long sleeve shirts, they were still exposed to acetochlor at high concentrations.

The urinary EMA concentrations of backpack sprayers on the day after spraying were slightly higher than that of the tractor sprayers but not significantly different. However, Gustin et al. [33] showed that the open-cabin applicators had significantly higher EMA concentrations compared to the closed-cabin applicators. However, in this study, the tractor sprayers were not normally in a closed cabin.

The acetochlor concentrations of the breathing zone air samples were significantly correlated with those of the total body dermal patch samples and the urinary EMA concentrations after spraying (Table 5). However, the acetochlor concentrations of the total body dermal patch samples were not significantly correlated with urinary EMA concentrations after spraying. As the skin is an effective barrier, it may take a longer time to pass through it and, thus, the amount of acetochlor entering the bloodstream is limited.

5. Conclusions

For backpack and tractor sprayers, exposure to the herbicide acetochlor was not significantly different for breathing zone air samples, total body dermal samples, and urinary EMA concentrations on the day after spraying. In addition, the urinary EMA concentrations on the day before spraying were significantly lower than those at the end of the spraying task as well as the day after spraying (which had the highest levels). The two groups of sprayers also used only regular clothing for PPE, resulting in inhalation and dermal exposures that increased their EMA levels when they sprayed. Furthermore, the breathing zone air samples were significantly correlated with those of the levels of the total body dermal patch samples and with the urinary EMA concentrations after spraying. Thus, both air and dermal exposure contribute to urinary EMA levels, and personal protective equipment improvements are needed. Future education of farmers should focus on the need to wear improved PPE, such as plastic aprons, coveralls, and rubber gloves during mixing and spraying.

Author Contributions: Conceptualization, S.R.W., P.K., N.K., S.P. and R.P.; methodology, S.R.W., R.P., N.K. and N.N.; formal analysis, R.P. and N.K.; investigation, R.P., N.N. and A.K.; resources, S.P., N.N. and A.K.; data curation, N.K., S.P. and R.P.; writing—original draft preparation, N.K.; writing—review and editing, N.K., S.P., P.K., R.P., N.N., A.K. and S.R.W.; supervision, S.R.W., P.K. and A.K.; funding acquisition, S.R.W. and P.K. All authors have read and agreed to the published version of the manuscript.

Funding: This research was financially supported by NIH, the Fogarty International Center, the National Institute of Environmental Health Science, and the Center for Disease Control under the award numbers U01 TW010091 and U2R TW010088.

Institutional Review Board Statement: This study was conducted according to the guidelines of the Declaration of Helsinki and was approved by the Institutional Review Board (or Ethics Committee) of the Faculty of Public Health, Mahidol University (protocol code MUPH 2015-146 and date of approval, 28 August 2015).

Informed Consent Statement: Informed consent was obtained from all subjects involved in the study.

Data Availability Statement: The data presented in this study are available on request from the corresponding author.

Acknowledgments: We would also like to thank all the participants and health-promoting hospital staff for helping in the recruitment of the subjects in the Nakhonsawan Provinces.

Conflicts of Interest: The authors declare no conflict of interest. The funders had no role in the design of the study; in the collection, analyses, or interpretation of data; in the writing of the manuscript; or in the decision to publish the results.

References

1. National Statistical Office. Statistics of Land, Whole Kingdom: 2012–2022. Available online: http://statbbi.nso.go.th/staticreport/Page/sector/TH/report/sector_11_18_TH_.xls (accessed on 1 April 2023). (In Thai)
2. National Statistical Office. The Informal Employment Survey 2022. 2022. Available online: http://www.nso.go.th/sites/2014/Pages/%E0%B8%AA%E0%B8%B3%E0%B8%A3%E0%B8%A7%E0%B8%88/%E0%B8%94%E0%B9%89%E0%B8%B2%E0%B8%99%E0%B8%AA%E0%B8%B1%E0%B8%87%E0%B8%84%E0%B8%A1/%E0%B9%81%E0%B8%A3%E0%B8%87%E0%B8%87%E0%B8%B2%E0%B8%99/%E0%B9%81%E0%B8%A3%E0%B8%87%E0%B8%87%E0%B8%B2%E0%B8%99%E0%B8%99%E0%B8%AD%E0%B8%81%E0%B8%A3%E0%B8%B0%E0%B8%9A%E0%B8%9A.aspx (accessed on 1 April 2023). (In Thai)
3. Bureau of Occupational and Environmental Diseases, Ministry of Public Health. Report Occupational and Environmental Diseases and Health Hazards in 2017. Available online: https://ddc.moph.go.th/uploads/ckeditor2//files/01_envocc_situation_60.pdf (accessed on 1 April 2023). (In Thai)
4. Office of Agricultural Economics. Quantity and Value of Imports of Agricultural Pesticides 2018–2022. Available online: http://www.oae.go.th/view/1/%E0%B8%9B%E0%B8%B1%E0%B8%88%E0%B8%88%E0%B8%B1%E0%B8%A2%E0%B8%81%E0%B8%B2%E0%B8%A3%E0%B8%9C%E0%B8%A5%E0%B8%B4%E0%B8%95/TH--TH (accessed on 1 April 2023). (In Thai)
5. Heydens, W.F.; Lamb, I.C.; Wilson, A.G. Chloracetanilides. In *Hayes' Handbook of Pesticide Toxicology*; Elsevier: Amsterdam, The Netherlands, 2010; pp. 1753–1769.
6. Environmental Protection Agency. *Federal Register Office of the Federal Register*; National Archives and Records Administration: Washington, DC, USA, 1994.
7. Department of Agriculture. Report The Herbicide Imported in 2007–2021. Available online: https://data.go.th/dataset/importherbicidevol (accessed on 1 April 2023). (In Thai)
8. Office of the Cane and Sugar Board. Sugarcane Area Report 2021–2022. Available online: http://www.ocsb.go.th/upload/journal/fileupload/13813--1585.pdf (accessed on 1 April 2023). (In Thai)
9. Ashby, J.; Kier, L.; Wilson, A.; Green, T.; Lefevre, P.; Tinwell, H. Evaluation of the potential carcinogenicity and genetic toxicity to humans of the herbicide acetochlor. *Hum. Exp. Toxicol.* **1996**, *15*, 702–735. [CrossRef]
10. Dearfield, K.L.; McCarroll, N.E.; Protzel, A.; Stack, H.F.; Jackson, M.A.; Waters, M.D. A survey of EPA/OPP and open literature on selected pesticide chemicals: II. Mutagenicity and carcinogenicity of selected chloroacetanilides and related compounds. *Mutat. Res. Genet. Toxicol. Environ. Mutagen.* **1999**, *443*, 183–221. [CrossRef] [PubMed]
11. Lerro, C.C.; Koutros, S.; Andreotti, G.; Hines, C.J.; Blair, A.; Lubin, J. Use of acetochlor and cancer incidence in the Agricultural Health Study. *Int. J. Cancer* **2015**, *137*, 1167–1175. [CrossRef] [PubMed]
12. Crump, D.; Werry, K.; Veldhoen, N.; Van, A.G.; Helbing, C.C. Exposure to the herbicide acetochlor alters thyroid hormone-dependent gene expression and metamorphosis in Xenopus Laevis. *Environ. Health Perspect.* **2002**, *110*, 1199–1205. [CrossRef] [PubMed]
13. Li, L.; Wang, M.; Chen, S.; Zhao, W.; Zhao, Y.; Wang, X. A urinary metabonomics analysis of long-term effect of acetochlor exposure on rats by ultra-performance liquid chromatography/mass spectrometry. *Pestic. Biochem. Physiol.* **2016**, *128*, 82–88. [CrossRef] [PubMed]
14. European Food Safety Authority. Conclusion on the peer review of the pesticide risk assessment of the active substance acetochlor. *EFSA J.* **2011**, *9*, 2143.
15. Das, N.; Maske, N.; Khawas, V.; Chaudhary, S.; Dhete, R. Agricultural fertilizers and pesticides sprayers—A review. *Int. J. Innov. Res. Sci. Technol.* **2015**, *1*, 44–47.
16. Bhatkar, A.K.; Khope, P.; Chaudhari, P. A Review: Development of Pesticide Spraying Machine. *IJRET Int. J. Res. Eng. Technol.* **2016**, *5*, 470–472.
17. Franke, A.; Kempenaar, C.; Holterman, H.; Van, Z.J. *Spray Drift from Knapsack Sprayers: A Study Conducted within the Framework of the Sino-Dutch Pesticide Environmental Risk Assessment Project PERAP*; Plant Research International: Wageningen, The Netherlands, 2010.
18. Kongtip, P.; Nankongnab, N.; Pundee, R.; Kallayanatham, N.; Pengpumkiat, S.; Chungcharoen, J.; Phommalachai, C.; Choochouy, N.; Sowanthip, P.; Khangkhun, P.; et al. Acute changes in thyroid hormone levels among Thai pesticide sprayers. *Toxics* **2021**, *9*, 16. [CrossRef]
19. NIOSH Manual of Analytical Methods. *Chlorinated and Organonitrogen Herbicides: Method 5602 (Air Sampling)*; DHHS (NIOSH) Pub.: Cincinnati, OH, USA, 1994; pp. 94–113.
20. Mahaboonpeeti, R.; Kongtip, P.; Nankongnab, N.; Tipayamongkholgul, M.; Bunngamchairat, A.; Yoosook, W. Evaluation of dermal exposure to the herbicide alachlor among vegetable farmers in Thailand. *Ann. Work Expo. Health* **2018**, *62*, 1147–1158. [CrossRef]
21. United States Environmental Protection Agency. Occupational and Residential Exposure Test Guidelines: OPPTS 875.2400 Dermal Exposure. Available online: https://www.regulations.gov/document?D=EPAHQ--OPPT--2009--0157--0012 (accessed on 1 April 2023).

22. Shah, P.; Moretto, A. Acetochlor. In Proceedings of the Pesticide Residues in Food 2015, Joint FAO/WHO Meeting on Pesticide Residues, Geneva, Switzerland, 15–24 September 2015.
23. Driskell, W.; Hill, R., Jr.; Shealy, D.; Hull, R.; Hines, C. Identification of a major human urinary metabolite of alachlor by LC–MS/MS. *Bull. Environ. Contam. Toxicol.* **1996**, *56*, 853–859. [CrossRef] [PubMed]
24. Beckmann Coulter, Instruction for Use. Creatinine. Available online: https://www.beckmancoulter.com/wsrportal/techdocs?docname=/cis/A69463/%%/EN (accessed on 1 April 2023).
25. Hornung, R.W.; Reed, L.D. Estimation of average concentration in the presence of nondetectable values. *Appl. Occup. Environ. Hyg.* **1990**, *5*, 46–51. [CrossRef]
26. Bootsikeaw, S.; Kongtip, P.; Nankongnab, N.; Chantanakul, S.; Sujirarat, D.; Mahaboonpeeti, R. Urinary glyphosate biomonitoring of sprayers in vegetable farm in Thailand. *Hum. Ecol. Risk Assess. Int. J.* **2021**, *27*, 1019–1036. [CrossRef]
27. Jallow, M.F.; Awadh, D.G.; Albaho, M.S.; Devi, V.Y.; Thomas, B.M. Pesticide knowledge and safety practices among farm workers in Kuwait: Results of a survey. *Int. J. Environ. Res. Public Health* **2017**, *14*, 340. [CrossRef] [PubMed]
28. Mergia, M.T.; Weldemariam, E.D.; Eklo, O.M.; Yimer, G.T. Small-scale farmer pesticide knowledge and practice and impacts on the environment and human health in Ethiopia. *J. Health Pollut.* **2021**, *11*, 210607. [CrossRef]
29. Wang, W.; Jin, J.; He, R.; Gong, H. Gender differences in pesticide use knowledge, risk awareness and practices in Chinese farmers. *Sci. Total Environ.* **2017**, *590*, 22–28. [CrossRef]
30. Jitnarin, N.; Kosulwat, V.; Rojroongwasinkul, N.; Boonpraderm, A.; Haddock, C.K.; Poston, W.S. Socioeconomic status and smoking among thai adults: Results of the National Thai Food Consumption Survey. *Asia Pac. J. Public Health* **2011**, *23*, 672–681. [CrossRef]
31. Wakabayashi, M.; McKetin, R.; Banwell, C.; Yiengprugsawan, V.; Kelly, M.; Seubsman, S. Alcohol consumption patterns in Thailand and their relationship with non-communicable disease. *BMC Public Health* **2015**, *15*, 1297. [CrossRef]
32. Barr, D.B.; Hines, C.J.; Olsson, A.O.; Deddens, J.A.; Bravo, R.; Striley, C.A. Identification of human urinary metabolites of acetochlor in exposed herbicide applicators by high-performance liquid chromatography-tandem mass spectrometry. *J. Expo. Sci. Environ. Epidemiol.* **2007**, *17*, 559–566. [CrossRef]
33. Gustin, C.A.; Moran, S.J.; Fuhrman, J.D.; Kurtzweil, M.L.; Kronenberg, J.M.; Gustafson, D.I. Applicator exposure to acetochlor based on biomonitoring. *Regul. Toxicol. Pharmacol.* **2005**, *43*, 141–149. [CrossRef]
34. Curwin, B.D.; Hein, M.J.; Sanderson, W.T.; Barr, D.B.; Heederik, D.; Reynolds, S.J. Urinary and hand wipe pesticide levels among farmers and nonfarmers in Iowa. *J. Expo. Sci. Environ. Epidemiol.* **2005**, *15*, 500–508. [CrossRef] [PubMed]
35. Hines, C.J.; Deddens, J.A.; Tucker, S.P.; Hornung, R.W. Distributions and determinants of pre-emergent herbicide exposures among custom applicators. *Ann. Occup. Hyg.* **2001**, *45*, 227–239. [CrossRef] [PubMed]
36. Hines, C.J.; Deddens, J.A.; Striley, C.A.; Biagini, R.E.; Shoemaker, D.A.; Brown, K.K.; MacKenzie, B.A.; Hull, R.D. Biological monitoring for selected herbicide biomarkers in the urine of exposed custom applicators: Application of mixed-effect models. *Ann. Occup. Hyg.* **2003**, *47*, 503–517. [PubMed]

International Journal of
Environmental Research and Public Health

MDPI

Article

Total Mercury Content in the Tissues of Freshwater Chelonium (*Podocnemis expansa*) and a Human Health Risk Assessment for the Amazon Population in Brazil

Fábio Júnior Targino [1], Joanna Damazio de Nunes Ribeiro [1], Julia Siqueira Simões [1], Carla Silva Carneiro [2], Stella Maris Lazzarini [3], Aline Ramos Souza [3,4], Micheli da Silva Ferreira [1], Sergio Borges Mano [1] and Eliane Teixeira Mársico [1,*]

[1] Faculty of Veterinary, Universidade Federal Fluminense (UFF), Niterói 24230-321, Brazil; fabiotargino@id.uff.br (F.J.T.); joannadnr@gmail.com (J.D.d.N.R.); julia_simoes@hotmail.com (J.S.S.); micheliferreira@id.uff.br (M.d.S.F.); sergiomano@id.uff.br (S.B.M.)

[2] Faculty of Pharmacy, Universidade Federal do Rio de Janeiro (UFRJ), Rio de Janeiro 21941-580, Brazil; carlasilvacarneiro@gmail.com

[3] Centro de Pesquisa e Preservação de Mamíferos Aquáticos (CPPMA), Eletronorte, Presidente Figueiredo 69736-000, Brazil; stellamarislazz@gmail.com (S.M.L.); alineramos.sza@gmail.com (A.R.S.)

[4] Institute of Biological Sciences, Universidade Federal do Amazonas (UFAM), Manaus 69077-000, Brazil

[*] Correspondence: etmarsico@id.uff.br

Abstract: Researchers recognize the silent, negative and deleterious effects caused by mercury pollution in gold mining areas. Freshwater turtles are culturally part of the diet of riverside populations in the Amazon region and this area presents mercury (Hg) pollution issues mainly due to gold mining activities. Thus, this research aimed to evaluate the total mercury (THg) content in the different organs of Amazonian giant river turtle (*Podocnemis expansa*) and carry out a human health risk assessment associated with the consumption of these animals. This study was conducted in the Vila Balbina, municipality of Presidente Figueiredo, state of Amazonas, Brazil. Skin (n = 28), muscle (n = 19) and brain (n = 2) samples were analyzed by Atomic Absorption Spectrometry (TDA-AAS) and a DMA-80™ mercury analyzer was used for the total mercury determinations. The average values found for THg in the skin, muscle and brain samples were, respectively, 0.1045 mg·kg^{-1}, 0.1092 mg·kg^{-1} and 0.0601 mg·kg^{-1}. Thus, THg was observed even though the *P. expansa* were kept in captivity, possibly due to previous contamination by air, water and food. The Hazard Quotient (HQ) was calculated considering a 9.07 g·day^{-1} intake dose of *P. expansa* and the consumption of turtles once a week showed an HQ = 2.45, which may cause long-term injuries to human health. Although the muscle concentrations were below the maximum limit established by the World Health Organization (WHO) and Brazilian regulatory agencies, it is important to evaluate consumption factors such as amount ingested, frequency and animal gender, which may cause a potential risk to regular consumers due to mercury bioaccumulation. The WHO may consider various aspects in order to warn the Amazon population about the severity and silent hazard of this metal, especially due to the importance of this matrix in the region. This region urgently needs government actions to inhibit clandestine mining and to prevent future serious, chronic health problems of the entire population.

Keywords: public health; food safety; environmental pollution; trace elements; biomonitoring; freshwater toxicology

Citation: Targino, F.J.; Ribeiro, J.D.d.N.; Simões, J.S.; Carneiro, C.S.; Lazzarini, S.M.; Souza, A.R.; Ferreira, M.d.S.; Mano, S.B.; Mársico, E.T. Total Mercury Content in the Tissues of Freshwater Chelonium (*Podocnemis expansa*) and a Human Health Risk Assessment for the Amazon Population in Brazil. *Int. J. Environ. Res. Public Health* **2023**, *20*, 6489. https://doi.org/10.3390/ijerph20156489

Academic Editors: Roel Vermeulen, Lauren Petrick and Maaike van Gerwen

Received: 19 April 2023
Revised: 21 June 2023
Accepted: 29 June 2023
Published: 1 August 2023

1. Introduction

Mercury (Hg) is a toxic trace element found in the Amazonian ecosystem due to its use in the amalgamation stage of gold extraction [1–3]. This toxic contaminant requires public health attention due to its toxicity, high absorption and low excretion after ingestion [4]. The toxic element is present in the aquatic environment where it is methylated, bioaccumulates and biomagnifies, providing the possibility for incorporation by human beings, who are at the top of the food chain [5]. Metal pollution, although less perceptible compared to other forms of pollution or contamination, can have profound and far-reaching impacts on ecosystems and human health [6]. Researchers have conducted extensive studies on mercury contamination in water, sediment and aquatic fauna across various countries and regions, including China [7], India [8], Italy [9] and Japan [10], as well as the Peruvian [11] and Brazilian Amazon [2].

Mercury can have toxic effects on tissues and systems such as the nervous, digestive and immune systems, and organs such as the lungs, kidneys, skin and eyes [12]. The organic form, methylmercury (MeHg), is a potent neurotoxin inducing oxidative stress activity, neuroinflammatory action and changes in heart rate [13]. Mercury is eliminated into the environment in the inorganic form and may undergo sediment methylation by physical, chemical and biological mechanisms, turning into MeHg, the most toxic form of this metal [14,15]. Due to its geochemical cycle, mercury can be found in areas far away from its mobilized source [16]. Inorganic Hg reaches the atmosphere, and, by precipitation, is deposited on the bottom of rivers, lakes and other watercourses [17]. Furthermore, humans may demonstrate pathognomonic effects [18].

To protect human health and limit dietary exposure to toxic metals, international agencies and Brazilian regulations have provided guidelines for the human intake of trace elements [19–21]. According to the World Health Organization (WHO), to estimate the amount of mercury per unit of body weight that can be ingested over a lifetime without health risks, the joint FAO/WHO Expert Committee recommended Provisional Tolerable Weekly Intakes (PTWIs) for MeHg of 1.6 $mg{\cdot}kg^{-1}$ body weight [22].

In recent years, particularly from 2016 to 2022, there has been a significant increase in illegal mining activity in the Brazilian Amazon, with a focus on indigenous territories. According to data from the Instituto Nacional de Pesquisa Espacial (INPE), there has been a staggering 787% rise in illegal mining activity within indigenous territories during this period [23]. This surge can be attributed to various factors, including the encouragement of such activities, inadequate monitoring and enforcement measures and the weakening of environmental regulations in the country [24,25]. Economic crisis and unemployment during the COVID-19 pandemic forced thousands of people to work in illegal mining in the Amazon [25].

The Amazon population's diet varies according to the season, but aquatic animals are the primary protein source for most riverside people, especially children. The consumption of turtles by natives and settlers is historically and culturally developed by Amazon communities [26], and the Amazonian giant river turtle (*Podocnemis expansa)* is the most consumed species. After colonization, chelonian commerce improved and the local population could explore the meat, eggs, oil and carapace as an additional financial source. Due to their long lifespan, aquatic turtles can accumulate Hg over extended periods and therefore it could be of interest to monitor environmental contamination [26,27]. The history of gold mining and hydroelectric power plants has influenced Hg availability in the environment and living organisms [5,28,29].

Furthermore, the consumption of chelonium meat from illegal hunting can lead to environmental imbalance, given the important ecological role turtles play in the Amazon's biodiversity [30]. Additionally, consumers are at risk of exposure to infectious agents, such as Enterobacteriaceae, notably *Salmonella* sp. [31,32]. Therefore, it is essential to develop research and environmental biomonitoring programs to measure the Hg concentration in *P. expansa* and assess the risk of exposure to this population. Thus, this research aimed to evaluate the presence of total mercury (THg) in different organs of the *P. expansa* chelonian species and carry out a human health risk assessment associated with the consumption of these animals.

2. Materials and Methods

2.1. Experimental Area

The location studied was Vila Balbina, in the municipality of Presidente Figueiredo, located in the North of Amazonas State (latitude 1°55′07.0 S and longitude 59°28′05.9 W), Brazil. This location was considered for study due to the extensive mining activity that has occurred there for decades. The *P. expansa* samples (Figure 1) were obtained from the Centro de Preservação e Pesquisa de Quelônios Aquáticos (CPPQA) (Figure 2) located near the River Uatumã mouth, a tributary of the black waters inserted in an area with anthropogenic actions such as metal extractions (niobium-Nb and tin-Sn).

Figure 1. Amazonian giant river turtle (*Podocnemis expansa*) in the herd of the Centro de Preservação e Pesquisa de Quelônios Aquáticos (CPPQA), Vila Balbina, Presidente Figueiredo municipality, Amazonas state, Brazil.

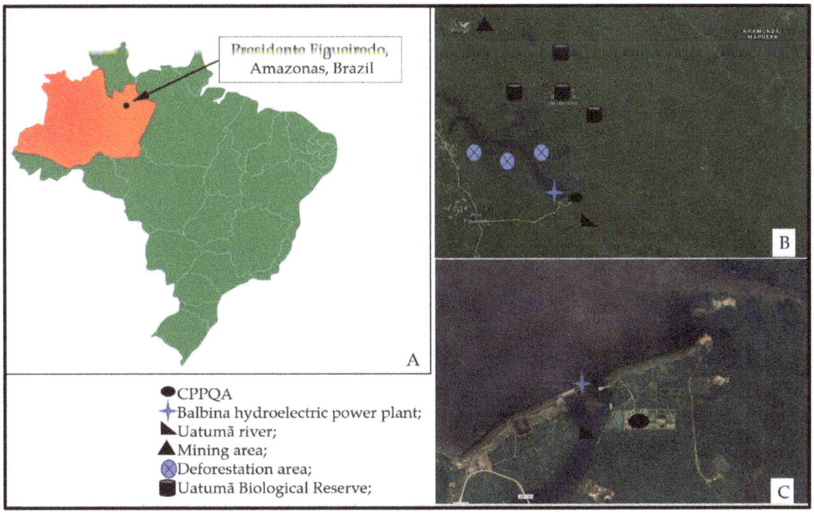

Figure 2. Location of sample collection in Centro de Preservação e Pesquisa de Quelônios Aquáticos (CPPQA), Vila Balbina, Presidente Figueiredo municipality, Amazonas state, Brazil. (**A**): Location in Brazil; Red: Amazonas state; Arrow and dot black: Presidente Figueiredo municipality; (**B**): 200 miles; (**C**): 200 m.

2.2. Chelonian Sampling

Twenty-eight samples of *P. expansa* were collected in June 2018 from the CPPQA captivity area, and the carapace length of each animal was measured. The evaluated animals (Figure 1) lived in confinement, kept in captivity at the CPPQA. These animals consisted of breeding males and female matrices, which are part of the restocking program on the beaches of the Uatumã river. Despite being kept in captivity, the food supply and water used in their enclosures and macrophyte production tanks came from the Uatumã river, downstream from the Balbina Hydroelectric Power Plant.

Skin and muscle samples (~2 cm^3) were obtained by biopsy punch (3 mm) from the pelvic member using a previous local anesthesia. Considering the difficulty of animal handling, 28 skin samples and 19 muscle samples were obtained, but only two brain samples were collected by the CPPQA when the animals died.

The animals underwent a biopsy using a punch, and they received subsequent treatment from veterinary doctors and research veterinarians at CPPQA. The treatment involved applying povidone-iodine and a healing ointment containing gentamicin, sulfanilamide, urea, vitamin A and excipients. This treatment regimen was administered daily for seven days to the wound area. The animals were housed in their original enclosure, which was in a designated area with net protection. This protocol was approved by the Comitê de Ética na Utilização de Animais (CEUA) of Universidade Federal Fluminense (UFF) under protocol number: 1012300319.

The samples were frozen (-18 °C $\pm$ 2 °C), stored in polypropylene packing and flown to a chemical laboratory for analysis at the Federal Fluminense University. The animals were gender-identified by observation of the reproductive organ dimorphism [33].

2.3. Chelonian Consumption Data

Ninety-one Balbina residents answered a questionnaire about the consumption of chelonians in the area. A semi-structured questionnaire was administered in an individual interview format. The inclusion criteria required participants to be over eighteen years of age and residents of Vila Balbina, located in the municipality of Presidente Figueiredo. The questionnaire consisted of two sections: (1). Socioeconomic factors (including gender, age, occupation, history of involvement in mining activities and education level); (2). Chelonian consumption patterns (including consumption of chelonian meat and turtle eggs, preferred species for consumption, frequency of consumption and annual amount of consumption of chelonian meat and eggs).

2.4. Instrumentation

The Direct mercury analyzer DMA80™ (Milestone Inc., Shelton, CT, USA) involves thermal decomposition, catalytic conversion, amalgamation and atomic absorption spectrophotometry. The heating stages are implemented to first dry and then thermally decompose the sample in a quartz cuvette, and a flow of oxygen carries the decomposition products through a hot catalyst bed. The mercury content is reduced to Hg (0) and then carried along with the reaction gases to an amalgamator, where the mercury is selectively trapped. Subsequently, the amalgamator is heated and releases the trapped mercury to the single beam, fixed wavelength atomic absorption spectrophotometer. The absorbance is measured at 253.7 nm as a function of the mercury content. All non-mercury vapors and decomposition products are flushed from the system by the continuous flow of gas (Milestone Inc., Shelton, CT, USA).

2.5. Analytical Procedures

The analyses were carried out according to the analytical procedures of the US Environmental Protection Agency (US EPA) 7473 (2007) [34]. An aliquot of 0.210 g of the chelonian sample was directly and accurately weighed into a quartz cuvette. The DMA80™ used does not require pretreated samples. According to the WHO/FAO recommendations [22], the samples were weighed on a wet weight basis (wwb).

The samples were submitted to the following heating stages: starting temperature of 200 ± 10 °C; drying stage for 60 s at 250 ± 10 °C; thermal decomposition at 650 ± 10 °C for 120 s and finally 100 s for the analytical reading and cooling. An amalgamator selectively trapped the Hg, and the oxygen flow removed any remaining gases or decomposition products (24 s); the amalgamator was rapidly heated (12 s), releasing the Hg vapor. The atomic absorbance was read at 253.7 nm.

The accuracy of the analytical procedure was evaluated by analyzing the Certified Reference Material (CRM) of DORM-2 (dogfish muscle tissue) obtained from the National Research Council Canada (NRCC). The CRM control samples were analyzed without any treatment, and non-analyzed aliquots were used to determine the moisture content according to the certificate instructions for analysis. The certified material concentration was determined in triplicate and reached 4.470 $\mu g \cdot g^{-1}$. The mean recovery was 94%.

2.6. Limits of Detection and Quantification

The limits of detection (LOD) and quantification (LOQ) were calculated from the mean concentration and relative standard deviation of ten blanks and the slope of the analytical curve (1.5×10^{-6} $mg \cdot kg^{-1}$ of THg). These limits were calculated as 3 and 10 times the standard deviation of 10 blanks and multiplied by the dilution factor used for sample preparation (1 g of sample/25 mL) using the following equations:

$$\text{LOD} = 3 \times \text{s and LOQ} = 10 \times \text{s}$$

where: s = the average standard deviation in units of instrumental response, of at least ten independent determinations of analytical blanks.

The limits of detection (LOD) and quantification (LOQ) were 94 $\mu g \cdot kg^{-1}$ and 314 $\mu g \cdot kg^{-1}$, respectively. The recovery value mean was 84.93%. The value of the LOQ was confirmed using ten independent replicates.

2.7. Human Health Risk Assessment

The human health risk was assessed by the US EPA/WHO method (1989) [35], assessing the exposure and characterizing the risk associated with the consumption of *P. expansa*. The exposure assessment was represented by I (Intake dose), which involves the description of the nature and size of the populations studied exposed to the chemical agent (Hg in this study), and the magnitude and duration of the exposure, calculated according to the following equation:

$$\text{I} = (\text{CF} \times \text{IR} \times \text{FI} \times \text{EF} \times \text{ED})/(\text{BW} \times \text{AT})$$

The Intake dose (I) was defined by the ratio between the sample and the estimated weight of the exposed population. The following factors were taken into account: the chemical concentration (CF) in *P. expansa* muscle (male, female and total mean); the intake rate (IR), taken as the average daily consumption of turtle of 9.07 $g \cdot person \cdot day^{-1}$ as estimated by Isaac et al. (2015) [36]; the fraction of food ingested (FI) obtained from a contaminated source (considered as the total food fraction in this study) [35,37]; the exposure frequency (EF) consumption considered annually, monthly and weekly; the exposure duration (ED), established from the average life expectancy for Brazilian people of 75.8 years [38]; body weight (BW), considered as 70 kg [35]; and average time (AT) as the annual exposure frequency by individuals during their life expectancy (AT = ED × 365 days).

For the risk characterization, the intake dose (I) was compared with the RfD (Reference dose of Hg), which, according to US EPA (2011) [20], is 3.0×10^{-4} $mg \cdot kg \cdot day^{-1}$, calculated using the following equation.

$$\text{HQ} = \text{I}/\text{RfD}$$

HQ greater than 1 (HQ > 1) order of magnitude indicating health risks.

2.8. Statistical Analysis

The normality of the data was confirmed with a Shapiro–Wilk test ($p > 0.05$). The one-way ANOVA and Tukey's test were applied using XLSTAT™ 2014.1 (Addinsoft Inc., New York, NY, USA) with a significance level of 95% ($p < 0.05$) to establish any significant differences between the means. A correlation analysis was applied, and Pearson's linear coefficient was calculated amongst the muscle and skin tissues. All analyses were carried out in triplicate.

3. Results and Discussion

3.1. Results

3.1.1. Total Mercury (THg) Content

The average carapace length for females was 74.42 cm ($n = 14$), and it was 47.18 cm ($n = 16$) for males. The present results show that although the female carapaces were larger than the males, higher THg concentrations were not observed in the female animals. The muscle tissue samples showed the highest concentrations of THg (0.6929 mg·kg^{-1}) as compared to the maximum values in the skin of (0.5457 mg·kg^{-1}) and in the brain of (0.0610 mg·kg^{-1}). The mean THg concentrations found in the skin ($n = 28$), muscle ($n = 19$) and brain ($n = 2$) were, respectively, 0.1045 mg·kg^{-1}, 0.1092 mg·kg^{-1} and 0.0601 mg·kg^{-1}. The results are shown in Table 1.

Table 1. Mean, standard deviation, minimum and maximum of total mercury (THg) concentrations in different tissues of Amazonian giant river turtle (*Podocnemis expansa*).

Tissues	Gender	*n*	Carapace (cm)	THg (mg·kg^{-1})	
				Mean ± SD	Min.–Max.
Skin	F	12	75.00	0.0817 ± 0.0165 [a]	ND–0.0406
	M	16	46.64	0.1215 ± 0.1401 [a]	ND–0.5457
Total Skin		28		0.1044 ± 0.2160 [a]	ND–0.5457
Muscle	F	5	74.83	0.0152 ± 0.1019 [b]	ND–0.6929
	M	14	47.18	0.1427 ± 0.1132 [a]	ND–0.5677
Total Muscle		19		0.1092 ± 0.1710 [a]	ND–0.6929
Brain *	F	2	77.00	0.0600 ± 0.0009	0.0591–0.0610

ND: not detected; *n* = number of samples; * Considering scarce samples statistics were not applied. [a,b] Means followed by the same letter in the same column do not statistically differ by the Tuckey test at 5% probability ($p < 0.05$).

Regarding the tissues, the muscle should be considered the most important because of its higher accumulation of THg in *P. expansa* compared to the skin, but the present research found no significant difference ($p < 0.05$). However, a notable difference was detected in the tissue concentrations, considering a minimum of "not detected" in the muscle and skin and a maximum of 0.6929 mg·kg^{-1} in the muscle. The average values found in the male samples demonstrated greater THg bioaccumulation in the male muscle than in the female muscle, and a significant difference ($p < 0.05$) was observed between the female muscle samples and the samples of the other tissues.

A positive correlation, regardless of gender, was observed between the skin and muscle THg concentrations (r = 0.87). The present research did not find a correlation between carapace length and muscle or skin THg concentrations for males (r = 0.011) or of females (r = 0.15), possibly due to the high variation between the tissue concentrations in the animal diet and the diversity of their origins. Furthermore, regarding the coefficient of determination (R^2), which measures the proportion in which the variables are related, a slight difference was observed when comparing tissues (skin and muscles) and sex, with $R^2 = 0.50$ for males and 0.43 for females.

3.1.2. Human Health Risk Assessment

Amongst those interviewed, 18% (15/91) consumed chelonian meat, especially tracajá (*Podocnemis unifilis*) and *P. expansa*. However, only 8% (7/91) reported consuming the eggs. Regarding the consumption data used to calculate the HQ, 8/91 ate chelonian meat at least once a year, 5/91 ate it once a month, and 2/91 ate it once a week during the egg-laying months (September to December). The people interviewed confirmed that these animals and eggs were caught in Vila Balbina, Presidente Figueiredo, Amazonas, Brazil.

The Hazard Quotient (HQ) was calculated for the different consumption intervals (annually, monthly and weekly) with respect to the THg concentrations in male plus female *P. expansa* and the mean THg values for both genders separately. Table 2 shows the HQ results obtained for the human risk assessment of *P. expansa* muscle.

Table 2. Human health risk assessment for the meat of Amazonian giant river turtle (*Podocnemis expansa*) by gender and ingestion frequency.

Frequency Consumption	Gender of *P. expansa*	Daily Intake (I) (mg/kg/day)	Hazard Quotient (HQ)
Annually	Male	1.85×10^{-5}	0.061
	Female	2.00×10^{-6}	0.006
	Mean	1.41×10^{-5}	0.047
Monthly	Male	2.21×10^{-4}	0.739
	Female	2.36×10^{-5}	0.078
	Mean	1.70×10^{-4}	0.565
Weekly	Male	9.61×10^{-4}	3.204
	Female	1.02×10^{-4}	0.341
	Mean	7.35×10^{-4}	2.452

3.2. *Discussion*

3.2.1. THg Content

The total mercury content in the different turtle tissues is related to Hg kinetics. Researchers have indicated that the total mercury content in freshwater fish is related to the species and habitat evaluated [5,39,40]. Brazilian studies have reported a THg content in Amazon fish muscle of over 0.500 mg·kg^{-1}, above the limit recommended by World Health Organization [19,41,42]. Although there are few studies with Amazonian freshwater chelonians, it is essential to consider the influence of the environment on the contents of trace elements.

The *P. unifilis* caught in the lower River Xingu accumulated lower concentrations of THg throughout its life when compared to other species of Amazon turtles. Thus, the Amazon basin rivers have mercury in a bioavailable form that organisms can accumulate. Recent studies have shown that *P. unifilis* and *P. expansa* are predominantly herbivorous, but may consume insects, crustaceans and mollusks [43,44].

The present authors observed that some studies indicate a difference in feeding preference between genders, with the females consuming more fruits and seeds and the males more stalks and shoots [45]. Thus, it was postulated that the lack of significant difference in Hg concentrations between males and females in this study could be attributed to the absence of variation in trophic levels among species gender [46]. Due to the maintenance of the animals analyzed in this research in captivity, the diet of males and females is similar, consisting of offering vegetables, fruits and macrophytes grown in the CPPQA itself. It assumes that in free-living animals at the mouth of the Uatumã river, the levels of Hg can be significantly different in males and females, due to eating habits, the period of prolonged fasting that females have before the laying phase and the transfer of inorganic chemical trace elements to eggs. Research has identified a maternal transfer of metals (essential or not) to the eggs in tests with Testudines nesting eggs, triggering genotoxic effects in the hatchlings [47].

Research carried out in the Amazon evaluated non-lethal and non-invasive ways of collecting samples from Amazonian chelonian species for Hg monitoring [48]. They found no significant differences between the carapace and internal organs of *Podocnemis erythrocephala* and *Podocnemis sextuberculata*.

Research conducted in a sustainable development reserve in Amazonas, Brazil evaluated THg concentration in *P. unifilis*. The study found an average THg concentration mean of 0.011 mg·kg^{-1} in six muscle samples [49].

Different THg concentrations may be observed between males and females due to changes in eating habits [50] and the elimination of Hg via the eggs during the reproductive period [51]. However, in contrast, the present study diverged from this observation and the THg concentration for males did not significantly differ from that of the females. This study found 0.0624 mg·kg^{-1} THg in *P. expansa* muscle samples, 0.432 mg·kg^{-1} THg in *Chelus fimbriatus* muscle and 0.033 mg·kg^{-1} THg in *P. erythrocephala* muscle [50,52]. These animals present distinct eating behaviors between the genders and life periods. Small amounts of fish and crab were observed in studies that analyzed the chelonian gastrointestinal content, but their primary food source is seeds and fruits, especially those from riverside trees and flooded forests. However, the literature also reported that juvenile *P. expansa*, in captivity, prefer to eat fish, insects and other meat [52,53].

3.2.2. Mechanism for THg Distribution

Hg is a metal that occurs naturally in the Amazonian environment in small quantities but constitutes an important pollutant derived from anthropogenic activities [40,54]. Once released into the environment, Hg is oxidized and will be available to be methylated, producing methylmercury (MeHg), a neurotoxic and teratogenic substance. MeHg undergoes bioaccumulation in aquatic organisms and is biomagnified through the trophic chain, constituting a source of exposure to man via the food chain, mainly due to fish consumption [18]. Traditional populations of fishermen, indigenous and riverside, are more exposed to this type of contamination due to the higher frequency of fish intake [55].

Elemental Hg in reptiles is well-absorbed by inhalation through the lungs and through the skin and is absorbed little by the gastrointestinal tract [27]. The interaction between Hg and the skin has relationships with affinity between Hg and the sulphide group of proteins [56,57]. In a study conducted by Schneider et al. (2022) that evaluated skin samples of Natator depressus collected off the coast of Australia, which present likely eating behaviors, Hg concentrations were 0.017 mg·kg^{-1}; in this study, the average concentrations of HgT in *P. expansa* skin were 0.1045 mg·kg^{-1} [58].

3.2.3. THg Bioaccumulation

Toxic metals entering the river body would be absorbed in sediments and migrate to the sediment and biota through a biological and chemical process [59,60], so researchers began to pay more attention to pollution by toxic metals [61,62].

The bioaccumulation of toxic metals by reptiles results from surface contact with the water and food chain [47,63,64], and the concentrations depend on the environmental levels of trace metals in the habitat, since they are absorbed and stored in the tissues [65]. Researchers have shown that the mercury concentrations in the muscles of freshwater animals varies widely depending on the capture location [59,66].

Captive breeding possibly contributed to reduced mercury concentrations compared to the reported data [57]. The researchers evaluated the THg concentrations in the muscles and keratin samples from the carapaces of free-living *P. expansa* in the Amazonas region and found THg concentrations in the muscle of 0.1 mg·kg^{-1} and in the carapaces of 2.8 mg·kg^{-1}. However, the diet and food consumption were human-controlled, thus interfering with the Hg kinetics.

In addition, the muscle THg concentration is related to eating habits, which interfere with the biomagnification of this trace element. Previous research regarding the THg concentrations in Amazon chelonian muscles from different species found concentrations of 0.106 mg·kg^{-1} in *P. expansa*, corroborating the results of the present research [51].

C. fimbriatus, a predatory carnivorous animal, is adapted for collecting fish and other animals in the aquatic environment [67] via a well-developed suction system and head movements toward the prey [68]. In contrast, when raised in captivity, young *P. expansa* prefer meat and fish [53], which may explain the higher average THg concentrations found in *C. fimbriatus* specimens compared to *P. expansa*.

Different Amazonian reptiles have been used to monitor THg bioaccumulation. Researchers evaluated this trace metal in various *C. crocodilus* tissues, this being an alligator from the river Purus (Amazonas state), and found a THg concentration of 0.389 mg·kg^{-1} [69]. Correia et al. (2014) evaluated THg in alligators from the Mamirauá reserve (Tefé, Amazonas State) and found THg concentrations of 0.407 mg·kg^{-1}; these animals showed predatory eating behavior [70]. Thus, the THg concentrations represent the biomagnification principle, which is associated with the consumption of fish, chelonians, small mammals and birds.

3.2.4. Human Health Risk Assessment

Although the established Provisional Tolerable Weekly Intake (PTWI) limits [12,19,20] suggest safe mercury intake values, researchers and global health organizations highlight that this metal is dangerous at any concentration due to its bioaccumulation characteristics, and thus there is no healthy mercury level for humans [35]. Bioavailability can be affected by controlling the biotic and abiotic factors. The most critical tool that can be used to reduce mercury in the environment and in animals is the prevention of pollution, which is considered a long-term goal [71]. In the state of Amazon, Balbina village (Vila Balbina) is highly influenced by mining and damming, both of which are anthropogenic activities on the River Uatumã causing environmental impacts and interfering with the Hg biological cycle [28].

The maximum tolerable limit of 0.5 mg MeHg·kg^{-1} for fish (except for predatory fish) [22] is established for seafood based on a weekly consumption. The Brazilian regulatory agencies also set the acceptable THg level for non-predatory fish, crustaceans, mollusks, cephalopods and bivalve mollusks at 0.5 mg·kg^{-1} and at 1.0 mg·kg^{-1} for predatory fish [21]. Although regulatory agencies have not established a THg limit for chelonians, it is important to note that, when comparing to the limit set for non-predatory fish, the THg concentration detected in the muscle of *P. expansa* (0.1045 mg·kg^{-1}) remained below the legal thresholds. However, it should be highlighted that this study identified two (2/19) *P. expansa* with THg concentrations exceeding 0.5 mg·kg^{-1}, which surpasses the levels established for non-predatory fish.

A Hazard Quotient (HQ) above 1 (HQ > 1) represents a potential health risk to consumers [20]. To determine the HQ for the consumption of *P. expansa*, it is crucial to consider the weekly consumption of the male categories. When considering the mean THg concentration in males and females for monthly or annual consumption, the HQ was <1 in all categories. Research carried out by Correia et al. (2014) [70] evaluated the HQ for alligator consumption by riverside populations and determined a value of 7.19, considering this quotient due to the higher Hg concentrations in these animals as compared to *P. expansa*, since they show predatory feeding behavior and are piscivores.

One of the characteristics of the consumption of turtles in the Amazon area is its seasonality, with the hunting of turtles and other animals for consumption being more intense in preserved and isolated areas [35]. In the present study, 18% (15/91) of the chelonian meat consumers interviewed considered it rare and occasional, only occurring in months with greater availability. Turtle consumption is taboo among the Amazonian population due to preservation practices of these species and constant supervision by environmental agencies [26].

The proximity and coexistence of Vila Balbina residents with environmental control bodies such as the Instituto Chico Mendes de Conservação da Biodiversidade (ICMBio) and the Instituto Brasileiro do Meio Ambiente e dos Recursos Naturais Renováveis (IBAMA) may have influenced the responses regarding the consumption of chelonian in the questionnaire administered to residents.

Furthermore, considering the accessibility of the study area, which is close to the state capital, Manaus (distance 107 km), it is reasonable to assume that consumption levels may vary in more isolated regions, such as indigenous, riverside and rural communities, where access to alternative sources of animal protein production is limited [26,71].

Although in this research only a small percentage of individuals reported consuming turtle meat and eggs, the true consumption might be higher. Female specimens are more consumed since they are easier to capture, their meat is more appreciated and the yield is higher [72]. The increased ease of capturing female *P. expansa* is associated with the time of egg-laying when these animals are more vulnerable. During this period, they leave the aquatic environment and deposit their eggs in nests on riverbeds. They are often captured, and the collected eggs, aside from being valued for their flavor, also hold medicinal purposes in ethnoknowledge, with claims of aphrodisiac properties.

Little information is available on the risk evaluation associated with chelonian consumption. Studies emphasized that total mercury concentrations in river fish from Malaysia and Ghana were influenced by the habitat, feeding habits, species and body length, and the increased HQ was due to the nearby gold mining area [73]. Brazilian researchers have reported HQ values ranging from 1.5 to 28.5 in the river Amazon and concluded that the elevated risk was associated with artisanal gold mining areas, where people also catch fish [1]. They suggested population risks at almost the sampling points due to consumption. It appears that there is considerable contamination of living beings that share the same environment with turtles due to their consumption without considering food safety [74]. Hence, determining the allowable limit for the consumption of these chelonians (daily, weekly or monthly) is crucial to preserve public health.

4. Conclusions

P. expansa chelonians, kept in captivity in Vila Balbina (Balbina village) at the mouth of the River Uatumã, accumulated THg in the muscles, skin and brain, with higher concentrations in the skin and muscle. Regarding gender differences, the males showed higher THg concentrations than the females. The concentration found in the *P. expansa* muscles were generally lower than the limits established for fish consumption.

The mercury content is mainly associated with animals' diets and is considered to accumulate in organisms. It is essential to notice that depending on the consumption characteristics, the frequency and animals' gender may be a risk to public health. Although the THg concentrations are below the acute metal toxicity level, the bioaccumulation characteristic deserves attention because of its long-term effects on human health.

Research is essential to assess the risk of Hg exposure resulting from the consumption of game meat, including chelonians meat, which holds a significant place in the diets of indigenous and riverside communities. It is imperative to evaluate mercury contamination in both wild-caught game animals and those raised in captivity through commercial breeding for meat production. Furthermore, analyzing a range of game animals from diverse origins and studying the consumption habits of different populations will contribute to understanding the significance of these meats as a potential source of Hg exposure.

A comprehensive study on this trace element in the Amazonian area is recommended to determine allowable consumption limits and identify the location most involved with contaminants. There are clear benefits and risks from the consumption of fishery products, so the population should be provided with enough information to maximize the positive health benefits while minimizing the risks from contaminants. General and scientific knowledge is scarce, reinforcing the importance of developing communication tools involving scientists, health professionals, regulators and the general public.

In addition to the need for further studies on the presence of Hg in the meat of chelonians from different species, locations and sexes, it is also important to investigate the Hg content in eggs. Furthermore, research on the consumption of these products is necessary to conduct risk assessments. In this study, *P. expansa* eggs were not collected and analyzed due to the seasonal nature of egg-laying, and the availability of this material was limited in the researched area.

Efforts should be established to address illegal mining in Brazil, especially in the Amazon region. These efforts include establishing controls in unlawful mining areas, increased surveillance of borders and indigenous territories, and strengthening environmental enforcement agencies such as ICMBio and IBAMA. Additionally, there is a need for continuous monitoring of mercury concentrations in fish and other food sources, along with regular risk assessments, to prevent the consumption of foods with high mercury levels.

Author Contributions: F.J.T.: methodology, validation, formal analysis, investigation, data curation, writing—original draft, visualization. J.D.d.N.R.: methodology, validation, formal analysis, investigation, data curation, visualization. J.S.S.: writing—original draft. C.S.C.: formal analysis. S.M.L.: supervision, investigation. A.R.S.: investigation. M.d.S.F.: conceptualization, resources, supervision, project administration. S.B.M.: writing—original draft. E.T.M.: resources, writing—original draft, visualization, supervision, project administration. All authors have read and agreed to the published version of the manuscript.

Funding: This research was financed in part by the Coordenação de Aperfeiçoamento de Pessoal de Nível Superior—Brazil (CAPES) (financing code 001), Brazil and by the Conselho Nacional de Desenvolvimento Científico e Tecnológico (CNPQ), Brazil (Grant number 311974/2019–8).

Institutional Review Board Statement: This work was evaluated and accepted by the UFF Human Ethics Committee (number 2899092), SisBio (number 63587-3) and the UFF Animal Ethics Committee (number 1012300319).

Informed Consent Statement: Informed consent was obtained from all subjects involved in the study.

Data Availability Statement: All data generated during analyzed this study are included in this article.

Conflicts of Interest: The authors declare no conflict of interest.

References

1. Castilhos, Z.; Rodrigues Filho, S.; Cesar, R.; Rodrigues, A.P.; Villas Bôas, R.; de Jesus, I.; Lima, M.; Faial, K.; Miranda, A.; Brabo, E.; et al. Human exposure and risk assessment associated with mercury contamination in artisanal gold mining areas in the Brazilian Amazon. *Environ. Sci. Pollut. Res.* **2015**, *22*, 11255–11264. [CrossRef] [PubMed]
2. Lino, A.S.; Kasper, D.; Guida, Y.S.; Malm, T.O. Total and methyl mercury distribution in water, sediment, plankton, and fish along the Tapajós River basin in the Brazilian Amazon. *Chemosphere* **2019**, *235*, 690–700. [CrossRef] [PubMed]
3. Pinto, L.C.M.; Dórea, J.G.; Bernardi, J.V.E.; Gomes, L.F. Mapping the Evolution of Mercury (Hg) Research in the Amazon (1991–2017): A Scientometric Analysis. *Int. J. Environ. Res. Public Health* **2019**, *16*, 1111. [CrossRef] [PubMed]
4. Okpala, C.O.R.; Sardo, G.; Vitale, S.; Bono, G.; Arukwe, A. Hazardous properties and toxicological update of mercury: From fish food to human health safety perspective. *Food Sci. Nutr.* **2017**, *58*, 1986–2001. [CrossRef] [PubMed]
5. Kasper, D.; Forsberg, B.R.; Amaral, J.H.F.; Leitão, R.P.; Py-Daniel, S.S.; Bastos, W.R.; Malm, O. Reservoir stratification affects methylmercury levels in River Water, Plankton, and Fish Downstream from Balbina Hydroelectric Dam, Amazonas, Brazil. *Environ. Sci. Technol.* **2014**, *48*, 1032–1040. [CrossRef]
6. Vardhan, K.H.; Kumar, O.S.; Panda, R.C. A review on heavy metal pollution, toxicity and remedial measures: Current trends and future perspectives. *J. Mol. Liq.* **2019**, *290*, 111197. [CrossRef]
7. Xu, Z.; Wu, S.; Christie, P.; Gao, X.; Xu, J.; Xu, S.; Liang, P. Impacts of estuarine dissolved organic matter and suspended particles from fish farming on the biogeochemical cycling of mercury in Zhoushan island, eastern China Sea. *Sci. Total Environ.* **2020**, *705*, 135921. [CrossRef]
8. Kumari, P.; Chowdhury, A.; Maiti, S.K. Assesment of heavy metal in the water, sediment, and two edible fish species of Jamshedpur Urban Agglomeration, India with special emphasis on human health risk. *Hum. Ecol. Risk Assess. Int. J.* **2018**, *24*, 1477–1500. [CrossRef]
9. Bettoso, N.; Pittaluga, F.; Predonzani, S.; Zanello, A.; Acquavita, A. Mercury Levels in Sediment, Water and Selected Organisms Collected in a Coastal Contaminated Environment: The Marano and Grado Lagoon (Northern Adriatic Sea, Italy). *Appl. Sci.* **2023**, *13*, 3064. [CrossRef]

10. Tomiyasu, T.; Shitauchi, R.; Kasai, D.; Kitagawa, T.; Kodamatani, H.; Kanzaki, R. The distribution and dynamics of residual mercury from the Chisso chemical plant in sediments of the Yatsushiro Sea, western Kyushu, Japan: Have recent sedimentations lowered surface mercury concentrations? *Environ. Sci. Pollut. Res.* **2023**, *30*, 72769–72781. [CrossRef]
11. Martinez, G.; McCord, S.A.; Driscoll, C.T.; Todorova, S.; Wu, S.; Araújo, J.F.; Vega, C.M.; Fernandez, L.E. Mercury Contamination in Riverine Sediments and Fish Associated with Artisanal and Small-Scale Gold Mining in Madre de Dios, Peru. *Int. J. Environ. Res. Public Health* **2018**, *15*, 1584. [CrossRef] [PubMed]
12. World Health Organization (WHO). *Elemental Mercury and Inorganic Mercury Compounds: Human Health Aspects; Concise International Chemical Assessment Document*; World Health Organization: Geneva, Switzerland, 2017; Volume 50.
13. Alexander, J.; Benford, D.; Boobis, A.; Bergman, Å. Scientific Opinion on Tetrabromobisphenol A (TBBPA) and its derivatives in food: EFSA Panel on Contaminants in the Food Chain (CONTAM). *EFSA J.* **2011**, *9*, 2477. [CrossRef]
14. Bonotto, D.M.; Wijesiri, B.; Vergotti, M.; Silveira, E.G.; Goonetilleke, A. Assessing mercury pollution in Amazon River tributaries using a Bayesian Network approach. *Ecotoxicol. Environ. Saf.* **2018**, *166*, 354–358. [CrossRef] [PubMed]
15. Vieira, M.; Bernardi, J.V.; Dórea, J.G.; Rocha, B.C.; Ribeiro, R.; Zara, L.F. Distribution and availability of mercury and methylmercury in different waters from the Rio Madeira Basin, Amazon. *Environ. Pollut.* **2018**, *235*, 771–779. [CrossRef] [PubMed]
16. Ma, M.; Du, H.; Wang, D. A New Perspective is Required to Understand the Role of Forest Ecosystems in Global Mercury Cycle: A Review. *Bull. Environ. Contam. Toxicol.* **2019**, *102*, 650–656. [CrossRef] [PubMed]
17. Beckers, F.; Rinklebe, J. Cycling of mercury in the environment: Sources, fate, and human health implications: A review. *Crit. Rev. Environ. Sci. Technol.* **2017**, *47*, 693–794. [CrossRef]
18. Kim, K.H.; Kabir, E.; Jahan, S.A. A review on the distribution of Hg in the environment and its human health impacts. *J. Hazard. Mater.* **2016**, *306*, 376–385. [CrossRef]
19. World Health Organization (WHO). *Guidence for Identifying Populations at Risk from Mercury Exposure*; World Health Organization: Geneva, Switzerland, 2008; p. 170.
20. United States Environmental Protection Agency (US EPA). Technical Support Document: National-Scale Mercury Risk Assessment Supporting the Appropriate and Necessary Finding for Coal- and Oil-Fired Electric Generating Unit 2011. Available online: https://www3.epa.gov/ttn/atw/utility/pro/hg_risk_tsd_3-17-11.pdf (accessed on 20 January 2023).
21. BRASIL. Ministério da Agricultura, Pecuária e Abastecimento (MAPA). Instrução Normativa nº. 160, de 1 de julho de 2022. Estabelece os Limites Máximos Toleráveis (LMT) de contaminantes em alimentos. Diário Oficial [da] República Federativa do Brasil, 6 de julho de 2022. Available online: https://www.gov.br/agricultura/pt-br/assuntos/inspecao/produtos-animal/plano-de-nacional-de-controle-de-residuos-e-contaminantes/1-in_160_2022_comp.pdf (accessed on 20 January 2023).
22. Food and Agriculture Organization of the United Nations (FAO). *Food Standards Programme Codex Committee on Contaminants in Foods 2016*; Food and Agriculture Organization/WHO: Roma, Italy, 2016.
23. Instituto Nacional de Pesquisa Espacial (INPE). Amazon Deforestation Monitoring Project (PRODES). 2022. Available online: http://terrabrasilis.dpi.inpe.br/downloads/ (accessed on 4 May 2023).
24. Gerson, J.R.; Szponar, N.; Zambrano, A.A.; Bergquist, B.; Broadbent, E.; Driscoll, C.T.; Erkenswick, G.; Evers, D.C.; Fernandez, L.E.; Hsu-Kim, H.; et al. Amazon forests capture high levels of atmospheric mercury pollution from artisanal gold mining. *Nat. Commun.* **2022**, *13*, 559. [CrossRef]
25. de Castro, F.F.; Goes, G.S.; Sena do Nascimento, J.A.; Tardin, M.M. Incidences of COVID-19 in major mining municipalities in the Brazilian Amazon: Economic impacts, risks and lessons. *Extr. Ind. Soc.* **2022**, *9*, 101033. [CrossRef]
26. Silva, A.L.; Begossi, A. Biodiversity, food consumption, and ecological niche dimension: A study case of the riverine populations from the Rio Negro, Amazonia, Brazil. *Environ. Dev. Sustain.* **2009**, *11*, 489–507. [CrossRef]
27. Grillitsch, B.; Schiesari, L. The ecotoxicology of metals in reptiles. In *Ecotoxicology Amphibians and Reptiles*, 2nd ed.; Sparling, D.W., Linder, G., Bishop, C.A., Krest, S., Eds.; CRC Press: Boca Raton, FL, USA, 2010.
28. Kehrig, H.A.; Malm, O.; Akagi, H.; Guimarães, J.R.D.; Torres, J.P.M. Methylmercury in Fish and Hair Samples from the Balbina Reservoir, Brazilian Amazon. *Environ. Res.* **1998**, *77*, 84–90. [CrossRef] [PubMed]
29. Silva Oliveira, C.A.; Kasper, D.; Sargentini Junior, E.; Bolson, M.A.; Torrezani, L.; Zara, L.F. Influence of environmental conditions on the mercury levels of the sediment along the Balbina Reservoir, Brazilian Amazon. *Environ. Monit. Assess.* **2021**, *193*, 50. [CrossRef] [PubMed]
30. Ellwanger, J.H.; Chies, J.A.B. Beyond diversity loss and climate change: Impacts of Amazon deforestation on infectious diseases and public health. *An. Acad. Bras. Ciênc.* **2020**, *92*, e20191375. [CrossRef] [PubMed]
31. Vliet, N.V.; Moreno, J.; Gómez, J.; Zhou, W.; Fa, J.E.; Golden, C.; Alves, R.R.N.; Nasi, R. Bushmeat and human health: Assessing the evidence in tropical and sub-tropical florests. *Ethnobiol. Conserv.* **2017**, *6*, 3. [CrossRef]
32. Carneiro, B.F. Isolamento e Identificação de *Salmonella* sp. e *Campylobacter* spp. em Amostras de Carne e Swab Cloacal, de Tartaruga da Amazônia (*Podocnemis expansa*). Master's Thesis, Universidade Federal de Goiás, Goiânia, Brazil, 2016.
33. Ceballos, C.P.; Hernández, O.; Morales-Betacourt, M.A.; Trujillo, F. *Podocnemis expansa*. In *Biología y Conservación de las Tortugas Continentales de Colombia*; Páez, V.P., Morales-Betacourt, M.A., Lasso, C.A., Castano-Mora, O.V., Bock, B., Eds.; Instituto de Investigación de Recursos Biológicos Alexander von Humboldt: Bogotá, Colombia, 2012; pp. 367–374.
34. United States Environmental Protection Agency (US EPA). Method 7473: Mercury in solids and Solutions by Thermal Decomposition, Amalgamation, and Atomic Absorption Spectro-Photometry. 2007. Available online: https://cfpub.epa.gov/si/si_public_record_report.cfm?Lab=NHSRC&dirEntryId=339252 (accessed on 20 January 2023).

35. *EPA/540/1-89/002*; Risk Assessment Guidance for Superfund: Human Health Evaluation Manual. United States Environmental Protection Agency (US EPA): Washington, DC, USA, 1989; p. 291.
36. Isaac, V.J.; Almeida, M.C.; Giarrizzo, T.; Deus, C.P.; Vale, R.; Klein, G.; Begossi, A. Food consumption as an indicator of the conservation of natural resources in riverine communities of the Brazilian Amazon. *An. Da Acad. Bras. De Ciências* **2015**, *87*, 2229–2242. [CrossRef]
37. Mieiro, C.L.; Pereira, M.E.; Coelho, J.P. Mercury bioaccessibility in fish and seafood: Effect of method, cooking and trofic level on consumption risk assessment. *Mar. Pollut. Bull.* **2022**, *179*, 113736. [CrossRef]
38. Instituto Brasileiro de Geografia e Estatística (IBGE). *Síntese de Indicadores Sociais: Uma Análise das Condições de vida da População Brasileira*; Coordenação de População e Indicadores Sociais: Rio de Janeiro, Brasil, 2016.
39. Yoshimoto, K.; Anh, H.T.V.; Yamamoto, A.; Koriyama, C.; Ishibashi, Y.; Tabata, M.; Nakano, A.; Yamamoto, M. Simple analysis of total mercury and methylmercury in seafood using heating vaporization atomic absorption spectrometry. *J. Toxicol. Sci.* **2016**, *41*, 489–500. [CrossRef]
40. Bastos, W.R.; Dorea, J.G.; Bernardi, J.V.E.; Manzatto, A.G.; Mussy, M.H.; Lauthartte, L.C.; Malm, O. Sex-related mercury bioaccumulation in fish from the Madeira River, Amazon. *Environ. Res.* **2015**, *144*, 73–80. [CrossRef]
41. Custódio, F.B.; Andrade, A.M.G.F.; Guidi, L.R.; Leal, C.A.G.; Gloria, M.B.A. Total mercury in commercial fishes and estimation of Brazilian dietary exposure to methylmercury. *J. Trace Elem. Med. Biol.* **2020**, *62*, 126641. [CrossRef]
42. Queiroz, J.V.; Vieira, J.C.S.; Bataglioli, I.C.; Bittarello, A.C.; Braga, C.P.; Oliveira, G.; Padilha, C.C.F.; Padilha, P.M. Total Mercury Determination in Muscle and Liver Tissue Samples from Brazilian Amazon Fish Using Slurry Sampling. *Biol. Trace Elem. Res.* **2018**, *184*, 517–522. [CrossRef]
43. Balensiefer, D.C.; Vogt, R.C. Diet of *Podocnemis unifilis* (Testudines, Podocnemididae) during the dry season in the Mamirauá Sustainable development Reserve, Amazonas, Brazil. *Chelonian Conserv. Biol.* **2006**, *5*, 312–317. [CrossRef]
44. Cunha, F.L.; Bernhard, R.; Vogt, R.C. Diet of an assemblage of four species of turtles (*Podocnemis*) in the Rio Uatumã, Amazonas, Brazil. *Copeia* **2020**, *108*, 103–115. [CrossRef]
45. Teran, A.F.; Vogt, R.C.; Gomez, M.D.F.S. Food habits of an assemblage of five species of turtles in the Rio Guapore, Rondonia, Brazil. *J. Herpetol.* **1995**, *29*, 536–547. [CrossRef]
46. Lara, N.R.F.; Marques, T.S.; Montelo, K.M.; de Ataídes, Á.G.; Verdade, L.M.; Malvásio, A.; de Camargo, P.B. A trophic study of the sympatric Amazonian freshwater turtles *Podocnemis unifilis* and *Podocnemis expansa* (Testudines, Podocnemidae) using carbon and nitrogen stable isotope analyses. *Can. J. Zool.* **2012**, *90*, 1394–1401. [CrossRef]
47. Frossard, A.; Coppo, G.C.; Lourenço, A.T.; Heringer, O.A.; Chippari-Gomes, A.R. Metal bioaccumulation and its genotoxic effects on eggs and hatchlings of giant Amazon river turtle (*Podocnemis expansa*). *Ecotoxicology* **2021**, *30*, 643–657. [CrossRef]
48. Schneider, L.; Belger, L.; Burger, J.; Vogt, R.C.; Jeitner, C.; Peleja, J.R.P. Assessment of non-invasive techniques for monitoring mercury concentrations in species of Amazon turtles. *Toxicol. Environ. Chem.* **2011**, *93*, 238–250. [CrossRef]
49. Borges, A.O.; Erickson, J.; Silva, L.A.; Fantin, C.; Domingos-Moreira, F.X.V. Mercury bioaccumulation, genotoxic and biochemical biomarkers reveal the health status of yellow-spotted Amazon River turtles (*Podocnemis unifilis*) in an environmental protection area in the Amazon. *Acta Amaz.* **2022**, *52*, 254–263. [CrossRef]
50. Schneider, L.; Belger, L.; Burger, J.; Vogt, R.C. Mercury bioaccumulation in four tissues of *Podocnemis erytrocephala* (*Podocnemididae*: Testudines) as a function of water parameters. *Sci. Total Environ.* **2009**, *407*, 1048–1054. [CrossRef]
51. Jagoe, C.H.; Brant, H.A. Maternal transfer of mercury in fish and reptiles: Concentrations n eggs versus somatic tissues. In Proceedings of the 24th Annual Meeting in North America of the Society of Environmental Toxicology and Chemistry, Austin, TX, USA, 9–13 November 2003.
52. Schneider, L.; Belger, L.; Burger, J.; Vogt, R.C.; Ferrara, C.R. Mercury levels in muscle of six species of turtles eaten by people along the Rio Negro of the Amazon Basin. *Arch. Environ. Contam. Toxicol.* **2010**, *58*, 444–450. [CrossRef]
53. Ojast, J. La tortuga arrau del Orinoco. *Def. De La Nat.* **1971**, *2*, 3–9.
54. Silva Montes, C.; Ferreira, M.A.P.; Giarrizzo, T.; Amado, L.L.; Rocha, R.M. The legacy of artisanal gold mining and its impact on fish health from Tapajós Amazonian region: A multi-biomarker approach. *Chemosphere* **2022**, *287*, 132263. [CrossRef] [PubMed]
55. Oliveira, R.C.; Dórea, J.G.; Bernardi, J.V.E.; Bastos, W.R.; Almeida, R.; Manzatto, A.G. Fish consumption by traditional subsistence villagers of the Rio Madeira (Amazon): Impact on hair mercury. *Ann. Hum. Biol.* **2010**, *37*, 629–642. [CrossRef]
56. Appelquist, H.; Asbirk, S.; Drabaek, I. Mercury monitoring: Mercury stability in bird feathers. *Mar. Pollut. Bull.* **1984**, *15*, 22–24. [CrossRef]
57. Schneider, L.; Eggins, S.; Maher, W.; Vogt, R.C.; Krikowa, F.; Kinsley, L.; Eggins, S.M.; Silvera, R. An evaluation of the use of reptile dermal scutes as a non-invasive method to monitor mercury concentrations in the environment. *Chemosphere* **2015**, *119*, 163–170. [CrossRef] [PubMed]
58. Schneider, L.; Tucker, A.D.; Vincent, K.; Fossette, S.; Young, E.J.; Whiting, S.D. First Assessment of Mercury (Hg) Concentrations in Skin and Carapace of Flatback Turtles (*Natator depressus*) (Garman) From Western Australia. *Front. Environ. Sci.* **2022**, *10*, 150. [CrossRef]
59. Eagles-Smith, C.A.; Ackerman, J.T.; Willacker, J.J.; Tate, M.T.; Lutz, M.A.; Fleck, J.A.; RobinStewart, A.; Wiener, J.G.; Evers, D.C.; Lepak, J.M.; et al. Spatial and temporal patterns of mercury concentrations in freshwater fish across the Western United States and Canada. *Sci. Total Environ.* **2016**, *568*, 1171–1184. [CrossRef] [PubMed]

60. Windsor, F.M.; Pereira, M.G.; Tyler, C.R.; Ormero, S.J. River organisms as indicators of the distribution and sources of persistent organic pollutants in contrasting catchments. *Environ. Pollut.* **2019**, *255*, 113144. [CrossRef]
61. Kronberg, R.M.; Drott, A.; Jiskra, M.; Wiederhold, J.G.; Björn, E.; Skyllberg, U. Forest harvest contribution to Boreal freshwater methyl mercury load. *Glob. Biogeochem. Cycles* **2016**, *30*, 825–843. [CrossRef]
62. Pignati, M.T.; Pezzuti, J.C.B.; Souza, L.C.; Lima, M.D.O.; Pignati, W.A.; Mendes, R.D.A. Assessment of Mercury Concentration in Turtles (*Podocnemis unifilis*) in the Xingu River Basin, Brazil. *Int. J. Environ. Res. Public Health* **2018**, *15*, 1185–1195. [CrossRef]
63. Zupo, V.; Graber, G.; Kamel, S.; Plicht, V.; Granitzer, S.; Gundacker, C.; Wittmann, K.J. Mercury accumulation in freshwater and marine fish from the wild and from aquaculture ponds. *Environ. Pollut.* **2019**, *255*, 112975. [CrossRef]
64. Santos, R.L.; Sousa Correia, J.M.; Santos, E.M. Freshwater aquatic reptiles (Testudines and Crocodylia) as biomonitor models in assessing environmental contamination by inorganic elements and the main analytical techniques used: A review. *Environ. Monit. Assess.* **2021**, *193*, 498. [CrossRef]
65. Cai, S.; Gan, J.; Liu, B.; Wang, J. Metal Concentrations in Tissues of Two Fish Species (*Mystus macropterus* and *Cyprinus carpio*) from the Chishui River, a Protected Area in China. *Fresen. Environ. Bull* **2017**, *26*, 4549–4554.
66. Zhang, H.; Guo, C.; Feng, H.; Shen, Y.; Wang, Y.; Zeng, T.; Song, S. Total mercury, methylmercury, and selenium in aquatic products from coastal cities of China: Distribution characteristics and risk assessment. *Sci. Total Environ.* **2020**, *739*, 140034. [CrossRef] [PubMed]
67. Lemell, P.; Beisser, C.J.; Gumpenberger, M.; Snelderwaard, P.; Gemel, R.; Weisgram, J. The feeding apparatus of *Chelus fimbriatus* (Pleurodira; *Chelidae*)–adaptation perfected? *Amphib. Reptil.* **2010**, *31*, 97–107. [CrossRef]
68. Lemell, P.; Natchev, N.; Beisser, C.J.; Heiss, E. Feeding in Turtles: Understanding Terrestrial and Aquatic Feeding in a Diverse but Monophyletic Group. In *Feeding in Vertebrates*; Springer: Berlin/Heidelberg, Germany, 2019; pp. 611–642.
69. Eggins, S.; Schneider, L.; Krikowa, F.; Vogt, R.C.; Silveira, R.; Maher, W. Mercury concentrations in different tissues of turtle and caiman species from the Rio Purus, Amazonas, Brazil. *Environ. Toxicol. Chem.* **2015**, *34*, 2771–2781. [CrossRef]
70. Correia, J.; Cesar, R.; Marsico, E.T.; Diniz, G.T.N.; Zorro, M.C.; Castilhos, Z. Mercury contamination in alligators (*Melanosuchus niger*) from Mamirauá Reservoir (Brazilian Amazon) and human health risk assessment. *Environ. Sci. Pollut. Res.* **2014**, *21*, 13522–13527. [CrossRef]
71. Burger, J.; Gochfeld, M. Mercury and selenium levels in 19 species of saltwater fish from New Jersey as a function of species, size, and season. *Sci. Total Environ.* **2011**, *409*, 1418–1429. [CrossRef]
72. Nunes, A.V.; Peres, C.A.; Constantino, P.D.A.L.; Santos, B.A.; Fischer, E. Irreplaceable socioeconomic value of wild meat extraction to local food security in rural Amazonia. *Biol. Conserv.* **2019**, *236*, 171–179. [CrossRef]
73. Doke, D.A.; Gohlke, J.M. Estimation of human health risk from exposure to methylmercury via fish consumption in Ghana. *J. Health Pollut.* **2014**, *4*, 18–25. [CrossRef]
74. Manavi, P.N.; Mazumder, A. Potential risk of mercury to human health in three species of fish from the southern Caspian Sea. *Mar. Pollut. Bull.* **2018**, *130*, 1–5. [CrossRef]

toxins

MDPI

Article

Evaluation of Statistical Treatment of Left-Censored Contamination Data: Example involving Deoxynivalenol Occurrence in Pasta and Pasta Substitute Products

Alessandro Feraldi [1,*], Barbara De Santis [2,*], Marco Finocchietti [1], Francesca Debegnach [2], Antonio Mandile [1,2] and Marco Alfò [1]

1 Department of Statistical Science, University of Rome "La Sapienza", 00185 Rome, Italy; marco.finocchietti94@gmail.com (M.F.); mandile.1709269@studenti.uniroma1.it (A.M.); marco.alfo@uniroma1.it (M.A.)
2 Department of Food Safety, Nutrition and Veterinary Public Health, Italian National Institute of Health, 00161 Rome, Italy; francesca.debegnach@iss.it
* Correspondence: alessandro.feraldi@uniroma1.it (A.F.); barbara.desantis@iss.it (B.D.S.)

Abstract: The handling of data on food contamination frequently represents a challenge because these are often left-censored, being composed of both positive and non-detected values. The latter observations are not quantified and provide only the information that they are below a laboratory-specific threshold value. Besides deterministic approaches, which simplify the treatment through the substitution of non-detected values with fixed threshold or null values, a growing interest has been shown in the application of stochastic approaches to the treatment of unquantified values. In this study, a multiple imputation procedure was applied in order to analyze contamination data on deoxynivalenol, a mycotoxin that may be present in pasta and pasta substitute products. An application of the proposed technique to censored deoxynivalenol occurrence data is presented. The results were compared to those attained using deterministic techniques (substitution methods). In this context, the stochastic approach seemed to provide a more accurate, unbiased and realistic solution to the problem of left-censored occurrence data. The complete sample of values could then be used to estimate the exposure of the general population to deoxynivalenol based on consumption data.

Keywords: left-censored data; limit of quantification; chemical contaminants; dietary exposure

Key Contribution: With the aim of minimizing the influence of left-censored data and reducing the associated uncertainty in dietary exposure estimations, the proposed approach is considered appropriate for the purpose of processing complex occurrence datasets of certain contaminants.

Citation: Feraldi, A.; De Santis, B.; Finocchietti, M.; Debegnach, F.; Mandile, A.; Alfò, M. Evaluation of Statistical Treatment of Left-Censored Contamination Data: Example involving Deoxynivalenol Occurrence in Pasta and Pasta Substitute Products. *Toxins* **2023**, *15*, 521. https://doi.org/10.3390/toxins15090521

Received: 13 July 2023
Revised: 18 August 2023
Accepted: 21 August 2023
Published: 24 August 2023

1. Introduction

In order to assess the dietary exposure of population groups to a contaminant that can pose a risk for human health, information regarding contamination values for food commodities and food consumption data need to be collected. When performing chemical risk assessment for dietary exposure, food analyses provide positive contamination values (VAL, values in μg/kg) and non-detected, left-censored values that, for certain classes of contaminants, can be high in number [1,2]. With regard to the latter, the numerical value is not known, and the only information available is that it is below a given threshold, often representing a laboratory-specific feature. The parameters that describe non-detected values are particularly important for exposure estimations, as these correspond to the lower tail of occurrence data distribution. These parameters, which are indicative of the threshold value, are the limit of detection (LOD), defined as the lowest concentration level that can be determined to be different from zero [1–3], and the limit of quantification (LOQ), defined as the minimum concentration or mass of an analyte that can be quantified with a given confidence and a given analytical procedure [3–5]. Non-detected values are a real challenge for any statistical analysis involving occurrence data. Ignoring the undetected values and analyzing only the quantified ones will inevitably result in extensively biased and therefore unrepresentative estimates. In addition to the analytical thresholds of the LOD or LOQ, the percentages of non-detected values in a sample also depend on the type of contaminant and on the food category. Since a large portion of non-detected values may lead to computational difficulties in calculating percentile or mean values [6,7], decision tree recommendations are provided to guide the statistical treatment of datasets containing various proportions of non-quantified results [7].

Currently, different statistical methods are available to interpret non-detected values and model combinations of positive (VALs) and non-detected values. Traditionally, exposure assessment is performed through a deterministic substitution method, which is widely used in food risk assessment [8–10]. The substitution method is based on the replacement of non-detected values with a value of zero, the LOQ, or the LOQ/2 to obtain and represent the lower-bound (LB) and upper-bound (UP) or to approximate a medium-bound (MB) occurrence scenario, respectively [6,7]. This substitution approach has the advantage of being simple in its implementation. However, as any deterministic procedure for simple imputation, it also has some disadvantages. For instance, it may produce biased estimates of summary statistics (even when the maximum extent of such biases is assessed and does not exceed the difference between the upper and the lower bounds) [11]. Additionally, it does not consider the intrinsic variability of non-detected values that, through the process of imputation, are treated as observed, thus deflating the variability of any estimator based on such completed data.

Such drawbacks have been an object of study, and more recently the international interest has moved towards the application of stochastic approaches to impute (multiple) non-detected values in food contamination data. The literature includes several efforts to deal with the issue of non-detected values, especially via maximum likelihood (ML) estimation applied to parametric models with censored data [11,12]. For this purpose, several attempts have been made using marginal and pseudo-likelihood-based methods [13] by fitting truncated parametric (e.g., lognormal) distributions to observed data [14]; log-probit regression (LPR) [15]; or non-parametric methods, such as the non-parametric estimator of the survival function [12,16–18].

Quantification limits (and all the associated issues) are likely to be found in data obtained from analyses performed in different environmental contexts such as water, air, food, and even biological specimens like urine samples. For example, Jones (2018) applied complex linear regression models, which allowed outcomes and covariates to be linear combinations of left-censored data on polychlorinated biphenyls (a class of tocix industrial chemicals), for the estimation of means and covariance matrices [13]. A further study proposed a transformation to convert (multiple) left-censored data on water quality to right-censored data enabling the use of common survival analysis techniques [18]. Stochastic approaches impute non-detected values via a number of plausible values drawn from a given statistical model, and all approaches are implemented to understand the variability in the resulting estimates and overcome the uncertainty arising in the imputation of non-detected values. In this study, alternative approaches were employed to deal with non-detected values and with the presence of multiple LOQ values, aiming to analyze and handle contamination data related to deoxynivalenol (DON), a mycotoxin that is an agricultural contaminant occurring in cereals and all derived products (including pasta and pasta substitutes) produced by toxigenic fungi from the *Fusarium* genus. Mycotoxins are toxic substances produced by toxigenic micromycetes, such as those of the *Aspergillus*, *Penicillium*, and *Fusarium* genera, and when they occur in food or feed they represent a risk to human and animal health [19].

Contamination levels are highly influenced by weather conditions and agricultural practices [20,21], which, when unfavorable, may lead to severe *Fusarium* spp. infections affecting cereals, such as *Fusarium* head blight and European corn borer. Humid conditions at the time of flowering favor fungal proliferation on the spikelet and promote high levels of DON and other *Fusarium* toxins on crops. In its opinion published in 2017 [22], the European Food Safety Authority (EFSA) identified vomiting as the critical acute adverse effect of the sum of DON, its 3-glucoside, and the acetylated forms (3-Ac-DON and 88 15-Ac-DON) for pigs, farmed minks, dogs, cats, and human risk assessment.

Despite the limitations in the available human data, the EFSA established a group acute reference dose (ARfD) of 8 μg/kg body weight (bw) per eating occasion for the sum of DON, 3-Ac-DON, 15-Ac-DON, and DON-3-glucoside. Moreover, in the absence of data on chronic effects in humans, the EFSA recognized the reduced body weight gain in experimental animals as the critical chronic effect for human risk assessment and established a group tolerable daily intake (TDI) of 1 μg/kg bw per day for the sum of DON and its metabolites [22].

Conventionally, the analytical detection of DON is carried out by liquid chromatography (LC) coupled with UV detection (typically, wave length λ = 220 nm) [23,24]. In order to gain better selectivity, a purification step with an immunoaffinity column (IAC) is recommended to be used during the extraction of the mycotoxin. Nowadays, tandem mass spectrometry (LC-MS/MS) techniques are the most commonly used for the detection of mycotoxins in general, and DON in particular. The literature includes extensive examples of analytical methods applied to the analysis of DON, its metabolites, and even multiple mycotoxins in cereal-based foods [25–29].

In this study, DON contamination values detected in pasta and pasta substitutes were produced by a battery of laboratories that used several analytical techniques and/or different analytical methods, thus providing a range of LOQ/LOD values in the dataset. Therefore, the available data were not only censored, but the level and "quality" of censoring depended on the specific laboratory. To cope with this complexity, three different multiple imputation (MI) techniques for imputing non-detected values with multiple LOQ values were considered. Through the replacement of each non-detected value with multiple plausible contamination values, these techniques allowed us to obtain multiple sets of plausible contaminations. In order to achieve a completed contamination dataset, each set obtained by a given imputation procedure was merged with the VALs. The estimates obtained by imputation were compared to those derived from the application of standard, deterministic, and substitution methods on the same dataset, which could be used to define boundaries for the former.

As will be shown in the present paper, the proposed approach is appropriate for processing complex datasets with a high number of non-detected values and a range of different LOD/LOQ values.

2. Results

2.1. Exploratory Statistics

The number of LOQ values present in the analyzed dataset and the exploratory statistics for the data on DON contamination are reported in Table 1. For each LOQ stratum, the number of contaminated samples; the corresponding percentage, mean, median, and standard deviation for the measured values; the number and percentage of non-detected values; and the size and percentage (with respect to the global sample) of the corresponding subsample, are reported.

Considering VALs only (i.e., 131 values), the mean values in the total sample as well as in each subsample associated with each LOQ value were higher than the medians, denoting a positive skewness in the distribution of the contamination data related to DON in pasta and pasta substitute products. This result empirically supported the selection of lognormal, Weibull, and gamma as appropriate candidate distributions. To provide a clearer representation of the data at hand, Figure 1a,b show the empirical distribution of the contamination data. In particular, Figure 1a shows the CDF for the contamination data, with the crosses representing observed contamination data and each bin representing the cumulative frequency of non-detected values according to the value of the laboratory-specific LOQ. Figure 1b shows the distribution of VALs only, and the positive skewness in the distribution of data on DON contamination in pasta and pasta substitute products can be clearly observed. Prior to the imputation, the average contamination in the total sample was 207.11 μg/kg.

Table 1. Contamination values (μg/kg) of pasta and pasta substitute products by LOQ. Exploratory statistics and frequency distribution.

	Detected					Non-Detected		Total	
LOQ (μg/kg)	n	%	Mean (μg/kg)	Median (μg/kg)	SD	n	%	n	%
26	23	59.0	111.5	101.0	56.9	16	41.0	39	18.8
50	94	70.7	225.6	169.2	230.8	39	29.3	133	63.9
92.5	3	75.0	371.0	220.5	288.4	1	25.0	4	1.9
100	9	45.0	203.2	164.0	99.6	11	55.0	20	9.6
150	2	28.6	207.0	207.0	35.3	5	71.4	7	3.4
500	0	0.0	-	-	-	5	100.0	5	2.4

Abbreviations: n—number, %—percentage, SD—standard deviation.

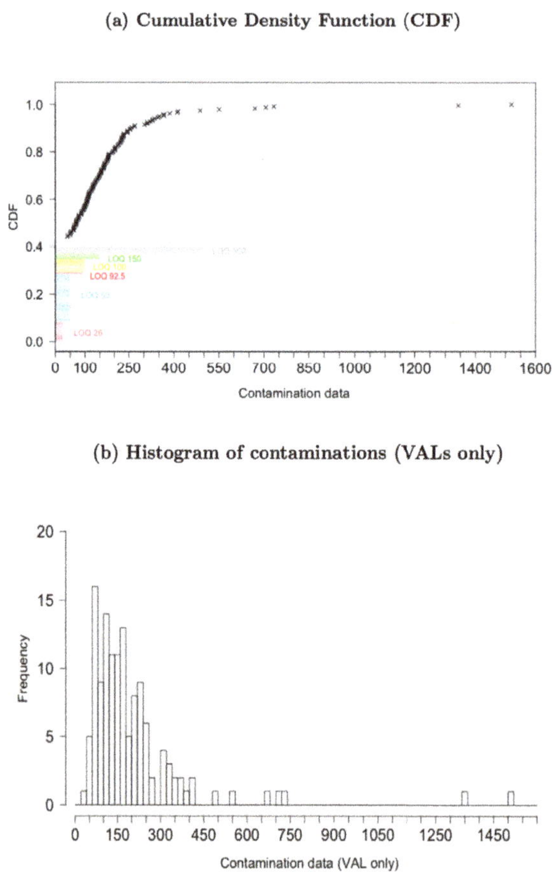

Figure 1. Distribution of observed data on DON contamination in pasta and pasta substitute products: (**a**) cumulative density function (CDF), (**b**) histogram of contaminations (VALs only).

Before proceeding to exposure estimates, a choice had to be made as to the three candidates and the MA distributions. The last one was preferred because it summarized different behaviors that were specific to each candidate distribution without the need to refer to a more complex (in terms of the number of parameters) single parametric form. This procedure seems to produce a more honest measure of precision with a reduced bias when compared to a best fit model [30]. Additionally, inference based on a model averaging (MA) procedure seems to outperform that obtained using the best model strategy [31].

2.2. Comparing Stochastic and Deterministic Methods

In Table 2, a comparison of the results obtained by applying the (deterministic) substitution method with the stochastic multiple imputation approach for the distribution obtained by MA is shown in terms of some basic exploratory statistics. The application of both the proposed MI procedures and the standard deterministic substitution method in order to replace the 77 non-detected values led to a decrease in the estimate of the average contamination value with respect to that based on the VALs only (207.11 μg/kg). This was to be expected, as the non-detected values lay in the left tail of the contamination data distribution.

Moreover, Table 2 shows that the contamination distributions estimated through the three procedures were similar. More specifically, since the values to be imputed were below a certain threshold, as expected, the estimates of contamination mainly differed in the right tails of the distributions. A statistical comparison of the three distributions was performed using the Kullback–Leibler measure, a method that is widely used in order to quantify the dissimilarity between probability distributions [32]. The calculated Kullback–Leibler measures were consistently small, ranging from 0.03 to 0.96, which provided a confirmation of the similarity between the distributions obtained through the three procedures.

Table 2. Exploratory statistics of the completed distribution for DON in pasta and pasta substitute products (μg/kg).

	CONTAMINATION VALUES					
Procedure	**Min**	**1st Quartile**	**Median**	**Mean**	**3rd Quartile**	**Max**
All	0.3	29.7	98.9	139.4	180.1	1519.6
Gold-standard	5.1	44.0	101.0	145.3	181.0	1519.6
Single	0.4	20.1	98.9	137.7	181.0	1519.6
LB	0.0	0.0	94.2	128.1	178.9	1519.6
UB	26.0	50.0	108.2	160.8	195.2	1519.6

Figure 2 shows the density estimates for the completed contamination data provided by the stochastic MI procedure (All). As for the Gold-Standard and Single procedures (see Appendix A, Figures A1 and A2, respectively), the positive skewness of the contamination data increased. This meant that DON contamination in pasta and pasta substitute products was effectively reduced when compared to the starting VAL distribution. With respect to the latter, after the MI and substitution methods, the distribution of contamination shifted to the left. This implied a reduction in the effective contamination scenario observed in pasta and pasta substitute products. In particular, a peak was shown between 0 and 50 μg/kg (see the top right portion of Figure 2), since the largest number of non-detected values (71%) belonged to the two largest strata, those associated with LOQ values equal to 26 and 50 μg/kg (see Table 1).

The estimated average contamination, the minimum and maximum values, and the quantiles provided by the suggested procedures were compared with the corresponding values obtained by the standard substitution method. This method is typically used because it is easy to understand and apply. It includes the LB, where the true contaminations are set equal to the minimum value that a non-detected value may assume (in this case zero), as well as the UB, where the true contaminations are all set equal to the corresponding LOQ, which represents their possible maximum value.

As shown in Table 2, the distributions of data on DON contamination in pasta and pasta substitute products derived from the stochastic multiple imputation approach (All, Gold-Standard, and Single) were included in LB and UB boundaries.

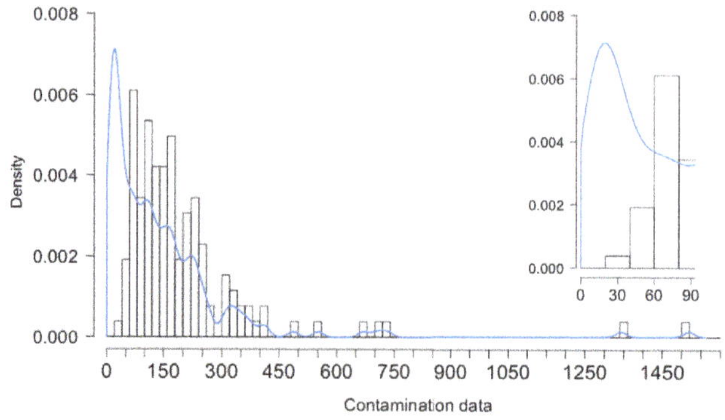

Figure 2. Density estimates of data on DON contamination in pasta and pasta substitute products applying All (MA only).

Figures 3 and 4 show the estimated average DON contamination in pasta and pasta substitute products and the corresponding 95% confidence interval (CI) based on the B = 100 imputed datasets, according to the three procedures. Estimated using the All procedure, the average level of DON contamination in pasta and pasta substitute products was approximately 139.4 μg/kg, with a 95% confidence interval ranging from 137.0 μg/kg to 141.9 μg/kg. The Gold-Standard procedure yielded a slightly higher mean estimate of 145.3 μg/kg. The estimates from the Single procedure (mean of 137.7 μg/kg) closely aligned with those obtained from the All procedure. Finally, the lower- and upper-bound scenarios (of the substitution methods) were included in this 95% CI obtained by the proposed stochastic procedures.

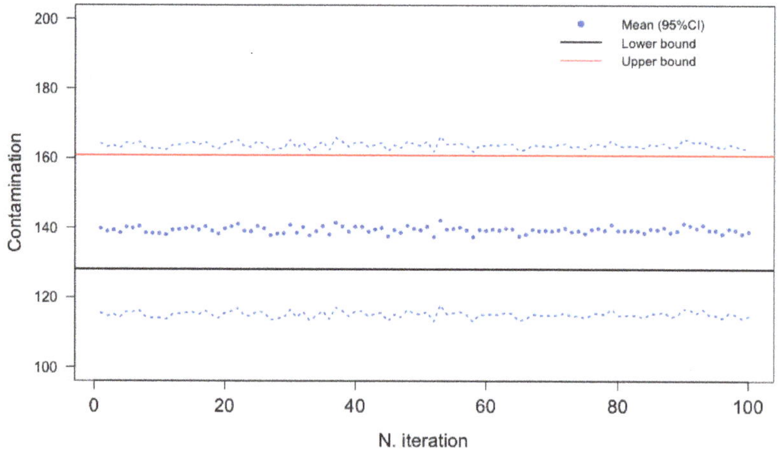

Figure 3. Mean estimate and 95% confidence interval (CI) for the mean of data on DON contamination in pasta and pasta substitute products across iterations (multiple imputation) (All procedure).

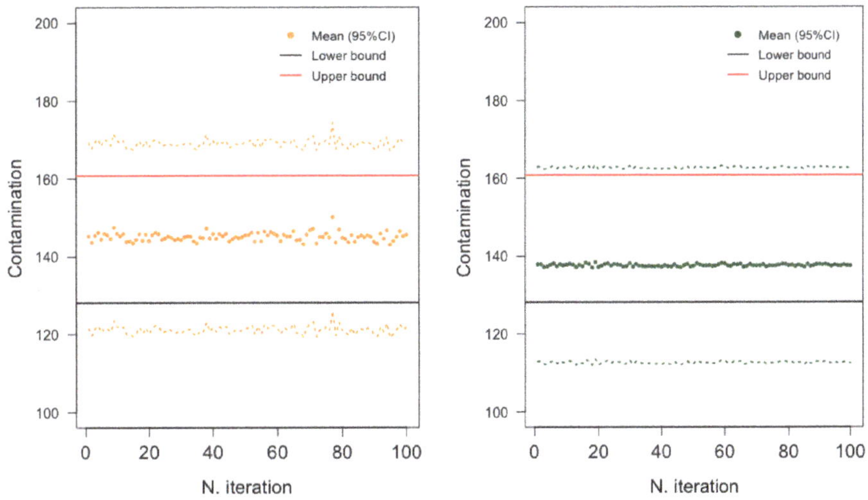

Figure 4. Mean estimate and 95% confidence interval (CI) for the mean of data on DON contamination in pasta and pasta substitute products across iterations (multiple imputation) (**left**: Gold-Standard procedure, **right**: Single procedure).

3. Discussion

The strength of the substitution methods lies in the fact that they can be easily used to screen the levels and evaluate which of the components in the assessment path are likely critical. However, it has been widely recognized that the substitution method is biased. The bias is a function of the true data variability, the percentage of non-detected values, and the sample size. Moreover, the substitution method ignores the variability due to uncertainty in the real non-detected values that are imputed. In other words, whether a sample contains 1% or 90% detected values does not drive the way the non-detected values are treated. This applies even in cases where the two types of samples have different features [11], with the most critical situation being that with the highest number of non-detected values and the presence of multiple LOQ values [11,12].

Stochastic approaches introduce more realism, since they are based on fitting distributions to VALs and thus, as opposed to the deterministic approach, take the variability of the data into account when imputing non-detected values. Further, they may help to provide estimates of the variability ascribed to each imputation, whereas in the substitution method this is completely neglected. By using a stochastic approach, each missing value is imputed by B potential values taken from the selected distribution. Therefore, the B estimated values of contamination and incidence represent a set of scenarios that allow one to derive measures of synthesis (e.g., averages) and variability, such as 95% CIs.

The Best Stochastic Procedure

The results of the exploratory statistics obtained by the procedure All, for which the imputation of non-detected values was based on the whole contamination sample, were the most reliable, as they took into account the whole left-censored contamination dataset. In the Gold-Standard procedure, the results depended on the stratum that was selected to be the best-fitted to the observed distribution. The selection could have limited the ability of the estimates to capture the real variability in the data on DON contamination in pasta and pasta substitute products, since it may not have reflect the variability and number of VALs in those strata that were characterized by a different LOQ value. Therefore, through the selection of the stratum with an LOQ equal to 50 μg/kg, the exploratory statistics for the contamination estimates provided by the Gold-Standard procedure proved to be slightly different from those obtained through the other two stochastic procedures (see Table 2). The Single and All procedures provided quite similar outputs. This was probably due to the fact that these two techniques had similarities, as they both considered all the contamination data to fit the candidate distributions. However, the Single procedure did not allow the imputation of non-detected values for the strata with no VALs (whose presence was mandatory in order to estimate the density distribution), and it did not make it possible to compute the likelihood for the stratum with an LOQ equal to 500 μg/kg. In addition, when the number of VALs was lower than 4 (as occurred in the stratum with an LOQ = 92.5 μg/kg), the estimated distribution was not informative and it was not sufficient for drawing imputations for non-detected values.

4. Conclusions

Three different procedures for the multiple imputation (MI) of left-censored data on DON contamination in pasta and pasta substitute products, which included a number of both detected (VALs) and non-detected values, were proposed. As a first output, the MI carried out through model averaging was to be preferred to the best fit model. This was due to its ability to produce estimates with a reduced bias as well as to summarize different behaviors that were specific to each candidate distribution. The procedure provided suitable imputations for pasta and pasta substitute products that were in agreement with each other and with those obtained by the deterministic estimates (LB and UB). In fact, the substitution method provided extreme scenarios for data on DON contamination in pasta and pasta substitute products, whereas MI provided contamination scenarios that were between the deterministic boundaries, which were too high (UB) or low (LB), respectively. Additionally, MI, which was based on fitting a specific parametric distribution, allowed us to account for (and estimate) the variability in the obtained contamination scenario due to imputation, which could not be achieved via deterministic methods. Among those proposed in this paper, the All procedure seemed to be the most appropriate.

In conclusion, MI provided suitable imputations and reliable contamination scenarios, especially for the All procedure performed via MA. However, the stochastic approach, which supplied distributions to be used for probabilistic calculations, should be considered complementary to the deterministic method rather than a replacement, since the latter led to conservative estimate calculations and offered an immediate assessment of the contamination boundaries.

Finally, although distributions and related mixtures other than lognormal, Weibull, gamma, and MA could be considered, the proposed approach was found to be appropriate for processing complex datasets of occurrence for contaminants with a high number of non-detected values, where the additional issue of a range of LOD/LOQ values makes the data handling even more challenging.

5. Materials and Methods

5.1. Occurrence Data

The available occurrence data for DON in pasta and pasta substitutes were gathered from the outcomes of the Italian National Official Control Plan for mycotoxins, which is the annual control plan (CP) requested by the European Commission from Member States. The main purpose of this is to verify the application of the rules and the functioning of national control systems. The CP is set each year by the Italian Ministry of Health to monitor and survey certain contaminants in food products throughout the national territory. The food sampling plans are defined with the support and collaboration of regions and provincial authorities, the "Direzione Generale per l'Igiene e la Sicurezza degli Alimenti e la Nutrizione" (DGISAN, Italian Ministry of Health, Rome, Italy) and the Italian National Reference Laboratory for mycotoxins in food and feed (NRL, Italian National Institute of Health, Istituto Superiore di Sanità), and they are approved by the "Coordinamento Tecnico Interregionale" (Technical Committee for the Heath Commission of the State-Region Conference). The annual CP provides guidance to regional and autonomous provincial authorities on the activities to be carried out by the official controlling body for the detection of mycotoxin occurrence (including DON) in food. It outlines and coordinates the actions for the verification of compliance with EU legislation and promotes the assessment of consumer exposure to contaminants.

The available dataset included data gathered and registered in the years 2016, 2017, and 2018. The pasta and pasta substitute samples were drawn at the final stage of the production and distribution of foodstuffs. The concentration levels of the contaminant in food were reported in terms of μg of DON per kg of food (μg/kg).

The samples were analyzed and the data were recorded by several official laboratories, distributed over the whole Italian territory. All laboratories operated under a quality control system following the ISO EN 17025 standard; they were equipped with different instruments for the detection and quantification of DON and used different analytical methods, for which different LOD/LOQ values were applied. The sample consisted of 208 contamination data for DON in pasta and pasta substitute products. Among these, 77 values (37.0%) were non-detected. The data collected came from laboratories distributed over eleven Italian regions and included six different LOQ values. Table 3 reports the distribution of DON occurrence, the production methods, the LOQ values, and the number of detected and non-detected contamination values reported by each laboratory.

Table 3. Distribution of DON occurrence, production methods, LOQ values, and detected/non-detected values across different laboratories.

Region	Production Method	LOQ	N. VAL	%VAL	N. LOQ
Basilicata	Unknown	50	1	100	0
		92.5	3	75	1
Emilia Romagna	Non-organic production	50	39	87	6
Friulia-Venezia Giulia	Non-organic production	100	8	73	3
Lazio	Unknown	26	23	59	16
Liguria	Unknown	50	2	67	1
Lombardia	Organic production	100	0	0	5
Piemonte	Unknown	50	2	50	2
		500	0	0	5
Puglia	Unknown	50	50	63	30
Sicilia	Unknown	100	0	0	3
Umbria	Unknown	150	2	29	5
Veneto	Unknown	100	1	100	0

Abbreviation: %—percentage.

5.2. The Proposed Stochastic Approach

In order to impute non-detected values, specific candidate parametric distributions were considered and fitted by maximum likelihood to the left-censored contamination data. After obtaining estimates for the parameters that indexed the parametric distributions, these were plugged in and used to draw plausible contamination values for the non-detected values. The whole (multiple) imputation procedure can be described by the following steps:

1. Consider a (possibly wide) family of candidate distributions;
2. Estimate parameters for each candidate distribution;
3. Assess the quality of fit and select the best distribution in terms of fit;
4. Model-average the candidate distributions with weights proportional to penalized likelihood criteria;
5. Impute non-detected values (three techniques) by drawing several (i.e., 100) values from the model-averaged distribution for each non-detected value.

STEP 1. Specify the (possibly wide) family of candidate distributions

The contamination values of DON in pasta and pasta substitute products were considered to form an independent and identically distributed (iid) sample drawn from a known parametric distribution. The contamination values were assumed to be independent across food commodities, and therefore they could be modeled by univariate distributions. Contamination measurements usually show a skewed distribution. These distributions are particularly common when VAL data are at hand, as they are characterized by large variances, a high proportion of non-detected values, and multiple LOQ values. The lognormal distribution properly describes VALs, and it is parsimonious in the sense that it is indexed by a two-dimensional (mean, variance) parameter [33,34]. For these reasons, it was selected as the starting point while defining the set of candidate distributions [35–38]. In addition to the lognormal model, the Weibull and gamma distributions were also considered in the analysis, as they represent two further parametric models that are often used to model contaminant data in food [39–42].

STEP 2. Estimate parameters for each candidate distribution

For each candidate distribution, the problem of estimating the parameters can be summarised as follows: a sample of observations $y, y_2, \ldots, y_n$ of a quantitative variable (representing data on DON contamination) was given, and it was assumed to be drawn from a probability density function (PDF) $f(x, \theta)$, which could either describe a lognormal, Weibull, or gamma random variable, where θ is the vector of parameters that indexed such a density. Since the parametric form of the distribution was known, in order to use such a distribution, the vector of parameters needed to be estimated. In this context, estimation was carried out based on the maximum likelihood (ML). The likelihood function, which represents the objective function to be maximized, is defined as the joint probability density function evaluated at the sample points:

$$L(y_1, y_2, \ldots, y_n, \theta) = \prod_{i=1}^{n} f(y_i; \theta), \tag{1}$$

while the maximum likelihood estimate (MLe) is defined as the value θ that maximizes the likelihood function.

In order to find this estimate, standard calculus methods (solving likelihood equations obtained by setting partial derivatives equal to 0 and checking that the matrix of second derivatives is negative-definite) are usually employed. However, when left-censored data are at hand, more complex methods are needed. ML estimation in the context of data left-censored by an LOQ [11,12] should consider the fraction of values above the LOQ and, therefore, the fraction of data below the LOQ, as well as the distribution for measured VALs. To summarise, let δ denote an indicator that is equal to 1 when the observation is noted and 0 when it is not. When $\delta = 0$, the information is that the value is below the threshold, i.e., $y <$ LOQ. Further, let $F(y; \theta)$ denote the cumulative distribution function (CDF), associated with $f(x, \theta)$, for the selected parametric distribution with parameter vector θ. The MLe for θ is the value that maximizes the likelihood function L for such left-censored data:

$$L(\theta) = \prod_{i=1}^{n} (f(y_i; \theta))^{\delta_i} (F(LOQ_i; \theta))^{(1-\delta_i)} \tag{2}$$

where LOQ_i denotes the LOQ for the $i - th$ value. The censored observations could, in fact, contribute only through the CDF, as the only information they gather is that the corresponding values are not greater than the LOQ. Maximizing the likelihood function is equivalent to maximizing the corresponding log-likelihood function:

$$log(L(\theta)) = \sum_{i=1}^{n} \delta_i log + (1 - \delta_i) log(F(LOQ_i; \theta)) \tag{3}$$

Finally, the first and second derivatives of the log-likelihood with respect to θ are used to calculate the estimate for θ, usually via iterative Newton-type algorithms [43].

STEP 3. Assessing the quality of fit and selecting the best distribution in terms of fit

A number of model selection criteria can be used to choose the "best fitting" distribution, such as the Bayesian information criterion (BIC) [44] and the Akaike information criterion (AIC) [45]. While inspired by different motivations, they are both based on the idea that model fit can be measured by the maximized value of the log-likelihood function, once this has been penalized according to the model complexity. Therefore, the AIC is defined as:

$$AIC = -2logL(\theta y; \delta) + 2K \tag{4}$$

where $logL(\theta y; \delta)$ is the maximized value of the log-likelihood function for a candidate distribution, and K represents the corresponding number of parameters. The lower the AIC value, the better the fit. In this study, the AIC index was used as the model selection criterion.

STEP 4. Model averaging

The MA step was performed using AIC-based weights [30]. MA starts from a set of plausible candidate statistical models (here, lognormal, Weibull, and gamma distributions were the considered distributions), and it combines such models in order to obtain a weighted average model where the weight associated with each candidate model is proportional to the corresponding measure of fit.

In detail, let us denote by M the family of candidate models, by $M_j, j = 1, \ldots, K$ the individual model belonging to this family, and by F_j the corresponding CDF. The procedure starts with the estimation of the natural parameters θ_j by ML; let us denote the corresponding covariance matrix $COV(\theta_j)$ and the estimate of the $p-th$ quantile ξ_p, $p \in (0,1)$. In a direct approach to MA, the quantile ξ_p is estimated for each candidate model, and the MA estimate is a weighted average of the K estimates. Based on model M_j, the MLe for ξ_p is $\xi_{p,j} = F_j(p; \theta_j)$ [30]. The variance of the averaged estimate is approximated by the delta method:

$$VAR(\hat{\xi}) \approx \Delta F_j^{-1}(p;\theta)^T COV(\hat{\theta}) \Delta F_j^{-1}(p;\theta) \tag{5}$$

where the gradient $\Delta F_j^{-1}(p;\theta_j)$ can be estimated by $\Delta F_j^{-1}(p;\hat{\theta}_j)$. The MA estimate is defined as the weighted average $\hat{\xi}_{p,MA_1} = \sum_{i=1}^{K} w_i \hat{\xi}_{p,i}$, where the weights w_i associated with each candidate model are defined by:

$$w_i = \frac{exp(-\frac{1}{2}\Delta_i)}{\sum_{j=1}^{K} exp(-\frac{1}{2}\Delta_j)} \tag{6}$$

and

$$\Delta_j = AIC_j - AIC_{min} \tag{7}$$

The model closest to the best fitting one is associated with the highest weights.

STEP 5. Impute non-detected values (All, Gold-Standard, Single)

In the presence of a single LOQ, it is assumed that a specific q-th percentile corresponds to the LOQ value or each candidate model. For each non-detected value B (i.e., B = 100), random values $(u_1, \ldots, u_B)$ were drawn from a continuous uniform distribution defined on the segment 0 to q, representing the tail of the distributions associated with non-detected values. Each value $u_1, \ldots, u_B$ represents a potential quantile $q_1, \ldots, q_B \leq q$ for the distributions (lognormal, Weibull, gamma, and MA) associated with the contamination values. In order to perform imputation, such quantiles were back-transformed to values between 0 and the corresponding LOQ value. A number (B) of completed datasets were obtained, each of them corresponding to a set of draws for the non-detected values. However, as remarked above, contamination data were characterized by different LOQ values, as these were specific to the recording laboratory. In order to handle the problem of multiple LOQ values, three procedures for the MI of non-detected values were proposed (namely, All, Gold-Standard, and Single). The All and Gold-Standard procedures imputed non-detected values by taking all the different LOQ values collectively, whereas the Single procedure imputed non-detected contaminations stratified by each LOQ value taken individually (i.e., taking 6 strata, see Table 1).

Procedure 1 (All).

The parameters for the candidate distributions were estimated using ML for left-censored data on the incomplete contamination sample, considered as a whole. This method did not take into account the laboratory from which the data originated. The basic idea of this procedure was to impute non-detected values according to the best fit to all the observed data and all considered distributions (including the MA distribution). The distributions were estimated on all contaminations, considered as a unique sample.

Procedure 2 (Gold-Standard).

After stratifying the contamination data according to the LOQ values, the stratum with the largest number of data (VALs and non-detected values) was selected as the Gold-Standard stratum, as this stratum provided the most accurate major empirical evidence. Therefore, the parameters indexing the candidate distributions were estimated through ML on the contamination data belonging to the Gold-Standard stratum. This procedure based the imputation of all non-detected contamination on the distributions of the contamination estimated using data from the Gold-Standard stratum only. In pasta and pasta substitute products, the Gold-Standard stratum had an LOQ equal to 50 μg/kg, with n = 133 contaminations in total, (94 VALs, and 39 non-detected values). Thus, all candidate distributions (including MA) were estimated based on observed contamination data with an LOQ of 50 μg/kg.

Procedure 3 (Single).

After stratifying the contamination data according to the LOQ values, the parameters indexing the candidate distributions were estimated via ML applied to contamination data from each stratum separately. The basic idea of this procedure was to impute the non-detected values, lower than the LOQ, considering the potential variability of the contamination for that stratum when compared with those of other strata. Then, imputed contamination data corresponding to each LOQ-specific stratum were joined as a unique completed contamination dataset.

Obviously, the imputation procedure was repeated in each case B = 100 times, and B completed datasets were obtained by the following procedure. Once parameters for candidate distributions were estimated according to one of the procedures previously discussed, the percentiles corresponding to each LOQ (i.e., 26 μg/kg, 50 μg/kg, 92.5 μg/kg, 100 μg/kg, 150 μg/kg, and 500 μg/kg) were computed for each candidate distribution. Thus, considering, for example, the stratum with LOQ = 26 μg/kg, which contained 16 non-detected values to be imputed, 100 values between 0 and the (distribution-specific) quantile corresponding to 26 μg/kg were randomly drawn for each non-detected value from the three candidate distributions. Afterwards, the corresponding quantiles as well as those obtained with a step of MA were computed. These quantiles were back-transformed (using the inverse sampling method) to contamination values in order to replace non-detected values below LOQ = 26 μg/kg. In this way, pseudo-random contamination values were drawn below the corresponding LOQ. This procedure was carried out for all the strata, each having a different number of non-detected values. Regardless of the imputation procedure selected (All, Gold-Standard, and Single), the set of imputed contaminations was then joined to the VALs. The multiple completed contamination datasets obtained by applying the All and Gold-Standard procedures consisted in 208 rows and 100 columns; each column referred to a completed dataset, where each of the 77 non-detected values was replaced by a set of reliable/plausible values drawn as discussed before from the candidate distributions, whereas the subset of 131 VALs was kept fixed. The procedure was repeated B = 100 times, and so the columns were filled in. As opposed to others, the Single procedure provided a smaller multiple complete contamination dataset (203 rows and 100 columns). Indeed, it was not possible to impute 5 non-detected values with an LOQ equal to 500 μg/kg, since the estimation of the candidate distributions is not possible for strata with no VALs. The phase of drawing several values lower than the corresponding LOQ for each non-detected value of DON in pasta and pasta substitute products is coherent with the Monte Carlo approaches used for sampling from a given theoretical (parametric) probability distributions [46].

5.3. Comparison between Deterministic Substitution Methods and Stochastic Approaches

To obtain contamination references as boundary levels to be compared with the results provided by MI, a substitution method (deterministic approach) was also applied. The contamination boundaries were defined as the LB and UB.

Author Contributions: All the authors contributed to the study concept and design. A.F., M.F. and A.M. conducted the analyses and wrote the first draft of the manuscript, with input from the other authors. B.D.S., F.D. and M.A. provided edits and feedback on subsequent versions of the manuscript. All authors have read and agreed to the published version of the manuscript.

Funding: This research received no external funding.

Institutional Review Board Statement: Not applicable.

Informed Consent Statement: Not applicable.

Data Availability Statement: The data that support the findings of this study are available on request from the corresponding authors, Alessandro Feraldi and Barbara De Santis.

Conflicts of Interest: The authors declare no conflict of interest.

Abbreviations

The following abbreviations are used in this manuscript:

iid	Independent and identically distributed
MLE	Maximum likelihood estimation
LOD	Limit of detection
LOQ	Limit of quantification
VAL	Detected values
CP	Control plan
DON	Deoxynivalenol
CDF	Cumulative density function
AIC	Akaike information criterion
MA	Model averaging
MI	Multiple imputation
UB	Upper bound
LB	Lower bound

Appendix A

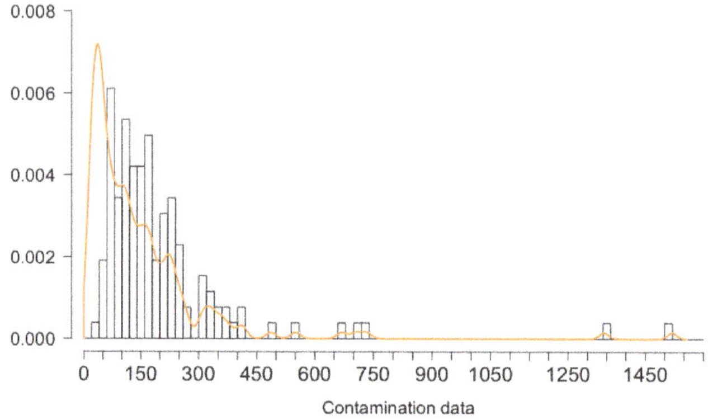

Figure A1. Density of DON contamination in pasta and pasta substitute products applying the Gold-Standard procedure (MA only).

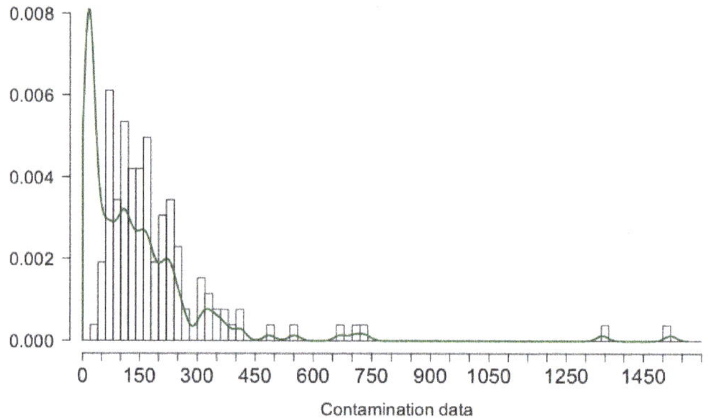

Figure A2. Density of DON contamination in pasta and pasta substitute products applying the Single procedure (MA only).

References

1. Keith, L.H.; Crummett, W.; Deegan, J.; Libby, R.A.; Taylor, J.K.; Wentler, G. Principles of environmental analysis. *Anal. Chem.* **1983**, *55*, 2210–2218. [CrossRef]
2. Girard, J. *Principles of Environmental Chemistry*; Jones & Bartlett Publishers: Burlington, MA, USA 2013; ISBN 1449693520.
3. MacDougall, D.; Crummett, W.B. Guidelines for data acquisition and data quality evaluation in environmental chemistry. *Anal. Chem.* **1980**, *52*, 2242–2249. [CrossRef]
4. Armbruster, D.A.; Pry, T. Limit of blank, limit of detection and limit of quantitation. *Clin. Biochem. Rev.* **2008**, *29*, S49.
5. Wenzl, T.; Haedrich, J.; Schaechtele, A.; Piotr, R.; Stroka, J.; Eppe, G.; Scholl, G. *Guidance Document on the Estimation of LOD and LOQ for Measurements in the Field of Contaminants in Food and Feed*; Publications Office of the European Union: Luxembourg, 2016.
6. EFSA Panel on Contaminants in the Food Chain. Management of left-censored data in dietary exposure assessment of chemical substances. *EFSA J.* **2010**, *8*, 1557.
7. Joint FAO/WHO Expert Meeting on Dietary Exposure Assessment Methodologies for Residues of Veterinary Drugs. In Proceedings of the Project to Review and Update the Principles and Methodology to Assess Dietary Exposure to Residues of Veterinary Drugs, Rome, Italy, 7–11 November 2011.
8. Hornung, R.W.; Reed, L.D. Estimation of average concentration in the presence of nondetectable values. *Appl. Occup. Environ. Hyg.* **1990**, *5*, 46–51. [CrossRef]
9. Glass, D.C.; Gray, C.N. Estimating mean exposures from censored data: Exposure to benzene in the Australian petroleum industry. *Ann. Occup. Hyg.* **2001**, *45*, 275–282. [CrossRef]
10. Ganser, G.H.; Hewett, P. An accurate substitution method for analyzing censored data. *J. Occup. Environ. Hyg.* **2010**, *7*, 233–244. [CrossRef]

11. Helsel, D.R. Fabricating data: How substituting values for nondetects can ruin results, and what can be done about it. *Chemosphere* **2006**, *65*, 2434–2439. [CrossRef]
12. Hwang, M.; Lee, S.C.; Park, J.H.; Choi, J.; Lee, H.J. Statistical methods for handling nondetected results in food chemical monitoring data to improve food risk assessments. *Food Sci. Nutr.* **2023**, 1–13. [CrossRef]
13. Jones, M.P. Linear regression with left-censored covariates and outcome using a pseudolikelihood approach. *Environmetrics* **2018**, *29*, e2536. [CrossRef]
14. Kuiper-Goodman, T.; Hilts, C.; Billiard, S.; Kiparissis, Y.; Richard, I.; Hayward, S. Health risk assessment of ochratoxin A for all age-sex strata in a market economy. *Food Addit. Contam.* **2010**, *27*, 212–240. [CrossRef] [PubMed]
15. Hawkins, N.; Norwood, S.; Rock, J. *A Strategy for Occupational Exposure Assessment*; American Industrial Hygiene Association: Fairview, VA, USA, 1991; ISBN 0932627463.
16. Kaplan, E.L.; Meier, P. Nonparametric estimation from incomplete observations. *J. Am. Stat. Assoc.* **1958**, *53*, 457–481. [CrossRef]
17. Schmoyeri, R.; Beauchamp, J.; Brandt, C.; Hoffman, F. Difficulties with the lognormal model in mean estimation and testing. *Environ. Ecol. Stat.* **1996**, *3*, 81–97. [CrossRef]
18. She, N. Analyzing censored water quality data using a non-parametric approach. *JAWRA J. Am. Water Resour. Assoc.* **1997**, *33*, 615–624. [CrossRef]
19. Čonková, E.; Laciakova, A.; Kováč, G.; Seidel, H. Fusarial toxins and their role in animal diseases. *Vet. J.* **2003**, *165*, 214–220. [CrossRef] [PubMed]
20. De Boevre, M.; Jacxsens, L.; Lachat, C.; Eeckhout, M.; Di Mavungu, J.D.; Audenaert, K.; Maene, P.; Haesaert, G.; Kolsteren, P.; De Meulenaer, B.; et al. Human exposure to mycotoxins and their masked forms through cereal-based foods in Belgium. *Toxicol. Lett.* **2013**, *218*, 281–292. [CrossRef]
21. Sumarah, M.W. The deoxynivalenol challenge. *J. Agric. Food Chem.* **2022**, *70*, 9619–9624. [CrossRef]
22. EFSA Panel on Contaminants in the Food Chain. Risks to human and animal health related to the presence of deoxynivalenol and its acetylated and modified forms in food and feed. *EFSA J.* **2017**, *15*, e04718.
23. Klötzel, M.; Schmidt, S.; Lauber, U.; Thielert, G.; Humpf, H.U. Comparison of different clean-up procedures for the analysis of deoxynivalenol in cereal-based food and validation of a reliable HPLC method. *Chromatographia* **2005**, *62*, 41–48. [CrossRef]
24. MacDonald, S.J.; Chan, D.; Brereton, P.; Damant, A.; Wood, R.; Dao Duy, K.; Felgueiras, I.; Feron, T.; Herry, M.; Ioannou-Kakouri, E.; et al. Determination of deoxynivalenol in cereals and cereal products by immunoaffinity column cleanup with liquid chromatography: Interlaboratory study. *J. AOAC Int.* **2005**, *88*, 1197–1204. [CrossRef]
25. Sulyok, M.; Berthiller, F.; Krska, R.; Schuhmacher, R. Development and validation of a liquid chromatography/tandem mass spectrometric method for the determination of 39 mycotoxins in wheat and maize. *Rapid Commun. Mass Spectrom.* **2006**, *20*, 2649–2659. [CrossRef]
26. Sulyok, M.; Krska, R.; Schuhmacher, R. A liquid chromatography/tandem mass spectrometric multi-mycotoxin method for the quantification of 87 analytes and its application to semi-quantitative screening of moldy food samples. *Anal. Bioanal. Chem.* **2007**, *389*, 1505–1523. [CrossRef]
27. Varga, E.; Glauner, T.; Berthiller, F.; Krska, R.; Schuhmacher, R.; Sulyok, M. Development and validation of a (semi-) quantitative UHPLC-MS/MS method for the determination of 191 mycotoxins and other fungal metabolites in almonds, hazelnuts, peanuts and pistachios. *Anal. Bioanal. Chem.* **2013**, *405*, 5087–5104. [CrossRef]
28. Malachová, A.; Sulyok, M.; Beltrán, E.; Berthiller, F.; Krska, R. Optimization and validation of a quantitative liquid chromatography–tandem mass spectrometric method covering 295 bacterial and fungal metabolites including all regulated mycotoxins in four model food matrices. *J. Chromatogr. A* **2014**, *1362*, 145–156. [CrossRef]
29. De Santis, B.; Debegnach, F.; Gregori, E.; Russo, S.; Marchegiani, F.; Moracci, G.; Brera, C. Development of a LC-MS/MS method for the multi-mycotoxin determination in composite cereal-based samples. *Toxins* **2017**, *9*, 169. [CrossRef] [PubMed]
30. Burnham, K.P. *Model Selection and Multimodel Inference. A Practical Information-Theoretic Approach*; Springer: New York, NY, USA, 2002.
31. Hansen, M.H.; Kooperberg, C. Spline adaptation in extended linear models (with comments and a rejoinder by the authors. *Stat. Sci.* **2002**, *17*, 2–51. [CrossRef]
32. Belov, D.I.; Armstrong, R.D. Distributions of the Kullback–Leibler divergence with applications. *Br. J. Math. Stat. Psychol.* **2011**, *64*, 291–309. [CrossRef] [PubMed]
33. Paulo, M.J.; van der Voet, H.; Jansen, M.J.; ter Braak, C.J.; van Klaveren, J.D. Risk assessment of dietary exposure to pesticides using a Bayesian method. *Pest Manag. Sci. Former. Pestic. Sci.* **2005**, *61*, 759–766. [CrossRef]
34. Shoari, N.; Dubé, J.S. Toward improved analysis of concentration data: Embracing nondetects. *Environ. Toxicol. Chem.* **2018**, *37*, 643–656. [CrossRef] [PubMed]
35. Aitchison, J.; Brown, J.A.C. *The Lognormal Distribution, with Special Reference to Its Uses in Economics*; Cambridge University Press: Cambridge, UK, 1957; Volume 16, pp. 228–230.
36. Shimizu, K. *History, Genesis, and Properties, Chap. 1 in Lognormal Distributions: Theory and Applications*; Crow, E.L., Shimizu, K., Eds.; CRC Press: Boca Raton, FL, USA, 1988; ISBN 9780367580278.
37. Lee, E.T. *Statistical Methods for Survival Data Analysis*; Wiley: Hoboken, NJ, USA, 1992; ISBN 1118095022.
38. Johnson, N.L.; Kotz, S.; Balakrishnan, N. Beta distributions. *Contin. Univariate Distrib.* **1994**, *2*, 210–275.
39. Hui, Y.H. *Handbook of Food Science, Technology, and Engineering*; CRC Press: Boca Raton, FL, USA, 2006; Volume 149, ISBN 9780849398476.

40. Jongenburger, I.; Reij, M.; Boer, E.; Zwietering, M.; Gorris, L. Modelling homogeneous and heterogeneous microbial contaminations in a powdered food product. *Int. J. Food Microbiol.* **2012**, *157*, 35–44. [CrossRef] [PubMed]
41. De Oliveira, T.L.C.; de Araújo Soares, R.; Piccoli, R.H. A Weibull model to describe antimicrobial kinetics of oregano and lemongrass essential oils against Salmonella Enteritidis in ground beef during refrigerated storage. *Meat Sci.* **2013**, *93*, 645–651. [CrossRef] [PubMed]
42. Shoari, N.; Dubé, J.S. Estimating the mean and standard deviation of environmental data with below detection limit observations: Considering highly skewed data and model misspecification. *Chemosphere* **2015**, *138*, 599–608. [CrossRef] [PubMed]
43. Ben-Israel, A. A Newton-Raphson method for the solution of systems of equations. *J. Math. Anal. Appl.* **1966**, *15*, 243–252. [CrossRef]
44. Schwarz, G. Estimating the dimension of a model. *Ann. Stat.* **1978**, *6*, 461–464. [CrossRef]
45. Akaike, H. Information Theory and an Extension of the Maximum Likelihood Principle. In Proceedings of the International Symposium on Information Theory, Tsahkadsor, Armenia, 2–8 September 1971; Petrov, B.N., Csaki, F., Eds.; Akadémiai Kiadó: Budapest, Hungary, 1973; pp. 267–281.
46. Hastings, W.K. Monte Carlo sampling methods using Markov chains and their applications. *Biometrika* **1970**, *57*, 97–109. [CrossRef]

toxins

MDPI

Article

B-Type Fumonisins in Post-Fermented Tea: Occurrence and Consumer Dietary Exposure in Guangxi, China

Taotao Qiu [1], Jialin Zhu [1], Huayi Zhang [1], Biyun Xu [1], Yanju Guo [1], Jingrong Li [1], Xin Xu [1], Fenglin Peng [1], Weiguo Liu [1], Shengmei Zhao [1], Zuocheng Yin [1] and Shihong Mao [2,*]

[1] College of Physical Education and Health, Guangxi Normal University, Guilin 541004, China; gxqiutaotao@163.com (T.Q.); shengmeizhao123@163.com (S.Z.)
[2] College of Tourism & Landscape Architecture, Guilin University of Technology, Guilin 541006, China
* Correspondence: glmsh@glut.edu.cn; Tel.: +86-0773-5821659

Abstract: Post-fermented tea (PFT), a commonly consumed beverage worldwide, is characterized by the rapid growth of its microbial groups and the substantial changes they undergo. Consequently, PFT may contain mycotoxins such as B-type fumonisins (FBs). This study aimed to assess the intake of FBs through the consumption of PFT among consumers in Guangxi, China. A novel quantitative method using high-performance liquid chromatography-mass spectrometry was used to determine the FB concentration in PFT products. Additionally, a PFT consumption survey was conducted using a face-to-face questionnaire, recording their body weight and PFT consumption patterns based on a three-day dietary recall method. Finally, hazard index was calculated to estimate the health risk of FBs from the consumption of PFT products in Guangxi. The results revealed that the occurrence of FBs in PFT was 20% (24/120), with a concentration ranging from 2.14 to 18.28 μg/kg. The results of the survey showed that the average daily consumption of PFT by consumers was 9.19 ± 11.14 g. The deterministic risk assessment revealed that only 0.026% of the provisional maximum tolerable daily intake of FBs was consumed through PFT, indicating that FB contamination in PFT is not a public health risk.

Keywords: post-fermented tea; exposure assessment; risk characterization; fumonisins

Key Contribution: Fumonisins are a group of mycotoxins that pose a threat to public health and can be ingested through contaminated post-fermented tea. However, results indicate that B-type fumonisins exposure from post-fermented tea consumption is not a serious health risk in Guangxi, China.

Citation: Qiu, T.; Zhu, J.; Zhang, H.; Xu, B.; Guo, Y.; Li, J.; Xu, X.; Peng, F.; Liu, W.; Zhao, S.; et al. B-Type Fumonisins in Post-Fermented Tea: Occurrence and Consumer Dietary Exposure in Guangxi, China. *Toxins* **2023**, *15*, 534. https://doi.org/10.3390/toxins15090534

Received: 27 July 2023
Revised: 26 August 2023
Accepted: 28 August 2023
Published: 30 August 2023

1. Introduction

Fumonisins (FUMs) are a group of mycotoxins that pose a threat to public health [1]. They are natural toxins with low molecular weight and are produced as a water-soluble secondary metabolite by *Fusarium verticillioides*, *Fusarium proliferatum*, and other *Fusarium* species, which can subsequently contaminate farm products in the field or during processing, transportation, or storage [2]. For example, *Fusarium* is the principal indigenous fungus found in the soils of subtropical tea plantations in China [3] and is a potential source of contamination in the production of tea. There are four categories of FUMs: A, B, C, and P. Notably, the B-type fumonisins (FBs) are the most abundant and toxic family, including fumonisin B_1 (FB_1), fumonisin B_2 (FB_2), and fumonisin B_3 (FB_3). The occurrence of FBs is often directly related to climatic conditions. Guangxi, one of the cities in China where post-fermented tea (PFT) is commonly consumed, has a typical subtropical monsoon climate with short winters, long summers, and abundant annual precipitation. The climatic conditions in Guangxi are extremely favorable for the growth of toxigenic fungi. Moreover, FBs are mainly found in cereals and other agricultural products and have many toxic effects, such as carcinogenicity, immunotoxicity, teratogenicity, and hepatotoxicity, as demonstrated through tests conducted on rodents [4,5]. In addition, FB_1 has been classified as belonging to the International Agency for Research on Cancer (IARC) Group 2B substances, implying that this mycotoxin is a possible human carcinogen [6] and has been shown to induce human esophageal cancer in many parts of Central America, Asia, and South Africa [7,8]. Additionally, FB_2 and FB_3 are almost structurally identical to FB_1 at a molecular level; therefore, the magnitude of their toxicity may be at the same level as that of FB_1 [9]. Due to their toxicity and risk to human health, the Joint Expert Committee on Food Additives evaluated FBs and allocated this mycotoxin a provisional maximum tolerable daily intake (PMTDI) of 2 μg/kg·bw/day [10]. Therefore, the need to detect mycotoxin contamination in food and assess their exposure to the human population is increasingly necessary for safeguarding human health.

As a healthy beverage known for its characteristic flavor, PFT, including the Pu-erh, Liupao, and Fu brick varieties, is a staple beverage that is consumed daily worldwide [11]. The components of PFT have unique biological activities, including antioxidant, antibacterial, antimutagenic, and anti-cancer properties, as well as health benefits that may aid in the prevention of cardiovascular disease [12]. The long aging process and pile fermentation method that is integral to the production of PFT are key to the formation of its characteristic flavors and active ingredients and involve the rapid growth of and substantial changes in the microbial groups present in the tea. Microbial groups can promote the transformation of the original tea compounds, improve the flavor of tea, and generate new active ingredients, including polyphenols, caffeine, and aroma-forming substances [13]. However, compared with the aging conditions of raw Pu-erh tea produced from non-fermented tea leaves, pile fermentation can enhance the formation of mycotoxins [14]. In addition, FB contamination can occur at any stage of commodity production, especially under warm and humid climatic environmental conditions [15]. Notably, PFT is packaged in various forms, such as baskets, cans, paper packaging, or aluminum foil bags, in Guangxi, China. Poor storage and processing conditions may lead to the growth of toxigenic microfungi, particularly under high temperature and humidity conditions. The presence of toxigenic microfungi in PFT is a risk factor for FB contamination. Furthermore, considering the water solubility of FBs, PFT consumption may be an important factor in the exposure of tea consumers to FBs. Considering the health risks associated with FBs, the risk assessment of PFT consumption has become increasingly important.

Consequently, in the last decade, the risk assessment of mycotoxin exposure from PFT consumption has received heightened attention. Studies on the presence of mycotoxins in PFT have been conducted in several countries. For example, a preliminary study on mycotoxins in PFT reported that all 36 Pu-erh tea samples purchased from Italian markets tested positive for aflatoxins (AFs) and ochratoxin A (OTA) [16]. In a recent study conducted in China, OTA and zearalenone were detected in PFT samples [17,18]. In another study, the mycotoxin citrinin was detected in Chinese Liupao tea [19]. However, FB contamination of PFT has not been widely studied in China, and there are gaps in the existing literature related to FB residues and the risk assessment of PFT consumption.

To avoid the toxic effects of FB exposure, an accurate and specific mycotoxin detection method is critical for evaluating the safety of tea samples. In the current study, high-performance liquid chromatography-tandem mass spectrometry (HPLC-MS/MS) was the preferred method for detecting mycotoxins due to its high accuracy, stability, and reproducibility. Currently, HPLC-MS/MS is widely used for the detection of FBs in human food and animal feed [20–23]. Moreover, considering the unique mycotoxin profile of PFT products compared with those of other food products and animal feed, a quantitative method of FB detection with high accuracy and broad applicability is required. Additionally, we calculated the dietary risk of FB exposure by comparing the estimated daily intake (EDI) with the PMTDI. The EDI comprises a combination of food consumption data, typical contamination levels, and consumer body weight. Although previous surveys have shown that FB exposure levels in many countries, including Tanzania, Somalia, Mexico, and Brazil, exceeded the PMTDI [24–27], indicating potential health risks from dietary exposure to different food products, few studies have assessed the potential human health risks of consuming PFT contaminated with FBs. To the best of our knowledge, the risk assessment of FBs in PFT in Guangxi has not been previously reported. Therefore, the aims of this study were to develop a detection method to monitor FBs in PFT, assess the occurrence of FBs in PFT in this geographical region for the first time, and conduct a risk assessment of FBs exposure for PFT consumers in Guangxi. We hypothesize that the FBs contaminating PFT are completely ingested by tea consumers.

2. Results and Discussion

2.1. Method Development and Validation

Post-fermented tea products are rich in fiber, pigments, polyphenols, and caffeine, which are easily coextracted with mycotoxins and strongly influence the accurate quantitation of FBs. Conventionally, an acetonitrile-water mixture (at different percentages) is used as the extraction solvent for mycotoxin analysis [20]. Acetonitrile, as a polar extraction solvent, can reduce the extraction of pigments and lipophilic materials such as chlorophylls and fats and has a high capacity to extract plant ingredients with different polarities [28]. Furthermore, acidified solutions have been shown to reduce the interference of proteins and sugars in mycotoxin extraction [29], and acetic acid has been widely used for the extraction of FBs [30,31]. Moreover, FBs are hydrophilic mycotoxins that are soluble in acetonitrile and water [32]. Therefore, in this study, an acetonitrile-water mixture (50:50, v/v) containing 5% acetic acid was used as the extraction solvent for the determination of FBs in PFT. In addition, multifunctional cleanup column (MFC) have been shown to be efficient in eliminating pigments, lipids, proteins, caffeine, and polyphenols from tea extraction solutions. Hence, an MFC was used in this study to eliminate PFT matrix interference for FB determination.

In this study, an accurate analytical method for the determination of FBs in PFT was developed using HPLC-MS/MS and validated based on recovery, linearity, limit of detection (LOD), limit of quantification (LOQ), and repeatability criteria. The retention times of FB_1, FB_2, and FB_3 were found to be about 11.02 min, 12.45 min, and 11.76 min, respectively (Figure 1(B1–B3)). The HPLC-MS/MS method showed good linearity for the target FBs, with correlation coefficients ranging from 0.9989 to 0.9996, as shown in Table 1. The LOD and LOQ were calculated using the method described below. In this study, the LOD and LOQ for FBs were determined to be in the ranges of 0.7–3.0 and 2–9.0 μg/kg (Table 1), respectively. The LOD and LOQ values were lower than those previously reported, demonstrating that the HPLC-MS/MS method used in this study had a high sensitivity for FB determination [33,34]. Previous studies have used HPLC to detect FBs; however, few HPLC-MS/MS methods have been developed for FB detection in PFT [35,36]. In particular, the potential for phenolic compounds to interfere with HPLC-fluorescence or HPLC-UV methods must be considered [18]. Therefore, we utilized an MFC to eliminate polyphenols from the tea extraction solution, thereby improving the sensitivity and accuracy of the HPLC-MS/MS method that we developed for FB determination in PFT products.

The accuracy of the method was calculated by comparing the recovery of FBs from spiked control blanks to those of PFT samples. Each recovery experiment was carried out three times over three days, and the average recovery for FBs in the PFT samples is presented in Table 1. The mean recovery values for different FBs ranged from 67.7 to 77.13%, and the RSD was 5.81%–7.9%. These results indicate that the PFT exhibited the strongest matrix effect, which is consistent with the results of previous studies [37]. Furthermore, the recovery values for each FB determined by the newly developed method satisfied the analysis requirements of Commission Regulation (EC) No. 401/2006 (recovery values: 60–120%, RSD $\leq$ 30%) [38].

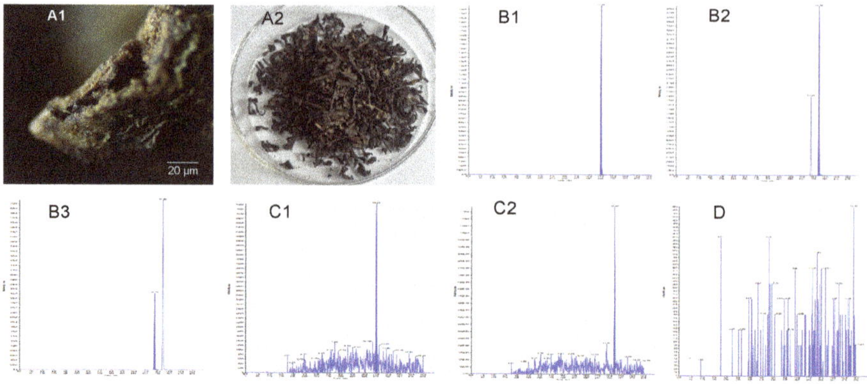

Figure 1. The microscope observation (**A1**) and photograph (**A2**) of the PFT samples; HPLC-MS/MS chromatogram of the FB_1 (**B1**), FB_2 (**B2**), and FB_3 (**B3**); Total ion chromatogram of PFT products contaminated Liupao tea with FB_1 (**C1**) and FB_2 (**C2**); a blank Liupao tea (**D**).

Table 1. Average recoveries, relative standard deviations (RSDs), limit of detection (LOD), limit of quantification (LOQ), linear regression equations, correlation coefficients (R^2) for FB determination by HPLC-MS/MS in PFT.

Mycotoxins	Fortification Level (μg/kg)	Recovery (%)	RSD (%)	LOQ (μg/kg)	LOD (μg/kg)	Regression Equation	R^2
Fumonisin B_1	250	67.7	5.81	3	9	y = 11,020x + 4705.7	0.9994
Fumonisin B_2	250	76.47	7.9	0.7	2	y = 20,797x + 116,020	0.9989
Fumonisin B_3	62.5	77.13	5.78	0.7	2	y = 26,700x + 53,722	0.9996

2.2. Occurrence and Concentration of FBs in PFT

The newly developed HPLC-MS/MS method was applied to analyze 120 commercial PFT samples, including Pu-erh, Liupao, and Fu brick teas. The HPLC-MS/MS chromatograms of Liupao tea contaminated with FB_1 and FB_2 and a blank sample without FBs contamination are shown in Figure 1(C1,C2,D). The present study revealed an incidence of positive samples of 20.00% (24/120), with levels of FB_1 and FB_2 contamination in the range of 2.00–18.00 μg/kg (Table 2). Of these, 80% of samples showed FB concentrations lower than the LOD, 10% of samples were contaminated with FB_1, 16.67% of samples were contaminated with FB_2, 6.67% of samples were co-contaminated with FB_1 and FB_2, and FB_3 was not detected in any sample. Notably, fungal hyphae were not observed in the positive samples by direct visual inspection, even when using optical microscopy (Figure 1(A1,A2)). Therefore, it would be difficult for consumers to assess FB contamination in PFT with direct visual inspection when buying PFT products. The mean FB_1 of all positive samples and total samples investigated was 5.42 and 0.54 μg/kg, respectively, and the mean FB_2 of all positive samples and total samples investigated was 5.70 and 0.95 μg/kg, respectively. The results demonstrated that the highest levels of FB_1 and FB_2 contamination were found in Fu brick tea at 11.81 and 18.28 μg/kg, respectively. Meanwhile, Liupao and Pu-erh teas showed the highest levels of FB_1 contamination at 7.07 and 10.44 μg/kg, respectively, and FB_2 contamination at 11.00 and 15.00 μg/kg, respectively. The lack of detectable FB_3 content in the 120 PFT samples analyzed was in accordance with results reported in several other studies on Pu-erh, green, and black teas [39,40]. However, the FB_1 and FB_2 values in this study were higher than those found in Pu-erh tea by Haas et al. [16] and PFT by Ye et al. [18], yet they were below the detection value of black tea and medicinal plants in Portugal [41]. The differences in the levels of FB contamination may be attributed to the different sample sources, purification techniques, and quantification methods.

Table 2. Occurrence and concentrations of FBs in PFT products sold in Guangxi.

PFT Products	Package Type	Positive/Total	Positive (%)	Number of Samples					FB_1 Maximum (μg/kg)	FB_2 Maximum (μg/kg)	FB_1 M1 (μg/kg)	FB_1 M2 (μg/kg)	FB_2 M1 (μg/kg)	FB_2 M2 (μg/kg)
				<LOD (μg/kg)	FB_1	FB_2	FB_3	FB_1 and FB_2						
Liupao teas	Basket packing, and paper packaging	8/38	21.05	30	4	8	0	4	7.07 ± 0.24	11.00 ± 0.32	0.55	5.25	0.87	4.13
	Aluminum foil bag packaging, plastic bag packaging, and canned	1/8	12.5	7	0	1	0	0	/	3.11 ± 0.14	/	/	0.39	3.11
Pu-erh teas	Basket packing, and paper packaging	11/50	22	39	6	7	0	2	10.44 ± 0.11	15.00 ± 0.68	0.6	5	0.92	6.57
	Aluminum foil bag packaging, plastic bag packaging, and canned	1/10	10	9	0	1	0	0	/	6.04 ± 0.78	/	/	0.6	6.04
Fu brick teas	Basket packing, and paper packaging	3/14	21.43	11	2	3	0	2	11.81 ± 0.54	18.28 ± 0.34	1	7	1.86	8.67
	Total	24/120	20	96	12	20	0	8	11.81 ± 0.54	18.28 ± 0.34	0.54	5.42	0.95	5.7

/: not detected. M1: mean value of total samples. M2: mean value of all positive samples.

Several previous studies have compared FB concentrations in food products, including tea. Table 2 shows that FB_1 and FB_2 contamination was detected in Fu brick, Liupao, and Pu-erh teas in Guangxi, which is consistent with the Pu-erh tea contamination data of Guangzhou, China [42]. The risk of FB contamination in PFT could be due to the high temperature and humidity environmental conditions in PFT production, sales, and storage areas (Guangdong and Guangxi in China), providing an ideal condition for the growth of *Fusarium* sp. [43]. Notably, the percentage of positive samples for Liupao, Pu-erh, and Fu brick teas packaged in baskets or paper was 21.05%, 22.00%, and 21.43%, respectively. In comparison, the percentage of positive samples for Liupao and Pu-erh tea packaged in aluminum foil bags, plastic bags, or cans was 12.50% and 10.00%, respectively (Table 2). Furthermore, the presence of FB_1 and FB_2 concurrent contamination in Liupao, Pu-erh, and Fu brick teas packaged in baskets or paper and the mean FB_1 and FB_2 in both the total samples investigated and all positive samples in PFT packaged with baskets or paper were higher than those of PFT packaged with aluminum foil bags, plastic bags, or cans. These results imply that utilizing aluminum foil bags, plastic bags, and cans as packaging can effectively prevent FB contamination during PFT storage. Scant data exists regarding the mechanisms of FB contamination of PFT [15]. We theorize that FB contamination may occur in all aspects of tea planting, harvesting, production processing, distribution, and storage; however, as rapid microbial growth is a part of the lengthy aging procedure of PFT production, methods for preventing the growth of and contamination by FBs are critical.

2.3. Demographic Profile and PFT Consumption Patterns in Guangxi

Exposure assessment is an important yet challenging step in risk assessment and mainly describes the hazard level of a substance based on factors such as exposure routes, dose, and frequency, along with population characteristics [44]. Based on the existing literature, limited data are available on the accurate consumption of PFT in Guangxi. To obtain more accurate consumption data, average PFT product consumption data were obtained from major cities in Guangxi using a face-to-face questionnaire method (Table 3). The demographic profile shows that the male consumer population (63.18%) was higher than that of female consumers, according to the survey results presented in Table 3. The PFT consumption data were consistent with the results reported by Yao et al. [45]. Similar to our findings, Guan et al. reported that the distribution of respondents according to age showed that more than half of the participants were older than 45 years [46]. The average body weight of PFT consumers was 62 ± 11.01 kg, which aligned with the results of previous reports designating an average body weight of 60 kg for Chinese adults [47]. Moreover, we found that 55.16% of participants preferred Liupao tea, 21.65% preferred Pu-erh tea, and 18.68% typically consumed two or more types of PFT (Table 4). In terms of brewing method preference, 40.77% of consumers chose multiple brewing for preparing PFT. Table 4 displays the frequency of PFT consumption in Guangxi, which confirmed that the majority of study participants drink PFT 1–7 times a week. In addition, Figure 2A shows no statistically significant differences in PFT consumption based on sex ($p > 0.05$), while significant differences were observed based on age ($p < 0.05$). The 45- to 60-year-old group showed an overall higher level of PFT consumption relative to the <45- and >60-year-old groups. The average consumption of PFT was 9.19 ± 11.14 g/day (Figure 2A), while the average consumption of PFT by unit weight was 0.15 ± 0.18 g/kg·bw/day for PFT consumers (Figure 2B).

Table 3. Demographic profile of PFT consumption participants sampled in Guangxi (N = 910).

Demographic Characteristics		Population Ratio
Gender	Male	575 (63.18%)
	Female	335 (36.81%)
Age	18–44	372 (40.88%)
	45–60	437 (48.02%)
	>60	101 (11.10%)
Respondents distribution	Nanning	96 (10.55%)
	Liuzhou	174 (19.12%)
	Guilin	228 (25.06%)
	Wuzhou	100 (10.99%)
	Other cities	312 (34.36%)
Total		910 (100%)
Average body weight (kg)		62.11 ± 11.01

Table 4. Preferences and frequency of PFT consumption in Guangxi. (N = 910).

	PFT Products	Sample Population Ratio		Brewing Method	Sample Population Ratio
Tea preference	Liupao teas	502 (55.16%)	Brewing method preference	Multiple brewing for drinking	371 (40.77%)
	Pu-erh teas	197 (21.65%)		Single brewing for drinking	338 (36.70%)
	Fu brick teas	41 (4.51%)		Long-term steaming for drinking	141 (15.49%)
	More than one type of tea	170 (18.68%)		More than one brewing method	64 (7.03%)
Times/day	Sample population ratio	Times/week	Sample population ratio	Times/month	Sample population ratio
1–2	639 (70.22%)	1–7	656 (72.09%)	1–15	398 (43.74%)
3–4	252 (27.70%)	8–14	197 (21.65%)	16–30	377 (41.43%)
>4	19 (2.09%)	>14	57 (6.26%)	>30	135 (14.84%)

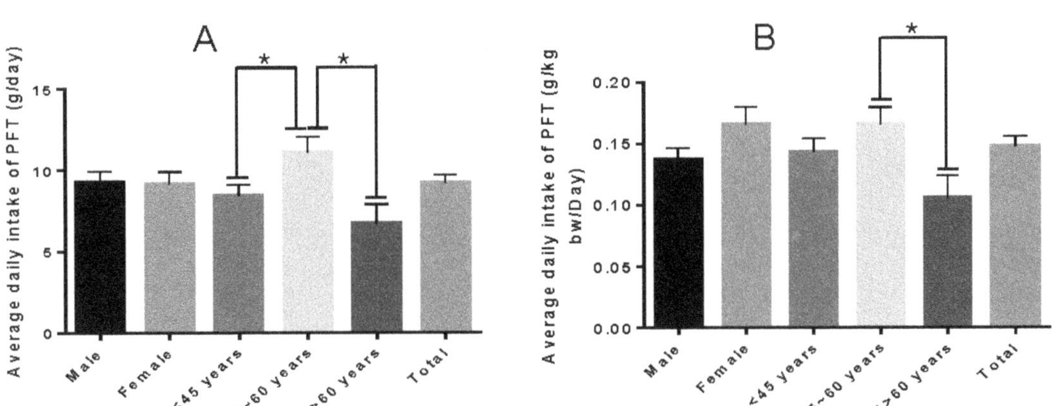

Figure 2. Average daily intake of PFT (**A**) and average daily intake of PFT by unit weight (**B**) in Guangxi (one-way ANOVA). * Mean significant difference ($p < 0.05$).

2.4. Exposure Assessment and Risk Characterization

Dietary exposure is commonly assessed using the point and probability assessment methods. Point assessment uses fixed values to determine each value in the mycotoxin assessment model, such as using the mean value to represent the average level of exposure in the population [48]. In this study, point assessment was calculated using the evaluation of the risk assessment of FBs in PFT from the chronic toxicity approach, based on methods recommended by the World Health Organization [49]. The data for levels of FB contamination and PFT consumption were combined to estimate the FB exposure level. The chronic PFT intake assessment of Guangxi consumers indicated that the level of FB exposure through the consumption of Pu-erh, Liupao, and Fu brick teas showed no observable risk (Table 5). The exposure to FBs in average PFT consumers amounted to 0.026% of the PMTDI of FBs (PMTDI, 2 μg/kg·bw/day), consumed via the teas included in this study. The average EDI of FB_1 was 0.000274, 0.000278, and 0.000343 μg/kg·bw/day for Liupao, Pu-erh, and Fu brick teas, respectively. The average EDI of FB_2 in this population was 0.00015, 0.000176, and 0.00032 μg/kg·bw/day for Liupao, Pu-erh, and Fu brick teas, respectively. These values are considerably lower than the PMTDI of FBs, indicating a low risk of FB toxicity for Guangxi PFT consumers. Similarly, Ye et al. detected the contaminant level of FBs in Chinese dark tea, and the hazard index (HI) values for FBs were far below 1.0 [18], indicating no risk concern for consumers.

Table 5. Dietary exposure assessment of FB consumption for PFT consumers in Guangxi.

PFT Products	FB Mean		FB Intake		HQ_1 (%)	HQ_2 (%)	HI (%)
	FB_1 Mean (μg/kg)	FB_2 Mean (μg/kg)	FB_1 Intake (μg/kg·bw/day)	FB_2 Intake (μg/kg·bw/day)			
Liupao tea	1.83	1	0.000274	0.00015	0.0137	0.00749	0.02119
Pu-erh tea	1.85	1.17	0.000278	0.000176	0.01388	0.00878	0.02265
Fu brick tea	2.29	2.13	0.000343	0.00032	0.01714	0.01599	0.03313
Average	1.99	1.43	0.000298	0.000215	0.0149	0.01075	0.02566

FBs mean: FB mean value of total samples. PMTDI = 2 μg/kg·bw/day. Average body weight = 62.11 kg. FB intake = (FB mean × PFT consumption)/average body weight. HQ = FB intake × 100/PMTDI. HI = HQ_1 + HQ_2.

The maximum regulatory limits for mycotoxins in tea have been established based on residual information on dry tea. However, as water infusion is the general method of tea consumption, it is necessary to monitor mycotoxin levels in PFT, estimate the average human intake, and perform health risk assessments. Recent studies have reported that the concentration of OTA and FB_2 in tea infusions depends on the infusion pH, and the total dissolved solids in the water do not substantially affect the transfer of mycotoxins from tea to infusions [50]. One limitation of this study was that, although FBs are water-soluble toxins, their migration rates during brewing were not addressed. Future studies are required to assess the migration rate of FBs during tea brewing to accurately determine FB intake. In this study, we hypothesized that the FBs contaminating PFT are completely ingested by tea consumers. Similar to the results of previous studies, those of the current study indicate that the actual risk of exposure to FBs through PFT consumption is lower than the estimated exposure level.

Another limitation of this study was that FB levels were only investigated in PFT. It is well known that the population of Guangxi consumes a great variety of foods on a daily basis, and the presence of contamination with multiple mycotoxins is common, especially in agricultural products [51]. One survey showed that 70% of Pu-erh tea samples tested positive for aflatoxin B_1 (AFB_1) [52], while another showed that AFB_2, AFG_1, AFG_2, deoxynivalenol, and enniatin B_1 were detected in tea samples [53]. Most foods, such as dairy products, cereals, beans, dried fruits, fresh fruits, and vegetable juices, are potentially contaminated with multiple mycotoxins [54–56]. The risk of mycotoxin contamination in human food and animal feed is widespread, owing to environmental and socioeconomic factors, agricultural practices, and production processes [24,57]. Although the results of our study show that FB contamination in PFT was not a public health risk, future studies are required to determine the combined synergistic or additive harmful effects caused by cumulative exposure to multiple mycotoxin contaminants, and an assessment of the risk of potential mycotoxin exposure from the daily intake of all foods is necessary to accurately capture the risk of intake in a population.

3. Conclusions

Post-fermented tea has a unique flavor and a clear bioactive effect, which are aligned with consumer preference. Consequently, the risk assessment of mycotoxin exposure in PFT has become increasingly important. A reliable and accurate method using purification steps coupled with HPLC-MS/MS was developed for determining FBs in PFT products. Overall, the incidence of FBs in the PFT samples was 20%, and the maximum level of FB_2 was 18 μg/kg. The risk assessment shows that the EDI ranged from 0.00015 to 0.000343 μg/kg·bw/day, which was far below the PMTDI value. Therefore, FB exposure through PFT beverage consumption does not represent a health risk for Guangxi consumers. Nevertheless, considering the potential for mycotoxin contamination of multiple food and beverage products, further detection of comprehensive dietary exposure to multiple mycotoxin contaminants is necessary. This study elucidates the risk of FB hazards in PFT products sold in Guangxi and provides a reference for further research on the detection and assessment of the mycotoxin contamination of other teas in other tropical and subtropical regions. Furthermore, the study results support the implementation of local initiatives aimed at preventing mycotoxin contamination.

4. Materials and Methods

4.1. Sources of PFT Samples, Chemicals, and Reagents

A total of 120 representative commercial PFT samples (60 Pu-erh, 46 Liupao, and 14 Fu brick teas) were randomly selected and purchased from various local markets in Guangxi, China, between January and December 2021. We obtained the FB_1, FB_2, and FB_3 reference standards from Pribolab Pte. Ltd. (Qingdao, China). FB_2 and FB_3 are mixed-standard substances. We purchased HPLC-grade methanol and acetonitrile (99% purity) from Merck (Darmstadt, Germany). All other chemicals were purchased from commercial sources. Magnesium sulfate was purchased from Shanghai Macklin Biochemical Technology Co., Ltd. (Shanghai, China) and baked at 120 °C for 12 h to remove any residual water before use. A MycoSpin^{TM} 400 MFC was purchased from Romer Labs Singapore Pte. Ltd. (Singapore). Ultrapure water was obtained using a Milli-Q system (MilliporeSigma, Billerica, MA, USA). Standard stock solutions (50 mg/L) were prepared by diluting commercial FBs in acetonitrile and water (50:50, v/v) containing 5% acetic acid. Interval standard solutions were prepared by pipetting 1.0 mL of FB_1, FB_2, and FB_3 into separate ampoules. The standard stock and interval standard solutions were stored at −20 and 4 °C, respectively.

4.2. Sample Pre-Treatment

Acetonitrile and water (50:50, v/v) containing 5% acetic acid were selected as the extraction solvents. A MycoSpin^{TM} 400 MFC was used for purification. The detailed procedure for the sample pretreatment was as follows: the ground PFT sample (10 g) was placed in a 250 mL triangular flask, to which 100 mL of acetonitrile and water (50:50, v/v) containing 5% acetic acid were added. After 2 h of shaking followed by centrifugation, the supernatant was decanted, and 2 mL of the supernatant was placed into a 5 mL centrifuge tube containing 250 mg of magnesium sulfate. Furthermore, 100 μL of acetic acid was added, and the mixture was vortexed for 3 min. After centrifugation, 750 μL of supernatant extract was passed through the MFC for purification. After the above extract was centrifuged for 1 min at 6793× g, 75 μL each of supernatant and internal standard solution were added. Finally, 120 μL of the supernatant was filtered through a 0.22 μm pore membrane prior to HPLC-MS/MS analysis.

4.3. Sample Analysis

The FBs in the PFT samples were determined using an Agilent 1290-Sciex QTRAP 5500 HPLC-MS/MS system equipped with a Dual Agilent Jet Stream ESI source operating in the positive mode and an Agilent 7683 B autosampler. For HPLC-MS/MS analysis, FBs were separated using a Phenomenex Gemini HPLC C18 column (4.6 mm × 150 mm) (Phenomenex Inc., Torrance, CA, USA). The separation was performed at 40 °C, and the injection volume was 20 μL. The mobile phase was composed of eluent A (water with 2 mmol/L ammonium acetate and 0.5% acetic acid) and eluent B (methanol with 2 mmol/L ammonium acetate and 0.5% acetic acid), and gradient elution was as follows: 10% B for 0–1 min, 10–97% B for 13 min, 97% B over 1 min, 10–97% B over 0.1 min, and 10% B for 5 min, with a flow rate of 1 mL/min and a total run time of 20 min. For the MS analysis, the ion source temperature was 650 °C, while the collision gas was of medium strength, the air curtain gas was 35 psi, ion source gas 1 was 60 psi, and ion source gas 2 was 65 psi.

Recovery rates were evaluated by adding concentrations of FB_1 (250 μg/kg), FB_2 (250 μg/kg), and FB_3 (62.50 μg/kg) into FBs-free PFT samples. Analyses were conducted once a day for three days, and recovery was determined by comparing the average peak areas of the spiked samples to those of the matrix-matched standard solutions. The LOD and LOQ were calculated based on signal-to-noise ratios (S/N) of 3 (3:1) and 10 times (10:1) the background chromatographic noise, respectively.

4.4. PFT Consumption Survey

A PFT consumption survey was conducted in Guangxi, China, during the first half of 2021. The population sample was predetermined as PFT consumers aged 18 years or older. Participants were informed (orally and via written instructions) about the general purpose of the survey and the basic principles of anonymity, confidentiality, and data protection. The data was collected through personal interviews. As all participants were aged 18 years old or older, interviews were conducted following verbal consent. A three-section structured questionnaire was developed and used in this study. The questionnaire was designed for direct interviews according to the China Health and Nutrition Survey and general guidelines for data collection [58]. The first section included general, self-reported demographic information, including gender, age, and weight. The second section examined PFT consumption patterns, including the frequency and quantity consumed. The third section prompted respondents to report their consumption of PFT the day before and in the last three days, including the type of product and the amount of consumed products (in g). The 24 h dietary recall is the most commonly used recall method. In addition to the 24 h dietary recall, a three-day dietary recall was included to cross-check the results and capture the habitual intake of PFT. To accurately capture the quantity and type of PFT intake, the participants were provided with 5 g each of Pu-erh, Liupao, and Fu brick tea as visual aids. These products are the most commonly consumed types of PFT in Guangxi; therefore, other PFT varieties (such as Qingzhuan and Kangzhuan) were not included in this study.

4.5. Exposure Assessment

An exposure assessment was developed to predict the severity and probability of adverse health effects from the intake of FBs via PFT. To characterize the FB exposure from PFT consumption, the amount of PFT consumed during a specific period and the typical FB contamination levels in PFT products were considered. In this study, the Guangxi PFT product consumption data were mathematically treated to represent the average amount (g) of PFT consumed per day based on the daily and three-day consumption reported during field research. Quantitative MS was performed to obtain preliminary estimates of the average concentration of FBs in PFT products sold in Guangxi.

The average concentrations used in the FB exposure calculations were determined using the following formula:

$$Ct = (M_1 + M_2 + \ldots + M_n)/N$$

where Ct is the average FB concentration, N is the number of testing samples, and M is the detected mean FB concentration in the PFT. The exposure to FB through PFT consumption was calculated as EDI using data on PFT consumption (μg/kg·bw/day), FB concentration, and body weight, according to the following formula:

$$\text{EDI} = \frac{M_{pc\times Cp}}{b\omega}$$

where M_{pc} is the average amount of PFT consumed (g), bw is body weight (kg), and C_p is the average concentration of FBs in PFT (μg/g). The hazard quotient (HQ) for exposure assessment was determined by comparing the EDI of FBs with the PMTDI of FBs set at 2 μg/kg·bw/day. To assess the risk resulting from dietary exposure to FBs in PFT, the HI for multiple mycotoxin exposures was calculated as the sum of the HQ using the following formula:

$$\text{HI} = \text{HQ}_1 + \text{HQ}_2 + \ldots + \text{HQ}_n \quad (1)$$

4.6. Statistical Analysis

Each experiment was performed in triplicate. All values are presented as the mean ± the standard error of the mean. Differences within the groups were analyzed using repeated-measures one-way analysis of variance. When the level of FBs in a PFT sample was higher than the LOD, the sample was considered positive, whereas an FB level lower than the LOD was considered negative. The PFT samples with FB concentrations below the LOD were assumed to be equal to LOD/2 according to the EU guidelines and were used for exposure calculations [59]. IBM SPSS 19.0 software (Armonk, NY, USA) was used for statistical analysis.

Author Contributions: Experiments and manuscript preparation, S.M. and T.Q.; conceived and designed the experiments, T.Q., F.P., W.L. and S.M.; performed the experiments, J.Z., H.Z., B.X. and Z.Y.; analyzed the data and wrote the manuscript, J.L., X.X., Y.G. and S.Z. All authors have read and agreed to the published version of the manuscript.

Funding: This work was supported by the Guangxi Natural Science Fundation (2020GXNSFAA159064, 2019GXNSFBA24506), Innovation Project of Guangxi Graduate Education(GY2023067, YCSW2023202) and the Guilin Science and Technology Development Program (Project) (20210202-2).

Institutional Review Board Statement: Not applicable.

Informed Consent Statement: Informed consent was obtained from all subjects involved in the study.

Data Availability Statement: The data presented in this study are available on request from the corresponding author.

Conflicts of Interest: The authors declare no competing financial interest.

References

1. International Agency of Research on Cancer. Mycotoxins and human health. In *Improving Public Health through Mycotoxin Control*; IARC: Lyon, France, 2012; pp. 87–104.
2. Li, R.; Tao, B.; Pang, M.; Liu, Y.; Dong, J. Natural occurrence of fumonisins B_1 and B_2 in maize from three main maize-producing provinces in China. *Food Control* **2015**, *50*, 838–842. [CrossRef]
3. Wang, L.M.; Huang, D.F.; Fang, Y.; Wang, F.; Li, F.L.; Liao, M. Soil fungal communities in tea plantation after 10 years of chemical vs integrated fertilization. *Chil. J. Agric. Res.* **2017**, *77*, 355–364. [CrossRef]
4. Chen, J.; Wen, J.; Tang, Y.; Shi, J.; Mu, G.; Yan, R.; Cai, J.; Long, M. Research progress on fumonisin B_1 contamination and toxicity: A review. *Molecules* **2021**, *26*, 5238–5259. [CrossRef] [PubMed]
5. Abdul, N.S.; Marnewick, J.L. Fumonisin B_1-induced mitochondrial toxicity and hepatoprotective potential of rooibos: An update. *J. Appl. Toxicol.* **2020**, *40*, 1602–1613. [CrossRef]
6. IARC Working Group on the Evaluation of Carcinogenic Risks to Humans. Some traditional herbal medicines, some mycotoxins, naphthalene and styrene. *IARC Monogr. Eval. Carcinog. Risks Hum.* **2002**, *82*, 1–556.
7. Chu, F.S.; Li, G.Y. Simultaneous occurrence of fumonisin B_1 and other mycotoxins in moldy corn collected from the People's Republic of China in regions with high incidences of esophageal cancer. *Appl. Environ. Microbiol.* **1994**, *60*, 847–852. [CrossRef]
8. Khan, R.B.; Phulukdaree, A.; Chuturgoon, A.A. Fumonisin B_1 induces oxidative stress in oesophageal (SNO) cancer cells. *Toxicon* **2018**, *141*, 104–111. [CrossRef]

9. Wall-Martínez, H.A.; Ramírez-Martínez, A.; Wesolek, N.; Brabet, C.; Durand, N.; Rodríguez-Jimenes, G.C.; García-Alvarado, M.A.; Salgado-Cervantes, M.A.; Robles-Olvera, V.J.; Roudot, A.C. Risk assessment of exposure to mycotoxins (aflatoxins and fumonisins) through corn tortilla intake in veracruz city (Mexico). *Food Addit. Contam. Part A* **2019**, *36*, 929–939. [CrossRef]
10. Marin, S.; Ramos, A.J.; Cano-Sancho, G.; Sanchis, V. Mycotoxins: Occurrence, toxicology, and exposure assessment. *Food Chem. Toxicol.* **2013**, *60*, 218–237. [CrossRef]
11. Zhang, M.; Otake, K.; Miyauchi, Y.; Yagi, M.; Yonei, Y.; Miyakawa, T.; Tanokura, M. Comprehensive NMR analysis of two kinds of post-fermented tea and their anti-glycation activities in vitro. *Food Chem.* **2019**, *277*, 735–743. [CrossRef]
12. Jiang, H.Y.; Shii, T.; Matsuo, Y.; Tanaka, T.; Jiang, Z.H.; Kouno, I. A new catechin oxidation product and polymeric polyphenols of post-fermented tea. *Food Chem.* **2011**, *129*, 830–8306. [CrossRef] [PubMed]
13. Abd El-Aty, A.M.; Choi, J.H.; Rahman, M.M.; Kim, S.W.; Tosun, A.; Shim, J.H. Residues and contaminants in tea and tea infusions: A review. *Food Addit. Contam. Part A Chem. Anal. Control Expo. Risk Assess.* **2014**, *31*, 1794–1804. [CrossRef] [PubMed]
14. World Health Organization. *Safety Evaluation of Certain Food Additives and Contaminants*; World Health Organization Press: Geneva, Switzerland, 2011; pp. 146–160.
15. Sedova, I.; Kiseleva, M.; Tutelyan, V. Mycotoxins in tea: Occurrence, methods of determination and risk evaluation. *Toxins* **2018**, *10*, 444. [CrossRef]
16. Haas, D.; Pfeifer, B.; Reiterich, C.; Partenheimer, R.; Reck, B.; Buzina, W. Identification and quantification of fungi and mycotoxins from pu-erh tea. *Int. J. Food Microbiol.* **2013**, *166*, 316–322. [CrossRef] [PubMed]
17. Cui, P.; Yan, H.; Granato, D.; Ho, C.T.; Ye, Z.; Wang, Y.; Zhang, L.; Zhou, Y. Quantitative analysis and dietary risk assessment of aflatoxins in Chinese post-fermented dark tea. *Food Chem. Toxicol.* **2020**, *146*, 111830. [CrossRef]
18. Ye, Z.; Wang, X.; Fu, R.; Yan, H.; Han, S.; Gerelt, K.; Cui, P.; Chen, J.; Qi, K.; Zhou, Y. Determination of six groups of mycotoxins in Chinese dark tea and the associated risk assessment. *Environ. Pollut.* **2020**, *261*, 114180. [CrossRef]
19. Li, Z.; Mao, Y.; Teng, J.; Xia, N.; Huang, L.; Wei, B.; Chen, Q. Evaluation of mycoflora and citrinin occurrence in Chinese Liupao Tea. *J. Agric. Food Chem.* **2020**, *68*, 12116–12123. [CrossRef]
20. Oplatowska-Stachowiak, M.; Haughey, S.A.; Chevallier, O.P.; Galvin-King, P.; Campbell, K.; Magowan, E.; Adam, G.; Berthiller, F.; Krska, R.; Elliott, C.T. Determination of the mycotoxin content in distiller' s dried grain with solubles using a multianalyte UHPLCMS/MS method. *J. Agric. Food Chem.* **2015**, *63*, 9441–9451. [CrossRef]
21. Pantano, L.; La Scala, L.; Olibrio, F.; Galluzzo, F.G.; Bongiorno, C.; Buscemi, M.D.; Macaluso, A.; Vella, A. QuEChERS LC-MS/MS screening method for mycotoxin detection in cereal products and spices. *Int. J. Environ. Res. Public. Health* **2021**, *18*, 3774. [CrossRef]
22. Mokubedi, S.M.; Phoku, J.Z.; Changwa, R.N.; Gbashi, S.; Njobeh, P.B. Analysis of mycotoxins contamination in poultry feeds manufactured in selected provinces of South Africa using UHPLC-MS/MS. *Toxins* **2019**, *11*, 452. [CrossRef]
23. Osaili, T.M.; Al-Abboodi, A.R.; Awawdeh, M.A.; Jbour, S.A.M.A. Affiliations expandassessment of mycotoxins (deoxynivalenol, zearalenone, aflatoxin B_1 and fumonisin B1) in hen's eggs in Jordan. *Heliyon* **2022**, *8*, e11017. [CrossRef]
24. Wielogorska, E.; Mooney, M.; Eskola, M.; Ezekiel, C.N.; Stranska, M.; Krska, R.; Elliott, C. Occurrence and Human-Health impacts of mycotoxins in Somalia. *J. Agric. Food Chem.* **2019**, *67*, 2052–2060. [CrossRef]
25. Kamala, A.; Kimanya, M.; Lachat, C.; Jacxsens, L.; Haesaert, G.; Kolsteren, P.; Ortiz, J.; Tiisekwa, B.; De Meulenaer, B. Risk of exposure to multiple mycotoxins from maize-based complementary foods in Tanzania. *J. Agric. Food Chem.* **2017**, *65*, 7106–7114. [CrossRef] [PubMed]
26. Piacentini, K.C.; Rocha, L.; Fontes, L.C.; Carnielli, L.; Reis, T.; Corrêa, B. Mycotoxin analysis of industrial beers from Brazil: The influence of fumonisin B_1 and deoxynivalenol in beer quality. *Food Chem.* **2017**, *218*, 64–69. [CrossRef] [PubMed]
27. Gilbert-Sandoval, I.; Wesseling, S.; Rietjens, I.M.C.M. Occurrence and probabilistic risk assessment of fumonisin B_1, fumonisin B_2 and deoxynivalenol in nixtamalized maize in Mexico city. *Toxins* **2020**, *12*, 644. [CrossRef] [PubMed]
28. González-Curbelo, M.Á.; Socas-Rodríguez, B.; Herrera-Herrera, A.V.; González-Sálamo, J.; Hernández-Borges, J.; RodríguezDelgado, M.Á. Evolution and applications of the QuEChERS method. *Trac. Trends Anal. Chem.* **2015**, *71*, 169–185. [CrossRef]
29. Rahmani, A.; Jinap, S.; Soleimany, F. Qualitative and quantitative analysis of mycotoxins. *Compr. Rev. Food Sci. Food Saf.* **2009**, *8*, 202–251. [CrossRef]
30. Bordin, K.; Rottinghaus, G.E.; Landers, B.R.; Ledoux, D.R.; Kobashigawa, E.; Corassin, C.H.; Oliveira, C.A.F. Evaluation of fumonisin exposure by determination of fumonisin B_1 in human hair and in Brazilian corn products. *Food Control* **2015**, *53*, 67–71. [CrossRef]
31. Yapo, A.E.; Strub, C.; Durand, N.; Ahoua, A.R.C.; Schorr-Galindo, S.; Bonfoh, B.; Fontana, A.; Koussémon, M. Mass spectrometry-based detection and risk assessment of mycotoxin contamination of 'kankankan' used for roasted meat consumption in Abidjan, Côte d'Ivoire. *Food Addit. Contam.* **2020**, *37*, 1564–1578. [CrossRef]
32. Visconti, A.; Doko, M.B.; Bottalico, C.; Schurer, B.; Boenke, A. Stability of fumonisins (FB_1 and FB_2) in solution. *Food Addit. Contam.* **1994**, *11*, 427–431. [CrossRef]
33. Sulyok, M.; Krska, R.; Senyuva, H. Profiles of fungal metabolites including regulated mycotoxins in individual dried Turkish figs by LC-MS/MS. *Mycotoxin Res.* **2020**, *36*, 381–387. [CrossRef] [PubMed]
34. Du, L.J.; Chu, C.; Warner, E.; Wang, Q.Y.; Hu, Y.H.; Chai, K.J.; Cao, J.; Peng, L.Q.; Chen, Y.B.; Yang, J.; et al. Rapid microwave-assisted dispersive microsolid phase extraction of mycotoxins in food using zirconia nanoparticles. *J. Chromatogr. A* **2018**, *1561*, 1–12. [CrossRef] [PubMed]

35. Monbaliu, S.; Wu, A.; Zhang, D.; Van Peteghem, C.; De Saeger, S. Multimycotoxin UPLC−MS/MS for tea, herbal infusions and the derived drinkable products. *J. Agric. Food Chem.* **2010**, *58*, 12664–12671. [CrossRef]
36. Mirón-Mérida, V.A.; Gong, Y.Y.; Goycoolea, F.M. Aptamer-based detection of fumonisin B_1: A critical review. *Anal. Chim. Acta* **2021**, *1160*, 338395–338427. [CrossRef]
37. Dzuman, Z.; Zachariasova, M.; Veprikova, Z.; Godula, M.; Hajslova, J. Multi-analyte high-performance liquid chromatography coupled to high-resolution tandem mass spectrometry method for control of pesticide residues, mycotoxins, and pyrrolizidine alkaloids. *Anal. Chim. Acta* **2015**, *863*, 29–40. [CrossRef] [PubMed]
38. European Commission. Commission Regulation (EC) No. 401/2006 of 23 February 2006 laying down the methods of sampling and analysis for the official control of the levels of mycotoxins in foodstuffs. *J. Eur. Communities* **2006**, *70*, 12–34.
39. Santos, L.; Marin, S.; Sanchis, V.; Ramos, A.J. Screening of mycotoxin multi contamination in medicinal and aromatic herbs sampled in Spain. *J. Sci. Food Agric.* **2009**, *89*, 1802–1807. [CrossRef]
40. Omurtag, G.Z.; Yazicioglu, D. Determination of fumonisin B_1 and B_2 in herbal tea and medicinal plants in Turkey by high-performance liquid chromatography. *J. Food Prot.* **2004**, *67*, 1782–1786. [CrossRef]
41. Martins, M.L.; Martins, H.M.; Bernardo, F. Fumonisins B_1 and B_2 in black tea and medicinal plants. *J. Food Prot.* **2001**, *64*, 1268–1270. [CrossRef]
42. Wu, J.Y.; Yang, G.Y.; Chen, J.L.; Li, W.X.; Li, J.T.; Fu, C.X.; Jiang, G.F.; Zhu, W. Investigation for Pu-Erh Tea contamination caused by mycotoxins in a tea market in Guangzhou. *J. Basic Appl. Sci.* **2014**, *10*, 349–356.
43. Nakagawa, H.; Hashimoto, R.; Matsuo, Y.; Sago, Y.; Yokoyama, K.; Takahashi, H. Detection and determination of fumonisins B_1, B_2, and B_3 contaminating Japanese domestic wine by liquid chromatography coupled to tandem mass spectrometry (LC–MS/MS). *Curr. Microbiol.* **2022**, *77*, 3057–3064. [CrossRef] [PubMed]
44. Stephenson, C.L.; Harris, C.A.; Clarke, R. An assessment of dietary exposure to glyphosate using refined deterministic and probabilistic methods. *Food Chem. Toxicol.* **2016**, *95*, 28–41. [CrossRef] [PubMed]
45. Yao, Y.; Chen, H.; Chen, L.; Ju, S.Y.; Yang, H.; Zeng, Y.; Gu, D.; Ng, T.P. Type of tea consumption and depressive symptoms in Chinese older adults. *BMC Geriatr.* **2021**, *21*, 331. [CrossRef] [PubMed]
46. Guan, X.; Yang, J.F.; Xie, X.Y.; Lin, C.; Li, J.Y. Research on the behavior of tea consumption in China with the CKB data. *J. Tea Sci.* **2018**, *38*, 287–295.
47. Jin, Y.L.; Liu, P.; Sun, J.F.; Wang, C.N.; Min, J.; Zhang, Y.F.; Wang, S.Y.; Wu, Y.N. Dietary exposure and risk assessment to lead of the population of Jiangsu province, China. *Food Addit. Contam. Part A Chem. Anal. Control Expo. Risk Assess.* **2014**, *31*, 1187–1195. [CrossRef]
48. Claeys, W.L.; Schmit, J.F.; Bragard, C.; Rogister, G.M.; Pussemier, L.; Schiffers, B. Exposure of several Belgian consumer groups to pesticide residues through fresh fruit and vegetable consumption. *Food Control* **2011**, *22*, 508–516. [CrossRef]
49. GEMS/FOOD. Global environment monitoring system-food contamination monitoring and assessment programme. In *GEM/Food Cluster Diets*; World Health Organization: Geneva, Switzerland, 2012.
50. Kiseleva, M.; Chalyy, Z.; Sedova, I. Tea: Transfer of mycotoxins from the spiked matrix into an infusion. *Toxins* **2021**, *13*, 404. [CrossRef]
51. Ibáñez-Vea, M.; Martínez, R.; González-Peñas, E.; Lizarraga, E.; Cerain, A.L. Co-occurrence of aflatoxins, ochratoxin a and zearalenone in breakfast cereals from Spanish market. *Food Control* **2011**, *22*, 1949–1955. [CrossRef]
52. Li, W.G.; Xu, K.L.; Xiao, R.; Yin, G.F.; Liu, W.W. Development of an HPLC-based method for the detection of aflatoxins in Pu-erh tea. *Int. J. Food Prop.* **2015**, *18*, 842–848. [CrossRef]
53. Pallarés, N.; Font, G.; Mañes, J.; Ferrer, E. Multimycotoxin LC-MS/MS analysis in tea beverages after dispersive liquid-liquid microextraction (DLLME). *J. Agric. Food Chem.* **2017**, *65*, 10282–10289. [CrossRef] [PubMed]
54. Ji, X.F.; Li, R.; Yang, H.; Qi, P.P.; Xiao, Y.P.; Qian, M.R. Occurrence of patulin in various fruit products and dietary exposure assessment for consumers in China. *Food Control* **2017**, *78*, 100–107. [CrossRef]
55. Kamle, M.; Mahato, D.K.; Gupta, A.; Pandhi, S.; Sharma, N.; Sharma, B.; Mishra, S.; Arora, S.; Selvakumar, R.; Saurabh, V.; et al. Citrinin mycotoxin contamination in food and feed: Impact on agriculture, human health, and detection and management strategies. *Toxins* **2022**, *14*, 85. [CrossRef] [PubMed]
56. Fusilier, K.; Chilvers, M.I.; Limay-Rios, V.; Singh, M.P. Mycotoxin co-occurrence in Michigan harvested maize grain. *Toxins* **2022**, *24*, 431. [CrossRef]
57. Magembe, K.S.; Mwatawala, M.W.; Mamiro, D.P. Mycotoxin contamination in stored maize and groundnuts based on storage practices and conditions in subhumid tropical Africa: The case of Kilosa district, Tanzania. *J. Food Prot.* **2016**, *79*, 2160–2166. [CrossRef]
58. Zhang, Y.B.; Li, Y.; Geng, T.T.; Pan, X.F.; Zhou, Y.F.; Liu, G.; Pan, A. Overall lifestyles and socioeconomic inequity in mortality and life expectancy in China: The China health and nutrition survey. *Age Ageing* **2022**, *51*, 167. [CrossRef] [PubMed]
59. Burstyn, I.; Teschke, K. Studying the determinants of ex-posure: A review of methods. *Ann. Occup. Hyg.* **1999**, *60*, 57–72.

pollutants

MDPI

Article

The Burden of Air Pollution: A Perspective on Global Health Inequalities

Agnieszka Jakubowska

Faculty of Economic Sciences, Koszalin University of Technology, 75-343 Koszalin, Poland; agnieszka.jakubowska@tu.koszalin.pl

Abstract: The aim of the research presented here is to assess the magnitude of the burden of health limitations due to air pollution on the world's economies. This burden was determined by the estimated number of premature years of life lost (YLLs) or years of health lost (YLDs) due to air-pollution-related diseases in the working-age population. Attention was drawn to the problem of existing inequalities in the global burden between economies with different income levels. The hypothesis of persistently high levels of inequality was tested by analysing the convergence process in a group of 204 countries over the period 1990–2019. The results indicate a high degree of variation in the level of health problems caused by air pollution. The analysis of the catching-up process of the most disadvantaged countries (with the highest level of health impacts from air pollution) did not show a positive convergence effect in the study group.

Keywords: air pollution; burden of disease; YLL; YLD; health inequalities; lost productivity

Citation: Jakubowska, A. The Burden of Air Pollution: A Perspective on Global Health Inequalities. *Pollutants* **2023**, 3, 419–436. https://doi.org/10.3390/pollutants3030029

Academic Editors: Roel Vermeulen, Lauren Petrick and Maaike van Gerwen

Received: 6 May 2023
Revised: 22 June 2023
Accepted: 10 July 2023
Published: 5 September 2023

1. Introduction

Air pollution is no longer just an environmental issue. It must also be considered in economic terms, as it causes a large global burden of disease. Although the consequences of poor air quality are particularly evident in the elderly (as a result of long-term exposure), the measurable effects of the burden of this risk factor can also be estimated for the working-age population. In terms of economic (social) costs, years of health and life lost prematurely due to illness or death significantly reduce the human resource potential of an economy [1–3] and are a source of lost productivity [4–7]. There is epidemiological evidence that particulate matter less than 2.5 μm in diameter (PM2.5) is an extremely hazardous factor for human health [8], indicating the need to study the impact of air pollution on morbidity and mortality, including the so-called indirect costs of disease. Reduced physical and mental fitness are now recognised as major causes of inactivity among people of working age [9,10], and the mechanisms by which health affects economic productivity are the subjects of numerous studies [11–14]. Although air pollution is commonly associated with areas of high production intensity, data on morbidity and mortality from this risk factor indicate that the burden of disease from air pollution is greatest in low- and middle-income countries [15]. It is in the lowest-producing countries that air pollution is now one of the major (or main) risk factors for the burden of disability or premature death.

The aim of the research presented here is to assess the magnitude of the burden of health limitations resulting from air pollution on the world's economies. This burden was determined by the estimated number of years of life lost prematurely (YLLs) or health lost (YLDs) due to air-pollution-related diseases in the working-age population. The analysis used Institute for Health Metrics and Evaluation (IHME) data from the Global Burden of Disease Study 2019 (GBD 2019) [16]. Attention was drawn to the problem of existing inequalities in the level of global burden on national economies determined by their income levels. The hypothesis of persistently high levels of inequality was tested by analysing the convergence process in a group of 204 countries over the period 1990–2019.

2. Literature Review

Evidence of the global burden of disease from environmental risk factors shows that the death rate from air pollution tends to increase as countries industrialise and move from low to middle income status [17]. Once a country reaches a higher level of development, these rates return to lower levels as both air quality and access to health care improve. Many studies confirm that the phenomenon of rapid increases in air pollution with increasing industrialisation in middle-income countries results in the highest mortality from this risk factor in this group of countries. The concept of the unequal burden of air pollution is based on the relationship between income inequality and economic development, which was put forward by Simon Kuznets in 1955 [18]. This is the so-called Environmental Kuznets Curve (EKC), which is often used to describe the relationship between economic growth and environmental quality. It refers to the hypothesis that there is an inverted U-shaped relationship between the value of output per capita and certain measures of environmental quality. The EKC thus describes a hypothetical relationship between various indicators of environmental degradation and the level of per capita income. In the early stages of a country's economic growth, emissions increase and environmental quality deteriorates, but above a certain level of per capita income (different for different environmental indicators) this trend reverses, leading to environmental improvement at high income levels [19]. It is therefore reasonable to assume that the level of environmental pressure can be expressed as an inverted U-shaped function of income per capita. According to the EKC concept, the precondition for environmental improvement is therefore an increase in income, which can then be translated into effective environmental policies. In the literature, the model has been the subject of numerous proposals for modifications to the relationship between environmental parameters and economic growth, highlighting potential mechanisms for adapting emission reduction practices that may differ significantly from the theoretical assumptions of the EKC model [20–22].

Ambient and household air pollution is currently a major global health problem, contributing to approximately 7 million (potentially preventable) deaths per year, reduced life expectancy and significant direct and indirect social costs [23,24]. Currently, air pollution is a major environmental risk factor for premature death and is responsible for significant morbidity with short and long-term health consequences [25]. Analyses of the association between air pollution and cerebrovascular [25–28], neurological [29–31], and psychiatric [32–34] disorders are well documented in the literature. Exposure to various forms of air pollutants, especially particulate matter, which is readily absorbed by the body, may potentially contribute to stroke, dementia, Parkinson's disease, cognitive impairment, and neurodevelopmental disorders [35,36]. There is also evidence of a positive association between high ambient levels of PM2.5 and the incidence of depression [37]. Inflammation and oxidative stress are thought to be key pathophysiological mechanisms by which air pollution causes brain damage [38]. It also creates conditions for thrombus formation by increasing coagulation factors and platelet activation [39]. At the same time, studies confirm a strong association between air pollution and cardiovascular morbidity and mortality [40,41].

As a result of its multiple effects on the human body, air pollution is now a major contributor to the global burden of disease. In 2019, 6.67 million deaths were attributed to this risk factor, accounting for 11.8% of all deaths. Of these, 4.1 million deaths (more than 61%) were specifically attributed to outdoor air pollution. According to data from the Institute for Health Metrics and Evaluation (IHME) published in the Global Burden of Disease Study 2019 [16], air pollution (outdoor and indoor) is the fourth-most-common risk factor in terms of intensity of exposure resulting in death, after high blood pressure, smoking, and poor diet. It is also a major cause of the burden of non-communicable diseases (NCDs), mainly ischaemic heart disease, stroke, lower respiratory infections, and chronic obstructive pulmonary disease. Air pollution is the fourth-leading risk factor (after malnutrition, hypertension and smoking) contributing to the global burden of disease, measured in healthy life years lost to death or disability. The IMHE estimates that air pollution will contribute to

213.3 million healthy life years lost to premature death or disability in 2019. This represents 17.5% of the global burden of disease from all risk factors combined.

As an environmental risk factor, air pollution is also a major contributor to the health and life expectancy of people of working age, with a measurable impact on national economies in terms of lost productivity of labour resources. According to the IMHE estimates cited above, in 2019, air pollution contributed to 43.0 million healthy life years lost due to premature death or disability among people in the 20–54 age group, representing 12.7% of the global burden of disease in this age group due to all risk factors combined.

Between 1990 and 2019, the global average total air pollution mortality rate for the population aged 20–54 years fell from 29.6 to 22.2 deaths per 100,000 people, mainly due to improvements in indoor air quality (the indoor air pollution mortality rate fell from 19.2 to 7.9 deaths per 100,000 people). Over the same period, the mortality rate from outdoor air pollution increased from 10.2 to 14.1 deaths per 100,000 population and now accounts for more than 63% of the total burden of premature death from air pollution [16]. The same data show a persistent burden attributed to airborne ozone pollution over the study period. The estimated burden of premature mortality due to exposure to this risk factor in the 20–54 age group decreased from 0.5 in 1990 to 0.4 deaths per 100,000 people in 2019. In line with the reduction in mortality rates, the contribution of ambient ozone to the global air pollution burden for 20–54 year olds increased from 1.7 to 1.8% over the period.

3. Materials and Methods

Data from the Institute for Health Metrics and Evaluation (IHME) Global Burden of Disease Study 2019 were used to assess the magnitude of the burden of disease caused by air pollution in working-age populations in 204 countries worldwide [16]. The spatial distribution of the burden was assessed using indicators representing the number of years of life lost (YLL) and the number of potential years of life with disability (YLD) due to diseases associated with exposure to air pollution. The burden measures were chosen on the basis of methods proposed in the literature to assess the phenomenon under study [42–44]. The approach adopted makes it possible to estimate the potentially lost effective working time of people of working age, while at the same time indicating the level of disease burden caused by a given risk factor on national economies. In the GBD 2019 study, three groups of pollutants were considered in the estimation of air pollution levels: (1) ambient particulate matter pollution, defined as the annual average daily exposure to outdoor concentrations of particulate matter with an aerodynamic diameter ≤ 2.5 µm (PM2.5); (2) household air pollution (NDS) from solid fuels, measured as individual exposure to indoor PM2.5 concentrations due to the use of solid fuels for cooking and heating; and (3) ozone air pollution. The average exposure data by country are population-weighted, i.e., they take into account the geographical distribution of the population in relation to the exposure.

Given the objective of the study, i.e., determining the degree of inequality in the burden on national economies resulting from the loss of effectiveness of human resources during the productive period, and the availability of aggregated data in the IHME database for age groups representing the range of highly productive labour resources, the analysis was based on variables describing the phenomenon under study in the population aged 20–54. Assuming that the level of health burden from air pollution is a function of both long-term exposure and increasing susceptibility to health risks associated with underlying diseases with increasing age, the reference group thus adopted in the process studied allowed at the same time to reduce the impact of age and related other cumulative health determinants on the occurrence of health damage. Comparability of data between countries was achieved by using intensity ratios, which refer to the number of cases per population size (rate). Statistical measures were used to determine the degree of variation in the parameters analysed across the study group of 204 countries.

Given the demonstrated heterogeneity of the burden of air pollution and its variability over time, the assumed global convergence in the burden of lost productivity on national economies was measured separately for parameters based on estimates of the burden of premature death (YLL(AP) rate) and long-term disability (YLD(AP) rate) in the working-age population defined for the whole study. The expected effect of decreasing global inequalities in the level of the burden of the consequences of air pollution on the world's economies will be an apparent process of positive convergence, consisting of a decrease over time in the average level of the indicators studied as their level of dispersion decreases. In order to assess the progress of the system under study towards equilibrium, the concept of beta (β) unconditional convergence, which takes into account different initial conditions and predicts that outliers will catch up more quickly with the target, and the concept of sigma (σ) convergence, defined as a decrease in the dispersion of performance, have been used. To test the β convergence hypothesis using cross-sectional data, a model was used to explain the growth of the trait under study in each country ($I = 1 \ldots, N$) between period t_0 and $t_0 + T$, using the initial value of the trait in each member state according to the formula:

$$\ln\left(\frac{y_{it_0+T}}{y_{it_0}}\right) = a + b\ln(y_{it_0}) + u_{it_0,it_0+T} \quad (1)$$

where: y_{it}—value of characteristic y in area i in period t; u_{it}—random disturbance. The existence of beta convergence is confirmed by the negative and statistically significant value of the estimator b.

The analysis assumes that beta convergence takes place in the study area when countries with an initially lower value of a given variable (the highest level of the YLL(AP) rate or the YLD(AP) rate) catch up by showing a faster rate of transition to the expected state, i.e., the ratio of the parameters y_{it0+T}/y_{it0} reaches the lowest possible value. The inclusion of absolute (unconditional) convergence in the study implies the assumption that all countries move towards the same long-run equilibrium state and reach it at the same time. However, countries with a worse initial position have a longer way to go. The occurrence of sigma convergence implies the achievement of the expected reduction in the level of dispersion of the characteristic under consideration over time. In order to verify the hypothesis of the occurrence of sigma convergence in the system under study, the amount of variance of the population in period t (σt^2) is used as a measure of the dispersion of the characteristic in question.

Taking into account the specificity of the studied process (burden of the studied world economies with the consequences of diseases caused by air pollution), the expected positive effect of the convergence process is a decrease in the average level of the studied parameters over time ($\overline{y}_{t_{0+T}} < \overline{y}_{t_0}$) along with a simultaneous decrease in the level of their dispersion ($\sigma^2_{t_{0+T}} < \sigma^2_{t_0}$). The results obtained make it possible to evaluate the pattern of convergence in the system studied.

4. Results

Estimating the burden of disease due to air pollution required simultaneous consideration of both the area-specific intensity of the risk factor (mainly PM2.5 in μg/m^3, indoor and outdoor) and the sensitivity of individual diseases to long-term exposure to these factors. A set of risk curves constructed from GBD data [45] shows that chronic obstructive pulmonary disease is particularly sensitive to increases in PM2.5 intensity (Figure 1). In this case, the level of the relative risk (RR) burden increases virtually linearly, reaching RR = 5.6 at a potential PM2.5 concentration of 600 μg/m^3. This means that the probability of adverse effects of chronic obstructive pulmonary disease (loss of life or health) in a person with long-term exposure to this level of PM2.5 is more than 5.5 times higher than in a person not exposed to this risk factor. A significant increase in the relative risk level with increasing PM2.5 concentrations in m^3 is also observed for ischaemic heart disease—in this case, with long-term exposure to PM2.5 at 600 μg/m^3, the RR increases to 3.2. For the burden of diseases such as stroke, type 2 diabetes, cancers of the trachea, bronchus, and lung, and lower respiratory tract infections, there is a strong increase in relative risk at initial exceedances of the safe exposure standards, while exceedances of 100 μg/m^3 no longer have a significant effect on the increase in risk. In the case of stroke, long-term exposure to

PM2.5 concentrations in the order of 90 μg/m^3 makes a person exposed to such exposure more than 3.5 times more likely to suffer or die from the disease than a person not exposed (RR = 3.6).

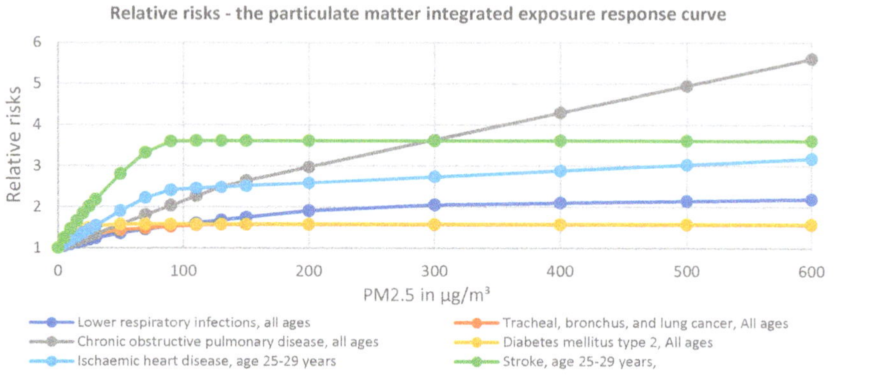

Figure 1. Integrated exposure–response curve for long-term exposure to PM2.5—relative risk (RR) values as a function of total PM2.5 intensity. Source: own elaboration based on: [45].

At the same time, GBD 2019 Summary Exposure Value (SEV) data [45] show a decrease in global exposure to total air pollutants over the past 30 years (Table 1), with the decrease driven by a significant reduction in exposure to household air pollution from solid fuels. Total exposure has increased for risk factors such as ambient particulate matter and ambient ozone.

Table 1. Summary exposure value (SEV) rate for 1990, 2019 and annualised rate of change (ARC) for 1990–2019.

Risks	SEV 1990 *	SEV 2019	ARC 1990–2019
I. Environmental/occupational risks	52.55 (48.66–55.92) **	45.36 (41.16–49.19)	−0.51 (−0.62–−0.40) *
A. Air pollution	45.37 (32.89–56.28)	34.72 (25.86–44.40)	−0.92 (−1.25–−0.61) *
A1. Particulate matter pollution	44.26 (31.87–55.07)	33.84 (25.08–43.43)	−0.93 (−1.26–−0.61) *
A1a. Ambient particulate matter pollution	15.65 (10.61–21.58)	26.19 (21.55–30.48)	1.78 (0.95–2.71) *
A1b. Household air pollution from solid fuels	27.33 (16.18–38.86)	12.04 (6.72–18.82)	−2.83 (−3.53–−2.20) *
A2. Ambient ozone pollution	50.67 (22.95–73.01)	61.19 (32.26–80.39)	0.65 (0.33–1.25) *

* Summary exposure value (SEV) is the RR-weighted prevalence of exposure, a univariate measure of risk weighted exposure, taking the value zero when no excess risk for a population exists and the value one when the population is at the highest level of risk. SEVs have been reported on a scale from 0% to 100%, on which a decline in SEV indicates reduced exposure to a given risk factor and an increase in SEV indicates increased exposure. ** 95% uncertainty interval (UI). Source: [45].

At present, air pollution levels vary widely between regions and countries. A greater burden of air pollution is observed in low- and middle-income countries, due to both the high levels of indoor pollution characteristic of low-income countries (due to reliance on solid fuels for cooking and heating) and the increasing levels of outdoor pollution as developing countries become more industrialised [46]. Global data on air pollution levels confirm the inequalities that still exist in this area. In fact, high average PM2.5 concentrations mainly affect African and Asian countries (Figure 2). Countries with the highest average PM2.5 concentrations include India (83 μg/m^3), Niger (80 μg/m^3), Nigeria (70 μg/m^3) and Egypt (68 μg/m^3). Their levels are almost double the global average and more than 500% higher than the European average. Between 1990 and 2019, while PM2.5 pollution levels in the European area decreased by almost 30%, the intensity of this risk factor increased by 9% in the Asian area. In the Indian area, the increase was 16%. However, the largest increase in PM2.5 intensity (43%) was recorded in Mongolia.

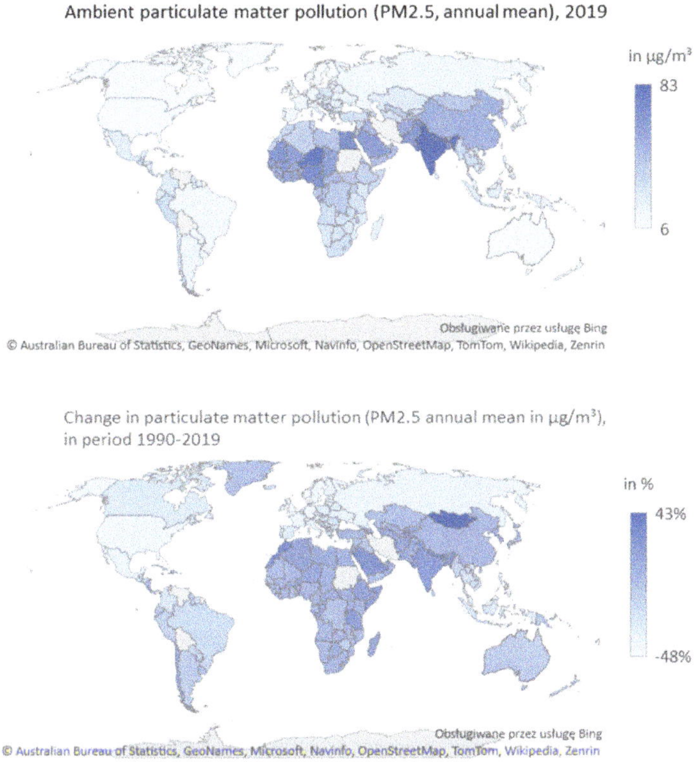

Figure 2. Spatial distribution of PM2.5 concentrations in 2019 and change from 1990 to 2019, N = 204. Source: own elaboration based on: [45].

An analysis of the relationship between a country's level of income (GDP per capita) and the average level of PM2.5 intensity (μg/m^3) in its territory shows that there is a persistent unfavourable relationship between air pollution levels and national product levels worldwide. This means that a large proportion of countries with low levels of production (low GDP per capita) have high levels of PM2.5 intensity (Figure 3). Over the period 1990–2019, the level of the correlation coefficient between the value of gross domestic product per capita and the intensity of PM2.5 in the air increased from −0.34 (1990) to −0.37 (2019), indicating a continuing stable trend in this respect.

According to IHME [16] data on the global burden of disease worldwide, in 2019, in the population aged 20–54 years, the average burden of disease resulting from long-term exposure to air pollution, as measured by the number of years of life and health lost due to these diseases (DALYs), was 1145.2 per 100,000 people in this age group. Taking into account the 204 global economies included in the study, this rate, depending on the country, ranged between 8672.6 (Solomon Islands) and 33.1 (Sweden). The key sources of air pollution burden in this age group were cardiovascular diseases (DALYrate = 786.5), chronic respiratory diseases (DALYrate = 108.7), diabetes and kidney disesse (DALYrate = 102.4), respiratory infections and tuberculososos (DALYrate = 84.1), and neoplasms (DALYrate = 54.7).

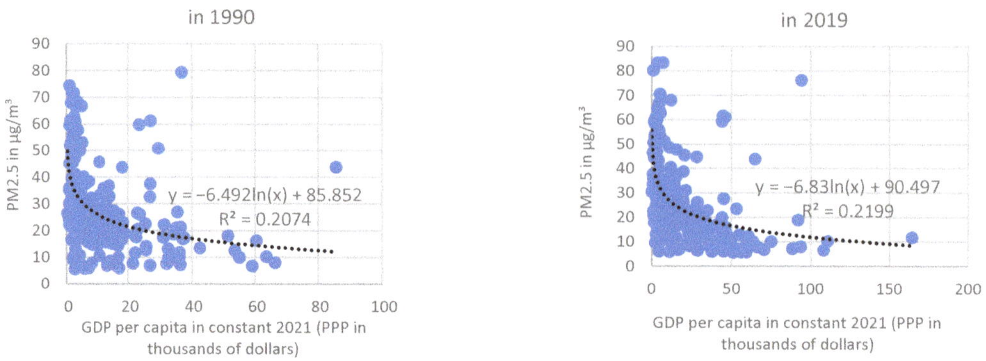

Figure 3. Relationship between the level of GDP per capita income level and PM2.5 air pollution levels. Source: GDP per capita [47] and PM2.5 in μg/m^3 [45].

A characteristic feature of the health burden of air pollution is the relatively high burden of premature mortality among those exposed. The ratio of years lived with disability (YLDs) to total years of healthy life lost (DALYs) due to exposure to air pollution in the population aged 20–54 years was on average 10.0%. In the study group it ranged from 4.3% (Solomon Islands) to 39.3% (Norway). A key issue in the burden on the potential of people of working age is therefore the high mortality rate of those affected by diseases related to air quality.

Over the period 1990–2019, the burden of premature mortality due to air pollution decreased significantly in the group of 204 countries studied—the average YLL(AP) rate fell from 879.8 to 1349.0 life years lost per 100,000 people aged 20–54 (the study also confirmed the statistical significance of these differences, assuming $\alpha = 0.05$, Wilcoxon test results $p = 0.000$). However, the scores remained highly dispersed, indicating persistent inequalities within the group (Figure 4). During the study period, the coefficient of variation of the YLL(AP) rate parameter increased from 81.0 to 100.4%. In contrast, the level of disability burden due to air pollution did not change significantly. The average YLD(AP)rate decreased from 144.6 to 143.0 life years lost per 100,000 persons aged 20–54. The level of the coefficient of variation decreased from 54.7 to 52.7 per cent over the period, indicating no apparent trend towards levelling out existing inequalities in this area (Table 2).

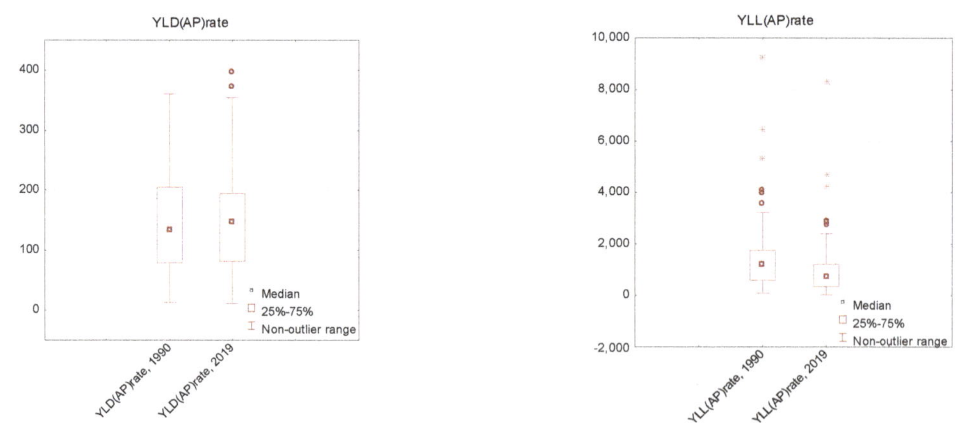

Figure 4. Distribution of YLL(AP)rate and YLD(AP)rate, population aged 20–54, 1990 and 2019, N = 204. Source: own study based on the data: [16].

Table 2. Descriptive statistics—YLL(AP)rate and YLD(AP)rate, population aged 20–54, 1990 and 2019, N = 204.

Variable	Descriptive Statistics						
	N Valid	Mean	Min.	Max.	Variance	SD	CV
YLL(AP)rate, 1990	204	1349.0	86.0	9249.9	1,193,365.6	1092.4	81.0
YLL(AP)rate, 2019	204	879.8	20.4	8299.1	780,868.3	883.7	100.4
YLD(AP)rate, 1990	204	144.6	12.9	360.3	6259.0	79.1	54.7
YLD(AP)rate, 2019	204	143.0	11.0	396.3	5674.7	75.3	52.7

Source: own study based on the data: [16].

Preliminary analysis of the IHME data confirms the assumption of inequalities in the global health burden of air pollution among the working-age population. This means that these people experience different levels of risk of morbidity and mortality from air pollution depending on the level of economic development of the area. As a result, most countries in Africa and Asia experience high levels of morbidity and mortality from exposure to air pollution, with a high contribution of this risk factor to the burden measured in healthy life years (HLY) lost due to death (YLL) or disability (YLD). The regional divide in this respect is clear, with most countries in Europe, North America and Latin America representing areas where the contribution of air pollution to total morbidity and mortality is low (Figure 5). In the study group of 204 countries, for the age group 20–54 years, the highest contribution of air pollution to total deaths occurred in Solomon Islands (20.0%), Vanuatu (17.7%), Kuwait (17.6%) and Egypt (16.9%). In contrast, the lowest contribution of air-pollution-related diseases to premature deaths was found in Finland (0.5%), Sweden (0.5%), Iceland (0.7%) and Estonia (0.7%). The highest contribution of air pollution to the burden measured in years of disability for the population aged 20 to 54 was found in Papua New Guinea (3.4%), Solomon Islands (3.4%) and Vanuatu (3.2%), and the lowest in Iceland (0.09%), New Zealand (0.10%) and Sweden (0.11%).

According to IHME data on the global burden of disease in 2019 [16], the YLL(AP) rate (i.e., YLLs per 100,000 people) in the population aged 20–54 years varied between 219.4 (high-income group countries) and 1428.5 (low-income group countries), depending on the level of economic development of the country (as measured by the World Bank). The level of the YLD(AP) rate (i.e., YLDs per 100,000 population) in the population aged 20–54 years varied during the study period between 68.7 (high-income countries) and 223.3 (lower-middle-income countries) (Table 3). The contribution of diseases caused by air pollution to the total number of years of life lost (YLL) in the 20–54 age group varied between 3.1% (high income group countries) and 9.5% (lower middle income group countries). It should be noted that in the high- and upper-middle-income countries, this share has decreased significantly over the period 1990–2019 (Table 3). In lower middle and low income countries, the share of air pollution as a risk factor for premature death increased by more than 12% on average. In terms of years of healthy life lost (YLDs), the contribution of air-pollution-related diseases to the total number of years lived with disability from all causes ranged from 0.5% (high-income countries) to 2.0% (lower-middle-income countries). Over the period 1990–2019, this share increased for high-income countries (an increase of 3.4%) and decreased for middle and low-income groups (a decrease of between 3% and 5%, depending on the income group).

The assessment of the correlation between the level of economic development and the level of the burden of disease caused by air pollution confirms the existence of a relatively strong relationship between these parameters in the group of 204 countries studied (Figure 6). The results show a clear correlation between the level of economic development of a country, measured by the value of GDP per capita, and the number of years of life (YLL(AP)rate) and health (YLD(AP)rate) lost due to air pollution. The relationships presented here confirm the unfavourable patterns observed globally—residents of countries with lower national product can statistically expect a higher burden of premature death and disease from air pollution. However, it is important to remember that globally, rising GDP is independently associated with greater urbanisation, and the increase in average exposure on the upward side of the EKC curve reflects both increasing industrialisation and a greater proportion of the total population living in urban areas, where air pollution levels are highest.

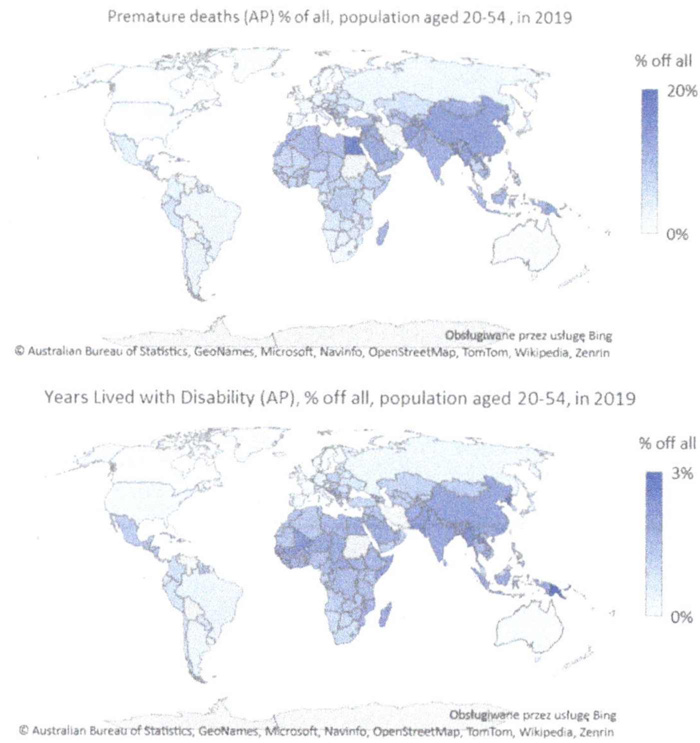

Figure 5. Contribution of air pollution (AP) to premature deaths and disability years, population aged 20–54, in 2019, N = 204.

Table 3. Air pollution (AP) burden of disease—YLLrate/YLD rate, population aged 20–54, by country group (breakdown by World Bank income level), in 2019 and change from 1990 to 2019.

World Bank Income Group	YLLs (AP) 2019			YLDs (AP) 2019		
	Rate *	% in All Causes	% in NCD	Rate *	% in All Causes	% in NCD
		2019				
High Income	219.4	3.1%	4.2%	68.7	0.5%	0.6%
Upper-Middle Income	771.5	7.6%	12.4%	163.6	1.6%	1.9%
Lower-Middle Income	1343.0	9.5%	16.3%	223.3	2.0%	2.5%
Low Income	1428.5	7.3%	15.5%	209.8	1.9%	2.5%
		Change over the period 1990–2019 (%)				
High Income	−50.0%	−31.5%	−40.3%	7.9%	3.4%	1.4%
Upper-Middle Income	−40.4%	−17.6%	−19.9%	−2.4%	−4.7%	−6.8%
Lower-Middle Income	−22.6%	12.6%	−7.5%	−7.6%	−4.1%	−7.2%
Low Income	−35.2%	12.6%	−11.0%	−10.8%	−3.5%	−11.9%

* YLLs/YLDs per 100,000 people.

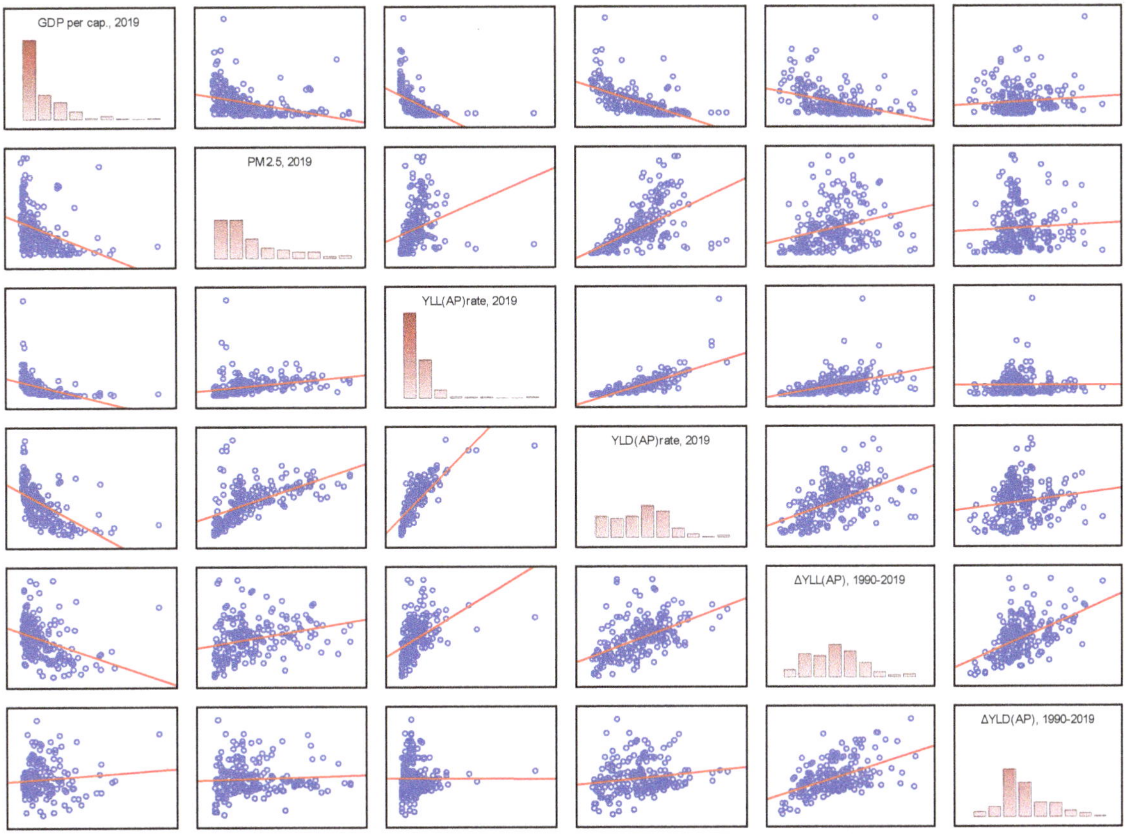

Figure 6. Correlation matrix.

Relatively strong ($r \geq 0.5$) and statistically significant ($p < 0.05$, for $\alpha = 0.05$) levels of negative correlation were obtained for both variable analyses (Table 4). The results obtained explain to some extent the reasons for the phenomenon of high air pollution levels in low-income economies. The statistically significant correlation ($r = -0.37$) shown between the level of national product (GDP per capita) and the level of PM2.5 pollution in its area indicates that a large part of low productivity economies are affected by higher levels of air pollution. At the same time, these results do not confirm the expected high positive correlation between PM2.5 intensity and the burden of disease from this risk factor on national economies. This suggests that there are other determinants of this process, such as access to health care. For the burden of premature mortality due to air pollution on the economy (YLLrate) in the 20–54 age group, the correlation coefficient was only 0.28. There was a stronger correlation between the intensity of exposure to PM2.5 and the burden of years of life lost due to disability (YLDrate). In this case, the correlation coefficient was 0.56 (with static significance) (Table 4).

In order to assess the persistence of global inequalities in the level of the burden of the consequences of air pollution on the world's economies, the correlation between the observed rate of change of the parameters under study (%ΔY, 1990–2019) and their level in 2019 was also assessed. The correlation coefficient obtained for the parameter YLLrate, $r = 0.46$ (at $p < 0.05$, for $\alpha = 0.05$), indicates the existence of a statistically significant, but not very strong, relationship between the current level of the burden of premature mortality due to air pollution and its rate of change over the study period. The results suggest that for a significant proportion of the countries studied, a high level of the burden indicator was associated with a higher rate of increase. This suggests that inequalities in the study group are being maintained over time. In the case of the burden of the effects of air pollution measured in years of disability (YLDrate), the correlation coefficient between the relative increase over time and the current level of burden was 0.16, which does not confirm the existing relationships in this area.

Table 4. Correlation results.

	Average	SD	GDP Per Cap., 2019 [1]	PM2.5 (Annual Mean, μg/m³), 2019 [2]	YLL(AP) Rate, 2019 [3]	YLD(AP) Rate, 2019 [3]	%ΔYLL(AP) Rate, 1990–2019	%ΔYLD(AP) Rate, 1990–2019
GDP per cap., 2019	22658	24332	1.000	−0.368 *	−0.489 *	−0.595 *	−0.362 *	0.092
PM2.5 (annual mean, μg/m³), 2019	26.1	17.7	−0.368 *	1.000	0.283 *	0.560 *	0.274 *	0.056
YLL(AP)rate, 2019	879.8	883.7	−0.489 *	0.283 *	1.000	0.785 *	0.458 *	−0.002
YLD(AP)rate, 2019	143.0	75.3	−0.595 *	0.560 *	0.785 *	1.000	0.516 *	0.163 *
ΔYLL(AP)rate, 1990–2019	−38%	26%	−0.362 *	0.274 *	0.458 *	0.516 *	1.000	0.522 *
ΔYLD(AP)rate, 1990–2019	3%	31%	0.092	0.056	−0.002	0.163 *	0.522 *	1.000

* Correlation coefficients determined are significant with $p < 0.05000$, N = 204. [1] Based on data: [47]. [2] Based on data: [45]. [3] Based on data: [16].

In order to verify the hypothesis assuming the existence of beta convergence in the studied process, an explanatory model of the growth of the studied characteristic in individual countries according to formula (1) was used. Assuming different initial conditions for individual world economies, determined by the level of the variables YLL(AP)rate and YLD(AP)rate in 1990, the rate of change of the analysed parameters in individual countries in the studied period (relation 2019/1990) was determined. It was assumed that in the case of countries with difficult initial conditions (high initial levels of the variables), this ratio should be as low as possible, indicating a rapid rate of decline of the parameter under study, giving the effect of catching up with the group (beta convergence).

Estimated from Formula (1), the regression models explaining the growth of the indicators studied, which determine the burden of premature death (YLLrate) or disability (YLDrate) due to air-pollution-related diseases in the study group of 204 countries, took the form:

(a) YLL (AP)rate variable:

$$\ln\left(\frac{\text{YLL (AP)rate}_{2019}}{\text{YLL (AP)rate}_{1990}}\right) = -1.7684 + 0.1716\ln(\text{YLL (AP)rate}_{1990}), R^2 = 0.0876;$$

(b) YLD variable (AP)rate:

$$\ln\left(\frac{\text{YLD (AP)rate}_{2019}}{\text{YLD (AP)rate}_{1990}}\right) = 0.2801 - 0.061\ln(\text{YLD (AP)rate}_{1990}), R^2 = 0.0199.$$

The resulting linear regression results of the parameters studied are shown in Figure 7.

In the case of the YLL(AP)rate parameter, the regression results obtained show that the value of the b-estimator is positive, indicating the absence of positive convergence effects within the group, despite a slight decrease in the degree of dispersion (variance) of the indicator under study between 1990 and 2019 (data in Table 2). This implies that disparities between the countries under study have decreased, but the observed lack of beta convergence does not warrant further conclusions in this respect. The expected effects of convergence, bringing the outlier economies (with the highest mortality burden) closer to the global average, are not visible.

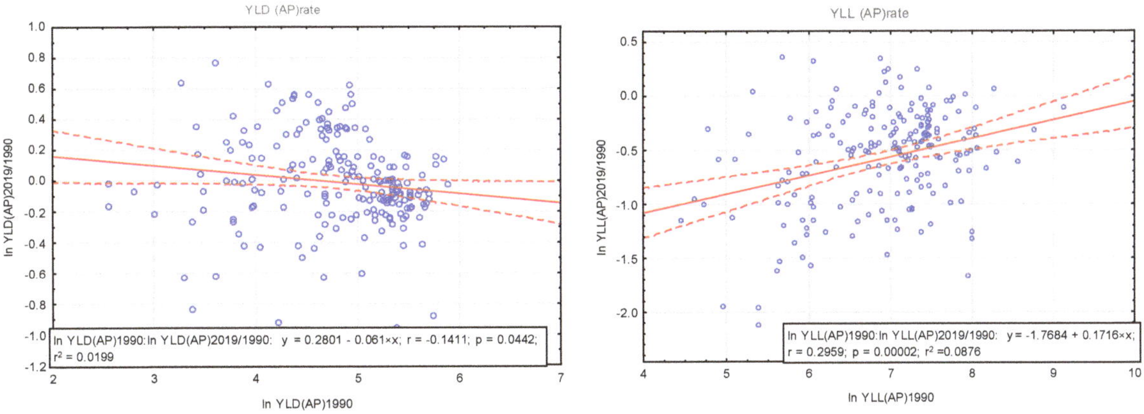

Figure 7. Patterns of convergence of YLLrate and YLDrate due to air pollution—beta convergence, over the period 1990–2019, N = 204.

In the case of the parameter YLD(AP)rate, the regression results obtained indicate that the value of the b-estimator is negative at −0.061 (at $p = 0.044$, for $\alpha = 0.05$), which could indicate a slow convergence effect in the group (however, the so-called half-life index derived from it indicates that, at the rate of convergence determined, the current disparities would only be halved in the next 330 years). The level of variance of this variable has also decreased slightly over time (data in Table 2), confirming the slow levelling of disparities associated with the convergence process. However, the lack of a measurable decrease in the YLD(AP) ratio over time in the group of world economies studied (the level of the variable decreased by only 7% over the period 1990–2019), as well as the small share of this burden in the total air pollution burden noted above, does not provide a sound basis for inferring a reduction in global inequalities in the health burden of air pollution.

The estimated regression model explaining the incremental total burden of premature death and disability (DALYrate) due to air-pollution-related diseases in the study group of 204 countries (Figure 8) was as follows:

$$\ln\left(\frac{\text{DALY (AP)rate}_{2019}}{\text{DALY (AP)rate}_{1990}}\right) = -1.3225 + 0.1184\ln(\text{DALY (AP)rate}_{1990}), R^2 = 0.0509.$$

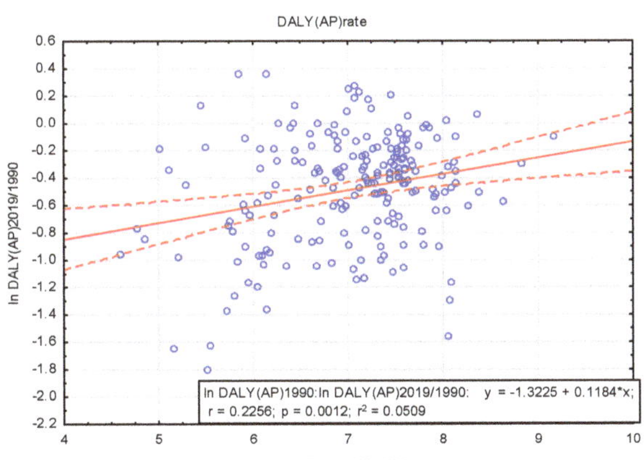

Figure 8. Patterns of DALYrate convergence due to air pollution—beta convergence, over the period 1990–2019, N = 204.

The results confirm that there is therefore no basis for inferring an ongoing process of convergence in the total health burden of air pollution worldwide. The parameter b (regression coefficient) is positive in this case b = 0.1184 and statistically significant ($p = 0.0012$, with an assumed $\alpha = 0.05$).

5. Discussion

Research reported in the literature indicates that adverse health effects of PM are strongly related to the size, composition and concentration of the pollutant particles and can occur even when relative risk estimates indicate that the pollutant parameters are within the specified range of values. The combination of species and diurnal variation in particle mass and composition are also major contributors to the toxicity of PM [48]. In practice, measuring the relevant parameters and quantifying their health effects is a difficult task, as many additional external factors strongly influence human morbidity and mortality [49]. Identifying and correcting for the influence of confounding factors remains a challenging task, although scientific advances are continually providing new information on the toxicity of trace elements in ambient PM [50,51]. It is expected that in the future, multi-element analysis will be required for PM determination, allowing the use of models to identify and partition sources [52,53]. On the other hand, a growing number of studies indicate that air pollution is more hazardous to human health than previously thought [54]. The current limit values for total mass concentrations of PM10 and PM2.5 are considered to be indicators of potential risk to the environment and human health. The PM10 fraction is light enough to float in the air for long periods of time and can therefore be transported over long distances, while PM2.5 is of particular health concern because it is small enough to enter the bloodstream through the lungs and cause serious illness [55]. In September 2021, the WHO published updated global air quality guidelines, with new reference values for most air pollutants significantly lower than in the previous guidelines. The annual average values for PM2.5 were reduced from 10 to 5 $\mu g/m^3$, for PM10 from 20 to 15 $\mu g/m^3$ and for nitrogen dioxide from 40 to 10 $\mu g/m^3$ [56].

Air pollution is a major determinant of human well-being worldwide, and the effects of air pollution and socioeconomic position (SEP) on health are well documented in the literature. Research shows that many socially disadvantaged communities, including in particular, low-income and ethnic minority communities, are exposed to disproportionately high levels of air pollution and are likely to be more vulnerable to the health effects of these pollutants due to the additional effects of material deprivation and psychosocial stress. However, the issue of assessing the socioeconomic consequences of air pollution is particularly complicated by the mechanisms that determine both the level of exposure of an area to environmental risk factors and its capacity to protect society from their adverse effects. Recent studies increasingly suggest that the level of exposure to air pollution is significantly determined by the socioeconomic position of those exposed and may play a key role in the epidemiology of disease and death [57]. As a consequence of the above, it must be borne in mind that a single variable, or a single index of several variables, cannot fully capture the mechanisms responsible for the cumulative health effects observed in specific communities following exposure to adverse environmental factors. However, many studies rely on single measures of SES, such as income or educational attainment, for lack of other options, without being able to consider the broader mechanisms through which SES may influence the relationship between air pollution and health.

The findings presented in this study are part of a trend in the literature concerning the unequal exposure of certain populations to environmental hazards and the disproportionate distribution of the effects of these hazards across populations [58–62]. At the same time, they are related to theories of environmental inequalities that allow deeper insights into the pathways of SES effects on levels of health inequalities [63,64], as well as providing an analytical framework to further develop studies of interactions between air pollution and social inequalities that have been conducted in the literature [65,66]. Indeed, it appears that groups with high SES, even when exposed to higher concentrations of air pollution, do not necessarily have a higher risk of disease or death. The results presented in this paper confirm that, globally, there is a stronger correlation between the level of GDP than between the level of average exposure of an area and PM2.5. Socio-economic differences, which determine limited access to resources, thus appear to be the common denominator underlying environmental inequalities.

The interdisciplinary nature of the study of environmental inequalities allows the use of a range of perspectives and tools appropriate to the social, natural and health sciences. Economic inequalities and their consequences for health have long been the focus of research by both economists and researchers from other disciplines concerned with population health. These researchers seek to understand the pathways by which socio-economic status is inversely related to indicators of health and disease. In the models they propose, people's characteristics are linked to various environmental exposures and to access to and quality of health care [67–70]. These studies clearly confirm the influence of socioeconomic factors and air pollution on life expectancy [71,72].

Despite the remarkable popularity of the issues of social and environmental inequalities in terms of health burden, there is still a need both to seek answers and to formulate proposals for solutions. Few studies in the area of environmental risks have even looked at changes in inequalities over time, which this study seeks to fill to some extent. Essentially, its aim was to answer the question: assuming that a global clean air policy has been pursued over the past few decades, has it succeeded in reducing the existing differences between countries in this area? The results of the present study do not provide a positive answer to this question, because despite a reduction in the average burden of air pollution, measured in terms of years of life and health lost, there has been no convergence in this area leading to a reduction in the average differences between countries, as the worse-off countries catch up with the better-off ones.

It should be emphasised that considerable progress has been made in the last decade in assessing the economic burden of air pollution [73,74], but estimates based on the indirect costs of disease are still subject to a high degree of uncertainty. Indeed, there is still no consensus on appropriate methods for assessing both the value of life and the costs of mortality due to air pollution [75]. For example, attempts have been made to value the economic impact of premature deaths based on the value of statistical life expectancy (VSL) [76] or the impact of pollution, mortality and disease on GDP growth rates [77]. This makes it difficult to objectively assess the persistent inequalities in the health burden of air pollution between countries and to set public and environmental health priorities.

6. Conclusions

The economic costs of the health burden of air pollution vary considerably depending on the assessment methods used. This study shows that the health burden is unevenly distributed globally and correlates more with the level of national product than with the actual level of exposure of an area to PM2.5 concentrations. This suggests that reducing the health burden of air pollution will require addressing two issues simultaneously: reducing exposure to pollutants and improving general health and health care in low-income countries. The results do not support the assumption of an ongoing process of positive convergence in the global burden of health effects from air pollution. Despite the observed decrease in average levels of the burden of air pollution in terms of years of life and health lost over the period 1990–2019, there was no reduction in average differences between countries, as countries with a worse situation (higher burden) caught up with countries with better performance.

By focusing the analysis on the working-age population, it has also been possible to gain perspective on the so-called indirect costs of air-pollution-related ill-health due to lost economic productivity. In the case of low-income economies, this reveals an existing mechanism for amplifying negative environmental effects, as even relatively small increases in risk factors have serious health consequences. At the same time, the results obtained go beyond the environmental aspect by providing quantifiable ways of estimating the economic costs of air pollution in individual countries, taking into account the costs of lost economic productivity. The development of a counterfactual scenario in which a risk factor is reduced or eliminated in a given area is important in this case.

Funding: This research received no external funding.

Data Availability Statement: Global Burden of Disease Collaborative Network. Global Burden of Disease Study 2019 (GBD 2019) Results. Seattle, United States: Institute for Health Metrics and Evaluation (IHME), 2020. Available from https://vizhub.healthdata.org/gbd-results/ (accessed on 29 April 2023); Global Burden of Disease Collaborative Network. Global Burden of Disease Study 2019 (GBD 2019) Air Pollution Exposure Estimates 1990–2019. Seattle, United States of America: Institute for Health Metrics and Evaluation (IHME), 2021; Global Burden of Disease Health Financing Collaborator Network produced estimates for Gross Domestic Product (GDP) from 1960–2050. Estimates are reported as GDP per person in constant 2021 purchasing-power parity-adjusted (PPP) dollars. https://ghdx.healthdata.org/record/ihme-data/global-gdp-per-capita-1960-2050-fgh-2021 (accessed on 29 April 2023).

Conflicts of Interest: The authors declare no conflict of interest.

References

1. Cole, M.A.; Neumayer, E. The impact of poor health on total factor productivity. *J. Dev. Stud.* **2006**, *42*, 918–938. [CrossRef]
2. William, J.; Lewis, M. Health Investments and Economic Growth: Macroeconomic Evidence and Microeconomic Foundations. World Bank Policy Research Working Paper Series; The World Bank Group: Washington, DC, USA, 2009.
3. Jakubowska, A.; Bilan, S.; Werbiński, J. Chronic diseases and labour resources: "Old and new" European Union member states. *J. Int. Stud.* **2021**, *14*, 129–138. [CrossRef]
4. Bialowolski, P.; McNeely, E.; VanderWeele, T.J.; Weziak-Bialowolska, D. Ill health and distraction at work: Costs and drivers for productivity loss. *PLoS ONE* **2020**, *15*, e0230562. [CrossRef] [PubMed]
5. Zhang, W.; Sun, H. How to analyze work productivity loss due to health problems in randomized controlled trials? A simulation study. *BMC Med. Res. Methodol.* **2021**, *21*, 130. [CrossRef] [PubMed]
6. Krol, M.; Brouwer, W. How to estimate productivity costs in economic evaluations. *Pharmacoeconomics* **2014**, *32*, 335–344. [CrossRef]
7. Yoshimoto, T.; Oka, H.; Fujii, T.; Nagata, T.; Matsudaira, K. The Economic Burden of Lost Productivity due to Presenteeism Caused by Health Conditions Among Workers in Japan. *J. Occup. Environ. Med.* **2020**, *62*, 883–888. [CrossRef]
8. Miller, M.R.; Shaw, C.A.; Langrish, J.P. From particles to patients: Oxidative stress and the cardiovascular effects of air pollution. *Future Cardiol.* **2012**, *8*, 577–602. [CrossRef]
9. Van Rijn, R.M.; Robroek, S.J.; Brouwer, S.; Burdorf, A. Influence of poor health on exit from paid employment: A systematic review. *Occup. Environ. Med.* **2014**, *71*, 295–301. [CrossRef]
10. Rice, N.E.; Lang, I.A.; Henley, W.; Melzer, D. Common health predictors of early retirement: Findings from the English Longitudinal Study of Ageing. *Age Ageing* **2011**, *40*, 54–61. [CrossRef]
11. Brouwer, W.; Verbooy, K.; Hoefman, R.; van Exel, J. Production Losses due to Absenteeism and Presenteeism: The Influence of Compensation Mechanisms and Multiplier Effects. *PharmacoEcon.* **2023**. Available online: https://link.springer.com/article/10.1007/s40273-023-01253-y (accessed on 29 April 2023). [CrossRef]
12. Bloom, D.E.; Canning, D.; Fink, G. Disease and development revisited. *J. Political Econ.* **2014**, *122*, 1355–1366. [CrossRef]
13. Abegunde, D.; Stanciole, A. *An Estimation of the Economic Impact of Chronic Noncommunicable Diseases in Selected Countries, Working Paper*; WHO Department of Chronic Diseases and Health Promotion, CHP, World Health Organization: Geneva, Switzerland, 2006.
14. Brown, S.; Sessions, J.G. The economics of absence: Theory and evidence. *J. Econ. Surv.* **1996**, *10*, 23–53. [CrossRef]
15. GBD 2019 Risk Factors Collaborators. Global burden of 87 risk factors in 204 countries and territories, 1990–2019: A systematic analysis for the Global Burden of Disease Study 2019. *Lancet* **2020**, *396*, 1223–1249. [CrossRef]
16. Global Burden of Disease Collaborative Network. *Global Burden of Disease Study 2019 (GBD 2019) Results*; Institute for Health Metrics and Evaluation (IHME): Seattle, WA, USA, 2020; Available online: https://vizhub.healthdata.org/gbd-results/ (accessed on 29 April 2023).
17. Stern, D.I. The environmental Kuznets curve. *Companion Environ. Stud.* **2018**, *49*, 49–54.
18. Kuznets, S. Economic growth and income inequality. *Am. Econ. Rev.* **1955**, *45*, 1–28.
19. Dinda, S. Environmental Kuznets curve hypothesis: A survey. *Ecol. Econ.* **2004**, *49*, 431–455. [CrossRef]
20. Chen, J.; Hu, T.E.; Van Tulder, R. Is the environmental Kuznets curve still valid: A perspective of wicked problems. *Sustainability* **2019**, *11*, 4747. [CrossRef]
21. Leal, P.H.; Marques, A.C. The evolution of the environmental Kuznets curve hypothesis assessment: A literature review under a critical analysis perspective. *Heliyon* **2022**, *8*, e11521. [CrossRef]

22. Ojaghlou, M.; Ugurlu, E.; Kadłubek, M.; Thalassinos, E. Economic Activities and Management Issues for the Environment: An Environmental Kuznets Curve (EKC) and STIRPAT Analysis in Turkey. *Resources* **2023**, *12*, 57. [CrossRef]
23. Mannucci, P.M.; Franchini, M. Health Effects of Ambient Air Pollution in Developing Countries. *Int. J. Environ. Res. Public Health* **2017**, *14*, 1048. [CrossRef]
24. Mannucci, P.M.; Harari, S.; Franchini, M. Novel evidence for a greater burden of ambient air pollution on cardiovascular disease. *Haematologica* **2019**, *104*, 2349–2357. [CrossRef] [PubMed]
25. Roth, G.A.; Mensah, G.A.; Fuster, V. The Global Burden of Cardiovascular Diseases and Risks: A Compass for Global Action. *J. Am. Coll. Cardiol.* **2020**, *76*, 2980–2981. [CrossRef]
26. Rajagopalan, S.; Al-Kindi, S.; Brook, R. Air Pollution and Cardiovascular Disease. *J. Am. Coll. Cardiol.* **2018**, *72*, 2054–2070. [CrossRef] [PubMed]
27. Cai, J.; Yu, S.; Pei, Y.; Peng, C.; Liao, Y.; Liu, N.; Ji, J.; Cheng, J. Association between Airborne Fine Particulate Matter and Residents' Cardiovascular Diseases, Ischemic Heart Disease and Cerebral Vascular Disease Mortality in Areas with Lighter Air Pollution in China. *Int. J. Environ. Res. Public Health* **2018**, *15*, 1918. [CrossRef] [PubMed]
28. Sørensen, M.; Poulsen, A.H.; Hvidtfeldt, U.A.; Christensen, J.H.; Brandt, J.; Frohn, L.M.; Ketzel, M.; Andersen, C.; Valencia, V.H.; Lassen, C.F.; et al. Effects of Sociodemographic Characteristics, Comorbidity, and Coexposures on the Association between Air Pollution and Type 2 Diabetes: A Nationwide Cohort Study. *Environ. Health Perspect.* **2023**, *131*, 027008. [CrossRef]
29. Bandyopadhyay, A. Neurological Disorders from Ambient (Urban) Air Pollution Emphasizing UFPM and PM2.5. *Curr. Pollut. Rep.* **2016**, *2*, 203–211. [CrossRef]
30. Kim, H.; Kim, W.H.; Kim, Y.Y.; Park, H.Y. Air pollution and central nervous system disease: A review of the impact of fine particulate matter on neurological disorders. *Front. Public Health* **2020**, *8*, 575330. [CrossRef]
31. Aldraihem, M.O.; Al-Ghamdi, F.; Murtaza, G.; Bashir, S. Air pollution and performance of the brain. *Arab. J. Geosci.* **2020**, *13*, 1158. [CrossRef]
32. Buoli, M.; Grassi, S.; Caldiroli, A.; Carnevali, G.S.; Mucci, F.; Iodice, S.; Cantone, L.; Pergoli, L.; Bollati, V. Is there a link between air pollution and mental disorders? *Environ. Int.* **2018**, *118*, 154–168. [CrossRef]
33. Khan, A.; Plana-Ripoll, O.; Antonsen, S.; Brandt, J.; Geels, C.; Landecker, H.; Sullivan, P.F.; Pedersen, C.B.; Rzhetsky, A. Environmental pollution is associated with increased risk of psychiatric disorders in the US and Denmark. *PLoS Biol.* **2019**, *17*, e3000353. [CrossRef]
34. Attademo, L.; Bernardini, F.; Garinella, R.; Compton, M.T. Environmental pollution and risk of psychotic disorders: A review of the science to date. *Schizophr. Res.* **2017**, *181*, 55–59. [CrossRef]
35. Hahad, O.; Lelieveld, J.; Birklein, F.; Lieb, K.; Daiber, A.; Münzel, T. Ambient Air Pollution Increases the Risk of Cerebrovascular and Neuropsychiatric Disorders through Induction of Inflammation and Oxidative Stress. *Int. J. Mol. Sci.* **2020**, *21*, 4306. [CrossRef]
36. Wang, Y.; Duong, M.; Brauer, M.; Rangarajan, S.; Dans, A.; Lanas, F.; Lopez-Jaramillo, P.; Puoane, T.; Yeates, K.; Chifamba, J.; et al. Household Air Pollution and Adult Lung Function Change, Respiratory Disease, and Mortality across Eleven Low-and Middle-Income Countries from the PURE Study. *Environ. Health Perspect.* **2023**, *131*, 047015. [CrossRef]
37. Rajkumar, R.P. The Relationship between Ambient Fine Particulate Matter (PM2.5) Pollution and Depression: An Analysis of Data from 185 Countries. *Atmosphere* **2023**, *14*, 597. [CrossRef]
38. Migliore, L.; Coppedè, F. Environmental-induced oxidative stress in neurodegenerative disorders and aging. *Mutat. Res. Genet. Toxicol. Environ. Mutagen.* **2009**, *674*, 73–84. [CrossRef]
39. Argacha, J.F.; Bourdrel, T.; van de Borne, P. Ecology of the cardiovascular system: A focus on air-related environmental factors. *Trends Cardiovasc. Med.* **2018**, *28*, 112–126. [CrossRef]
40. Franchini, M.; Mannucci, P.M. Thrombogenicity and cardiovascular effects of ambient air pollution. *Blood* **2011**, *118*, 2405–2412. [CrossRef]
41. Laden, F.; Schwartz, J.; Speizer, F.E.; Dockery, D.W. Reduction in fine particulate air pollution and mortality: Extended follow-up of the Harvard Six Cities study. *Am. J. Respir. Crit. Care Med.* **2006**, *173*, 667–672. [CrossRef]
42. Williams, A. Calculating the global burden of disease: Time for a strategic reappraisal? *Health Econ.* **1999**, *8*, 1–8. [CrossRef]
43. Łyszczarz, B.; Sowa, K. Production losses due to mortality associated with modifiable health risk factors in Poland. *Eur. J. Health Econ.* **2022**, *23*, 33–45. [CrossRef]
44. Al-Hemoud, A.; Gasana, J.; Al-Dabbous, A.N.; Al-Shatti, A.; Al-Khayat, A. Disability Adjusted Life Years (DALYs) in Terms of Years of Life Lost (YLL) Due to Premature Adult Mortalities and Postneonatal Infant Mortalities Attributed to PM2.5 and PM10 Exposures in Kuwait. *Int. J. Environ. Res. Public Health* **2018**, *15*, 2609. [CrossRef] [PubMed]
45. Institute for Health Metrics and Evaluation (IHME). Global Burden of Disease Collaborative Network. In *Global Burden of Disease Study 2019 (GBD 2019) Air Pollution Exposure Estimates 1990–2019*; Institute for Health Metrics and Evaluation (IHME): Seattle, WA, USA, 2021.
46. Ritchie, H.; Roser, M. Air Pollution. Published Online at OurWorldInData.org. Available online: https://ourworldindata.org/air-pollution (accessed on 29 April 2023).

47. Global Burden of Disease Health Financing Collaborator Network Produced Estimates for Gross Domestic Product (GDP) from 1960–2050. Estimates Are Reported as GDP Per Person in Constant 2021 Purchasing-Power Parity-Adjusted (PPP) Dollars. Available online: https://ghdx.healthdata.org/record/ihme-data/global-gdp-per-capita-1960-2050-fgh-2021 (accessed on 29 April 2023).
48. Davidson, C.I.; Phalen, R.F.; Solomon, P.A. Airborne particulate matter and human health: A review. *Aerosol Sci. Technol.* **2005**, *39*, 737–749. [CrossRef]
49. Al-Kindi, S.G.; Brook, R.D.; Biswal, S.; Rajagopalan, S. Environmental determinants of cardiovascular disease: Lessons learned from air pollution. *Nat. Rev. Cardiol.* **2020**, *17*, 656–672. [CrossRef] [PubMed]
50. Agibayeva, A.; Guney, M.; Karaca, F.; Kumisbek, A.; Kim, J.R.; Avcu, E. Analytical Methods for Physicochemical Characterization and Toxicity Assessment of Atmospheric Particulate Matter: A Review. *Sustainability* **2022**, *14*, 13481. [CrossRef]
51. Hassan Bhat, T.; Jiawen, G.; Farzaneh, H. Air Pollution Health Risk Assessment (AP-HRA), Principles and Applications. *Int. J. Environ. Res. Public Health* **2021**, *18*, 1935. [CrossRef] [PubMed]
52. Ogrizek, M.; Kroflič, A.; Šala, M. Critical review on the development of analytical techniques for the elemental analysis of airborne particulate matter. *Trends Environ. Anal. Chem.* **2022**, *33*, e00155. [CrossRef]
53. Lavigne, E.; Burnett, R.T.; Weichenthal, S. Association of short-term exposure to fine particulate air pollution and mortality: Effect modification by oxidant gases. *Sci. Rep.* **2018**, *8*, 16097. [CrossRef]
54. Carvalho, H. New WHO global air quality guidelines: More pressure on nations to reduce air pollution levels. *Lancet Planet. Health* **2021**, *5*, e760–e761. [CrossRef]
55. Lelieveld, J.; Evans, J.; Fnais, M.; Giannadaki, D.; Pozzer, A. The contribution of outdoor air pollution sources to premature mortality on a global scale. *Nature* **2015**, *525*, 367–371. [CrossRef]
56. WHO. WHO Global Air Quality Guidelines. In *Particulate Matter (PM2.5 and PM10), Ozone, Nitrogen Dioxide, Sulfur Dioxide and Carbon Monoxide*; World Health Organization: Geneva, Switzerland, 2021.
57. O'Neill, M.S.; Jerrett, M.; Kawachi, I.; Levy, J.I.; Cohen, A.J.; Gouveia, N.; Wilkinson, P.; Fletcher, T.; Cifuentes, L.; Schwartz, J.; et al. Workshop on Air Pollution and Socioeconomic Conditions. Health, wealth, and air pollution: Advancing theory and methods. *Environ. Health Perspect.* **2003**, *111*, 1861–1870. [CrossRef]
58. Hajat, A.; Hsia, C.; O'Neill, M.S. Socioeconomic disparities and air pollution exposure: A global review. *Curr. Environ. Health Rep.* **2015**, *2*, 440–450. [CrossRef]
59. Gouveia, N. Addressing Environmental Health Inequalities. *Int. J. Environ. Res. Public Health* **2016**, *13*, 858. [CrossRef]
60. Goodman, A.; Wilkinson, P.; Stafford, M.; Tonne, C. Characterising socio-economic inequalities in exposure to air pollution: A comparison of socio-economic markers and scales of measurement. *Health Place* **2011**, *17*, 767–774. [CrossRef]
61. Molitor, J.; Su, J.; Molitor, N.T.; Rubio, V.G.; Richardson, S.; Hastie, D.; Morello-Frosch, R.; Jerrett, M. Identifying vulnerable populations through an examination of the association between multi-pollutant profiles and poverty. *Environ. Sci. Technol.* **2011**, *45*, 7754–7760. [CrossRef]
62. Jakubowska, A.; Rabe, M. Air Pollution and Limitations in Health: Identification of Inequalities in the Burdens of the Economies of the "Old" and "New" EU. *Energies* **2022**, *15*, 6225. [CrossRef]
63. Grandjean, P.; Bellanger, M. Calculation of the disease burden associated with environmental chemical exposures: Application of toxicological information in health economic estimation. *Environ. Health* **2017**, *16*, 123. [CrossRef] [PubMed]
64. Miranda, M.L.; Edwards, S.E.; Keating, M.H.; Paul, C.J. Making the Environmental Justice Grade: The Relative Burden of Air Pollution Exposure in the United States. *Int. J. Environ. Res. Public Health* **2011**, *8*, 1755–1771. [CrossRef]
65. Fairburn, J.; Schüle, S.A.; Dreger, S.; Karla Hilz, L.; Bolte, G. Social Inequalities in Exposure to Ambient Air Pollution: A Systematic Review in the WHO European Region. *Int. J. Environ. Res. Public Health* **2019**, *16*, 3127. [CrossRef] [PubMed]
66. Shen, W.; Srivastava, S.; Yang, L.; Jain, K.; Schröder, P. Understanding the impacts of outdoor air pollution on social inequality: Advancing a just transition framework. *Local Environ.* **2020**, *25*, 1–17. [CrossRef]
67. Santana, P.; Almendra, R.; Pilot, E.; Doreleijers, S.; Krafft, T. Environmental Inequalities in Global Health. In *Handbook of Global Health*; Kickbusch, I., Ganten, D., Moeti, M., Eds.; Springer: Cham, Switzerland, 2021. [CrossRef]
68. Spalt, E.W.; Curl, C.L.; Allen, R.W.; Cohen, M.; Williams, K.; Hirsch, J.; Adar, S.D.; Kaufman, J. Factors influencing time-location patterns and their impact on estimates of exposure: The Multi-Ethnic Study of Atherosclerosis and Air Pollution (MESA Air). *J. Expo. Sci. Environ. Epidemiol.* **2016**, *26*, 341–348. [CrossRef]
69. Samet, J.M. The environment and health inequalities: Problems and solutions. *J. Health Inequalities* **2019**, *5*, 21–27. [CrossRef]
70. Martinez, G.S.; Spadaro, J.V.; Chapizanis, D.; Kendrovski, V.; Kochubovski, M.; Mudu, P. Health Impacts and Economic Costs of Air Pollution in the Metropolitan Area of Skopje. *Int. J. Environ. Res. Public Health* **2018**, *15*, 626. [CrossRef] [PubMed]
71. Yang, H.; Wang, S.; Ren, Z.; Liu, H.; Tong, Y.; Wang, N. Life expectancy, air pollution, and socioeconomic factors: A multivariate time-series analysis of Beijing City, China. *Soc. Indic. Res.* **2022**, *162*, 979–994. [CrossRef]
72. Pu, H.; Wang, S.; Wang, Z.; Ran, Z.; Jiang, M. Non-linear relations between life expectancy, socio-economic, and air pollution factors: A global assessment with spatial disparities. *Environ. Sci. Pollut. Res.* **2022**, *29*, 53306–53318. [CrossRef]
73. Yin, H.; Brauer, M.; Zhang, J.; Cai, W.; Navrud, S.; Burnett, R.; Howard, C.; Deng, Z.; Kammen, D.M.; Schellnhuber, H.J.; et al. Population ageing and deaths attributable to ambient PM2·5 pollution: A global analysis of economic cost. *Lancet Planet Health* **2021**, *5*, e356–e367. [CrossRef]

74. Conti, S.; Ferrara, P.; D'Angiolella, L.S.; Lorelli, S.C.; Agazzi, G.; Fornari, C.; Cesana, G.; Mantovani, L.G. The economic impact of air pollution: A European assessment. *Eur. J. Public Health* **2020**, *30* (Suppl. S5), ckaa165-084. [CrossRef]
75. Nair, M.; Bherwani, H.; Mirza, S.; Anjum, S.; Kumar, R. Valuing burden of premature mortality attributable to air pollution in major million-plus non-attainment cities of India. *Sci. Rep.* **2021**, *11*, 22771. [CrossRef]
76. Safari, Z.; Fouladi-Fard, R.; Vahedian, M.; Mahmoudian, M.H.; Rahbar, A.; Fiore, M. Health impact assessment and evaluation of economic costs attributed to PM2.5 air pollution using BenMAP-CE. *Int. J. Biometeorol.* **2022**, *66*, 1891–1902. [CrossRef]
77. Maiti, M.; Jadhav, P. Impact of pollution level, death rate and illness on economic growth: Evidence from the global economy. *SN Bus. Econ.* **2021**, *1*, 109. [CrossRef]

toxics

MDPI

Article

Micronuclei, Pesticides, and Element Mixtures in Mining Contexts: The Hormetic Effect of Selenium

Marcela E. Varona-Uribe [1], Sonia M. Díaz [1], Ruth-Marien Palma [2], Leonardo Briceño-Ayala [1], Carlos Trillos-Peña [1], Eliana M. Téllez-Avila [2], Lyda Espitia-Pérez [3], Karina Pastor-Sierra [3], Pedro Juan Espitia-Pérez [3] and Alvaro J. Idrovo [4,*]

[1] School of Medicine and Health Sciences, Universidad del Rosario, Bogotá D.C. 111221, Colombia; marcela.varona@urosario.edu.co (M.E.V.-U.); sdiaz21@gmail.com (S.M.D.); leonardo.briceno@urosario.edu.co (L.B.-A.); carlos.trillos@urosario.edu.co (C.T.-P.)

[2] Environmental and Occupational Health Group, National Institute of Health, Bogotá D.C. 111321, Colombia; rpalma@ins.gov.co (R.-M.P.); etellez@ins.gov.co (E.M.T.-A.)

[3] Grupo de Investigación Biomédicas y Biología Molecular, Universidad del Sinú, Montería 230001, Colombia; lydaespitia@unisinu.edu.co (L.E.-P.); karinapastor@unisinu.edu.co (K.P.-S.); pedrojespitia@unisinu.edu.co (P.J.E.-P.)

[4] Public Health Department, School of Medicine, Universidad Industrial de Santander, Bucaramanga 680002, Colombia

* Correspondence: idrovoaj@yahoo.com.mx

Abstract: The contexts where there are mining and agriculture activities are potential sources of risk to human health due to contamination by chemical mixtures. These contexts are frequent in several Colombian regions. This study explored the potential association between the frequency of micronuclei and pesticides and elements in regions with ferronickel (Montelibano, Córdoba) and gold (Nechí, Antioquia) mining, and a closed native mercury mine (Aranzazu, Caldas), with an emphasis in the potential effect of selenium as a potential chelator. A cross-sectional study was carried out with 247 individuals. Sociodemographic, occupational, and toxicological variables were ascertained. Blood and urine samples were taken for pesticide analysis (5 organophosphates, 4 organochlorines, and 3 carbamates), 68 elements were quantified in hair, and micronuclei were quantified in lymphocytes. The mixtures of elements were grouped through principal component analysis. Prevalence ratios were estimated with robust variance Poisson regressions to explore associations. Interactions of selenium with toxic elements were explored. The highest concentrations of elements were in the active mines. The potentially most toxic chemical mixture was observed in the ferronickel mine. Pesticides were detected in a low proportion of participants (<2.5%), except paraoxon-methyl in blood (27.55%) in Montelibano and paraoxon-ethyl in blood (18.81%) in Aranzazu. The frequency of micronuclei was similar in the three mining contexts, with means between 4 to 7 (p = 0.1298). There was great heterogeneity in the exposure to pesticides and elements. The "hormetic effect" of selenium was described, in which, at low doses, it acts as a chelator in Montelibano and Aranzazu, and at high doses, it can enhance the toxic effects of other elements, maybe as in Nechí. Selenium can serve as a protective agent, but it requires adaptation to the available concentrations in each region to avoid its toxic effects.

Keywords: pesticides; mining; chemical mixtures; selenium; hormesis; genetic damage; Colombia

Citation: Varona-Uribe, M.E.; Díaz, S.M.; Palma, R.-M.; Briceño-Ayala, L.; Trillos-Peña, C.; Téllez-Avila, E.M.; Espitia-Pérez, L.; Pastor-Sierra, K.; Espitia-Pérez, P.J.; Idrovo, A.J. Micronuclei, Pesticides, and Element Mixtures in Mining Contexts: The Hormetic Effect of Selenium. *Toxics* **2023**, *11*, 821. https://doi.org/10.3390/toxics11100821

Academic Editors: Roel Vermeulen, Lauren Petrick, Maaike van Gerwen and Jose V. Tarazona

Received: 7 July 2023
Revised: 7 August 2023
Accepted: 16 August 2023
Published: 29 September 2023

1. Introduction

Pesticides are chemical substances used to prevent or control any pest, and their inadequate handling can have adverse effects on the health of agricultural workers, their families, the community, and the environment [1]. Exposure to these substances can be through skin contact, ingestion, or inhalation and can cause acute or chronic poisoning. Chronic adverse effects usually appear long after the first exposure, even at low doses, being more frequent among adults [2] since various pesticides accumulate in the body and the environment, causing disease after several years of exposure [3]. There is evidence of the relationship between chronic exposure to some pesticides and an increased occurrence of neurological, respiratory, dermatological, and renal diseases, reproductive disorders, cancer, and endocrine disorders, among others. Many of these disorders are associated with insecticides and herbicides, especially organophosphates, organochlorines, and triazine compounds [4,5]. Considering the availability of susceptibility biomarkers, carcinogenic effects of pesticides can be seen in early changes in genomic DNA, such as DNA-DNA and DNA-protein cross-linking, strand breaks, and DNA adduct formation, generating defective cells [6–9]. Moreover, the study of micronuclei has been useful in identifying agents associated with inflammation, obesity, adverse reproductive outcomes (infertility and pregnancy complications), chronic diseases (diabetes mellitus, cardiovascular, kidney, and neurodegenerative diseases), and accelerated aging syndromes [10].

Agriculture for human and animal consumption is also practiced in many places where mining activities are performed. The consequences of mining and agricultural activity in the same place are not very well known, but they undoubtedly have to do with the type of mining, the geochemical of soil, and the absorption capacity of crops. Open-pit mining is often associated with extensive environmental damage, as mining waste can spread via air and surrounding water sources. It contrasts with damage to water sources related to alluvial and underground mining, although it has higher deleterious effects in alluvial activities [6,11]. Concerning the geochemistry of soils and types of vegetation, generalizations are not easy since the chemical mixtures observed in rocks and soils can be highly variable. For this, specific studies are required for each mining-agricultural region. Research on exposure to pesticides, elements, and genetic damage has been explored in several studies, but mainly in isolation from other chemicals. Studies on mixtures of pesticides with toxic substances, trace elements, and substances with no known adverse effects in humans, animals, and plants are scarce, even though it is a common form of exposure.

In Colombia, there is a high consumption and market of pesticides, even though the records are inaccurate [12]. In this country, mining regions may coincide with crops cultivated years ago or fumigated to prevent vector-borne transmitted diseases [13]. When pesticides are persistent in the environment, such as organochlorines, the presence of toxic minerals may propitiate the occurrence of adverse effects not well-known in humans [14]. Although organochlorine pesticides have been forbidden in Colombia, there is evidence that they were used to control pests in coca, marijuana, and poppy crop regions [15]. In this regard, there are approximately 88,629 m^3 of soil contaminated with organochlorines in the country; however, there is a significant underreporting of contaminated sites [16].

The legal consumption of pesticides in Colombia includes agricultural and livestock uses, the control of vector-borne diseases, and the use of glyphosate in the illicit crop control program. It is estimated that ~13.17 kg of pesticide per hectare (kg/ha) are used in the country, much higher than the world average estimated at 2.57 kg/ha [17]. The largest amounts of pesticides are used in agriculture, where organophosphates, carbamates, dithiocarbamates, phosphonic acid, pyrethroids, and bipyridyls are more frequent. Pyrethroids and organophosphates with residual action are used to control vector-borne diseases in the places with the highest occurrence [18]. Glyphosate has been used in the illicit crop control program in two commercial presentations: Cuspide® 480SL has been applied for ground spraying and Roundup-Ultra® (glyphosate + Cosmoflux 411F®) for aerial spraying [19]. The exposure to glyphosate in the control of illicit crops is much higher than that observed in other uses around the world [20,21]. In addition, it should be noted that, in 2015, the use of glyphosate in these activities was abandoned due to the potential carcinogenic risk [22], but it continues to be used in other commercial presentations for agricultural uses.

In these mining-agricultural contexts, multiple and heterogeneous exposures can generate synergistic effects; that lead to greater damage or disease or antagonistic effects where the adverse effects of exposure are minimized by the presence of other chemical compounds. Perhaps the most notorious case in Colombia may be the presence of very high selenium concentrations in soils [23], which may reduce the adverse effects of metals and metalloids given their chelating action [24]. However, the presence of recognized toxics, such as mercury, lead, cadmium, and pesticides, with elements that are not widely studied in the country, such as arsenic, beryllium, and uranium [25,26], could have unpredictable effects. Potential adverse effects in humans due to exposure to complex mixtures of chemical substances are chronic diseases [7,8] and genomic damage [27,28]. In this context, this study explored the association between exposure to some pesticides and elements related to anthropogenic and non-anthropogenic activities with the occurrence of micronuclei among individuals residing in three mining areas in Colombia. Potential selenium effects were emphasized. It corresponds to a first approximation of potential adverse effects among humans in mining contaminated sites with agricultural or vector-borne disease control activities.

2. Materials and Methods

2.1. Study Design and Mining Contexts

A cross-sectional study was conducted with three populations potentially exposed to mining and agricultural activities. The populations were from Montelibano (Córdoba), Nechí (Antioquia), and Aranzazu (Caldas) (see Figure 1).

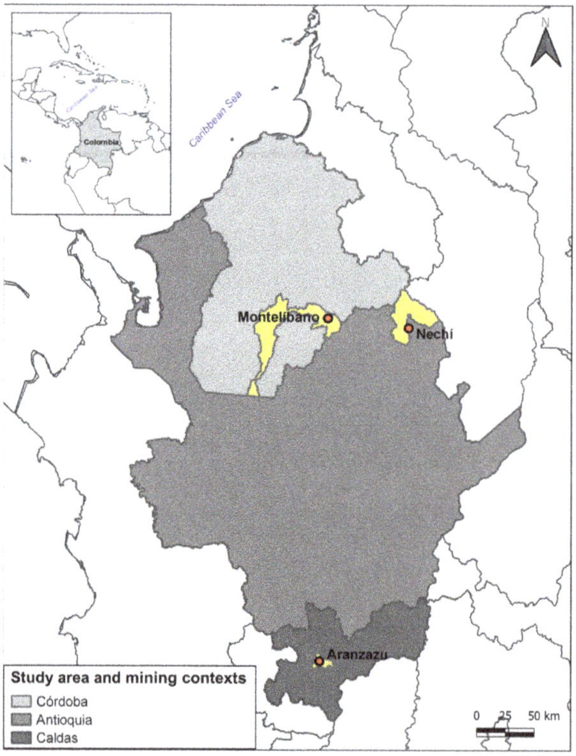

Figure 1. Mining contexts with agricultural activities included in the study.

The selection of these municipalities was based on the specific characteristics of each place. Montelibano is close to one of the largest ferronickel mines in the world. Additionally, there is a coal mine in the region, and gold is exploited in an artisanal way. Exposure to chemical mixtures (Ni, Fe, As, Pb, Cd, mainly) has been described [29,30]. In this region, *Plasmodium falciparum* infection is frequent [31]; thus, insecticides are used. Moreover, cassava and purple yam (*Dioscorea alata*) are common crops in this territory. Nechí is a municipality where gold is extracted from alluvium, which has led to the use of mercury for decades to separate and extract the precious metal from other rocks. According to previous studies, it is one of the most mercury-contaminated areas in the world. Mercury levels in the air are from 300 to 1 million ng Hg/m^3 in public spaces and ~10,000 ng Hg/m^3 in private domestic sites [32]. Moreover, in this region, pesticides are used in crops of rice, corn, cassava, banana, and purple yam and to control vector-borne diseases. In Aranzazu, there is a geological fault where native mercury presence is important; in fact, the only mercury mine in Colombia exists in this municipality. It was closed in 1975 due to financial problems and neurotoxicity among the workers [33]. Currently, it is a small municipality whose economic activities are related to avocado and coffee cultivation. It is an example of what can happen after a mercury mine is closed.

2.2. Participants and Data Collection

Potential participants from the three municipalities were selected by non-probabilistic sampling due to the lack of an updated sampling frame and the logistical limitations related to informal work and difficult access to mining regions. In the Nechí and Montelibano regions, for decades, there have been illegal armed groups, which has limited the presence of public and environmental health researchers due to the inherent security risks. Individuals $\geq$18 years old living in the place for a minimum of six months participated in the study. Those individuals who had presented illnesses that prevented them from adequately answering the questionnaire and who had undergone X-rays, radiotherapy, or chemotherapy during the previous year were excluded. Consequently, participants were formal and informal workers or individuals dedicated to housework activities. A questionnaire was applied that included sociodemographic variables (age, sex, education, health affiliation), labor, and toxicological (smoker status, alcohol consumption). Additionally, an adaptation of the World Health Organization's instrument, "Guidance for identifying populations at risk from mercury exposure", was applied [34].

2.3. Pesticides in Blood and Urine

A 5 mL venous blood sample and a 50 mL urine sample were extracted from each participant, which remained refrigerated until processing and analysis to determine pesticides. Concentrations (ppb) of carbamate insecticides: aldicarb, carbofuran, and propoxur; the organochlorines: α-endosulfan, β-endosulfan, endosulfan sulfate, and hexachlorobenzene, and the organophosphorus: malathion, ethyl paraoxon, methyl paraoxon, ethyl parathion, and methyl parathion, were determined in blood and urine. They were selected because they are widely used in Colombia and have high acute toxicity. The analyses were performed through Quick, Easy, Cheap, Effective, Rugged, and Safe (QuEChERS) extraction and high-performance liquid chromatography (HPLC) analyses with triple quadrupole (SQ) LC/MSD. We rely on the method published by Sciex (equipment supplier) for environmental applications [35]. The limits of detection are available in Appendix A (Tables A1 and A2).

2.4. Element Mixtures in Hair

Hair (in high proportion keratin rich in sulfur) was selected as an internal dose biomarker given the interest in quantifying metals and metalloids; the way of entering the body is via ingestion, although the concentration of some elements may be due to external contamination or compounds included in cosmetics [36]. Hair samples were taken from the occipital region of the scalp (~3 cm long) of all participants. The 68 elements included in the analysis were: lithium (Li), beryllium (Be), boron (B), sodium (Na), magnesium (Mg), aluminum (Al), silicon (Si), phosphorus (P), sulfur (S), potassium (K), calcium (Ca), scandium (Sc), titanium (Ti), vanadium (V), chromium (Cr), manganese (Mn), iron (Fe), cobalt (Co), nickel (Ni), copper (Cu), zinc (Zn), gallium (Ga), germanium (Ge), arsenic (As), bromine (Br), selenium (Se), rubidium (Rb), strontium (Sr), yttrium (Y), zirconium (Zr), niobium (Nb), molybdenum (Mo), ruthenium (Ru), rhodium (Rh), palladium (Pd), silver (Ag), cadmium (Cd), indium (In), tin (Sn), antimony (Sb), iodine (I), cesium (Cs), barium (Ba), lanthanum (La), cerium (Ce), praseodymium (Pr), neodymium (Nd), samarium (Sm), europium (Eu), gadolinium (Gd), terbium (Tb), dysprosium (Dy), holmium (Ho), erbium (Er), thulium (Tm), ytterbium (Yb), lutetium (Lu), tantalum (Ta), tungsten (W), rhenium (Re), platinum (Pt), gold (Au), mercury (Hg), thallium (Tl), lead (Pb), bismuth (Bi), thorium (Th), and uranium (U). Note that several elements (or related compounds) listed by IARC as carcinogenic (group 1), probable (2A), or possible (2B) carcinogenic to humans were included in the analysis. Group 1 are As, Be, Cd, Cr, Sr, Ni, and Th. Group 2A are Sb, Co, In, Mg, Mn, and Si. Group 2B are Hg, Mo, Pb, K, Na, Ti, and V.

The preparation and the sample analysis were carried out by the ICP-MS and ICP OES Laboratory (LABSPECTRO) of the Chemistry Department of the Pontificia Universidade Católica do Rio de Janeiro, in Brazil. Over there, the samples were cleaned with deionized water and acetone for three repetitions, with each solution inside an ultrasound bath to eliminate exogenous elements. Hair samples were oven-dried at 60 °C and weighed. Samples were then acidified with 2.50 mL of HNO_3 in 50 mL polypropylene tubes and digested at 70 °C for 4 h. Finally, they were diluted for elemental determination. Hair element concentrations were evaluated by inductively coupled plasma mass spectrometry using an Elan DRCII ICP-MS spectrometer (Analyst 200, PerkinElmer, Sciex, Norwalk, CT, USA) [37]. Each curve calibration point (blank, reagent blank, and sample) was analyzed with the internal rhodium standard. Argon was used as carrier gas at a flow rate of 50 mL min^{-1}.

2.5. Cytogenetic Analysis

Fenech's methodology was used to explore genetic instability [38]. First, heparinized whole blood (0.5 mL) was added to 4.5 mL of RPMI 1640 medium (Sigma R8758, St. Louis, MO, USA) supplemented with 2 mM L-glutamine (Sigma A5955, St. Louis, MO, USA), 10% fetal bovine fetal serum (Gibco 15000-044, Brazil), 100 µL/mL antibiotic-antimycotic (Sigma A5955, St. Louis, MO, USA), and 2% phytohemagglutinin (Sigma L8754, St. Louis, MO, USA). Cultures were incubated at 37 °C for 44 h under 5% CO_2. After this incubation period, 6 µg/mL cytochalasin B (Sigma C6762, St. Louis, MO, USA) was added. After incubation, lymphocytes were harvested by centrifugation at 1200 rpm for 8 min, centrifuged again, fixed in 25:1 (v/v) methanol/acetic acid, placed on a clean microscope slide, and stained with Diff-Quik stain (Medion diagnostics; 726443. Düdingen. CH). For each participant's blood sample, 2000 binucleated (BN) cells (1000 from each of two slides prepared from duplicate cultures) were evaluated for the presence of DNA damage indices—micronuclei (MN) using bright-field optical microscopy at 200–1000× magnification. All slides were coded for blinded analysis according to the criteria proposed by Fenech.

2.6. Statistical Methods

The databases were created using Epi Info 7.0, and the statistical analyses were performed with the Stata 17 program. Values reported by the laboratory as not detected or below the detection limit were considered zero in the statistical analyses. First, descriptions of participants with proportions or measures of central tendency and dispersion were used according to the observed distribution. Shapiro–Wilk test was used to explore the distribution of continuous variables. Comparisons between the three population groups were made with x^2 or Kruskal–Wallis tests. Then, a detailed exploration of possible element mixtures suggested that in each mining site, there are different distributions of elements. For this reason, the following analysis was performed for each mine independently.

To identify the mixtures of elements in each mining site, principal component analyses with the 68 element concentrations in hair were carried out. It was useful because it is a mathematical model that explains covariance or correlation among element concentrations in hair in terms of unobserved latent variables [39], in this case, element mixtures. One a priori criterion to consider a factor as relevant was that they exhibited a proportional contribution higher than 5%. In Montelibano, two common factors (FC1M and FC2M) were extracted after a varimax rotation (Figure 2). With the same method, in Nechí (FC1N, FC2N, and FC3N) (see Figure 3) and Aranzazu (FC1A and FC2A) (see Figure 4), three and two factors were extracted, respectively. In Montelibano, the factors explained 46.08% of the variance, whereas, in Nechí and Aranzazu, it was 41.43% and 35.15%, respectively. Rotated components and scree plots are available in Appendix A (Tables A3–A5, Figures A1–A3). In addition, factor scores were calculated for every element, and after these scores were normalized.

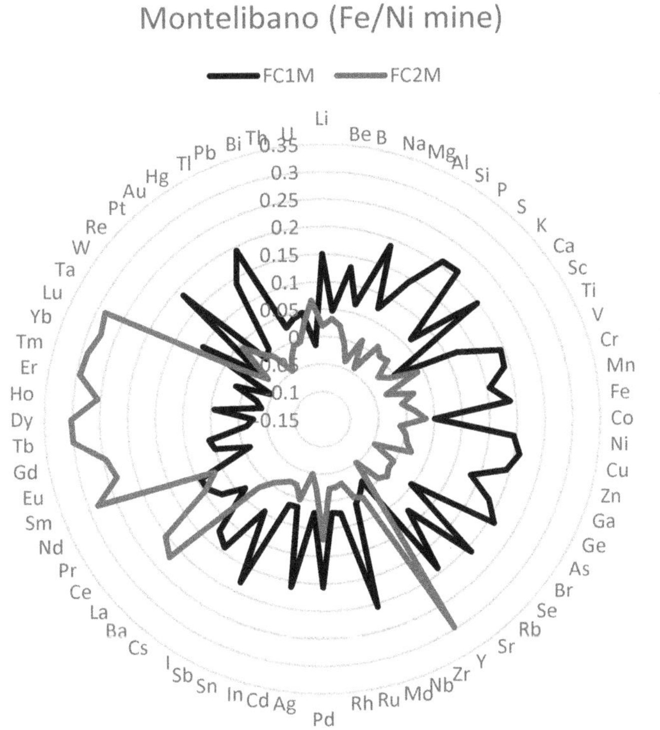

Figure 2. Main element mixtures in Montelibano (factor loadings).

To explore the associations with the frequency of micronuclei, prevalence ratios (PR) with 95% confidence intervals (95% CI) were estimated for each mining site. They were first estimated bivariate and then multiple, using Poisson regression models with robust variance [40]. In the analysis, special care was taken with some elements that could have adverse or antagonistic effects. It was considered whether it belonged to the chemical mixtures included in the analyses; if they were not included, they were analyzed independently in each mining site. Interactions between selenium and other elements were explored in Nechí and Aranzazu because of its potential protective effect [41].

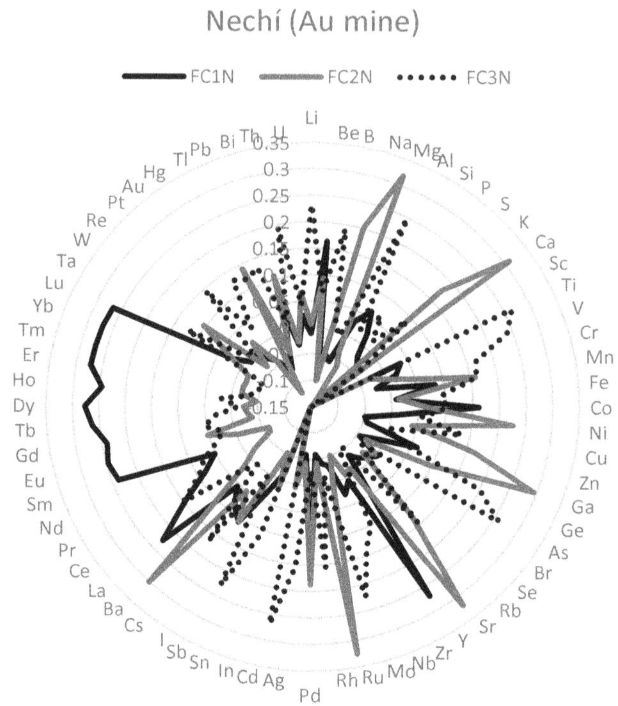

Figure 3. Main element mixtures in Nechí (factor loadings).

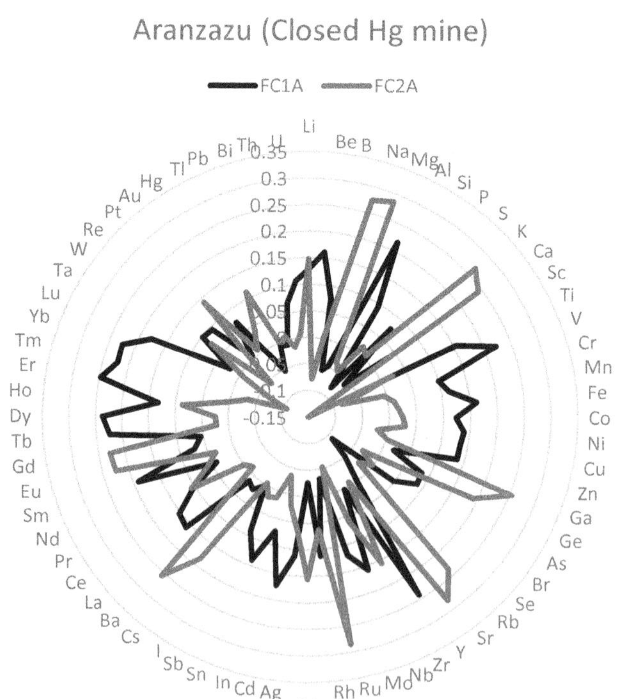

Figure 4. Main element mixtures in Aranzazu (factor loadings).

3. Results

3.1. Mean Characteristics of Participants

Information from 306 individuals was collected, but after excluding those with missing data, information from 247 individuals was analyzed. Of these, 30.77% were from Montelibano (n = 76), 34.01% from Nechí (n = 84), and 35.22% from Aranzazu (n = 87). It was evident that participants differed between the three locations (see Table 1). There was high participation of women, much higher in Montelibano and lower in Aranzazu. Age was also higher in Montelibano and lower in Aranzazu. There were many more married or cohabiting in Montelibano and more single in Nechí. Regarding education, it is evident that the lowest levels were found in Nechí, including a very high proportion of illiterates, which contrasts with the highest educational levels in Montelibano. As expected, there were more individuals carrying out mining-related activities in Montelibano and none in Aranzazu since the mercury mine has been closed since 1975; however, Aranzazu is where more individuals were engaged in work related to agriculture.

Cigarette consumption was higher in Aranzazu and lower in Montelibano, while alcoholic beverage consumption was higher in Montelibano and lower in Aranzazu. Regarding food consumption, it was striking that the only difference was in the consumption of fish, which was higher in Montelibano and lower in Aranzazu. In relation to the elements in the hair, it was evident that the highest concentrations are found in places where there are active mines; these include elements that, in some of their chemical forms, are carcinogenic or probably carcinogenic for humans, according to the IARC. Selenium, with its potential chelating effect, had a higher concentration in Nechí and a lower one in Aranzazu.

Table 1. Mean characteristics of individuals participating in the study (n = 247).

Site (Minerals)	**Montelibano (Fe/Ni)**	**Nechí (Au)**	**Aranzazu (Closed Hg)**	
Variables	**(n = 76) %/Median (Min–Max)**	**(n = 84) %/Median (Min–Max)**	**(n = 87) %/Median (Min–Max)**	***p* Value**
Sex (female)	80.26	72.62	60.92	0.023
Age (years)	54 (21–81)	48 (21–83)	40 (19–83)	<0.001 [a]
Civil status				
Married/Free union	85.53	65.48	70.11	0.029
Divorced/Widower	2.63	7.14	10.34	
Single	11.84	27.38	19.54	
Education				
Illiteracy	2.63	11.90	5.75	<0.001
Elementary (partial or full)	13.16	39.29	39.08	
Secondary (partial or full)	31.58	22.62	37.93	
Technic (partial or full)	35.53	19.05	4.60	
University (partial or full)	17.11	7.14	12.64	
Occupation				
In mining activities	52.63	44.05	0	<0.001
In agricultural activities	6.58	22.62	45.98	<0.001
Cigarette consumption (any moment)	21.05	39.29	43.68	0.007
Alcohol consumption (any moment)	61.84	44.05	40.23	0.015
Food consumption				
Fish	97.37	91.67	71.26	<0.001
Canned food	56.58	48.81	60.92	0.274
White meat	97.37	92.86	96.55	0.328

Table 1. *Cont.*

Site (Minerals)	Montelibano (Fe/Ni)	Nechí (Au)	Aranzazu (Closed Hg)	
Variables	**(n = 76) %/Median (Min–Max)**	**(n = 84) %/Median (Min–Max)**	**(n = 87) %/Median (Min–Max)**	***p* Value**
Red meat	94.74	94.05	96.55	0.734
Fruits	96.05	94.05	94.25	0.825
Vegetables	98.68	97.62	94.25	0.242
Carcinogenic elements * (IARC 1)				
Arsenic	0.14 (0.05–4.28)	0.18 (0.06–1.64)	0.12 (0.05–0.36)	<0.001 [a]
Beryllium	0 (0–16.06)	0 (0–0.07)	0 (0–0.02)	<0.001 [a]
Cadmium	0.05 (0–1.69)	0.04 (0.01–0.48)	0.05 (0.01–0.94)	0.395 [a]
Chromium	0.51 (0.18–19.06)	0.33 (0.18–7.25)	0.33 (0.20–3.53)	<0.001 [a]
Strontium	2.09 (0.1–55.42)	3.49 (0.33–46.21)	2.97 (0.19–46.13)	0.011 [a]
Nickel	0.45 (0.05–17.77)	0.25 (0.05–2.07)	0.14 (0.02–3.92)	<0.001 [a]
Thorium	0 (0–0.27)	0 (0–0.04)	0 (0–0.01)	<0.001 [a]
Probably carcinogenic * (IARC 2A)				
Antimony	0.03 (0.00–0.35)	0.02 (0.00–0.35)	0.01 (0.00–0.29)	0.040 [a]
Cobalt	0.04 (0.00–4.46)	0.03 (0.01–1.14)	0.01 (0.00–0.13)	<0.001 [a]
Indium	0 (0–0.002)	0	0 (0–0.004)	0.3484
Magnesium	77.90 (12.97–1449)	105.81 (19.44–2131)	49.73 (9.51–608.74)	<0.001 [a]
Manganese	1.66 (0.06–92.63)	3.09 (0.15–46.54)	0.80 (0.10–11.50)	<0.001 [a]
Silica	0 (0 -7598)	0 (0–4495)	1201 (0–3913)	<0.001 [a]
Possibly carcinogenic * (IARC 2B)				
Mercury	1.89 (0.06–12.29)	2.48 (0.17–17.14)	0.08 (0–1.63)	<0.001 [a]
Molybdenum	0.05 (0.01–1.40)	0.06 (0.02–0.36)	0.04 (0.02–0.24)	<0.001 [a]
Lead	0.92 (0.04–89.01)	0.64 (0.08–75.17)	0.55 (0.06–163.37)	0.761 [a]
Potassium	1.83 (0–155.47)	3.98 (0–106.10)	1.08 (0–129.28)	<0.001 [a]
Sodium	3.12 (0.47–238.26)	3.94 (0.49–701.25)	1.44 (0.30–379.54)	<0.001 [a]
Titanium	0.13 (0–3.16)	0.24 (0.09–1.24)	0.18 (0.08–0.60)	<0.001 [a]
Vanadium	0.05 (0.00–1.41)	0.12 (0.03–0.85)	0.05 (0.02–0.45)	<0.001 [a]
Other elements				
Phosphorus	219 (66–2883)	218 (111–342)	205 (148–297)	0.114 [a]
Selenium	1.35 (0.23–14.70)	1.40 (0.46–60.69)	1.01 (0.41–1.44)	<0.001 [a]

* Classification of the International Agency for Research on Cancer. [a] Kruskal-Wallis test.

3.2. Pesticide Exposure

In relation to the quantified pesticides (Table 2), the most relevant finding is that they were detected in only a low proportion of the participants, which suggests a low number of individuals with recent exposure. In the blood samples, the following were not detected: detected aldicarb, alpha-endosulfan, betha-endosulfan, endosulfan-sulfate, malathion, hexachlorobenzene, parathion-methyl, propoxur, and carbofuran. In the urine samples, the following were not detected: aldicarb, alpha-endosulfan, betha-endosulfan, endosulfan-sulfate, malathion, hexachlorobenzene, and carbofuran. In Montelibano, more pesticides were detected, where paraoxon-methyl in blood was the most frequent, whereas, in Aranzazu, the most common was paraoxon-ethyl in blood.

Table 2. Concentrations of selected pesticides in blood and urine among inhabitants near three mines in Colombia.

Montelibano (Fe/Ni Open-Pit Mine)		Positives	Concentrations (ppm)					
			Percentiles					
Pesticide	**Sample**	**%**	**Min**	**25**	**90**	**95**	**99**	**Max**
Paraoxon-methyl	Blood	27.55	ND	ND	0.24	0.29	37.26	37.26
Parathion-ethyl	Blood	1.02	ND	ND	ND	ND	ND	1.02
Paraoxon-ethyl	Urine	2.04	ND	ND	ND	ND	0.06	0.06
Parathion-ethyl	Urine	2.04	ND	ND	ND	ND	0.07	0.07
Parathion-methyl	Urine	2.04	ND	ND	ND	ND	0.07	0.07
Propoxur	Urine	2.04	ND	ND	ND	ND	0.13	0.44
Nechí (Au fluvial mines)		**Positives**	**Concentrations (ppm)**					
			Percentiles					
Pesticide	**Sample**	**%**	**Min**	**25**	**90**	**95**	**99**	**Max**
Paraoxon-ethyl	Blood	0.99	ND	ND	ND	ND	ND	0.75
Parathion-ethyl	Urine	0.99	ND	ND	ND	ND	ND	0.05
Aranzazu (Hg closed mine)		**Positives**	**Concentrations (ppm)**					
			Percentiles					
Pesticide	**Sample**	**%**	**Min**	**25**	**90**	**95**	**99**	**Max**
Paraoxon-ethyl	Blood	18.81	ND	ND	0.08	0.09	0.10	0.13
Paraoxon-methyl	Urine	0.99	ND	ND	ND	ND	ND	0.38

ND: Not Detectable.

3.3. Element Mixtures in Hair

There were very different mixtures of elements in the populations residing in the surroundings of the three mines (see Figures 2–4). In the FC1M mixture in Montelibano, the most important elements were P, S, Cu, As, Se, Mo, Li, Mg, Ca, V, Cr, Mn, Fe, Ni, Zn, Ge, Sr, Pd, Cd, Sb, Re, and Tl. This factor included the elements with economic interest (Fe/Ni) and recognized toxins, such as As, Cd, Mn, Ta and Cr, and Se as potential chelators. This mixture is highly dangerous to the health of those who are exposed, as it was described previously [25,26]. FC2M included Dy, Tb, Y, Er, Tm, Lu, Sm, Yb, Ho, Gd, Eu, La, and Ce. This mixture included rare earth elements (REEs) related to modern products, such as electronics, chemicals, medical, aviation, and defense technologies, and, more recently, agriculture [42]. Some studies suggest that REEs are associated with nephrotoxic, neurotoxic, and reproductive disorders, fibrotic tissue injury, oxidative stress, pneumoconiosis, and cytotoxicity [43]. Maybe REEs are the consequence of the smelting processes in the ferronickel mine. Note that Be, Hg, and Pb have no important participation in these mixtures.

In Nechí, FC1N included Dy, Er, Tm, Yb, Y, Lu, Tb, Eu, Ho, Gd, Sm, La, Pd, Be, and Co. This mixture included less frequent elements together with Be and Co. FC2N in Nechí included Ru, Sr, Mg, Ca, Ba, Ga, Ni, Na, Pd, K, Zn, and Mn, with Ni and Mn being the most toxic mixture. FC3N in Nechí included Cd, Ti, As, Al, V, Li, Sb, Mo, Ge, Th, B, Cs, Cr, Mn, Rh, and Ag. This is the most toxic mixture in Nechí. Note that Hg, Se, and Pb have no important participation in these mixtures. FC1A in Aranzazu included Er, Y, Dy, Yb, Tb, V, Tm, Al, Sm, Lu, Cd, Fe, Be, La, Nb, Ti, and Ce. In this mixture, the most toxic elements were Cd and Be. FC2A in Aranzazu included Mg, Ru, Sr, Na, K, Ba, Ga, Ca, Rb, Gd, Eu, Ge, Cs, Zr, and Pd. This element mixture only included Mg as a recognized toxic. Note that Hg, Se, As, Cr, Mn, P, and Pb have no important participation in these mixtures.

3.4. Micronuclei Frequency

The distribution of micronuclei observed in study participants by the municipality and in total is shown in Figure 5. As can be seen, in Montelibano, the median was 4 micronuclei with 25th and 75th percentiles of 2 and 6.5; in Nechí, the median was 5 micronuclei with 25th and 75th percentiles of 0 and 9, and in Aranzazu the median was 7 micronuclei with 25th and 75th percentiles of 2 and 10. However, there were no statistically significant differences between the three mining contexts ($p = 0.1298$, Kruskal–Wallis test).

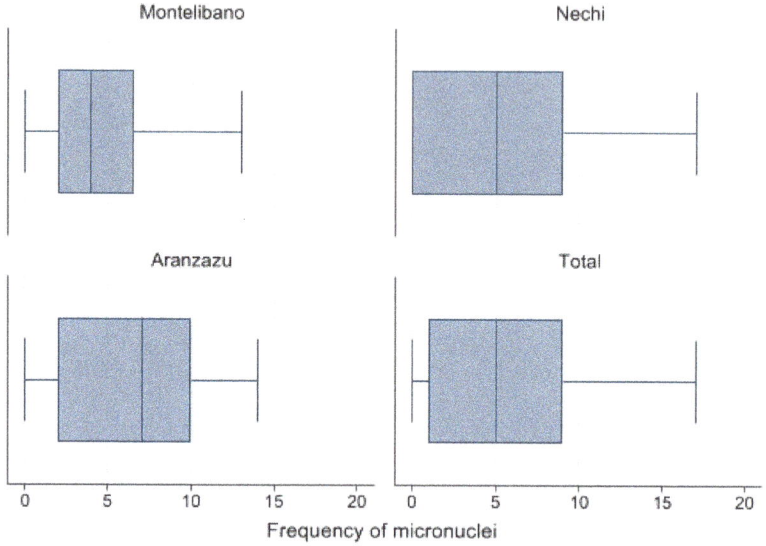

Figure 5. Frequency of micronuclei in the three mining contexts.

3.5. *Bivariate Analysis*

A first bivariate exploration (Table 3) of the possible associations with the number of micronuclei allowed us to identify that being a woman is protective in Montelibano, suggesting a differential occupational effect compared to men. Higher educational levels were protective, with their most noticeable effect being in Montelibano. Food consumption seems to mark routes of intake of toxic elements since the consumption of fish, white meat, and vegetables were risk factors in Montelibano, while the consumption of fruits and vegetables was protective in Nechí and Aranzazu, respectively. In relation to the mixtures of elements in hair, it was striking that no mixture showed an association in Montelibano, while in Nechí, a mixture (FC1N) with a protective effect was observed, and in Aranzazu there was a mixture (FC1A) that was a risk factor. Mercury in hair was a risk factor in Aranzazu, which was where the lowest concentration was observed among its inhabitants. Contrary to expectations, selenium in hair was a risk factor in Nechí, where the highest values of selenium were. Something similar happened with lead in hair, which was a protective factor in Nechí. Another unexpected finding is that beryllium in hair was a protective factor in Montelibano.

3.6. *Multiple Analysis*

Table 4 shows the most relevant variables associated with the number of micronuclei obtained in the multiple regression models. In Montelibano, it turned out that the consumption of fish and lead in hair were risk factors, as well as the pesticides parathion ethyl in blood and paraoxon ethyl in urine. The mixture of FC1M elements, beryllium in hair, and the consumption of vegetables and red meat were protective factors. Due to its importance, it should be clear that selenium is part of the FC1M mix.

Table 3. Crude prevalence ratios (PR) between pesticide and element exposure and micronuclei occurrence.

Site (Type of Mine)	Montelibano (n = 76) Fe/Ni		Nechí (n = 84) Au		Aranzazu (n = 87) Closed Hg	
Variables	**PR**	**95% CI**	**PR**	**95% CI**	**PR**	**95% CI**
Sex (female)	0.68	0.47–0.99	0.82	0.56–1.21	0.86	0.63–1.17
Age (years)	1.01	0.99–1.02	1.00	0.99–1.01	1.00	0.99–1.01
Civil status						
Married/Free union	1		1		1	
Divorced/Widower	1.26	0.51–3.11	1.31	0.64–2.68	1.14	0.68–1.91
Single	1.17	0.68–2.02	1.26	0.85 -1.87	1.01	0.70–1.45
Education						
Illiteracy	1		1		1	
Elementary (partial or full)	0.38	0.24–0.60	1.02	0.56–1.86	0.63	0.42–0.94
Secondary (partial or full)	0.48	0.34–0.66	0.85	0.46 -1.59	0.67	0.43–1.03
Technic (partial or full)	0.41	0.31–0.53	1.00	0.53–1.89	0.78	0.40–1.54
University (partial or full)	0.45	0.31–0.64	1.29	0.65–2.55	0.78	0.50–1.22
Occupation						
In mining activities	0.90	0.66–1.23	1.12	0.77–1.64	NA	
In agricultural activities	1.13	0.71–1.79	0.58	0.33–1.02	0.99	0.72–1.35
Consumption (any moment)						
Cigarette	1.33	0.94–1.87	0.88	0.58–1.33	1.15	0.84–1.57
Alcohol	0.81	0.59–1.12	1.18	0.81–1.71	0.86	0.62–1.17
Food consumption						
Fish	2.27	1.11–4.64	1.74	0.53–5.74	1.33	0.91–1.97
Canned food	1.14	0.84–1.54	1.44	0.99–2.08	1.09	0.78–1.54
White meat	1.29	1.00–1.66	1.93	0.80–4.65	1.05	0.67–1.64
Red meat	0.89	0.62–1.26	1.34	0.45–4.01	0.84	0.35–2.04
Fruits	1.12	0.38–3.30	0.48	0.32–0.71	0.73	0.52–1.01
Vegetables	1.50	1.28–1.75	0.52	0.19–1.39	0.75	0.58–0.96
Pesticides (ppb)						
Paraoxon-methyl in blood	1.00	0.99–1.01				
Paraoxon-ethyl in blood					18.43	0.71–476.02
Element mixtures in hair						
FC1	0.97	0.94–1.00	0.90	0.83–0.98	1.04	1.00–1.07
FC2	0.98	0.94–1.03	1.04	0.99–1.09	1.02	0.99–1.05
FC3	NA		0.96	0.90–1.03	NA	
Selected elements in hair						
Mercury	0.97	0.93–1.02	0.98	0.92–1.03	1.76	1.06–2.93
Selenium	NA		1.02	1.01–1.02	0.58	0.25–1.34
Lead	1.01	1.00–1.01	0.94	0.89–0.98	1.00	0.99–1.00
Beryllium	0.77	0.66–0.89	NA		NA	

NA: Not Applicable.

In Nechí, the FC2N mixture (which includes Mg, Mn, and Ni), lead, and selenium were identified as risk factors. FC1N (with beryllium and very rare elements) and fruit consumption appeared as protective factors. Additionally, an interaction between lead and selenium was identified. In Aranzazu, the main risk factor is mercury, followed by nickel. Selenium and fruit consumption were identified as protective factors. A strong interaction between mercury and selenium was identified.

These findings may seem strange at first approximation but can be better understood by considering the different exposure pathways involved in each type of mine. In Montelibano, the exposure can be food and aerial, while in Nechí, it is mainly food in the context of occupational and para-occupational aerial exposure to mercury. In Aranzazu, the exposure is mainly for food.

Table 4. Adjusted prevalence ratios (adj. PR) between pesticide and element exposure and micronuclei occurrence.

Site (Type of Mine)	Montelibano (n = 76) (Open-Pit Fe/Ni)		Nechí (n = 84) (Fluvial Au)		Aranzazu (n = 87) (Closed Hg)	
Variables	**Adj. PR**	**95% CI**	**Adj. PR**	**95% CI**	**Adj. PR**	**95% CI**
Pesticides (ppb)						
Parathion ethyl (blood)	0.04	0.01–0.14				
Paraoxon ethyl (urine)	0.01	0.00–0.04				
Element mixtures in hair (scores)						
Factor 1	0.97	0.94–0.99	0.88	0.79–0.98		
Factor 2			1.05	1.00–1.09		
Hair concentration (ppm)						
Beryllium	0.67	0.59–0.76				
Lead	1.01	1.00–1.01	1.37	1.06–1.77		
Mercury					36.02	2.69–380.55
Selenium			1.05	1.03–1.08	0.86	0.32–2.33
Nickel					1.28	1.13–1.45
Interactions						
Selenium X Lead			0.78	0.64–0.95		
Selenium X Mercury					0.11	0.02–0.77
Diet consumption						
Fish	4.73	3.94–5.68				
Vegetables	0.47	0.38–0.59				
Red meat	0.67	0.54–0.83				
Fruits			0.55	0.38–0.80	0.66	0.46–0.95

4. Discussion

The most important finding of this study is the great heterogeneity in the exposure to pesticides and elements in the three contexts that combine mining and the use of pesticides in agriculture or to control vector-borne diseases. Within the complexity, it was clear that where there is recent mining activity is where more pesticides and elements were detected, and higher concentrations were observed. This is logical since mining mobilizes large amounts of rocks and soils, facilitating the entry of various elements to humans; these mainly through food and water, although the respiratory tract is likely in places with open pit mining. In fact, chemical mixtures were associated with the number of micronuclei only in Montelibano and Nechí, where there are active mines. In Aranzazu, the associations are between isolated elements or in interaction with another element.

In addition to this expected finding, selenium showed two different faces. On the one hand, selenium acted as a protective factor and, on the other, as a risk factor framed in the context of statistical interaction. This seems to be strong evidence of the "hormetic response" [44], in which selenium at low doses is beneficial (Aranzazu case) or acts synergistically with other chemicals (Nechí case) when it is in high concentrations, increasing the adverse effect. In the case of Montelibano, the presence of selenium in a mixture with many toxic elements seems to be related to the mitigation of the adverse effect. Note that the observed concentrations of selenium in hair in Montelibano were intermediate between Nechí and Aranzazu. To the best of our knowledge, this is the first time that the "hormetic effect" of selenium in humans has been reported under non-experimental conditions. Hormetic dose response has been described for several substances in different animal species [45]. Identifying the "hormetic effect" of a substance, such as selenium, allows us to understand the contradictory effects observed in several studies. Unfortunately, it is not easy to define the adequate dose related to only beneficial effects.

Reduction of exposure to toxic metals and metalloids using Se is a priority among susceptible populations, such as children and pregnant and lactating women. In general, consumption of selenium-rich foods can be part of interventions to avoid the adverse effects of contaminants, such as Hg [46]. The intake of Se can come from foods that, by definition, have higher concentrations, such as nuts, or from the consumption of artificially enriched foods. Another option is to apply Se to crops, such as rice, reducing Hg accumulation in plants [47]. However, special care must be taken in places where the geology includes selenium in high concentrations because synergistic adverse effects, such as those observed in Nechí in this study, may occur. Knowing the history is important because, in the 1626, fray Pedro Simon described the toxicity of Se in corn and other vegetables that grew in specific places; poisoned animals and humans lost their hair and had dermal lesions [48]. In consequence, although there are some regional studies [49], Colombia needs to improve studies on Se in soils, especially in places where there are mining activities.

In Aranzazu, there is a health issue that must be pointed out due to the possible implications it has on the results, and that, unfortunately, could not be addressed in the present study. Among the inhabitants of Aranzazu, endogamy has been common for several generations; in fact, the occurrence of neuropsychiatric disorders, such as mood bipolar disorder, is so high that studies have been carried out that report its association with the genetic factor [50]. In this case, the question arises as to whether these genetic changes have an influence on the number of micronuclei observed, such as that of the other mining sites.

This study should be interpreted considering the limitations of the study. First, the population was not representative, so the findings should be understood as exploratory of important environmental health problems. Interest in the study was the main motivation of the participants, which explains the greater participation of women; men tend to be more cautious in places where there are illegal armed groups, as occurred in Montelibano and Nechí. Colombia is one of the countries with more socio-ecological conflicts [51] and where more assassinations of environmental leaders occur [52], which is evidence of the difficulties of carrying out environmental health studies in places with the greatest contamination.

Perhaps no pesticides were found because it is easier to find them in those who directly carry out fumigation activities, which by tradition is an activity associated with men. Given that the miners could be the ones most exposed to elements, it is possible that their low participation has the effect that their results are higher than reported. An important issue to note is that, due to the high number of elements included, several of these do not have certified standards. However, due to the exploratory and novel nature of the study approach, it was decided to include all the results. If there was a measurement error, we have no evidence that it was differential.

In conclusion, the results show the complex exposures that occur in mining contexts where there are also pesticides used in agricultural activities or to control vector-borne diseases. Each place is so different from the others that it is difficult to extrapolate the findings to other similar areas. In the Colombian case, it is a difficult task because the data on the use of pesticides do not have adequate information; it is even more complicated in places where illegal activities are carried out, such as coca, marijuana, or poppy cultivation [15]. In addition, soil geochemistry data is scarce, and much of what exists is only available to mining companies that collect data as part of their economic activities. For this reason, studies such as this one are essential to understand the effects of large extractive activities, especially if they were carried out before and after starting work.

Additionally, this study identified the region where the open pit ferronickel mine is located as the most polluted and which demands urgent attention from the environmental and health authorities. Since this may be happening at other large open pit mines, it is important to perform similar studies in regions with coal in La Guajira [53] and Cesar departments [54] and gold mining in La Colosa in Cajamarca, Tolima [55]. Periodic monitoring of environmental agents, mainly in the most polluted sites, is essential to have adequate vigilance of the potential adverse effects associated with chemical agents. Without a doubt, it is very important that Colombian authorities prioritize toxic chemicals because there is an exaggerated use of pesticides and many extractive activities with a high impact on the environment.

Author Contributions: Conceptualization and methodology, M.E.V.-U., S.M.D., R.-M.P., L.B.-A., C.T.-P., E.M.T.-A., L.E.-P., K.P.-S., P.J.E.-P. and A.J.I.; validation, S.M.D.; formal analysis, A.J.I. and S.M.D.; writing—original draft preparation, A.J.I.; writing—review and editing, M.E.V.-U., R.-M.P., E.M.T.-A., S.M.D., L.E.-P. and C.T.-P.; supervision, project administration, and funding acquisition, M.E.V.-U., R.-M.P., S.M.D. and L.E.-P. All authors have read and agreed to the published version of the manuscript.

Funding: This research was funded by Ministerio de Ciencia Tecnología e Innovación (Minciencias) in Colombia under contract No. 905–2019, Universidad del Rosario, Universidad del Sinú, the National Institute of Health, and Universidad Industrial de Santander.

Institutional Review Board Statement: The study was conducted in accordance with the Declaration of Helsinki and the regulations established in Resolution 8430 of 1993 of the Colombian Ministry of Health. It was approved by The Ethics Committee of the Colombian National Health Institute (registry #23-2019) and the Scientific Research Ethics Committee of the Universidad Industrial de Santander (act 23, 19 December 2019).

Informed Consent Statement: Informed consent was obtained from all subjects involved in the study.

Data Availability Statement: The data presented in this study are available on request from the corresponding author.

Acknowledgments: The authors thank the residents of the participating municipalities for their collaboration during the fieldwork; to the ICP-MS and ICP OES Laboratory (LABSPECTRO) of the Chemistry Department of the Pontificia Universidade Católica do Rio de Janeiro, Brazil, for the support with hair analyses, and to Dina Ricardo-Caldera who made the map for the article.

Conflicts of Interest: The authors declare no conflict of interest. The funders had no role in the design of the study, in the collection, analyses, or interpretation of data, in the writing of the manuscript, or in the decision to publish the results.

Appendix A

Validation data of pesticide analyses—blood/urine LC/Ms-Ms QUECHERS extraction; Main element mixtures in each mining site obtained with principal component analysis.

Table A1. Validation data—blood LC/Ms-Ms QuEChERS extraction.

Pesticide	Lineal Range (ppb)	Detection Limit (ppb)	Quantification Limit (ppb)
Aldicarb (ppb)	0.075 to 15.00	0.01	0.075
Propoxur (ppb)	0.075 to 15.00	0.01	0.075
Carbofuran (ppb)	0.075 to 15.00	0.01	0.075
α-endosulfan (ppb)	5.00 to 50.00	1.00	5.00
β-endosulfan (ppb)	5.00 to 50.00	1.00	5.00
Endosulfan sulfate (ppb)	2.00 to 75.00	0.05	2.00
Malathion (ppb)	0.075 to 15.00	0.01	0.075
hexachlorobenzene (ppb)	0.125 to 75.00	0.025	0.125
ethyl paraoxon (ppb)	0.075 to 15.00	0.01	0.075
methyl paraoxon (ppb)	0.075 to 15.00	0.01	0.075
ethyl parathion (ppb)	0.075 to 15.00	0.01	0.075
methyl parathion (ppb)	0.075 to 15.00	0.01	0.075

Table A2. Validation data—urine LC/Ms-Ms QuEChERS extraction.

Pesticide	Lineal Range (ppb)	Detection Limit (ppb)	Quantification Limit (ppb)
Aldicarb (ppb)	0.075 to 15.00	0.01	0.075
Propoxur (ppb)	0.075 to 15.00	0.01	0.075
Carbofuran (ppb)	0.075 to 15.00	0.01	0.075
α-endosulfan (ppb)	5.00 to 50.00	1.00	5.00
β-endosulfan (ppb)	5.00 to 50.00	1.00	5.00
Endosulfan sulfate (ppb)	2.00 to 75.00	0.05	2.00
Malathion (ppb)	0.075 to 15.00	0.01	0.075
hexachlorobenzene (ppb)	0.125 to 75.00	0.025	0.125
ethyl paraoxon (ppb)	0.075 to 15.00	0.01	0.075
methyl paraoxon (ppb)	0.075 to 15.00	0.01	0.075
ethyl parathion (ppb)	0.075 to 15.00	0.01	0.075
methyl parathion (ppb)	0.075 to 15.00	0.01	0.075

Table A3. Montelibano mine. Rotated components.

Variable	Comp1	Comp2	Unexplained
Li7ppm	0.1500	0.0191	0.52
Be9ppm	0.0473	0.0327	0.9327
B11ppm	0.1305	0.0220	0.63
Na23ppm	0.0662	−0.0063	0.9131
Mg24ppm	0.1870	−0.0380	0.3187
Al27ppm	0.0742	0.0091	0.8829
Si28ppm	0.1413	−0.0416	0.6132
P31ppm	0.2082	0.0145	0.09689
S34ppm	0.2107	0.0026	0.09373
K39ppm	0.0639	0.0067	0.9136
Ca44ppm	0.1978	−0.0258	0.2299
Sc45ppm	0.0012	−0.0089	0.9992
Ti47ppm	0.1200	0.0388	0.6604
V51ppm	0.1932	−0.0285	0.2674
Cr52ppm	0.1895	0.0186	0.2429
Mn55ppm	0.1549	−0.0064	0.5178
Fe57ppm	0.1885	0.0145	0.2573
Co59ppm	0.0504	0.0342	0.9244
Ni60ppm	0.1947	−0.0113	0.2421
Cu65ppm	0.2097	0.0007	0.1052
Zn66ppm	0.1931	0.0067	0.2328
Ga69ppm	0.1342	0.0188	0.6137
Ge74ppm	0.1828	−0.0469	0.3514
As75ppm	0.2112	−0.0023	0.09576
Br79ppm	0.0475	−0.0033	0.9551
Se82ppm	0.2097	0.0057	0.09745
Rb85ppm	0.0761	−0.0035	0.8838
Sr88ppm	0.1905	−0.0503	0.2963
Y89ppm	−0.0160	0.2961	0.06107

Table A3. *Cont.*

Variable	Comp1	Comp2	Unexplained
Zr90ppm	−0.0002	0.0105	0.9988
Nb93ppm	0.0153	0.0024	0.9949
Mo98ppm	0.2059	−0.0278	0.1661
Ru102ppm	0.0243	−0.0237	0.9856
Rh103ppm	0.0215	−0.0223	0.9882
Pd105ppm	0.1568	0.0694	0.3767
Ag107ppm	0.0215	−0.0223	0.9882
Cd114ppm	0.1610	−0.0487	0.4974
In115ppm	0.0162	0.0020	0.9944
Sn118ppm	0.0167	−0.0215	0.9916
Sb121ppm	0.1809	−0.0231	0.3559
I127ppm	0.0471	−0.0154	0.957
Cs133ppm	0.1414	−0.0053	0.5982
Ba137ppm	0.1245	0.0183	0.6661
La139ppm	0.0366	0.2215	0.3809
Ce140ppm	0.0761	0.2036	0.3263
Pr141ppm	0.0914	0.0952	0.6742
Nd143ppm	0.0930	0.0656	0.7374
Sm147ppm	−0.0111	0.2836	0.1337
Eu153ppm	0.0492	0.2336	0.2762
Gd157ppm	0.0561	0.2459	0.1821
Tb159ppm	−0.0142	0.3002	0.03225
Dy163ppm	−0.0265	0.3027	0.03045
Ho165ppm	0.0436	0.2556	0.1711
Er166ppm	−0.0390	0.2938	0.09388
Tm169ppm	−0.0274	0.2887	0.1194
Yb172ppm	0.0159	0.2768	0.1242
Lu175ppm	−0.0468	0.2851	0.1482
Ta181ppm	0.1006	0.0208	0.7759
W182ppm	−0.0107	−0.0288	0.9866
Re185ppm	0.1857	0.0421	0.2286
Pt194ppm	−0.0017	0.0101	0.9989
Au197ppm	0.0089	−0.0077	0.9982
Hg202ppm	0.1397	−0.0477	0.621
Tl205ppm	0.1923	−0.0274	0.2741
Pb208ppm	0.0280	−0.0081	0.9848
Bi209ppm	0.0389	−0.0043	0.97
Th232ppm	0.0463	0.0276	0.9398
U238ppm	−0.0153	0.0667	0.9529
	Component rotation matrix		
Comp1	0.9545	0.2982	
Comp2	−0.2982	0.9545	

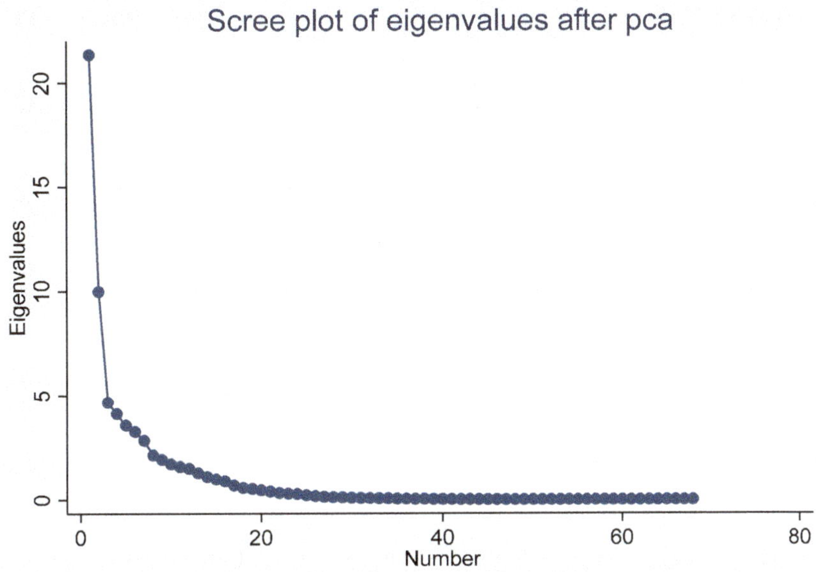

Figure A1. Montelibano mine. Scree plot of eigenvalues after principal component analysis.

Table A4. Nechí mine. Rotated components.

Variable	Comp1	Comp2	Comp3	Unexplained
Li7ppm	−0.0115	0.0063	0.2322	0.668
Be9ppm	0.1641	0.0943	0.0676	0.428
B11ppm	−0.0730	−0.0993	0.1918	0.6905
Na23ppm	−0.0492	0.2026	0.0141	0.6789
Mg24ppm	−0.0579	0.3185	−0.0070	0.2158
Al27ppm	0.0292	−0.0349	0.2480	0.5471
Si28ppm	0.0614	−0.0603	0.0216	0.9232
P31ppm	0.0210	−0.1330	0.0652	0.8326
S34ppm	−0.0232	−0.1452	0.0180	0.8187
K39ppm	−0.0366	0.1778	0.0907	0.7146
Ca44ppm	−0.0331	0.3095	−0.0346	0.256
Ti47ppm	0.0354	−0.0302	0.2649	0.4755
V51ppm	0.0194	−0.0054	0.2402	0.5968
Cr52ppm	−0.0241	0.0426	0.1676	0.8227
Mn55ppm	0.0832	0.1594	0.1615	0.4053
Fe57ppm	0.0088	0.0143	0.0925	0.9365
Co59ppm	0.1617	0.0669	0.1027	0.4149
Ni60ppm	0.0119	0.2265	0.0548	0.5576
Cu65ppm	−0.0485	0.0411	0.1353	0.8828
Zn66ppm	−0.0424	0.1659	−0.0004	0.7825
Ga69ppm	0.0612	0.2940	0.0238	0.2001

Table A4. *Cont.*

Variable	Comp1	Comp2	Comp3	Unexplained
Ge74ppm	−0.0107	0.0976	0.2001	0.6738
As75ppm	−0.0443	−0.0311	0.2606	0.6023
Br79ppm	0.0044	−0.0016	−0.0314	0.9943
Se82ppm	−0.0081	−0.0209	−0.0087	0.9942
Rb85ppm	−0.0186	0.1297	0.1449	0.7437
Sr88ppm	−0.0344	0.3195	−0.0553	0.1947
Y89ppm	0.2702	−0.0120	−0.0440	0.1154
Zr90ppm	0.0062	−0.0070	0.1009	0.9305
Nb93ppm	0.0274	−0.0513	0.1370	0.8336
Mo98ppm	−0.0339	−0.0011	0.2230	0.7128
Ru102ppm	−0.0402	0.3249	−0.0172	0.1875
Rh103ppm	−0.0585	−0.0429	0.1548	0.8387
Pd105ppm	0.1690	0.1871	−0.0274	0.2792
Ag107ppm	−0.0585	−0.0429	0.1548	0.8387
Cd114ppm	−0.0324	0.0052	0.2684	0.5776
Sn118ppm	−0.0175	0.0126	0.1077	0.9321
Sb121ppm	0.0264	−0.0518	0.2278	0.6098
I127ppm	0.1067	0.1061	0.0060	0.7229
Cs133ppm	0.0865	0.0496	0.1696	0.5898
Ba137ppm	0.0545	0.3004	−0.0011	0.2019
La139ppm	0.2261	0.0191	0.0360	0.2558
Ce140ppm	0.1444	−0.0180	0.1492	0.4578
Pr141ppm	0.0817	−0.0595	0.0962	0.7948
Nd143ppm	0.0515	−0.0620	0.0744	0.8882
Sm147ppm	0.2375	−0.0029	0.0403	0.1947
Eu153ppm	0.2467	0.0513	0.0218	0.107
Gd157ppm	0.2404	−0.0383	0.0542	0.1628
Tb159ppm	0.2629	−0.0297	0.0305	0.05878
Dy163ppm	0.2765	−0.0237	−0.0136	0.03806
Ho165ppm	0.2456	−0.0470	0.0194	0.1984
Er166ppm	0.2728	−0.0181	−0.0349	0.09089
Tm169ppm	0.2717	−0.0177	−0.0377	0.1013
Yb172ppm	0.2706	−0.0207	−0.0560	0.13
Lu175ppm	0.2661	−0.0185	−0.0436	0.1461
Ta181ppm	0.0219	0.0252	0.1231	0.8716
W182ppm	−0.0100	0.1034	0.0185	0.916
Re185ppm	−0.0068	−0.0379	−0.0067	0.9866
Pt194ppm	−0.0261	0.0131	0.1446	0.8793
Au197ppm	−0.0408	−0.0434	0.0348	0.9597
Hg202ppm	−0.0631	−0.1192	0.1304	0.7625

Table A4. *Cont.*

Variable	Comp1	Comp2	Comp3	Unexplained
Tl205ppm	−0.0717	0.1403	0.1189	0.7638
Pb208ppm	−0.0367	−0.0415	0.1321	0.885
Bi209ppm	0.0371	0.1073	0.0109	0.8753
Th232ppm	−0.0311	−0.0263	0.1933	0.7794
U238ppm	0.0228	0.0653	0.0465	0.9332
	Component rotation matrix			
Comp1	0.9102	0.2221	0.3497	
Comp2	−0.1539	0.9650	−0.2122	
Comp3	−0.3845	0.1393	0.9125	

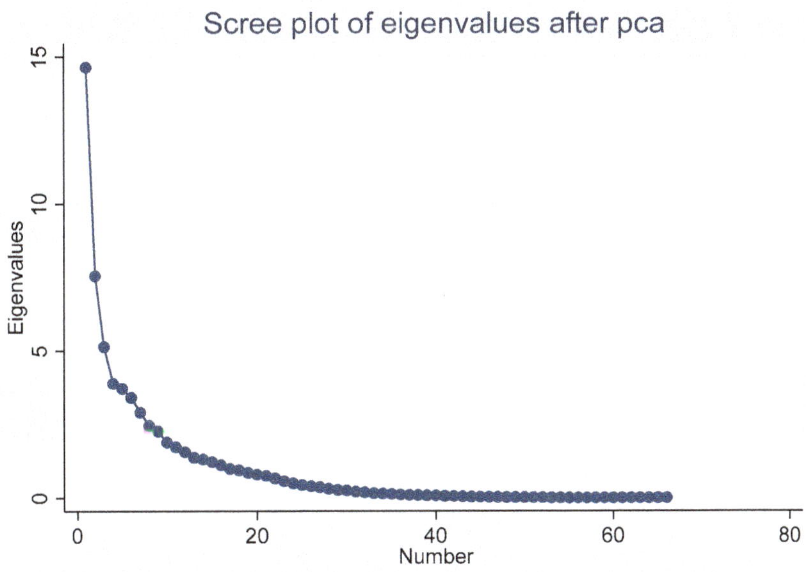

Figure A2. Nechí mine. Scree plot of eigenvalues after principal component analysis.

Table A5. Aranzazu (closed) mine. Rotated components.

Variable	Comp1	Comp2	Unexplained
Li7ppm	0.1282	0.1476	0.4032
Be9ppm	0.1615	−0.0780	0.6992
B11ppm	0.0763	0.0005	0.9241
Na23ppm	−0.0588	0.2748	0.2883
Mg24ppm	−0.0468	0.2847	0.2233
Al27ppm	0.2177	−0.0328	0.4314
Si28ppm	0.0929	−0.0485	0.8996
P31ppm	−0.0800	0.0166	0.9247

Table A5. *Cont.*

Variable	Comp1	Comp2	Unexplained
S34ppm	0.0755	0.0064	0.9215
K39ppm	−0.0603	0.2665	0.3328
Ca44ppm	0.0432	0.2435	0.2644
Ti47ppm	0.1547	0.0238	0.6532
V51ppm	0.2220	−0.0831	0.4365
Cr52ppm	0.1002	−0.0045	0.8731
Mn55ppm	0.1143	0.0114	0.8184
Fe57ppm	0.1626	0.0175	0.6304
Co59ppm	0.1233	0.0255	0.77
Ni60ppm	0.1389	0.0307	0.7051
Cu65ppm	0.1355	−0.0199	0.7793
Zn66ppm	0.1358	0.0027	0.7577
Ga69ppm	0.0198	0.2537	0.2743
Ge74ppm	0.0849	0.1918	0.3856
As75ppm	0.0877	−0.0314	0.912
Br79ppm	0.0488	−0.0019	0.9698
Se82ppm	−0.0894	−0.0231	0.8737
Rb85ppm	−0.0483	0.2360	0.4736
Sr88ppm	0.0024	0.2806	0.1626
Y89ppm	0.2415	−0.0064	0.2545
Zr90ppm	0.0019	0.1568	0.738
Nb93ppm	0.1582	0.0610	0.5575
Mo98ppm	0.1197	−0.0525	0.8358
Ru102ppm	−0.0322	0.2846	0.2052
Rh103ppm	0.1125	0.0636	0.735
Pd105ppm	−0.0263	0.1554	0.7689
Ag107ppm	0.1125	0.0636	0.735
Cd114ppm	0.1717	0.0207	0.5838
In115ppm	0.0733	−0.0367	0.9379
Sn118ppm	0.1385	0.0131	0.7344
Sb121ppm	0.0901	0.0179	0.8783
I127ppm	0.0012	−0.0047	0.9998
Cs133ppm	0.0212	0.1844	0.6031
Ba137ppm	0.0124	0.2555	0.2828
La139ppm	0.1589	−0.0109	0.6848
Ce140ppm	0.1520	−0.0003	0.7003
Pr141ppm	0.1014	0.1068	0.6583
Nd143ppm	0.0410	0.0606	0.9192
Sm147ppm	0.1897	0.0352	0.4652
Eu153ppm	0.0707	0.2241	0.276

Table A5. *Cont.*

Variable	Comp1	Comp2	Unexplained
Gd157ppm	0.0545	0.2279	0.312
Tb159ppm	0.2235	0.0222	0.306
Dy163ppm	0.2364	0.0241	0.2221
Ho165ppm	0.1325	0.0871	0.5982
Er166ppm	0.2470	0.0025	0.2027
Tm169ppm	0.2213	−0.0317	0.4104
Yb172ppm	0.2237	−0.1070	0.4231
Lu175ppm	0.1777	−0.0688	0.6391
Ta181ppm	0.0220	−0.0088	0.9945
W182ppm	0.0982	0.0779	0.7488
Re185ppm	0.0937	−0.0567	0.8952
Pt194ppm	0.0062	0.1416	0.7806
Au197ppm	0.0734	0.0080	0.9246
Hg202ppm	−0.0141	0.0413	0.9841
Tl205ppm	−0.0535	0.1120	0.8788
Pb208ppm	−0.0062	0.0018	0.9996
Bi209ppm	−0.0086	0.0077	0.999
Th232ppm	0.0674	−0.0158	0.947
U238ppm	0.1072	0.0069	0.8444
	Component rotation matrix		
Comp1	0.8020		
Comp2	0.5974	−0.8020	

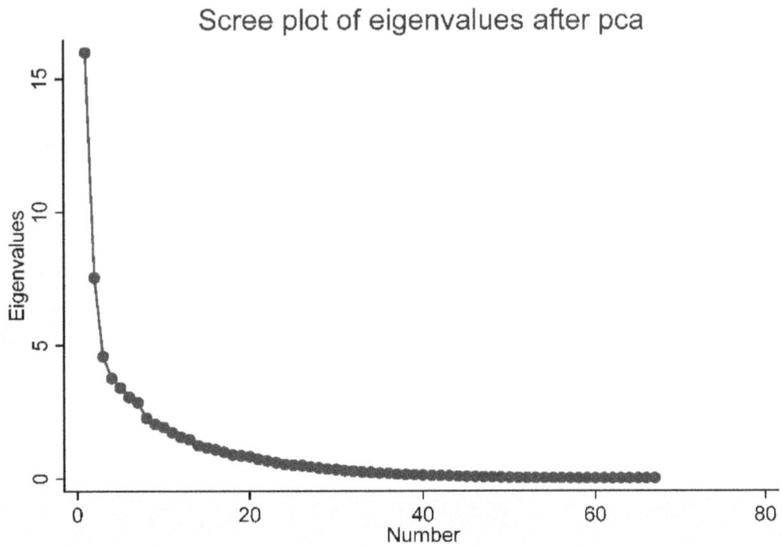

Figure A3. Aranzazu mine. Scree plot of eigenvalues after principal component analysis.

References

1. Kaur, K.; Kaur, R. Occupational Pesticide Exposure, Impaired DNA Repair, and Diseases. *Indian J. Occup. Environ. Med.* **2018**, *22*, 74. [CrossRef] [PubMed]
2. Stanaway, J.D.; Afshin, A.; Gakidou, E.; Lim, S.S.; Abate, D.; Abate, K.H.; Abbafati, C.; Abbasi, N.; Abbastabar, H.; Abd-Allah, F.; et al. Global, Regional, and National Comparative Risk Assessment of 84 Behavioural, Environmental and Occupational, and Metabolic Risks or Clusters of Risks for 195 Countries and Territories, 1990–2017: A Systematic Analysis for the Global Burden of Disease Study 2017. *Lancet* **2018**, *392*, 1923–1994. [CrossRef]
3. Damalas, C.; Koutroubas, S. Farmers' Exposure to Pesticides: Toxicity Types and Ways of Prevention. *Toxics* **2016**, *4*, 1. [CrossRef] [PubMed]
4. Kim, K.-H.; Kabir, E.; Jahan, S.A. Exposure to Pesticides and the Associated Human Health Effects. *Sci. Total Environ.* **2017**, *575*, 525–535. [CrossRef] [PubMed]
5. Rani, L.; Thapa, K.; Kanojia, N.; Sharma, N.; Singh, S.; Grewal, A.S.; Srivastav, A.L.; Kaushal, J. An Extensive Review on the Consequences of Chemical Pesticides on Human Health and Environment. *J. Clean. Prod.* **2021**, *283*, 124657. [CrossRef]
6. Cotton, J.; Edwards, J.; Rahman, M.A.; Brumby, S. Cholinesterase Research Outreach Project (CROP): Point of Care Cholinesterase Measurement in an Australian Agricultural Community. *Environ. Health* **2018**, *17*, 31. [CrossRef] [PubMed]
7. Imran, I.; Ansari, A.; Saleem, S.; Azhar, A.; Zehra, S. Insights of OPs and PYR Cytotoxic Potential In Vitro and Genotoxic Impact on PON1 Genetic Variant among Exposed Workers in Pakistan. *Sci. Rep.* **2022**, *12*, 9498. [CrossRef] [PubMed]
8. Marcelino, A.F.; Wachtel, C.C.; Ghisi, N.d.C. Are Our Farm Workers in Danger? Genetic Damage in Farmers Exposed to Pesticides. *Int. J. Environ. Res. Public Health* **2019**, *16*, 358. [CrossRef]
9. Mostafalou, S.; Abdollahi, M. Pesticides: An Update of Human Exposure and Toxicity. *Arch. Toxicol.* **2017**, *91*, 549–599. [CrossRef]
10. Fenech, M.; Holland, N.; Kirsch-Volders, M.; Knudsen, L.E.; Wagner, K.-H.; Stopper, H.; Knasmueller, S.; Bolognesi, C.; El-Zein, R.; Bonassi, S. Micronuclei and Disease—Report of HUMN Project Workshop at Rennes 2019 EEMGS Conference. *Mutat. Res. Toxicol. Environ. Mutagen.* **2020**, *850–851*, 503133. [CrossRef]
11. Ofosu, G.; Dittmann, A.; Sarpong, D.; Botchie, D. Socio-Economic and Environmental Implications of Artisanal and Small-Scale Mining (ASM) on Agriculture and Livelihoods. *Environ. Sci. Pol.* **2020**, *106*, 210–220. [CrossRef]
12. Valbuena, D.; Cely-Santos, M.; Obregón, D. Agrochemical Pesticide Production, Trade, and Hazard: Narrowing the Information Gap in Colombia. *J. Environ. Manag.* **2021**, *286*, 112141. [CrossRef] [PubMed]
13. Castellanos, A.; Chaparro-Narváez, P.; Morales-Plaza, C.D.; Alzate, A.; Padilla, J.; Arévalo, M.; Herrera, S. Malaria in Gold-Mining Areas in Colombia. *Mem. Inst. Oswaldo Cruz.* **2016**, *111*, 59–66. [CrossRef] [PubMed]
14. Zaragoza-Bastida, A.; Valladares-Carranza, B.; Ortega-Santana, C.; Zamora-Espinosa, J.; Velázquez-Ordoñez, V.; Aparicio-Burgos, J. Repercusiones del Uso de los Organoclorados sobre el Ambiente y Salud Pública. *AbanicoVet* **2016**, *6*, 43–55.
15. Varona, M.E.; Díaz-Criollo, S.M.; Lancheros-Bernal, A.R.; Murcia-Orjuela, A.M.; Henao-Londoño, G.L.; Idrovo, A.J. Organochlorine Pesticide Exposure among Agricultural Workers in Colombian Regions with Illegal Crops: An Exploration in a Hidden and Dangerous World. *Int. J. Environ. Health Res.* **2010**, *20*, 407–414. [CrossRef] [PubMed]
16. García Ubaque, C.A.; García Ubaque, J.C.; Vaca Bohórquez, M.L. Compuestos Orgánicos Persistentes en Colombia: Cuantificación y Diagnóstico para Pesticidas Organoclorados. *Tecnura* **2015**, *19*, 163. [CrossRef]
17. FAOSTAT. Organización de las Naciones Unidas Para la Alimentación y la Agricultura. Plaguicidas. 2019. Available online: http://www.fao.org/faostat/es/#data/EP/metadata (accessed on 31 May 2023).
18. Castro Cely, Y. Relación entre la Exposición a Plaguicidas con el Intento Suicida y Suicidio en Colombia durante el Periodo 2009–2018. Ph.D. Thesis, Universidad Nacional de Colombia, Bogotá, Colombia, 2023.
19. Idrovo, A.J. Plaguicidas usados en la Fumigación de Cultivos Ilícitos y Salud Humana: ¿Una Cuestión de Ciencia o Política? *Rev. Salud Pública* **2004**, *6*, 199–211. [CrossRef] [PubMed]
20. Varona, M.; Henao, G.L.; Diaz, S.; Lancheros, A.; Murcia, A.; Rodriguez, N.; Alvarez, V.H. Effects of Aerial Applications of the Herbicide Glyphosate and Insecticides on Human Health. *Biomedica* **2009**, *29*, 456–475. [CrossRef]
21. Gillezeau, C.; van Gerwen, M.; Shaffer, R.M.; Rana, I.; Zhang, L.; Sheppard, L.; Taioli, E. The Evidence of Human Exposure to Glyphosate: A Review. *Environ. Health* **2019**, *18*, 2. [CrossRef]
22. Idrovo, A.J.; Rodríguez-Villamizar, L.A. Moving Back in Policy Banning Glyphosate Use in Colombia. *Lancet* **2018**, *392*, 1194. [CrossRef]
23. Ancizar-Sordo, J. Occurrence of Selenium in Soils and Plants of Colombia, South America. *Soil Sci.* **1947**, *63*, 437–438. [CrossRef]
24. Spiller, H.A.; Hays, H.L.; Casavant, M.J. Rethinking Treatment of Mercury Poisoning: The Roles of Selenium, Acetylcysteine, and Thiol Chelators in the Treatment of Mercury Poisoning: A Narrative Review. *Toxicol. Commun.* **2021**, *5*, 19–59. [CrossRef]
25. Rodriguez-Villamizar, L.A.; Medina, O.M.; Flórez-Vargas, O.; Vilanova, E.; Idrovo, A.J.; Araque-Rodriguez, S.A.; Henao, J.A.; Sánchez-Rodríguez, L.H. Chemical Element Mixtures and Kidney Function in Mining and Non-Mining Settings in Northern Colombia. *Int. J. Environ. Res. Public Health* **2023**, *20*, 2321. [CrossRef] [PubMed]
26. Florez-Vargas, O.; Vilanova, E.; Alcaide, C.; Henao, J.A.; Villarreal-Jaimes, C.A.; Medina-Pérez, O.M.; Rodriguez-Villamizar, L.A.; Idrovo, A.J.; Sánchez-Rodríguez, L.H. Geological Context and Human Exposures to Element Mixtures in Mining and Agricultural Settings in Colombia. *Sci. Total. Environ.* **2023**, *898*, 165632. [CrossRef] [PubMed]
27. Braun, J.M.; Gennings, C.; Hauser, R.; Webster, T.F. What Can Epidemiological Studies Tell Us about the Impact of Chemical Mixtures on Human Health? *Environ. Health Perspect.* **2016**, *124*, A6–A9. [CrossRef] [PubMed]

28. Martin-Reina, J.; Casanova, A.G.; Dahiri, B.; Fernández, I.; Fernández-Palacín, A.; Bautista, J.; Morales, A.I.; Moreno, I. Adverse Health Effects in Women Farmers Indirectly Exposed to Pesticides. *Int. J. Environ. Res. Public Health* **2021**, *18*, 5909. [CrossRef] [PubMed]
29. Idrovo, A.J.; Rivero-Rubio, C.; Amaya-Castellanos, C.; Idrovo, A.J.; Rivero-Rubio, C.; Amaya-Castellanos, C. Perception of Pollution and Arsenic in Hair of Indigenous Living near a Ferronickel Open-Pit Mine (Córdoba, Colombia): Public Health Case Report. *Rev. Univ. Ind. Santander. Salud* **2017**, *49*, 115–123. [CrossRef]
30. Idrovo, A.J. Cerro Matoso Mine, Chemical Mixtures, and Environmental Justice in Colombia. *Lancet* **2018**, *391*, 2320. [CrossRef]
31. Fernández-Niño, J.A.; Idrovo, A.J.; Cucunubá, Z.M.; Reyes-Harker, P.; Guerra, Á.P.; Moncada, L.I.; López, M.C.; Barrera, S.M.; Cortés, L.J.; Olivera, M.; et al. Paradoxical Associations between Soil-Transmitted Helminths and Plasmodium Falciparum Infection. *Trans. R. Soc. Trop. Med. Hyg.* **2012**, *106*, 701–708. [CrossRef]
32. Cordy, P.; Veiga, M.M.; Salih, I.; Al-Saadi, S.; Console, S.; Garcia, O.; Mesa, L.A.; Velásquez-López, P.C.; Roeser, M. Mercury Contamination from Artisanal Gold Mining in Antioquia, Colombia: The World's Highest per Capita Mercury Pollution. *Sci. Total Environ.* **2011**, *410–411*, 154–160. [CrossRef]
33. Bonilla, J.S.; Figueroa, H.; Idrovo, A.J. *Historia de la mina de mercurio "La Esperanza", 1948–1983, Aranzazu, Caldas*; Un Desastre Olvidado: Bucaramanga, Colombia, 2023.
34. Word Health Organization. Guidance for Identifying Populations at Risk from Mercury Exposure. Available online: https://www.who.int/publications/m/item/guidance-for-identifying-populations-at-risk-from-mercury-exposure (accessed on 28 June 2023).
35. Shin, Y.; Lee, J.; Park, E.; Lee, J.; Lee, H.S.; Kim, J.-H. A Quantitative Tandem Mass Spectrometry and Scaled-Down QuEChERS Approach for Simultaneous Analysis of Pesticide Multiresidues in Human Urine. *Molecules* **2019**, *24*, 1330. [CrossRef] [PubMed]
36. Kales, S.; Christiani, D. Hair and Metal Toxicity. In *Hair in Toxicology*; Tobin, D.J., Ed.; The Royal Society of Chemistry: London, UK, 2005; pp. 125–158. ISBN 978-0-85404-587-7.
37. Nunes, J.A.; Batista, B.L.; Rodrigues, J.L.; Caldas, N.M.; Neto, J.A.G.; Barbosa, F. A Simple Method Based on ICP-MS for Estimation of Background Levels of Arsenic, Cadmium, Copper, Manganese, Nickel, Lead, and Selenium in Blood of the Brazilian Population. *J. Toxicol. Environ. Health A* **2010**, *73*, 878–887. [CrossRef] [PubMed]
38. Fenech, M. Cytokinesis-Block Micronucleus Cytome Assay. *Nat. Protoc.* **2007**, *2*, 1084–1104. [CrossRef]
39. Widaman, K.F. Common Factor Analysis Versus Principal Component Analysis: Differential Bias in Representing Model Parameters? *Multivar. Behav. Res.* **1993**, *28*, 263–311. [CrossRef] [PubMed]
40. Barros, A.J.D.; Hirakata, V.N. Alternatives for Logistic Regression in Cross-Sectional Studies: An Empirical Comparison of Models That Directly Estimate the Prevalence Ratio. *BMC Med. Res. Methodol.* **2003**, *3*, 21. [CrossRef] [PubMed]
41. Stafoggia, M.; Breitner, S.; Hampel, R.; Basagaña, X. Statistical Approaches to Address Multi-Pollutant Mixtures and Multiple Exposures: The State of the Science. *Curr. Environ. Health Rep.* **2017**, *4*, 481–490. [CrossRef] [PubMed]
42. Ramos, S.J.; Dinali, G.S.; Oliveira, C.; Martins, G.C.; Moreira, C.G.; Siqueira, J.O.; Guilherme, L.R.G. Rare Earth Elements in the Soil Environment. *Curr. Pollut. Rep.* **2016**, *2*, 28–50. [CrossRef]
43. Gwenzi, W.; Mangori, L.; Danha, C.; Chaukura, N.; Dunjana, N.; Sanganyado, E. Sources, Behaviour, and Environmental and Human Health Risks of High-Technology Rare Earth Elements as Emerging Contaminants. *Sci. Total Environ.* **2018**, *636*, 299–313. [CrossRef]
44. Calabrese, E.J.; Baldwin, L.A. Inorganics and Hormesis. *Crit. Rev. Toxicol.* **2003**, *33*, 215–304. [CrossRef]
45. Rix, R.R.; Guedes, R.N.C.; Cutler, G.C. Hormesis Dose–Response Contaminant-Induced Hormesis in Animals. *Curr. Opin. Toxicol.* **2022**, *30*, 100336. [CrossRef]
46. Spiller, H.A. Rethinking mercury: The role of selenium in the pathophysiology of mercury toxicity. *Clin. Toxicol.* **2018**, *56*, 313–326. [CrossRef] [PubMed]
47. Chen, J.; Hao, S.; Bañuelos, G.; Zhou, X. A Quantitative Review of the Effects of Se Application on the Reduction of Hg Concentration in Plant: A Meta-Analysis. *Front. Plant. Sci.* **2023**, *14*, 1199721. [CrossRef] [PubMed]
48. Simón, P. Noticias Historiales de las Conquistas de Tierra Firme en las Indias Occidentales. Available online: https://www.bibliotecadigital.uchile.cl/discovery/fulldisplay/alma991007267759703936/ (accessed on 2 August 2023).
49. Mendoza, O.G.; Castellanos, F.; Jiménez, J.F. *Selenio (Se). En: Atlas Geoquímico de Colombia, Versión 2020*; Servicio Geológico Colombiano: Bogotá, Colombia, 2020.
50. Bedoya, G.; García, J.; Montoya, P.; Rojas, W.; Amézquita, M.E.; Soto, I.; López, M.C.; Ospina-Duque, J.; Ruiz-Linares, A. Análisis de isonimia entre poblaciones del noroeste de Colombia. *Biomedica* **2006**, *26*, 538. [CrossRef] [PubMed]
51. Pérez-Rincón, M.P.; Peralta-Ardila, M.; Vélez-Torres, I.; Méndez, F. Conflicto Armado Interno y Ambiente en Colombia: Análisis desde los Conflictos Ecológicos, 1960–2016. *J. Pol. Ecol.* **2022**, *29*, 672–703. [CrossRef]
52. Albarracín, J.; Milanese, J.P.; Valencia, I.H.; Wolff, J. Local Competitive Authoritarianism and Post-Conflict Violence. An Analysis of the Assassination of Social Leaders in Colombia. *Int. Interact.* **2022**, *49*, 237–267. [CrossRef]
53. Espitia-Pérez, L.; Sosa, M.Q.; Salcedo-Arteaga, S.; León-Mejía, G.; Hoyos-Giraldo, L.S.; Brango, H.; Kvitko, K.; da Silva, J.; Henriques, J.A. Polymorphisms in Metabolism and Repair Genes Affects DNA Damage Caused by Open-Cast Coal Mining Exposure. *Mutat. Res. Toxicol. Environ. Mutagen.* **2016**, *808*, 38–51. [CrossRef] [PubMed]

54. Guerrero-Castilla, A.; Olivero-Verbel, J.; Marrugo-Negrete, J. Heavy Metals in Wild House Mice from Coal-Mining Areas of Colombia and Expression of Genes Related to Oxidative Stress, DNA Damage and Exposure to Metals. *Mutat. Res. Toxicol. Environ. Mutagen.* **2014**, *762*, 24–29. [CrossRef]
55. Vélez-Torres, I.; Vanegas, D. Contentious Environmental Governance in Polluted Gold Mining Geographies: The Case of La Toma, Colombia. *World Dev.* **2022**, *157*, 105953. [CrossRef]

MDPI AG
Grosspeteranlage 5
4052 Basel
Switzerland
Tel.: +41 61 683 77 34

MDPI Books Editorial Office
E-mail: books@mdpi.com
www.mdpi.com/books

www.ingramcontent.com/pod-product-compliance
Lightning Source LLC
Chambersburg PA
CBHW041940260926
51850CB00018B/349
* 9 7 8 3 7 2 5 8 1 7 4 3 6 *